International Conference on Physics and Energy (ICPAE 2021)

Journal of Physics: Conference Series Volume 2040

Kancheepuram, India
24 March 2021

ISBN: 978-1-7138-5011-3
ISSN: 1742-6588

Printed from e-media with permission by:

Curran Associates, Inc.
57 Morehouse Lane
Red Hook, NY 12571

Some format issues inherent in the e-media version may also appear in this print version.

Institute of Physics
Dirac House, Temple Back
Bristol BS1 6BE UK

Phone: 44 1 17 929 7481
Fax: 44 1 17 920 0979

techtracking@iop.org

Additional copies of this publication are available from:

Curran Associates, Inc.
57 Morehouse Lane
Red Hook, NY 12571 USA
Phone: 845-758-0400
Fax: 845-758-2633
Email: curran@proceedings.com
Web: www.proceedings.com

TABLE OF CONTENTS

Journal of Physics: Conference Series **2040** (2021) 011001 doi:10.1088/1742-6596/2040/1/011001

International Conference on Physics and Energy 2021(ICPAE 2021)

PREFACE

This special issue includes the proceedings of the First International Conference on Physics and Energy 2021 (ICPAE 2021), which was held during 21st April 2021 at RanBal Research Institute (RBRI) in Kancheepuram, Tamilnadu, India. ICPAE 2021 provided a platform to experts, researchers, scholars, and academicians across the globe to exhibit their recent innovative research results and the most recent trends in the diverse fields of Physics and Energy.

The first edition of ICPAE 2021 was represented by many potential authors. The program comprised of three plenary and four Invited lectures and eighty oral presentations covering emerging fields of Physics and Energy. The organizing committee sincerely acknowledges the support and continuous encouragement from Shri. E. Santhanam, Smt. S. Selvambal [Patrons-RBRI], Prof. P. Anandhi [Conference Chair] and Prof. P. Kumaran [Conference Secretary], Wolaita Sodo University, Ethiopia.

Also, the organizing committee would like to express thanks to Dr.D.S.Vijayan, Associate Professor/Dept .of Civil Engineering, Aarupadai Veedu , Paiyanoor, Chennai, India, for rendering his support to the publishing committee. We are grateful to the advisory committee and technical review committee (National/International) for their valuable suggestions and contributions in organizing the mammoth scientific event.

The first edition of the ICPAE 2021 proceedings includes fifty six papers eventually accepted after a careful peer-review process accomplished on the presented articles. The organizers sincerely thank to the guest editors and all the reviewers for their great efforts in bringing out the high standard scientific publication. The organizers of the conference are indebted to the publisher of Journal of Physics: Conference Sries (IOP Publisher), for giving consent to publish these papers under the special issue category.

Conference Secretary
Dr.P.Kumaran M.E., Ph.D
Wolaita Sodo University, Ethiopia

Dr. S. Harikrishnan M.E, Ph.D
Professor & Head
Department of Mechanical Engineering
Kings Engineering College
Irungattukottai, Chennai – 602117, India

Dr.P.Anandhi M.Tech., Ph.D
Associate Professor
Department of Electronics and Communication Engineering
P.B.College of Engineering
Irungattukottai, Chennai – 602117, India

Journal of Physics: Conference Series **2040** (2021) 011001 doi:10.1088/1742-6596/2040/1/011001

Advisory Committee

- Dr. R. Thangappan, Periyar University, Salem, India
- Dr. G. Murugadoss, Sathyabama University, Chennai, India
- Dr. S. Sankar, Sriram Engineering College, Chennai, India
- Dr. R. Vinayagamoorthy, SCSVMV University, Kancheepuram, India
- Dr. D.S. Vijayan, AVIT, Chennai, India
- Dr. S. Harikrishnan, Kings Engineering College, Chennai, India
- Dr. A. Rajaram, EGS Pillay Engineering College, Nagapattinam, India
- Dr. S. Madhu, Saveetha School of Engineering, Chennai, India
- Dr. K.C. Ramya, Sri Krishna College of Engineering and Technology, Coimbatore, India
- Dr. K. Anisha, SCSVMV University, Kancheepuram, India
- Dr. K. Sathishkumar, Executive Officer, Ranbal Research Institute, Kancheepuram, India
- Dr. K. Nimel Ross, Kings Engineering College, Chennai, India
- Dr. G. Chinnadurai, Rajiv Gandhi College of Engineering, Chennai, India

Technical Review Committee (International Members)

- Dr. Mohammed Hafiz Ali, KFUPM, KSA
- Dr. Muhammed Bin Mat Noor, Universiti Malaysia, Pahang
- Dr. Nidal AbuHamdeh, King Abdulaziz University, Saudi Arabia
- Dr. Tarek Bouktir, University of Ferhat Abbas Setif, Algeria
- Dr. Hamzah Ahmad, Universiti Malaysia, Pahang
- Dr. Tooraj Yousefi, Ryerson University, Canada
- Dr. Hakan F Oztop, Firat University, Turkey
- Dr. Robert A Taylor, University of New South Wales, Australia
- Dr. M. Shanmugam, Yeungnam University, South Korea
- Dr. Ayodele A Daniyan, Obafemi Awolowo University, Nigeria
- Dr. Aloysius F. Hepp, Chief Technologist Nanotech Innovations, USA
- Dr. Muhammad Sharif, The University of Alabama, USA
- Dr. Azura Che Soh, Universiti Putra Malaysia, Malaysia
- Dr. Mazin Zeki Othman, Dhofar University, Oman
- Dr. Isaac I. Akinwumi, Covenant University, Nigeria

- Dr. Balew Baye, Debre Tabor University, Ethiopia
- Dr. Alaa El Din Mahmoud, Environmental Sciences Department, Faculty of Science, Alexandria University, Egypt.
- Dr. Benlekkam Mohamed Lamine, Smart Structure Laboratory, University Center d'Ain Temouchent, Algeria.

Technical Review Committee (National Members)

- Dr. Pradip Lingfa, NERIST, Arunachel Pradesh, India
- Dr. K.M. Pandey, NIT Silchar, India
- Dr. A. Arivarasan, Kalasalingam University, Krishnankovil, India
- Dr. R. Prakash, VIT University, Vellore, Tamilnadu, India
- Dr. D. Dineshkumar, Sathyabama University, India
- Dr. A. Rajasekhar, SRM Institute of Science & Technology, India
- Dr. A.D. Dhass, Indus Institute of Technology and Engineering, Indus University, Gujarat, India

Journal of Physics: Conference Series **2040** (2021) 011002 doi:10.1088/1742-6596/2040/1/011002

Peer review declaration

All conference organisers/editors are required to declare details about their peer review. Therefore, please provide the following information:

- **Type of peer review: ~~Single-blind~~ / Double-blind / ~~Triple-blind~~ / ~~Open~~ / ~~Other~~ (please describe)**
  All the submitted papers were sent to the two reviewers, in order to evaluate the novelty and originality of the works attempted by the authors.

- **Conference submission management system:**
  Through conference email (icpae2019@gmail.com)

- **Number of submissions received:**
  176

- **Number of submissions sent for review:**
  147

- **Number of submissions accepted:**
  72

- **Acceptance Rate (Number of Submissions Accepted / Number of Submissions Received X 100):**
  48.97%

- **Average number of reviews per paper:**
  Two

- **Total number of reviewers involved:**
  45

- **Any additional info on review process:**
  First, the plagiarism on the papers was verified with Turnitin. Plagiarism reports showed anything above 20% on the papers was rejected before they were sent to the reviewers. The papers containing below 20% plagiarism were sent to the reviewers to evaluate the originality of the work. Two reviewers were invited to review each paper and duration of the review process was 4 – 6 weeks. Based on the reviewers' comments, the decisions on the papers were taken.

- **Contact person for queries (please include: name, affiliation, institutional email address)**
  Dr.S.Harikrishnan, Kings Engineering College, Irungattukottai, Chennai, India.
  ramhkn@gmail.com & harikrishnan@kingsedu.ac.in

Lung Cancer Detection using VGG NET 16 Architecture

Thanzeem Mohamed Sheriff S[1], Venkat Kumar J[2], Vigneshwaran S[3], Aida Jones[4], and Jose Anand[5]

Department of Electronics and Communication Engineering, KCG College of Technology, Chennai, Inida, [1]thanzeem4@gmail.com, [2]venkatvenras@gmail.com, [3]waran8870@gmail.com, [4]aidajones89@gmail.com, [5]joseanandme@yahoo.co.in

Corresponding author's e-mail: thanzeem4@gmail.com

Abstract. Cancer is one of the main reason for loss of human life across the world. All the medical practitioners and researchers are dealing with the demanding situations to fight against cancer. Based on the report in 2019 from American Cancer Society, 96,480 deaths are anticipated due to skin cancers, 142,670 deaths are from lung cancers, 42,260 deaths are from breast cancers, 31,620 deaths are from prostate cancers, and 17,760 deaths are from mind cancers. Initial detection of most cancers has the pinnacle precedence for saving the lives. This paper proposed a lung cancer detection using Deep Learning based on VEE NET architecture. This was one of the famous models submitted to ILSVRC-2014. Visual checkup and manual practices are used on this venture for the various types of cancer diagnoses. This guide interpretation of scientific images that needs massive time intake and is notably susceptible to mistakes. Thus, in this project, we apply deep learning algorithms to identify lung cancer and its presence without the need for several consultations from different doctors. This leads to an earlier prediction of the presence of the disease and allows us to take prior actions immediately to avoid further consequences in an effective and cheap manner avoiding human error rate. In this project lung cancer and its presence is determined. A web application is developed as a hospital application where an input x-ray image is given to detect lung cancer.

Keywords. Lung Cancer, Deep Learning, ILSVRC-2014.

1. Introduction

Lung cancer is also called lung carcinoma. This is a kind of harmful lung tumor considered as unrestrained cell growth in lung tissues. This development spread outside the lung by the method of metastasis into neighboring tissue or additional parts of the body [1]. Most cancers that twitch inside the lung, called primary lung cancers, are carcinomas. The common symptoms are persistent cough, weight problem, tininess of breath, and shoulder, arm, and chest cautions. The mainstream of approximately 85 % of people who cause lung cancer are owing to long-term tobacco smoking. Nearly 10–15 % of belongings happen in people that they not having smoking habit [2]. These belongings are frequently created by a combination of genetic influences and experience to radon gases, asbestos, second-hand smoking, or other kinds of pollution. The Computed Tomography (CT) images from the chest radiography is used to detect lung cancer [3]. The identification is inveterate by biopsy that is characteristically achieved by bronchoscopy or CT-guidance. In 2012, globally lung cancer is affected by 1.8 million people and 1.6 million people have died. This creates is the leading common description for cancer-related death in men and second shared in women after breast cancer [4]. The greatest shared age for lung cancer analysis is 70 years. In general, nearly 20 % of peoples are affected by lung cancer. The survive rate after five years of diseases occurred is very poor, while outcomes on average are worse in the developing world [5]. More about segmentation of image analysis for the detection of non-invasive anemia application is discussed [6] and for the same a ridge regression algorithm is used [7]. Image analysis on Matlab software in analyzed [8] and unsupervised prediction for medical treatment on lung cancer with stereotactic body radiation therapy is mentioned [9]. For the identification of boundaries in medical images and edge vector and mapping method is used [10]. Also for security system image analysis is done on the finger vein [11] and a content based image retrieval method is implemented for texture and shape analysis [12]. Lung cancer detection from CT images are discussed in [13], and CT is used for nodule classification of thorax images using SVM [14]. Image classification based on deep convolution networks is described in [15] and predictive variance for accelerating stochastic gradient detection [16] and deep leaning based algorithm [17] for the detection of images using convolution techniques [18]. Diagnosis of lung cancer using clinical and radiomic features was described [19], and CT based image analysis for predicting lung adenocarcinoma in [20] and a nomogram development of clinical staging using image feature analysis in non-small cell lung cancer is discussed [21].

2. Deep Learning

Deep learning is an algorithm that impersonates the network of neurons during a brain. This is a subset of machine learning which is named as deep learning and this uses deep neural networks. As shown in figure 1 the deep learning technique is built with connected layers. The primary layer is understood because of the input layer. The final layer is the output layer. The layers which is used between input and output layer is the hidden layers. The meaning for deep suggests that the network joins neurons in additional than the two layers.

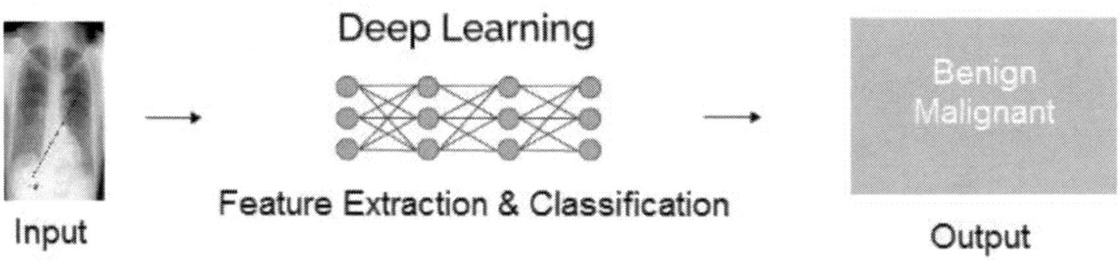

Figure 1. Deep Learning Architecture

Every hidden layer consists of number of neurons. The successive neurons are connected with one another. The function of neuron is to process and propagate the received input which is received from the above layer. Signal strength provided to neuron within the successive layer is based on factors such as bias, the load and the activation parameters. Huge volume of input data is consumed by the network for the various process to be performed by the multiple layers. Based on the inputs the network under consideration will learn themselves with the complex functions at each layer. Deep learning approach is an influential technique to obtain better prediction of an actionable output. Deep learning outperforms in pattern identification and knowledge based forecast. Combining both will enhance the performance with extraordinary leads on factors such as innovation, management and production. By the use of deep learning, traditional techniques are better performed. The deep learning approaches are 41 percent more accurate than other image classification machine learning techniques, and the same with 27 percent in face recognition applications and 25 percent better in voice recognition applications..

2.1. *Deep Learning Process*
A deep neural network delivers progressive accurateness in numerous tasks, from object recognition to speech recognition, and will acquire mechanically, without predefined information expressly coded by the application programmer.

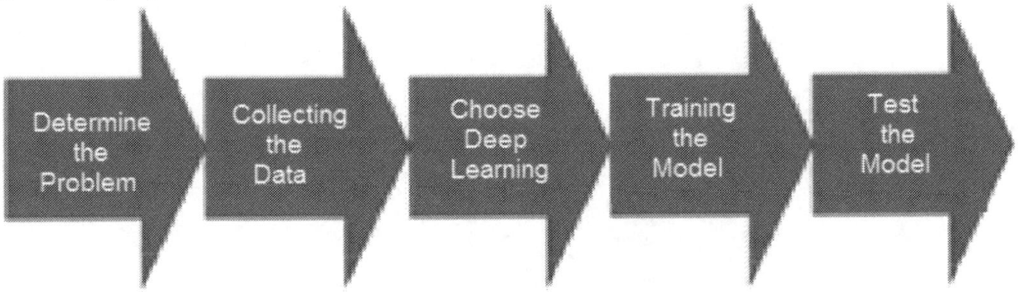

Figure 2. Learning Process

Figure 2 shows the learning process involved in the deep learning technique. Every layer has an interior layer of data, i.e., the grading of data. In a neural network application with 4 layers will learn additional composite features than that with two levels. Learning happens in two stages. The first stage contains nonlinear transformation inputs and produces a statistical prototypical as output. In the second level the algorithm intents to rise the model with a mathematical process stated as a derivative. Neural network concept repeats the 2 levels hundreds of times or even to thousands of times until it reaches the tolerance level of accuracy. This processes is repeated as iteration until an optimal solution is received.

3. VGG NET 16 Architecture
The trained model using VGG Net 16 architecture is shown in the figure 3. Convolution 1 input layer is a standard image size of 224 x 224 RGB. The input is conceded over all the 16 layers in order to find the best accurate results, where filters are applied with a exact minor reception field of size 3 ×3. In some layout, it also uses 1 × 1 convolution filters, that are seen as straight modification of input stations. Convolution line altered to 1 pixel, the spatial padding of convolution layer input is such that the spatial resolution is stored after the convolution, e.g., Padding is 1-pixel for 3 × 3 convolution layers. The collection is done with max-pooling lenses, which follow one of the convolution layers. Max-pooling is done with a 2 × 2 pixel convolution layer.

All the 3 fully connected layers will continue with the convolutional layer with different depths in various structures. The initial two layers have 4096 channels and the third layer connects to ILSVRC classification in 1000 channels making one channel per category. Soft-max layer is the retaining layer and the configuration of completely integrated layers are similar across the entire networks. The entire hidden layers are fitted to non-linear retrieval represented by ReLU. This is also known that there is no network other than one which contain the local response and such practice do not have any improved performance in the ILSVRC database thus leading to memory usage and computational time.

3.1. *ResNet Algorithm*
The construction of ResNet50 has 4 stages are there. This network will take an input image with width and height, as duplicate 32, and 3 as channel width. Aimed at the description, look at the input size as 224 x 224 x 3. All ResNet drawings make the first conversion and multi-integration using the dimensions of 7 × 7 and 3 × 3 respectively. After the phase 1 the network twitches and consists of 3 remaining blocks covering 3 layers individually. The maximum number of characters used to achieve the convolution function on all 3 layers of block 1 is 64, 64, and 128 respectively. Curved arrows point to the joining of the identity. The combined arrow signifies that the process of the convolution in

Journal of Physics: Conference Series **2040** (2021) 012001 doi:10.1088/1742-6596/2040/1/012001

Residual Block is done in stride 2, henceforth, the length and width of the input will be cut in half, but the channel width will be doubled. As we move from one section to another, the channel width is folded and the input size is minimized by half.

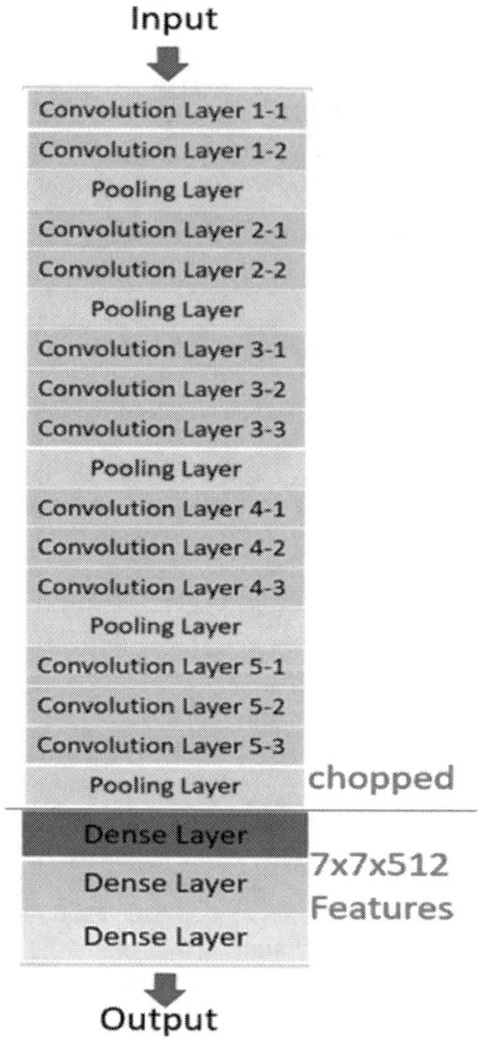

Figure 3. VGG16 Architecture

In deep networks like ResNet50, ResNet152, etc. and for each remaining F function, 3 layers are arranged in sequence. Convolutions 1×1, 3×3, 1×1 are used in the three levels. Layers of 1×1 convolution are responsible for dropping and reinstating size. The 3×3 layer is left as a bottle with a small size of input/output. Finally, the network has a Pooling layer followed by a fully connected layer of 1000 neurons (output of the ImageNet class).

4. Training Process

In this training procedure, we are going to train the model by performing image augmentation first. This image augmentation will help us to train the model effectively, by pre-processing all the images present in the data set and converting each image into multiple images of different angles. Various steps involved in our training are given below which are array pre-processing, rotation, flipping, etc. Then after augmentation train the datasets by extracting the features using a combination of two deep learning algorithms VGG-16 and Resnet algorithm. It undergoes a process called optimization which will optimize the model and loss minimization which will reduce the noises generated during training. In the last it will be undergoing a process is called model seriation which will be evaluated after generating model using the testing dataset and predict the presence of lung cancer Then python codes are executed using visual studio software. The image given in figure 4 is the flowchart of the training process in the proposed system. The modules represented in the training process are dataset collection, dataset pre-processing, dataset separation, training and testing the dataset using VGG -16. After the optimization technique finally the validation and evaluation process is carried out to obtain the optimized result.

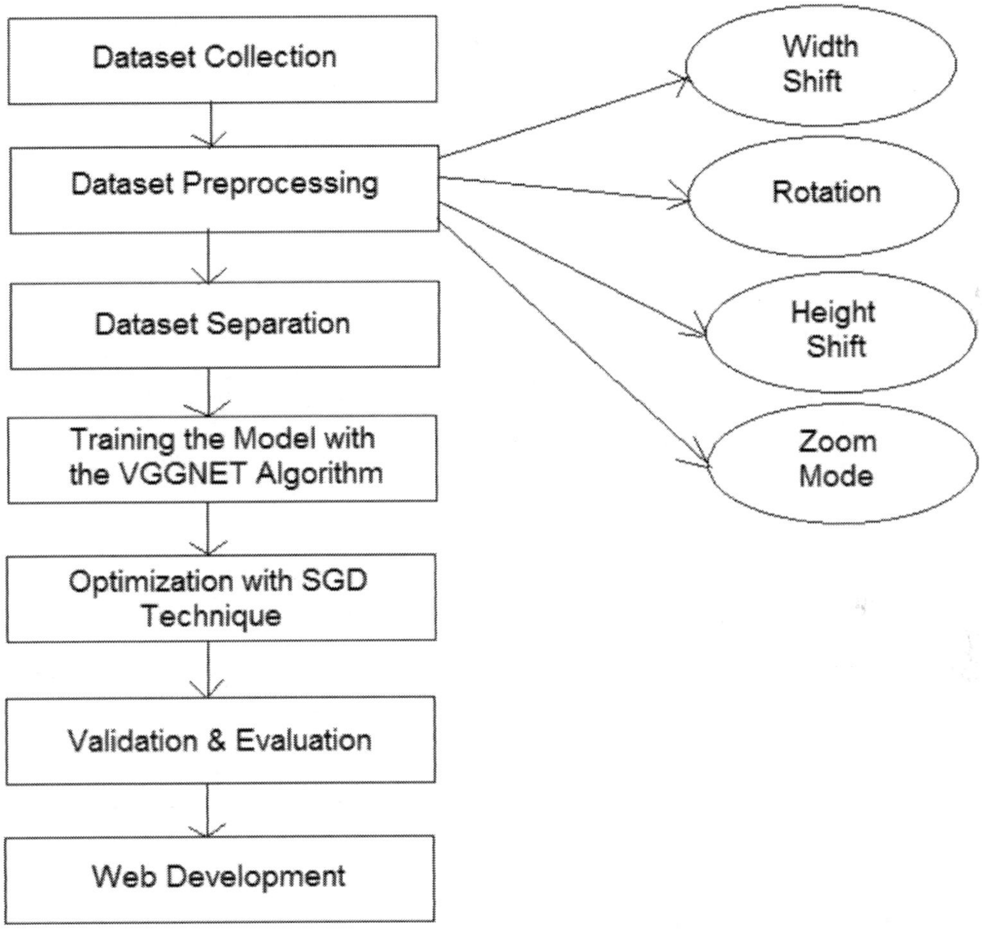

Figure 4. Flowchart Representation

4.1. Importing Necessary Libraries
The required libraries are imported for training the model. Some libraries like NumPy, OS, Adam, SGD are imported and the process is continued by pre-processing.

4.2. Performing Image to Array Pre-processing
The pixels of each image contain a 0:255 aspect ratio and 8bit data. If we give this huge value for training, it will not be trained accordingly. So, to avoid this accuracy problem we are reducing the aspect ratio from 0:255 to 0:1 so that the training will be more efficient [0:1] in the sense we are reducing it to 0:42, 0.25.

4.3. Aspect Aware Pre-processor
During the pre-processing and training of images of the lung in the image should not be affected by any of the operations like shifting, cropping, flipping, rotating, resizing, etc. So, to maintain the original features of an image we are doing Aspect Aware pre-processing.

4.4. Performing Image Augmentation
Here we have converted single image into the multiple images of different categories by Rotating, Flipping, Contrast adjustment, Inverting etc.

4.5. Training and Testing Dataset
The data is divided into two categories: training and testing, with 75% of the data being used for training and 25% for testing.

4.6. Label Binarizer
It converts the name of labels such as Normal, Severe and Mild into Binary format so that the machine can understand it so we can get accurate output, as the machine cannot understand human language it's advisable to convert human language to Binary.

4.7. Optimization
First an ADAM Optimizer is used and trained up to 10 epochs. Additionally, SGD Optimizer layer is introduced for adjusting the learning write and adding it as a layer to get more Accuracy.

5. Discussion and Evaluation
As the execution get completed the accuracy is plotted for the training using the deep learning algorithm. As discussed [1] the system is not very much accurate and the real-time process is quite complex, and this is overcome by the proposed model with ease implementation and accuracy. The cancer cannot be diagnosed with limited parameters [2], but in the system proposed presence of cancer can be identified at an early stage. There is no satisfactory accuracy in the detected image for cancer diagnosis [13], the system implemented using deep learning algorithm has complete accuracy. The evaluation results are shown in table 1.

Table 1. Evaluation after Fine Tuning

	Precision	Recall	F1-score	Support
Ignore	1.00	1.00	1.00	34
Lung_cancer	1.00	1.00	1.00	27
Lung_normal	1.00	1.00	1.00	49
Accuracy			1.00	110
Macro Average	1.00	1.00	1.00	110

Figure 5 shows the image for the training loss and the training accuracy validation graph obtained after the training process. In the first training process the accuracy was less and there would be more loss. After repeating the training process, the accuracy was increased and loss was reduced. The trained model graph also automatically stored in the file.

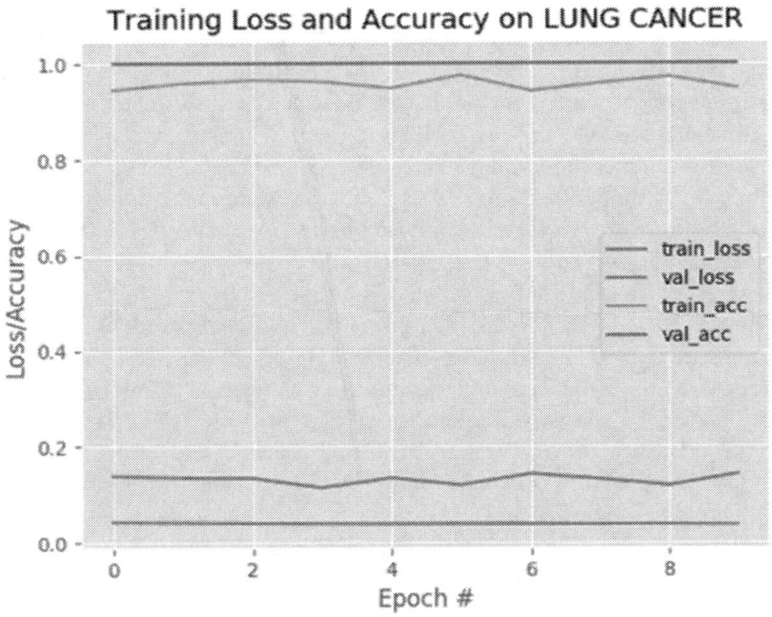

Figure 5 Training Loss and Accuracy

6. Conclusion

In this paper, the recognition of lung cancer is done in early phases. The architecture used is VGG which consist of 16 layers. The input is passed through all the 16 layers in order to find the best accurate results. VGG can also be compared with other architectures or else the other convolutional layers can be increases the accuracy and probability for better results.

References

[1] C Parmar, R T Leijenaar, P Grossmann, E R Velazquez, J Bussink, D Rietveld, M M Rietbergen, B Haibe-Kains, P Lambin and H J Aerts 2015 Radiomic feature clusters and prognostic signatures specific for lung and head & neck cancer *Scientific Reports* **5** 11044

[2] M G Vicente, A F H Martínez, Á M S Castrejon, G A J Londoño, J M Udias and P L. Atance 2018 Heterogeneity in [18F] Fluorodeoxyglucose positron emission tomography/computed tomography of non–small cell lung carcinoma and its relationship to metabolic parameters and pathologic staging. *Molecular Imaging* **13** (9) 7290.2014. 00032

[3] J Duchi, E Hazan and Y Singer 2011 Adaptive Subgradient Methods for Online Learning and Stochastic Optimization *Journal of Machine Learning Research* **2** 2121-2159

[4] Kuruvilla and K Gunavathi 2014 Lung cancer classification using neural networks for CT images *Computer Methods and Programs in Biomedicine* **113** (1) 202–209

[5] H C Shin, H R Roth, M Gao et al. 2016 Deep convolutional neural networks for computer-aided detection: CNN architectures, dataset characteristics and transfer learning *IEEE Transactions on Medical Imaging* **35** (5) 1285– 1298

[6] Sivachandar Kasiviswanathan, Thulasi Bai Vijayan, Lorenzo Simone, Giovanni Dimauro 2020 Semantic Segmentation of Conjunctiva Region for Non-Invasive Anemia Detection Applications. *Electronics 2020* **9** 1309 doi:10.3390/electronics9081309

[7] Sivachandar Kasiviswanathan, Thulasi Bai Vijayan, Sheila John 2020 Ridge regression algorithm based non-invasive anaemia screening using conjunctiva images *Journal of Ambient Intelligence and Humanized Computing* https://doi.org/10.1007/s12652-020-02618-3.

[8] Subbarayan Sadasivam, S Karthick Ramanathan 2009 Effective watermarking of digital audio and image using Matlab technique *International Conference on Machine Vision IEEE*

[9] H Li. M Galperin-Aizenberg, D Pryma, C B Simone II and Y Fan 2019 Unsupervised machine learning of radiomic features for predicting treatment response and overall survival of early-stage non-small cell lung cancer patients treated with stereotactic body radiation therapy *Radiotherapy and Oncology* **129** (2) 218-226

[10] J Anand, and K Sivachandar 2013 An Edge Vector and Edge Map Based Boundary Detection in Medical Images *International Journal of Innovative Research in Computer and Communication Engineering* **1** (4) 1050-1055

[11] J Anand, T G A Flora, and A S Philip 2013 Finger-Vein based Biometric Security System *International Journal of Research in Engineering and Technology* **2** (12) 197-200

[12] Sibia E V, Mareena G, and J Anand 2014 Content Based Image Retrieval Technique on Texture and Shape Analysis using Wavelet Feature and Clustering Model *International Journal of Enhanced Research in Science Technology & Engineering* **3** (8) 224-229

[13] Vaishali C Patil and Shrinivas R Dhotre 2016 Lung Cancer Detection from Images of Computer Tomography Scan *International Journal of Advanced Research in Computer and Communication Engineering* **5** (7)

[14] H M Orozco and O O V Villegas 2013 Lung nodule classification in CT thorax images using support vector machines *12th Mexican International Conference on Artificial Intelligence* 277–283

[15] Rizhevsky A, Sutskever I and Hinton G E 2012 Imagenet classification with deep convolutional neural networks *Advances in Neural Information Processing Systems* 1097–1105

[16] Johnson R and Zhang T 2013 Accelerating Stochastic Gradient Descent Using Predictive Variance Reduction *Proceedings of the 26th International Conference on Neural Information Processing Systems* 315-323

[17] Sutskever, J Martens, G Dahl and G Hinton 2013 On the importance of initialization and momentum in deep learning *PMLR* 1139-1147

[18] C Szegedy, W Liu, Y Jia, P Sermanet, S Reed, D Anguelov et al. 2014 Going deeper with convolutions *arxiv preprint* arxiv: 1409 4842

[19] H Zhou, D Dong, B Chen, M Fang, Y Cheng, Y Gan, R Zhang, L Zhang, Y Zang and Z Lin 2018 Diagnosis of distant metastasis of lung cancer based on clinical and radiomic features Translational *Oncology* **11** (1) 31-36

[20] T P Coroller, P Grossmann, Y Hou, E R Velazquez, R T Leijenaar, G Hermann, P Lambin, B Haibe Kains, R H Mak and H JAerts 2015 CT based radiomic signature predicts distant metastasis in lung adenocarcinoma *Radiotherapy and Oncology* **114** (3) 345-350

[21] M C Desseroit, D Visvikis, F Tixier, M Majdoub, R Perdrisot, R Guillevin, C C Le Rest and M Hatt 2016 Development of a nomogram combining clinical staging with 18-FDG PET/CT image features in non-small cell lung cancer stage I- III *European Journal of Nuclear Medicine and Molecular Imaging* **43** (8) 1477-1485

Wind Turbine Fault Monitoring System Using MQTT

G Sasikala [1], Yadala Poorna Sai Chandra [2], Nallamilli Siva[3], Avilala Sai Vinesh[4]

[1] ECE Department, Vel Tech Rangarajan Dr. Sagunthala R&D Institute of Science and Technology,Chennai,India.
[2]ECE Department, Vel Tech Rangarajan Dr. Sagunthala R&D Institute of Science and Technology,Chennai,India. Email : yadalapoornasaichandra@gmail.com
[3]ECE Department, Vel Tech Rangarajan Dr. Sagunthala R&D Institute of Science and Technology,Chennai,India. Email : aivinesh.avilala@gmail.com.
[4] ECE Department, Vel Tech Rangarajan Dr. Sagunthala R&D Institute of Science and Technology,Chennai,India. Email : saivinesh.avilala@gmail.com.
Corresponding Author Email ID: sasikalaeverest369@gmail.com,

Abstract. Electricity can be generated by many sources in different ways. The one best well-known resource is wind energy. This energy reduces the emission of carbon dioxide and this energy can be produced through wind turbines. The system which is going to be proposed consists of Node MCU, MQTT protocol and sensors like temperature sensor and wind speed sensor are used to detect the faults, which occur in the wind turbine such as temperature in generator and bearing faults respectively. Whenever the value of temperature in the generator exceeds threshold value, then sensor will give the alert. Whenever the sufficient wind speed is detected but the power generation is low with respective to actual energy produced by the turbine, then sensor will alert. These sensors are connected to micro controller unit ESP8266 Wi-Fi module for detection operation and these parameters will be sent to the main location through MQTT by which the client can monitor these detected parameters in a MQTT Client Cloud network through web page and also these parameters can directly send to the user mobile through SMS and Emails.

Keywords—MQTT, Sensors, Node MCU, Monitoring

1. Introduction

The resources like Sunlight and the wind energy are the best natural resources to produce electrical energy. The wind energy is considered as the good source of energy, which is used to rotate the wind turbines in order to produce the current energy. The wind turns 2 or 3 blades around a rotor, which is attached to the main shaft in the turbine, which is used to spins a generator is the one which responsible for creating electricity. Wind turbines are at a risk of fault, because of installation in harsh environment like sea shores, deserts as well as the plains which are far away from the control unit. Due to the hard conditions around the wind turbine, there might be a chance of occurrence of a trouble and sometimes which leads to the power off. In order to stop the occurrence of the trouble it is needed to maintain the fault diagnosis and monitoring devices to screen the execution time status and faults occurred to improve the life time and the efficiency of the wind turbine.

1.1. Problem Identification

Wind turbines are fault inclined, because of their installation in harsh environment like desert, plains aside from that they're complicated gadget, that are placed some distance far away from the manage center. in order that area the threat of fault occurrence and the facet results can be extra, even it ends in electricity off.

1.2. Proposed Solution

The parameters which are required to monitor the wind turbine will be collected by the monitoring system from different parts of the turbine. Based upon the data gathered from turbine the controlling system will compare the data with threshold data which was fed previously and it will alert through the message to the server. The monitoring system consists of various sensors installed at various places of the turbine. The sensors like temperature, humidity, and speed detecting sensor collects the information from the turbine and sends to the MQTT server using ESP8266 NodeMCU.

The micro-controller NODEMCU ESP8266 has ability to perform WIFI related activities hence it is also called as WIFI module. All the ESP models have same type of ESP processor varies only in the type of breakout board used. The board of ESP8266-01 has only two gpio pins whereas ESP-12 has 16 GPIO pins many devices surrounding cannot be connected to internet directly on its own. So, ESP8266 module enables these devices to connect with internet and makes to communicate and make them as IOT devices. This is low cost; small module and it is really simply and easy to use.

In our proposed model we are using NODEMCU ESP8266 and sensors like temperature, humidity and speed sensors. These sensors are situated in and out of the turbine which are used to monitor the temperature and humidity. The IR speed sensor senses the speed of the air. These data will be sends to the WIFI module. The code will be written in the Arduino IDE, here we will select the board as NODEMCU ESP8266 then we will connect the hardware component of ESP8266 through USB cable to the port of pc the compiled code will be dumped in the ESP8266, the resultant output will be seen in the serial monitor. The WIFI module is connected to the server through the MQTT protocol. The all data will be sends to the server and the output will be monitored with the live graphs.

Internet of Things devices have been a part of our daily lives. From smart homes, smart bulbs and other house hold appliances; the developers are including this technology to create a network of connected devices that make the daily life a little easier. All these been made possible for the easy communication. There are many ways to communicate among devices, one of the ways is using single protocol commonly used is MQTT. MQTT is known as machine-to-machine protocol designed for ease and low weight publish/subscribe message transportation.

Table 1. Supervisory and Control Data Acquisition Data Set

Time Stamp	Wind Speed(m/s)	Generator Temperature (°C)
2018-10-01 00:10:00	4.94	64.653
2018-10-01 00:20:00	5.361	64.632
2018-10-01 00:30:00	5.01	61.16

These values are the average different parameter values of wind turbine which is shown on Table 1. Whenever the sensor values exceed the above values a message will be sent to the receiver through E-mail or an SMS [6].

2. Literature Review

In this paper the research is carried out with the sensor issue in a Wind Turbine model with fault tolerant control [1]. Which is utilized to permits boundaries of the regulator to be reconfigured in agreement mistake information acquired online from sensors to improve the dependability and by and large execution of the framework when a blunder experience. The essential part is planned by sliding mode observer based fault detection filter to create a lingering sign to assess flaw. Fault detection filter is intended to amplify affectability shortcoming

The author proposed a 3-stage philosophy for discovery and disconnection of detecting component deficiencies, i.e., SHORT issue, CONSTANT flaw, and NOISE issue, in WSN-based turbine CMS [2]. The proposed sensor flaw location and seclusion incredibly will expand the exactness and dependability of wind turbine CMSs. Information gathered from wind turbines in the field are utilized to assess the viability of the proposed strategy.

This paper presents 2 novel methodologies arranged to the plan of flaw open minded control plans for solid guideline of generator twist in a breeze turbine that can be influenced by both model vulnerabilities and actuator issues in its generator/converter [3]. The essential methodology depends on fluffy model reference versatile control in which a fluffy deduction component is utilized for boundary variation with no information on the possible shortcomings in the framework. The optional methodology misuses fluffy displaying and recognizable proof approach to build up a coordinated model-based shortcoming recognition and finding, and a programmed signal amendment instrument is utilized to enjoy issues in the framework dependent on online information of analyzed issues.

To identify the wind turbine [4] and to change the fault components three types of curve are used such that power curve, blade curve, rotor curve.

This paper proposes a little example Wind Turbine flaw location technique with the manufactured deficiency information utilizing generative ill-disposed nets [5]. basically, in view of past information, an unpleasant shortcoming information age measure is created to change the typical information to the harsh deficiency information. Optionally, a harsh flaw information purifier is created by GANs to make the unpleasant issue information more acquainted with the genuine shortcoming information.

The author [7] suggested here MQTT protocol for both way communication between the console and robot. The work is carried out with cloud platform, control and tracking system

Publisher subscriber is a one such protocol used in surveillance IOT. A scheme to stream a lot of data from publisher to subscriber using an IoT device that is microcontroller NodeMCU 8266 using two different protocol, there are MQTT and CoAP [8]. The NodeMCU will act as responder or publisher and can act as subscriber or sending request to get data from publisher. By the experiment, we can see that the differences in accuracy, time spend and the convenience from using Message Queuing Telemetry Transport (MQTT) and CoAP schemes.

In the midtowns, underground links are utilized rather than of overhead transmission lines. It is difficult to go through the particular spot of the shortcomings [9]. we have utilized IOT based generally method with Google information base for the issue discovery with the assistance of Node .Node MCU Module with transformer and utilized Google information base to checking the situation with transformers.

3. Methodology

3.1. Proposed Methodology

When the simulation process gets started, initially the IR sensor checks for the speed of the blades, if the speed of air is high, then after sometimes it checks again, else if the speed of air is low, the temperature sensor senses the temperature. The detected temperature from sensor installed inside the generator will send the temperature value with the help of ESP8266 and through MQTT to the MQTT server. The range of the temperature is shown in the format of graph for every interval of time or whenever the values exceed the triggered values the receiver will be given with an alert through an Email or an SMS to their mobile.

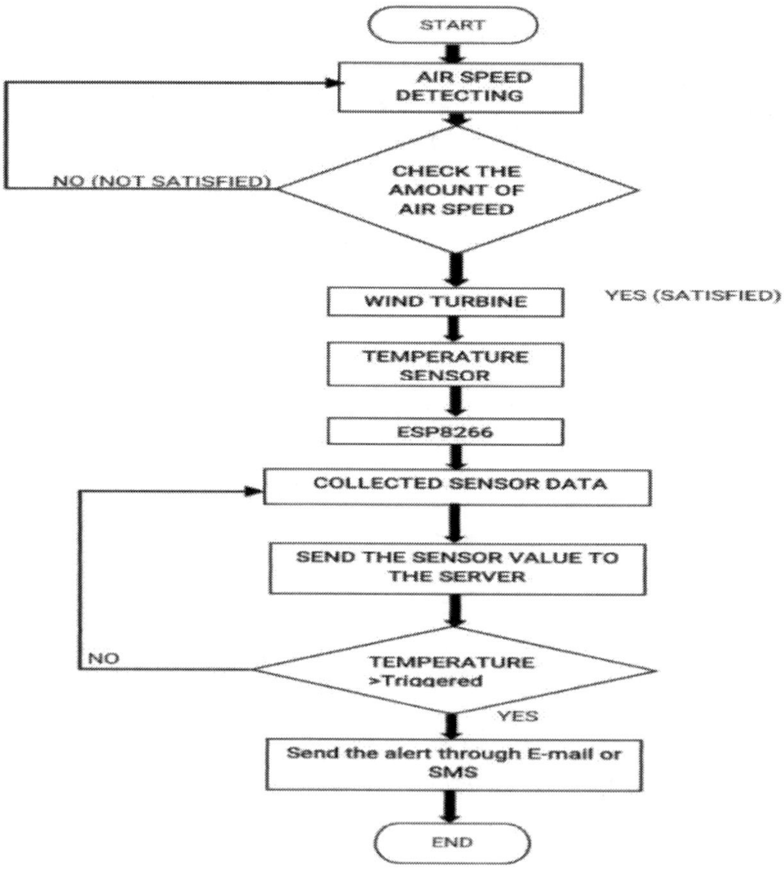

Figure 1. Flow Chart

By using above Figure 1 a circuit is designed and simulated code is written in the Arduino IDE.

3.2. Proposed Model

The proposed model for fault monitoring in wind turbine system using MQTT has been demonstrated using the block diagram figure 2 as follows.

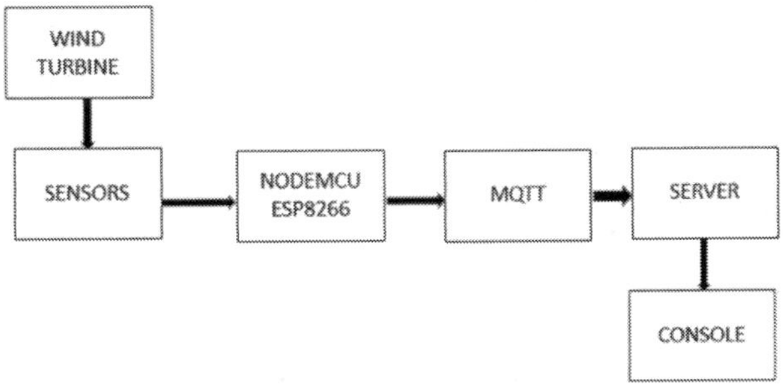

Figure 2. Block Diagram

Figure 2 describes about the block diagram of our proposed model. Firstly, from the wind turbine the faults are detected by the various types of sensors of different parameters. The sensors are connected to the NODEMCU ESP8266.The data which is published in the Node MCU transmitted to the MQTT server. From the MQTT server will be sent as an Email or an SMS.

4. Methodology

4.1. DHT 11 Sensor

It is a temperature and humidity based sensing module as shown in Figure 3 to give digital output. This module can be connected to any micro-controller. The suitable one is Arduino and Raspberry. With that we can find amount air present in surroundings.

Figure 3. DHT11 Sensor

4.2. Node MCU ESP8266

This is user-friendly very low cost WIFI module, which develops a simple TCP (Transmission control protocol)/IP (Internet Protocol) connection and can easily be interfaced with micro controllers via Serial Port. Normally it is used in IOT cloud based Wi Fi device with the operating range of 3 to 3.6 volts. ESP8266 WIFI module can easily be interfaced with micro controller's board (Arduino UNO) via Serial Port as shown in Figure 4.

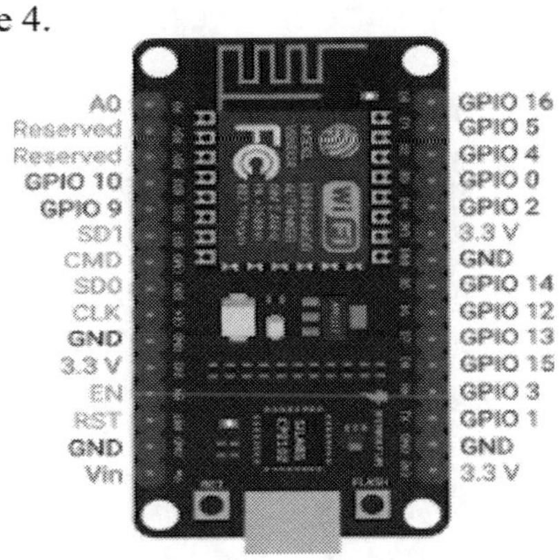

Figure 4. Node MCU ESP8266

4.2.1. Features of NODEMCU

- 32-bit micro-controller SOC works at 80 MHz having antenna switches, power amplifier, digital peripheral interfaces, low noise receive amplifier, power management module and with filter capability.
- It comprises of 64 Kilo Bytes of boot Read just Memory, 80 KB client information RAM and 32 KB guidance RAM.

4.3. Arduino IDE

- It is an open-source gadgets stage dependent on simple to utilize equipment and programming.
- This board is flexible runs on Mac, Windows, and Linux and able to access all type of inputs and as well as all type of outputs.
- It provides several development tools like, IDE (Integrated Development environment).
- In the Arduino 1.8.13 we can convert the any language code into hexadecimal code for uploading into Arduino.

The Figure 5 shows the output of the code where that Sketch uses 279972 bytes (29 percent) of program storage space. Maximum is 958448 bytes. Global variables use 27804 bytes (33 percent) of dynamic memory, leaving 54116 bytes for local variables. Maximum is 81920 bytes.

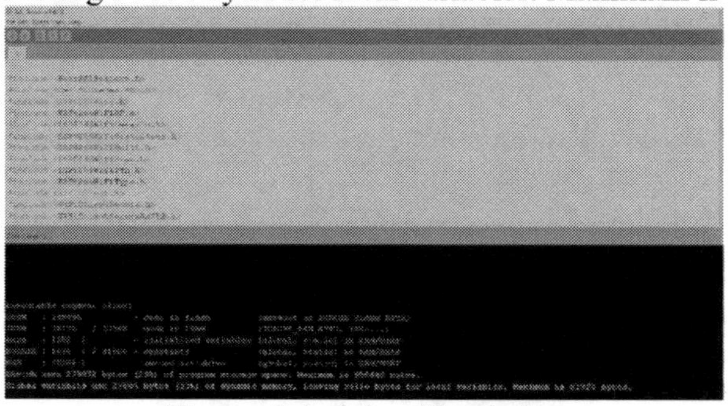

Figure 5. Arduino Simulation

4.4. Adafruit IO

It is a cloud administration free open source tool that simply implies to run it for you and you don't need to oversee it. This will connect to projects to many web services like Android SMS, E-Mail, RSS feeds, weather services, etc

5. Results and Discussion

5.1. Arduino Output

The simulated ESP8266 output is shown in Figure 6 using Arduino IDE. When the codes of sensors which are being used in the hardware were dumped into the Arduino IDE and after the successfully compilation, the codes are uploaded to the ESP8266 and the sensors will detect the parameters which will be sent to the receiver through MQTT protocol.

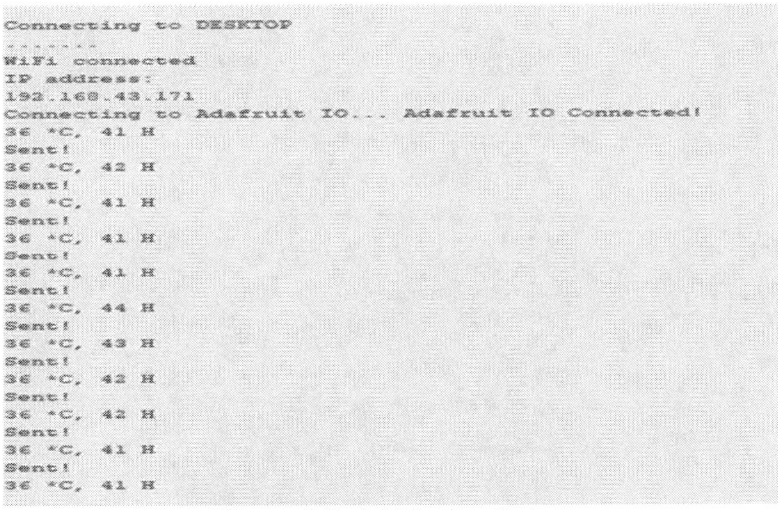

Figure 6. Serial Monitor Output

5.2. Server Output

The Figure 7 shows the results of the sensors used in the hardware in the MQTT server.

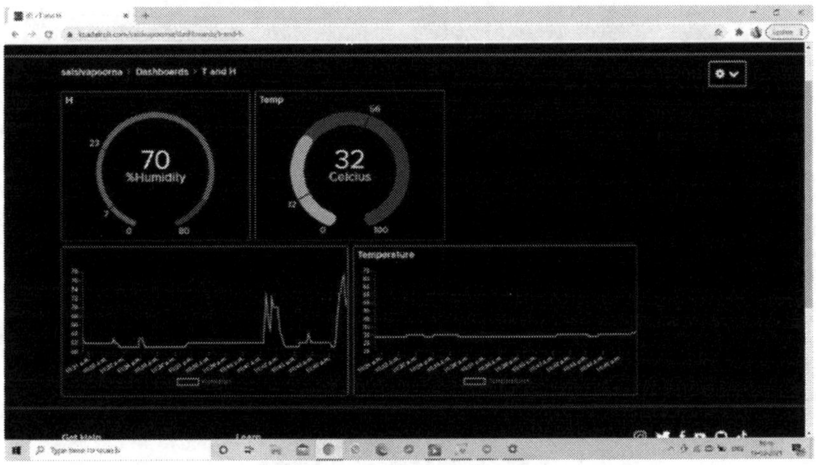

Figure 7. Server Output

5.3. Message Output

Figure 8 shows the output of the Wind Turbine whenever the temperature exceeds the threshold value a message will be sent to phone.

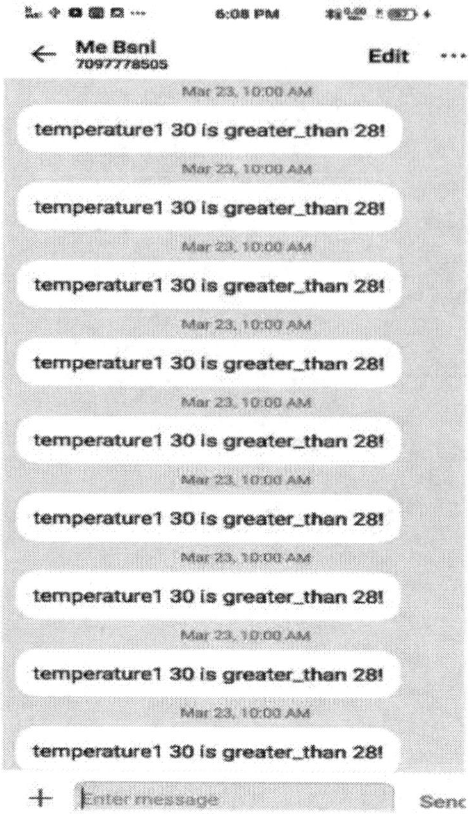

Figure 8. Message Output

5.4. E-Mail Output

The Figure 9 shows the results of the monitoring system through E-mails whenever the temperature exceeds the threshold level.

Figure 9. E-Mail Output

5.5. *Wind Turbine Model Output*
The Figure 10 shows the output in the wind Turbine consists of sensor and anemometer to checkthe speed of the air.

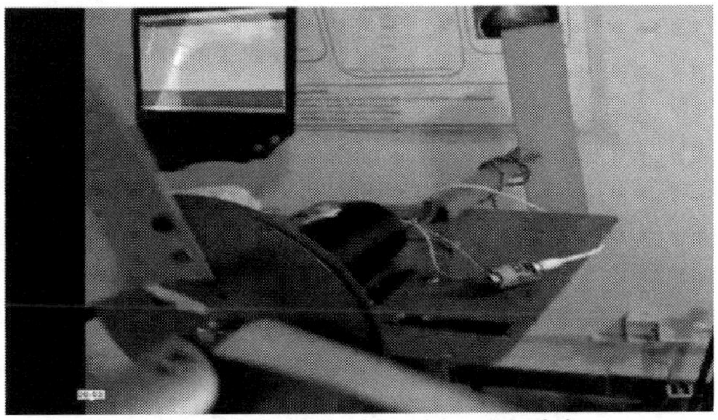

Figure 10. Wind Turbine Model

6. Conclusion
In Wind turbines there is a possibility of fault occurrence which is explained in problem statement. So, the chance of fault occurrence will be more, even it may lead to power off. The proposing remote monitoring system will detect the faults occurred in the turbine and sends using ESP8266 through the MQTT to receiver.

References
[1] "Active Fault Tolerance Control For Sensor Fault Problem in Wind Turbine Using SMO with LMI Approach", Nuralif Mardiyah, Novendra, Setyawani, Bella Retno and Zulfatman Has Department of Electrical Engineering, University of Muhammadiyah Malang, Indonesia, Proceeding of EECSI 2018, Malang – Indonesia.

[2] 2017-PSEC-0784, Aziz Altaf Khuwaja, "Sensor Fault Detection and Isolation for a Wireless Sensor Network-Based Remote Wind Turbine Condition Monitoring System",by Yayu Peng Student Member, IEEE Power and Energy Systems Lab Department of ECE,University of Nebraska-Lincoln Lincoln, NE 68588-0511 USA.

[3] IEEE Transactions on Control Systems Technology "Wind Turbine Fault Diagnosis and Fault-Tolerant Torque Load Control Against Actuator Faults" by Hamed Badihi, Student Member, IEEE, Youmin Zhang, Senior Member, IEEE, and Henry Hong.

[4] A Survey of "Wind Turbine Control Monitoring and Fault Detection on Wind Energy" by M.Kokila Research Scholar Department of Computer Science Ayya Nadar Janaki Ammal College,Sivakasi.

[5] IEEE Transactions on Industrial Informatics, VOL. 15, NO. 7, JULY 2019, "A Small-Sample Wind Turbine Fault Detection Method With Synthetic Fault Data Using Generative Adversarial Nets"by inhai Liu , Member, IEEE, Fuming Qu , Student Member, IEEE, Xiaowei Hong , and Huaguang Zhang , Fellow, IEEE.

[6] Research Article, "Wind Fleet Generator Fault Detection via SCADA Alarms and Autoencoders" by Mattia Beretta, Juan Jos´e C´ardena2, Cosmin Koch 2 and Jordi Cusid´o, Published on 3 December 2020.

[7] "Study of using MQTT Cloud Platform for Remotely Control Robot and GPS Tracking" by Nut Aroon,Department of Computer Engineering School of Engineering, BangkokUniversity Pathumthani, Thailand.

[8] "he Use and Performance of MQTT and CoAP as Internet of Things Application Protocol using NodeMCU ESP8266 " by, Sandy Suryo Prayogo, Yulisdin Mukhlis , dept of Electrical

Engineering,Gunadarma University ,Jakarta, Indonesia.
[9] "IOT Based Fault Detection of Underground Cables through Node MCU Module", by Laxmi Goswami ,Vinay Anand ,Department of EE, Sanskriti University,Uttar Pradesh, India..

Control of Environmental Pollution by Utilizing Wastes from Industry on Fly Ash Based Geopolymer Concrete

Ispara Xavier S[1*], Divahar R[2], Aravind Raj P S[2], Devi Saran M[3], Varsha Manohar[3], Athul Sudhakaran[3]
[1]Assistant Professor, Department of Civil Engineering, Aarupadai Veedu Institute of Technology, Vinayaka Mission's Research Foundation, Paiyanoor, Chennai.
[2]Associate Professor, Department of Civil Engineering, Aarupadai Veedu Institute of Technology, Vinayaka Mission's Research Foundation, Paiyanoor, Chennai.
[3]U.G. Student, Department of Civil Engineering, Aarupadai Veedu Institute of Technology, Vinayaka Mission's Research Foundation, Paiyanoor, Chennai.
(*Corresponding Author, Email:isparaxavier.civil@avit.ac.in)

Abstract. Geo-polymer Concrete is mainly hardened cementitious paste made from fly-ash and alkali solution. It helps to combines waste product into useful product. In this the major point is to investigate the solidity parameters of GPC utilize varying proportions of fly-ash, GGBS & Silica fume in addition to alkali solution (Na-OH and Na2SiO3), which reduces demand of OPC which leads to the emission of CO2 to the environment and to produce a more durable infrastructure. With various advantages of these materials as studied in detail in this project, better strength parameters for Geo-polymer concrete are expected by using varying proportions of Fly-ash, GGBS, Silica-fume along with an alkali solution (NaOH and Na2SiO3) than conventional concrete. Thus, Geo-polymer concrete helps in the complete replacement of Portland cement and act as an environmentally friendly material by the reduction of CO2 emission to the society which leads to control the environmental pollution. It is expected that centralizing over needs & examining in a way that includes a lot of careful detail in this area will help to materialize flue-ash based GPC as a trading & non-polluting material also set the seal on the sustainability of construction.

Keywords: Geo-polymer Concrete, Fly-ash, GGBS, Silica-fume.

1. Introduction

Expansion of construction shows to the usage of numerous method which sources, damaging reaction of discharge in the habitat. But the order of concrete is multiplying gradually for its development & manufacturing by entire types of suited forms. To avoid these issues, the concrete should be made with eco-friendly materials. To bring-up non-polluted concrete, the cement wanted to restore with some other binders which should not be creating any bad effect on the surroundings. The use of industrial byproducts as binders can reduce the problem. By the regard of the vanguard GPC is optimistic techniques. The GP's were the division of the genealogy of dead polymers undergo synthesized configuration alike to essential zeolite materials, but the multilayers were un-structured. So that the Geo-polymer (GP) technology could bring down the CO2 outflow to the atmosphere. The main aim to investigate the hardness limit of flue-ash formed GPC by varying proportions of flue-ash, GGBS & silica fume along with an activator solution (NaOH & Na2SiO3).

(V. Gopi & K. Shyam Chamberlin 2019) come-up with an experimental investigation on the strength and durability of concrete incorporated with silica fume and fly-ash; in this study, the utility of concrete enlarges with the increase in the proportion of silica fume and fly ash in concrete. (M. G. Kamaldeep & K. Shyam Chamberlin 2019) carried out a paper on studies the production of concrete by using a mixture of fly-ash and GGBS as combined material. In this the representation of M40 by comparatively restore Fly ash and GGBS to cement were radiate and analyzed for Compressive strength at 7, 28 days. Flexural strength merited were also attained at 28 days and obtained results were compared with conventional concrete. (Akshatha K. B 2018) carried out an experimental study of concrete using silica fume. In this examination, the introduction of silica fume to the materials directed to reduce the resilience of OPC and the bulk of the material with silica fume is close to the OPC. But in the compressive strength in the concrete, the silica fume gets increases in it. And also, flexural strength & split tensile strength are also close variants where the silica fume was added. So, the modulus of elasticity in addition to with silica fume concrete there was no specific trend of dissimilarity. (V. Hariharan et al. 2017) studied on implementation in fly-ash & GGBS as an entire renewal of the cement in Geo-polymer concrete. The experimental investigations executed, the automatic resources of Geo-polymer concrete such that the compressive strength, split tensile strength & flexural strength expand with expansion in the GGBS gratified. By that absorption of NaOH solution rises and the energy will expand. Compressive strength of GP enlarges within the expansion of era in concrete.

1.1 Geo-polymer Concrete

Geo-polymer concrete is the combination of squandering substance for example fly-ash, silica fume and also GGBS; were they mixed with the binding compound. Portland cement is not mandatory for the fabrication of GPC. The manufacturing of the binder is by the synthesizing of alkaline liquid with the original matter loaded with silica & alumina. The principal fragment of GPC are industrial products such as the silicon & aluminium and also alkaline start up this polymerizes this medium into the molecular pattern and a webbing to invent a desensitized binder. And well known as the inanimate polymer cement. Also, the major polymerization take place in geo-polymer concrete occurs while the heat obtruded to throughout the healing span. Lengthy healing time will amplify the polymerization action and the outcome will be high-level compressive strength. GPC have a marvellous defiance with chemical strike and display the capability of hostile habitats where the toughness of Portland cement will be analysed. It will relevant in aquatic regions, the place where high-level CO2 soils. Thus geo-polymer concrete helps in the complete replacement of Portland cement. . It is expected that centralizing over needs & examining in a way that includes a lot of careful detail in this area will help to materialized flue-ash based GPC as a trading & non-polluting material & set the seal on the sustainability of construction.

2. Materials and Objectives

The ingredients used for this work are same as those used in the conventional concrete. But here fly-ash, GGBS and silica fume are introduced as fully replacement for cement. The material used are cement, sand, coarse aggregate, fly-ash, GGBS, silica fume, super plasticizer and water.

The future developments of construction should be economy and environmental basis. Fly-ash based GPC is an economic alternative to ordinary concrete. The overall cost for geo-polymer concrete is under of conventional concretes, since it is utilizing easily and freely available for fly-ash, cheapest chemicals NaOH and Na2SiO3 solution.

Beyond such economic benefits, the major advantage of geo-polymer concrete is related to environment:

To construct a convenient thing and supply extreme level concrete besides OPC.

Mainly reduces CO2 emission and eco-friendly products.

Journal of Physics: Conference Series **2040** (2021) 012003 doi:10.1088/1742-6596/2040/1/012003

2.1 General

This output is comparing the solidity parameters of geo-polymer concretes with Ordinary Portland Cement (OPC). These describes quantity calculation, properties of materials used and casting of specimens. The total number of cubes, beams and cylinder are shown in Table 1.

Table 1. No. of cubes, beams and cylinders for GPC and OPC

	No. of cubes, beams, and cylinders for GPC	No. of cubes, beams, cylinders for OPC
7th day	70% flyash, 20% GGBS, 10% silica fume 60% flyash, 30% GGBS, 10% silica fume 50% flyash, 40% GGBS, 10% silica fume	9 cubes
28th day	70% flyash, 20% GGBS, 10% silica fume 70% flyash, 20% GGBS, 10% silica fume 70% flyash, 20% GGBS, 10% silica	6 cubes, 6 cylinders 4 beams

3. Result and Discussion

Test outcome of the geo-polymer concretes specimens were emulated with the normal concrete specimens of same size & grade and the results were presented using tables and graphs. The function of concrete by fully renewal of cement by varying the percentage of pozzolanic materials; (70% fly-ash, 20% GGBS, 10%silicafume), (60% fly-ash, 30% GGBS, 10% silica fume), (50% fly-ash, 40% GGBS, 10% silica fume). Workability then the compressive strength, flexural strength & the split tensile strength of the test specimens. The strength parameters of ordinary concretes specimens and Geo-polymer Concrete specimens.

The sieve analysis is done to both fine & coarse aggregates to measure the percentage of sieves. Fine aggregates pass through 4.75mm IS sieve & in other-hand coarse aggregates retains through 4.75 mm. In Fig. 1.is shown the findings of sieve analysis of both, fine & coarse aggregates.

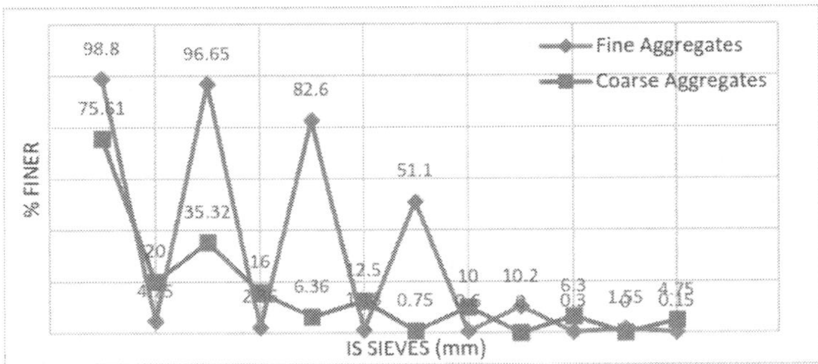

Figure 1. Sieve Analysis Result

Then the cast of specimens will takes-place through various test such as workability in the concrete, where the mixer material can be moulded into any shape. For maximum strength the compaction should be 100% in the concrete. In which slump & compaction tests were approved for inspect the workability of the concrete. After this test only the degree of workability will be finalized.

The conventional concrete specimens were prepared first & compressive strength tested for an age of seventh and twenty-eighth days after cast of the specimens. In this study, for conventional concrete, 150×150×150 mm cube specimens were developed. Three of these cubes are testing for compressive strength at the span of seventh and twenty-eighth days after cast the specimens. Fig. 2. given below shows the compressive strengths of conventional concretes at seventh day and twenty-eighth day as a result.

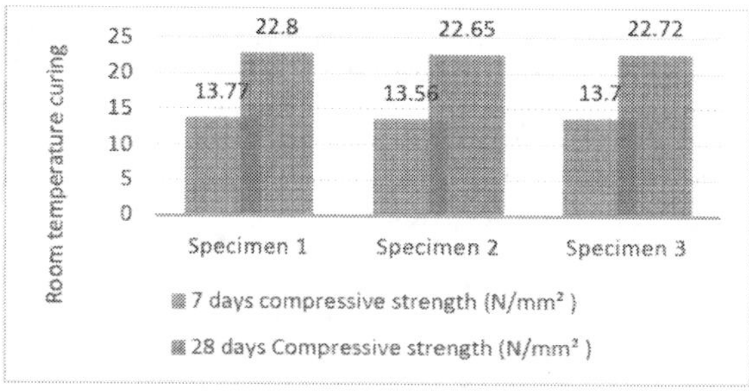

Figure 2. Compressive strength of specimens

Split tensile strength are indirect practice that determinism the tensile strength of concretes used in a cylinder which splits over the vertical diameter. The length of the specimen must not be below than that of the diameter & not more than double the diameter. The cylindrical specimens of size 150mm diameter × 300mm height were prepared and tested for split tensile strength on conventional concrete. The fill-up must be enacted in need of upset & enlarges interminably at a nominal rate within the span of 1.2 N to 2.4 N. Fig. 3.given below shows the Split tensile strengths of conventional concrete at 28th day.

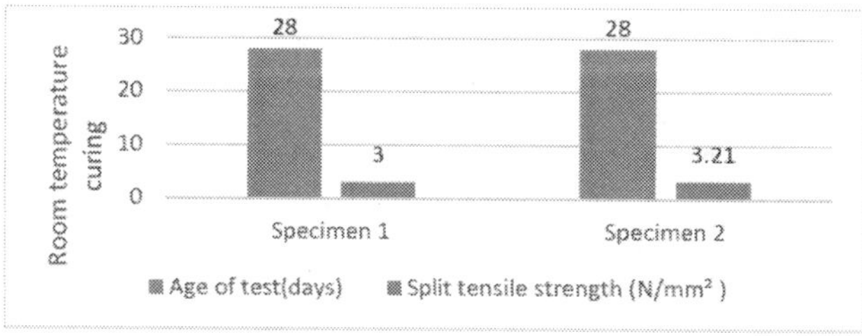

Figure 3. Split tensile strength of conventional concrete for 28th day

Flexural strength of a conventional concretes is deliberated by the load of 150×150 mm concrete beams with a width of three times. The flexural strength results were found in the span of 2.5N/mm2 - 2.38 N/mm2. Fig. 4. given below shows the Flexural strength on conventional concrete at 28th day.

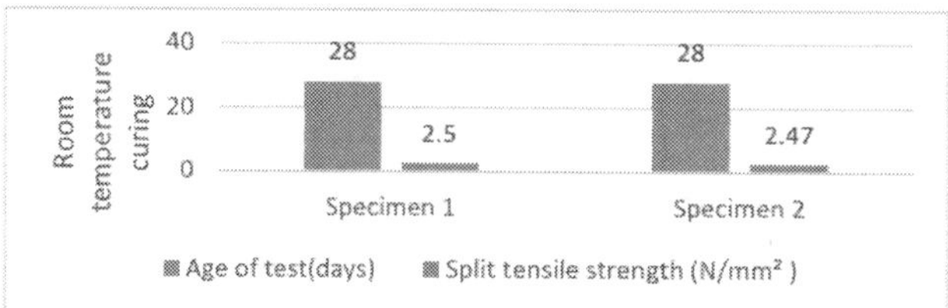

Figure 4. Flexural strength of conventional concrete for 28th day

In this work, the mix proportions used to make the geo-polymer concrete were based on the arrangement & the habitual property of flue-ash & GGBFS synthesised Alkali-activate concretes. Types of matter fraction for geo-polymer concrete mixers were shown in Table 2. The ratio from alkaline to the binder will be considered as 0.35-0.5, & also like that from sodium silicate to sodium hydroxide solution as 2.5-3. Sum of aggregate in the mass of concrete is 66-88%, & the content of fine aggregate in amount aggregate sum is adopted like 35%. Scale of plasticizer is filled like 2%-4.5% by the mass of binder. Additional water content on condition that lacking can be included in the span of 0.03%-0.07% by mass of bitumen substances.

Table 2. Materials used in geo-polymer concrete mixes

Materials Used	Range of materials
Alkaline liquid-Binder	0.35-0.5
Sodium silicate-Sodium hydroxide	2.5-3
Distilled Water-Binder	0.17-0.25
Sum of aggregate in mass of concrete	66-88%
Fine aggregate contented in amount of aggregate	35%
Water content	0.03%–0.07% of mass of bitumen substances
Super Plasticizer	2%–4.5% of mass of bitumen substances

In the GPC mix design the dissimilar materials used are flue-ash, GGBFS, sodium silicate, sodium hydroxide, fine & coarse aggregates, also water & synthetic polymers. Blend percentage of these medium are taken with the substructure of unit weight of plain concrete. Percentage of fine & coarse aggregates used were 70-30% as a result. Quotient of alkaline to bitumen substance was taken as 0.35 & ratio of sodium silicate & sodium hydroxide was 3. Synthetic polymers of 4.5% was utilize to enhance the viability of fresh GPC. All mix design values were prepared, and the values were shown in Table 3.

Table 3. Mix proportions weight of all materials

Materials(kg)	Quantity for 1m3	For each percentage proportions	Total quantity needed for all proportions.
Fly-ash	373.1	14.7	44.1
GGBS	106.6	4.2	12.6
Silica fume	53.3	2.1	6.3
Coarse aggregate	1176	46.339	139
Fine aggregate	504	19.85	59.55
Mass of Na2SiO3	65	2.56	7.68
Total water content in Na2SiO3	69	2.71	8.13
Mass of NaOH	13.9	0.547	1.64
Total water content in Na2SiO3	39.09	1.54	4.62
Super Plasticizer	24.48	0.96	2.88

4. Conclusion

Geo-polymers have much potential as construction materials as well as non-pollutant materials due to their excellent mechanical properties. They can be cured at ambient temperature and emit less CO2 compared to Portland cement which leads to control the environmental pollution. The mixture proportions and the manufacturing process used to make the Geo-polymer concrete was based on the arrangement & the habitual property of flue-ash & ground granulated blast-furnace slag synthesized alkali activate concretes.

- Fly-ash is acknowledged as the superior medium to composing the Geo-polymer concrete & announce to the peak of compressive strength to geo-polymer concretes.
- Fly-ash rooted Geo-polymer concrete indicated magnificent property & approved for structural developing by analyst & with the expanding merit of the NaOH molarity that the stability of Geo-polymer concrete gets grown-up.
- Rising the ratio of GGBS also expanded the fire resistance of concrete.
- By utilizing of silica fume in the concrete that expand the compressive strength. Avoid the breeze forces high corrosion & abrasion resistance, huge structures obtain the silica fume content to allow the high toughness concrete.
- Which is a breakthrough to the furtherance of using super-plasticizer in concrete-technology and here Ceraplast-300 helps to generate intense resilience in concrete no vibration in the course of set down.
- With various leads of these materials as studied detailed in this project, better strength parameters for Geo-polymer concrete are expected by using varying proportions of Fly-ash, GGBS, Silica-fume along with an alkali solution (NaOH and Na2SiO3) than conventional concrete. Thus, Geo-polymer concrete helps in the complete replacement of Portland cement. It is expected that centralizing over needs & examining in a way that includes a lot of careful detail in this area will help to materialize flue-ash based GPC as a trading & non-polluting material also set the seal on the sustainability of construction.

REFERENCES

[1]. Akshatha K.B (2018) Experimental Study of Concrete using Silica Fume, International Research Journal of Engineering and Technology (IRJET) e-ISSN: 2395-0056, Volume: 05 Issue: 05.
[2]. D. Lavanya, J. Guru Jawahar, and C. Sashidhar (2016), —Performance of Fly Ash and GGBS Based Geo-polymer Concrete in Acid Environment, ‖ International Journal of Research and Scientific Innovation (IJRSI), Vol. 3.
[3]. K. Chandrakar and V.V.Singh (2017), —An Experimental Study on High Performance Concrete using Mineral Fly Ash and GGBS with M-60 grade Concrete,‖ International Journal Of Advance Research, Ideas and Innovations In Technology (ISSN) , Vol. 3, no. 6.
[4]. Khushpreet Singh and Prakul Thakurl (2018), A Review on Geo-polymer Concrete: A Green Concrete, International Research Journal of Engineering and Technology (IRJET), Volume: 05 Issue: 05.
[5]. Khushal Chandra Kesharwani, Amit Kumar Biswas, Anesh Chaurasiya and Ahsan Rabbani (2017), —Experimental study on use of fly-ash in concrete‖, International Research Journal of Engineering and Technology (IRJET), Volume: 04 Issue: 09.
[6]. Manimaran E and Mohankumar G (2017) —Influence Of Sodium Hydroxide Concentration on the Strength of Fly ash Based GPC‖, International Journal of Engineering Sciences & Research Technology.
[7]. M .G. Kamaldeep and K. Shyam Chamberlin (2019) Experimental Study on Performance of Concrete by using Combination of Fly-ash and GGBS Blended Material, International Journal of Recent Technology and Engineering (IJRTE) ISSN: 2277-3878, Volume-7, Issue-6.
[8]. Patankar S.V, Ghugal Y.M and Jamkar S.S (2014) —Effect of Concentration of Sodium Hydroxide and Degree of Heat Curing on Fly ash Based Geo-polymer Mortar," Indian journal of materials science, vol. 2014, pp. 1–6
[9]. Prince Arulraj and G Shashikant and (2019), A Research Article on —Geo-polymer Concrete, International Journal of Innovative Technology and Exploring Engin1Zeering (IJITEE) ISSN: 2278-3075, Volume-8.

[10]. Reddy, S. Vijaya Bhaskar, and P. Srinivasa Rao (2016)"Experimental studies on compressive strength of ternary blended concretes at different levels of micro silica and GGBS.",International Journal of Advanced Engineering, Management and Science Vol. 3, Issue10.

[11]. S Kumar, J Pradeepa and P M Ravindra (2013), Experimental investigation on optimal strength parameters of fly ash based geo-polymer concrete, International Journal of Structural and Civil Engineering Research, Vol.2.

[12]. V. Gopi and K. Shyam Chamberlin (2019) Experimental Investigation on Strength and Durability of Concrete Incorporated with Silica Fume and Fly Ash, International Journal of Recent Technology and Engineering (IJRTE) ISSN: 2277-3878, Volume-7, Issue-6.

[13]. V.Hariharan, G.Silambarasan and N.Harishkumar (2017) Utilization of Fly-ash and GGBS as A Fully Replacement of Cement in Geo-polymer Concrete, International Conference on Emerging trends in Engineering, Science and Sustainable Technology (ICETSST-2017).

[14]. Wankhede, P. R., and V. A. Fulari (2017) "Effect of fly ash on properties of concrete," International Journal of Emerging Technology and Advanced Engineering, Vol. 4, Issue 7 pp. 284-289.

[15]. www.google.com

Fuzzy Logic Intelligent system for an Automatic medical waste segregation

V Ramani Bai Gopinath[1], S H Dhavarpanah[2], G Kangadharan[3], R Ruzaimah[1]

[1] Department of Civil Engineering, Faculty of Engineering, Technology and Built Environment, UCSI University Kuala Lumpur, Malaysia
[2] Intelligent Energy Systems PTY LTD, Sydney, Australia
[3] Alpha Cambridge International School, Tiruchirappalli, India

Email: ramaniucsi@gmail.com

Abstract. AUTOM is an acronym given to a commercially designed automatic waste disposal master tool for medical laboratories and clinical waste separation through this research study. In this research, a fuzzy rule-based system is designed which can segregate 24 different waste types which are selected from 8 more common medical waste groups. In the proposed system, after capturing each frame some pre-processing operations are done, features are extracted, fuzzy parameters, fuzzy terms and fuzzy rules are determined and finally a rule with a maximum certification degree is fired. In the designed fuzzy rule-based system 24 roles corresponding to the 24 detectable objects exist, 14 fuzzy parameters are defined, and a maximum of 3 fuzzy terms for each fuzzy parameter have been considered. The system is flexible on different light conditions, view degree of the camera, and 360° object rotation. Our experiments on a real environment and using both new and used wastes have shown up to 94% of better performance of the system developed.
Key words: Automation, Clinical waste segregation, Fuzzy intelligent system, Fuzzy rule-based system, Hazardous waste.

1. Introduction

Large hospitals, according to studies, generate 2.0 kg of garbage per bed every day. 0.5 kg of this can be classified as biological risk waste. Injuries from infected sharps and Hepatitis B, C, and HIV infection can occur if hospital trash is not handled appropriately and thrown off [1]. Burning, chlorination, baking soda, packaging, and UV irradiation are all options for treating this massive level of healthcare waste. Automatic trash classification is essential for all these systems.

The unit operations' optimal values of the output indicate that the integrated biological system in this study is ideal for the rapid growth of bacteria [2]. Massive plastics are mainly utilized to produce bottles, packaging, and bags. Intelligent sorting systems, based on image recognition and classification, have been developed [1]: (1) shape recognition[3], (2) linear discriminate analysis (LDA) applied to gray level histogram on images **Error! Reference source not found.**, (3) using other physical properties such as density, near-infrared spectroscopy (NIR) **Error! Reference source not found.**, or (4) studying the difference of in the transmittance (or absorption) between different plastics to an infrared InGaAsP laser [4]. Usually, the recycling process development investigation is focused on automatic sorting for massive plastics. Several reviews can be found for this issue. Further literature studies have proved the vulnerability of

waste disposal. Construction of buildings, waste disposal, and other practices have all led to the contamination of Sungai Kerayong. Most of the sediments that emerged from mines effluent, domestic waste, and soil erosion are raising the value in suspended solids [8].
This paper uses a diary methodology to study the impacts of air pollution ratings on employees' routine physical resources and behaviour. According to multilevel data obtained over 2 weeks from 155 employees in metropolitan China, ratings of air pollution severity deplete workers' self-control resources [10]. Hospitals are currently looking for disposable pharmaceutical products that can be disinfected and recycled to cut patient costs by 40% to 60% [11]. The algorithm, and a parallel version, are detailed in the Fuzzy module of the free computer algebra software Sage Math. Solve Fuzzy System, or SFS, is an efficient technique for finding actual solutions of computation systems with symmetrical L-R fuzzy numbers as coefficients and constrained support and objective distribution functions are L and R. The real solutions of a system can be derived from the solutions of some polynomial systems with real coefficients [12]. It presented a numerical approach for generating this index as well as a way of computing them. When comparing two fuzzy sets, the comparison should be available to respond where the order of the fuzzy sets cannot be identified, and hence their order is ignored. Two crisp relations with shared factors over a collection of fuzzy sets with trapezoidal classifiers were generated using the probability-based preferred intensity index. There exist a stringent order relation and the rest is a relation that reflects order indifference, based on the parameter value [13].

Environmental data can be organised and analysed using relational databases and GIS, the Geographical Information System. The coastal resource information system for change of climate has been built to be easy and versatile. Variation in the degree of information given for each attribute and country is possible thanks to the database structure [14]. A distance in between the sum of upper semicontinuous fuzzy subsets of a finite-dimensional Euclidean space but also its T-convex hull has an upper limitation. As a result of this discovery, we describe an iterative approach for generating a T-convex hull of an upper semicontinuous fuzzy set. As an extension of the conclusions about t-norm fuzzy sets, we offer an example from a game-theoretic situation where t-norm fuzzy sets were utilised to handle uncertainty in a recurring two-player game [15]. In this research, a fuzzy rule-based system is introduced. The input of the system is a 2D RGB online captured frame and the output of the system is an array that shows the certainty factor of belonging of each object in the frame to a waste type and a waste class. The considered classes and subclasses in this project include Microbiological Waste (Ampoules, PCR Tubes, Petri dishes, Sample Tubes), Non-Recyclable (Ceramic Mugs, Feeding Bottles, Food Containers), Pathological Waste (Bandage Spool, Bottles, Cotton Balls, Gauze Pads), Pharmaceutical Waste (Inhalers, Medicine Bottles, Pipettes, Vials), Recyclable (Aerosol Cans, Forks, Juice Bottles, Mineral Water, Paper Cups), Sharps (Razor Blades, Scalpel Blades, Scalpels, Syringes). To classify the waste types, a fuzzy intelligent system is utilized including 24 rules and several fuzzy terms which are based on the extracted features from the input frame.

2. Intelligent Model Development

A flowchart of the method which is followed in this research to segregate the wastes is shown in Figure 1. Generally, this process is combined with two main steps; first, some pre-processing operations are applied to the image to enhance image quality, separate objects, and eliminate the unwanted area of the image. In the rest of this section, after a brief view of the utilized fuzzy system, our method is described in detail.

2.1 Fuzzy Intelligent System

Fuzzy logic is a multi-valued logic that allows intermediate values to be defined between conventional evaluations like yes/no, true/false, black/white, etc. Notions like rather warm or pretty cold can be formulated mathematically and algorithmically processed. In this way, an attempt is made to apply a more human-like way of thinking in the programming of computers ("soft" computing). Fuzzy logic systems address the imprecision of the input and output

variables by defining fuzzy numbers and fuzzy sets that can be expressed in linguistic variables (e.g. small, medium, and large).
A fuzzy rule-based approach to modelling is based on verbally formulated rules overlapped throughout the parameter space. They use numerical interpolation to handle complex non-linear relationships. Many existing systems need the rules to be formulated by an expert. Fuzzy rules are linguistic IF-THEN- constructions that have the general form "IF A THEN B" where A and B are (collections of) propositions containing linguistic variables. A is called the premise and B is the consequence of the rule. In effect, the use of linguistic variables and fuzzy IF-THEN- rules exploits the tolerance for imprecision and uncertainty. In this respect, fuzzy logic mimics the crucial ability of the human mind to summarize data and focus on decision-relevant information.

For incorporating membership values representing ambiguous knowledge, fuzzy logic offers a variety of appealing aggregation operands. In functional form, a fuzzy set's membership function can be a normal distribution function, a triangular function, a trapezoidal-shaped function, and so on.
Because specifying the control rules in fuzzy systems is frequently quicker and better, needing fewer rules, the systems run faster than traditional systems. Fuzzy systems are frequently tractable, robust, and reduced.

The following is the method for achieving the fuzzy output of such an expert knowledge base [9].
The shooting level of the ith rule is evaluated by:

$$A_i(x_0) \times B_i(y_0) \quad (1)$$

The output of the ith rule is calculated by:

$$C_i(w) = A_i(x_0) \times B_i(y_0) \rightarrow C_i(w) for\ all\ w \in W \quad (2)$$

The overall system output, C, is obtained from the individual rule outputs C_i by:

$$C(w) = Agg\{C_1, \ldots, C_n\} for\ all\ w \in W \quad (3)$$

2.2 Waste Segregation Method
The model development has followed the flow chart as presented in Figure 1. Applying some pre-processing operations on the captured image. Each process and the effect of its application on the image is illustrated in **Error! Reference source not found.**. The steps involved in this method are explained below.

2.2.1 Normalizing RGB image and setting light coefficients.
To reduce the effect of different light conditions in the image capturing process, triple RBG elements are normalized. First, the image is binaries. The average of each R, G, and B element in the background pixels (In the binaries image pixels which are detected as non-object pixels) is calculated. The average of all background pixels is calculated. Finally based on the difference of average of each element and calculated total average, corresponding values are updated. Sometimes based on the light condition, the range of intensity values of different pixels (including the ones that belong to the various kinds of wastes and background) is too limited, so their separation is not possible. To solve this problem, this range should get widen. This process can be applied in two steps. First when the final binary image is extracted. (Extraction binary image is one of the most critical steps. By this process, object areas in the image are separated from background pixels). Second in decision making and while intensity is used as a feature to detect the waste type. In both steps based on the range of the average intensity values in background and object Gray-scale pixels, a coefficient is defined and applied to the corresponding intensity values.

2.2.2Converting RGB image to Gray-Scale image

2.2.3 Building binary image

2.2.4 Releasing unwanted borders of the binary image

Because of the inaccurate view of the camera, usually, there is an unwanted light border in the captured image as the image object. Thus, the border should be removed before classification. This process is done by sweeping the image on each side. Each vertical and horizontal line in which all of its pixels are detected as non-background pixels is removed from the image.

2.2.5 Applying noise reduction on the binary image

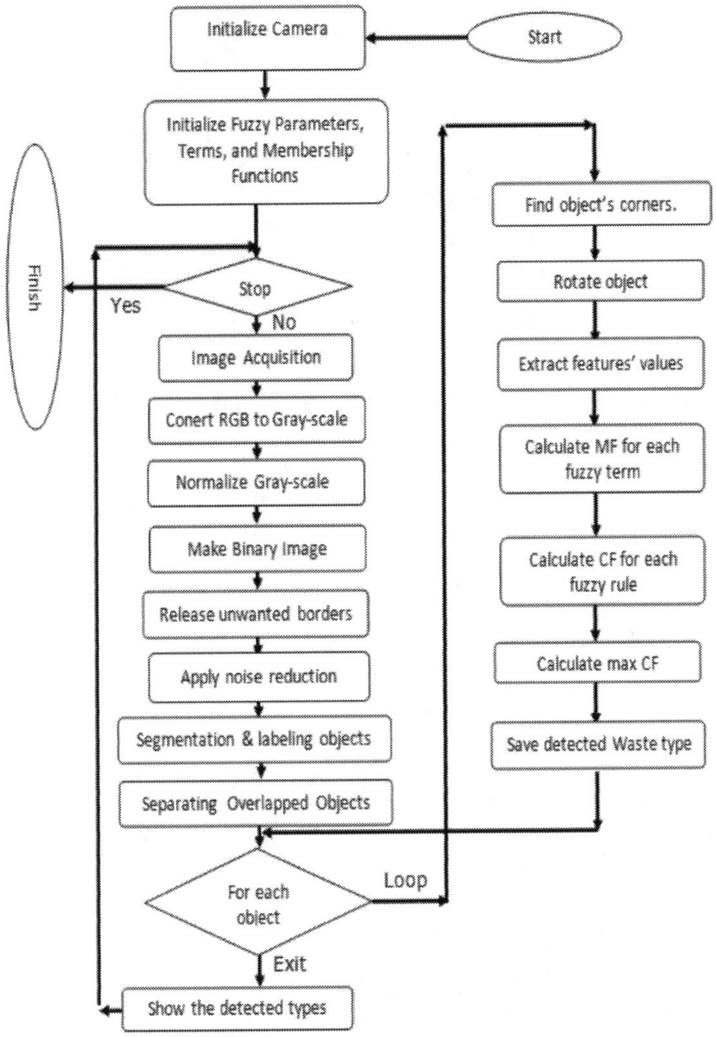

Figure 1. Fuzzy Medical Waste Segregation Flowchart

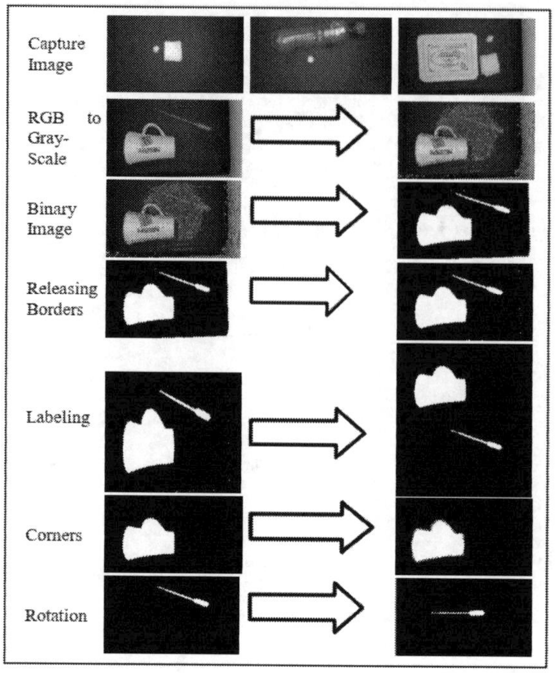

Figure 2. Sample Pre-processing Output from the model developed.

2.2.6 Segmentation and labelling existing objects
The labelling operator is applied, and objects are counted in the binary image. In the next step, small portions are ignored. To do so, several pixels of that section is counted and is compared with a threshold value. The threshold is calculated based on the size of the corresponding image.

2.2.7 Separating overlapped objects
After applying morphological operations on each object and separating different parts of results, they are compared with the original binary image, and overlapped objects are separated.

2.2.8 Finding the corners of objects and

2.2.9 Rotating each object to its horizontal position.

2.3 Extracting proper features for each object from the corresponding captured image
Classification is done based on the information achieved from the captured image. All the extracted information from the image cannot be investigated for classification and several more informative features should be selected. The utilized features or a combination of them should be able to distinguish objects from each other.

Since the designed classification is not a history-based method, it only uses features extracted from the current capture frame. The captured image is a 2D colour image. Therefore, only the features are accountable which are extracted from spatial information of each pixel or each object of this image. From all features, geometric features like length, width, the ratio of length on width, diameter, object geometric shape, and so on are more effective in this project. They are accountable because of their independence to the light and environment condition and similarity of different objects which are categorized as a waste type. RGB colour information is less utilized because of the light condition, proximity of camera and objects, and light effect on the colour and intensity of captured object. HSI colour model which is used in these kinds of projects is investigated but because of the above reasons, they are utilized less. The final utilized features include the following:

- How much the object looks like a square?

$$S_i = min\left\{\frac{1-\delta_{h_i}}{l_i}, \frac{1-\delta_{l_i}}{h_i}\right\} \times \beta \qquad (4)$$

Where$i$stands to ith object.$\delta_{h_i}$is the standard deviation of the height of different parts of an object. $\delta_{l_i}$is the standard deviation of the length of different parts of the object. $l_i$andh_i refer to the mean of length and height of different parts of the object. And β is a coefficient that is set to 0.8 for small objects and 1 for big objects.

- The ratio of max length to a max height of the object.
- The ratio of max length on mode height of the object.
- The ratio of waste's Length on its mode Height (Height with most frequent)
- The ratio of the height of the object to the height of the image.
- The ratio of the length of the object to the length of the image.
- How many degrees of having two lumps.
- Distance between two lumps on the object
- HSI color model (Hue, Saturation, Intensity) and mean of each element all over the object.
- How much the object looks like a trapezium?
- How many degrees of having two height levels.
- How many degrees of having a hole inside the object.
- How much the object looks like a circle?

$$C_i = \frac{\left(\frac{1-Ci_r}{Ci_r+Ni_r}\right) + 2 \times Ratio_r}{3}$$
$$Ratio_r = min\left\{\frac{1-|2\times r-h_i|}{2\times r}, \frac{1-|2\times r-l_i|}{2\times r}\right\} \qquad (5)$$

r increases step by step from zero to whenever the number of pixels at radius r which are belonged to the object, Ci_r, be less than the pixels which do not belong, Ni_r.
How much the object looks white?

$$W_i = \frac{mean\{R,G,B\}}{mean(|\{R,G,B\} - mean\{R,G,B\}|)} \qquad (6)$$

where R, G, B refer to an average value of red, green, and blue values respectively of all object pixels.
A ratio of mean I (Intensity) on mean H (Hue).

- Size of the object.

2.4 Determination of fuzzy parameters
Extracted fuzzy parameters and corresponding fuzzy terms are listed in Table .

2.5 Definition of fuzzy terms
In this project, many fuzzy terms are defined. This work is done because of two reasons. Firstly, there are 24 waste types and classes. Classification of this number of waste types is only possible using a lot number of rules and to distinct each rule from the others lots of fuzzy terms are required.
Secondly, each fuzzy term is used in different rules, and the definition of computational fuzzy terms instead of computing them is preferable.
Generally, this system is designed based on the definition of many fuzzy terms to increase the power of its ability to classify waste types that are coming from various numbers of fuzzy terms.

2.6 Initializing fuzzy membership function parameters

One of the most effective of the performance of the system is that values are assigned to different parameters of fuzzy membership functions. Considering the nature of the utilized fuzzy terms in this project, three types of fuzzy membership functions are used.

Advancing and Dropping Linear Fuzzy Membership Functions, Triangular Fuzzy Membership Functions, and Trapezoid shape Fuzzy Membership Functions. Linear FMF's have two parameters, Triangular FMFs have three parameters, and finally, Trapezium FMFs have four parameters.

According to a large number of utilized fuzzy terms, several fuzzy parameters which should be assigned by proper values are a lot. In this project, values are assigned to the parameters experimentally. Based on the nature of the corresponding fuzzy term, an initial set of values are selected. Fuzzy term values are calculated. Fuzzy rules are fired and based on the accuracy of detected objects these values are adapted.

Table 1. Fuzzy Parameters and their Fuzzy Terms

No.	Fuzzy Parameters and fuzzy sets
1.	The ratio of max length on the max height of the object (4 sets)
2.	The ratio of max length on mode height of the object (1 set)
3.	Degree of having two lumps (1 set)
4.	Size of the object (4 sets)
5.	How much the object looks like a square? (2 sets)
6.	How much the object looks like a trapezium? (2 sets)
7.	How much the object looks like a circle? (3 sets)
8.	Intensity on Hue (2 sets)
9.	Hue (2 sets)
10.	Saturation (2 sets)
11.	Mean value of white color (1 set)
12.	The ratio of the height of the object on the height of the image (1 set)
13.	The ratio of the length of the object on the length of the image (1 set)
14.	Degree of having two height levels (1 set)

2.7 Definition of fuzzy rules

The decision-maker or classifier of the system is an intelligent system including some rules. The rules are designed based on the fuzzy terms and membership degree that each fuzzy parameter has.

The twenty-four (24) rules are designed for the system. It means for each waste type, a separate rule is designed. Each of these rules calculates the membership degree of the mentioned object in the corresponding class of waste type. So the output of rules is 24-degree values. Finally, the object is classified as a waste type with a maximum value between all rules.

By using a more comprehensive rule more than a waste type can be classified. Using this method and by reducing the number of rules, the speed of the system increases. The system did not follow this method so for each waste type a separate rule exists.

This decision was made due to the low importance of execution time in this project compared to its accuracy. In addition, processing of 24 rules is not time-consumer among intelligent systems with tens of thousands of rules. Furthermore, the extension of the system to classify more waste types in the future is much easier.

3. Results and Discussion

The sample output from the AUTOM system is presented in Figure 3. As it is clear in this figure, all the samples are classified correctly. In addition, Table 2 shows the percentage of correct detection times for each waste type separately.

TABLE 2. ACCURARY OF DIFFERENT WASTE TYPES

Waste Type	Accuracy	Waste Type	Accuracy
Ampoules	92%	PCR Tubes	87%
Petridishes	100%	Sample Tubes	85.5%
Ceramic Mugs	96.2%	Feeding Bottles	86.4%
Food Containers	100%	Bandage Spool	94%
Bottles	94.3%	Cotton Balls	97%
Gauze Pads	100%	Inhalers	96%
Medicine Bottles	94.3%	Pipettes	93.45%
Vials	95.4%	Aerosol Cans	88%
Forks	96.5%	Juice Bottles	87%
Mineral Water	87.5%	Paper Cups	91.5%
Razor Blades	100%	Scalpel Blades	97.2%
Scalpels	98.2%	Syringes	100%
Average		94.06%	

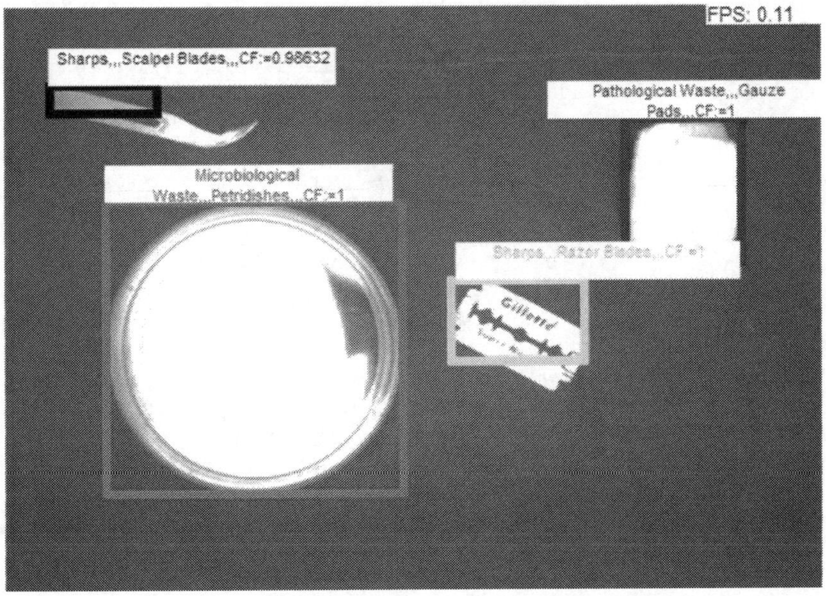

Figure 3. Output from the AUTOM fuzzy model

In some items such as Mineral Water whose shape is very near to Feeding Bottles and Juice Bottles or Aerosol Cans which do not follow a specific shape in different samples, the results are worse. Briefly average of the accuracy of the system is more than 94% which is acceptable.

- High accuracy of detection in the proposed method.
- There is no limitation on the number of waste types. By adding a new rule, a new waste type is detectable.
- The decision method is conceptual and understandable for human beings.
- Unlike methods such as ANN, detection of a big number of waste types is possible.
- It is not required to provide training samples.
- By using the fuzzy concept, the separation of similar objects is more meaningful.
- The transparency of the system is high. It is possible to explain why and how a decision has been made.
- Technical Challenges and Risks
- The high number of effective parameters. (Each fuzzy membership function has at least two parameters.)
- Adding a new type and new rule affects the other performances.
- Fuzzy parameters should be set manually.
- The unmanageable effect of environmental conditions on the achieved accuracy.
- Calculation of fuzzy membership values and sequential investigation of rules is time-consuming.

Like all systems, the proposed system in this research has advantages and some drawbacks. The selected number of these items are listed above respectively as an outcome of this study. The technical preferences and advantages of the new prototype developed will ensure for better waste segregation system than the existing method in practice.

4. Conclusions

In this research, a rule-based waste segregation system was introduced. The main contribution of the system was classifying 24 predefined waste types from 8 medical waste classes. Many fuzzy terms have been utilized to cover all similar states between these waste types and a fuzzy rule-based system is used to calculate the certainty factor of belonging each captured image object to one waste type. This system was run and tested in a real environment and using real waste samples. The results show that on different light conditions, different view degrees of the camera, and different positions of the objects, this system is capable to segregate objects. The technology is adaptable to a variety of lighting situations, camera view angles, and 360-degree object rotation. The tests in a real environment with both new and used wastes demonstrate that the system performs at a rate of 94%- 100% for shaped wastes, from 85-93% accuracy for shining objects and up to 97% for shapeless objects.

5. Acknowledgment

The authors acknowledge and thank the MOSTI (Ministry of Science, Technology, and Innovation), Malaysia for their support and the publication that has resulted in this paper. The authors would also thank the collaborating organizations especially the University of Nottingham Malaysia Campus and others for their excellent coordination among researchers and the assistance provided by postgraduate Research students.

REFERENCES

[1] Jawed A 2006 Hospital waste management issues and steps taken by the government of Pakistan. Ministry of environment. J of Hospital Infection, 14 (92), pp. 15-22.

[2] Assefa R et al 2019 Nitrogen removal in integrated anaerobic–aerobic sequencing batch reactors and constructed wetland system: a field experimental study, Applied Water Science, 9(5), pp.1-11.

[3] S Brunner et al 2012 Investigation of Algorithms for the Reliable Classification of Fluorescently Labelled Plastics, IEEE Intern. Instrumentation and Measurement Technology Conf. (I2MTC), pp 1659–1664.
[4] C. Amuthadevi, D. S. Vijayan, Varatharajan Ramachandran, "Development of air quality monitoring (AQM) models using different machine learning approaches", Journal of Ambient Intelligence and Humanized Computing, https://doi.org/10.1007/s12652-020-02724-2
[5] Amutha devi, Gayathri Monicka.J, "Emerging bio-medical applications and open research challenges in cognitive internet of things (CIOT)" International Journal of Pharmacy and Technology" Dec-2016 Vol. 8 Issue No.4 5049- 5054
[6] Brunner S, Kargel C 2011 Evaluation of potential emission spectra for the reliable classification of fluorescently coded materials, In Algorithms and Technologies for Multispectral, Hyperspectral, and Ultraspectral Imagery XVII, International Society for Optics and Photonics, 8048, pp. 80480I.
[7] Gerhardt et al 2011 Detection of single molecules illuminated by a light-emitting diode, Sensors, 11(1), pp. 905–916.
[8] Gopinath R B, M R Tamjis 2008 Water quality in healthcare, International Journal of Environmental Technology and Management, 9(1), pp. 125-140.
[9] S Sathish Kumar et al 2010 An Improved Method of Segmentation Using Fuzzy-Nero Logic, 2nd. Intern. Conf. on Computer Research and Development, pp. 671-675.
[10] Amutha devi, Gayathri Monicka.J, "Development of fuzzy approach to predict the fetus safety and growth using AFI" Journal of super computing" Dec-2019 DOI 10.1007/s11227-019-03099-8
[11] Bai et al 2013 Effective hospital waste classification to overcome occupational health issues and reduce waste disposal cost. Infection control & hospital epidemiology, 34(11), pp. 1234-1235.
[12] Aubry et al 2020 Computing real solutions of fuzzy polynomial systems, Fuzzy Sets and Systems, 399, pp. 55-76.
[13] Dombi J, Jonas T 2020 Ranking trapezoidal fuzzy numbers using a parametric relation pair, Fuzzy Sets and Systems, 399, pp. 20-43.
[14] Ramani Bai et al 2011 Towards a Database for an Information Management System on Climate Change: An Online Resource, Climate Change and the Sustainable Use of Water Resources, Springer Berlin Heidelberg, pp. 61–67.
[15] Bielawski J, Tabor J 2021 Convex hull of a fuzzy set and triangular norms. Fuzzy Sets and Systems, 417, pp. 93-109.

(CSM2-RA-R2-TI): Cyber Security Maturity Model for Risk Assessment Using Risk Register for Threat Intelligence

Lakshmi Prasanna B[1, 3], M SaidiReddy[2]

[1] Research Scholar, [2]Associate Professor
Department of Computer Science and Engineering, Koneru Lakshmaiah Education Foundation, Hyderabad, Telangana, India.
[3]Assistant Professor, Department of Computer Science, RBVRR Women's College, Hyderabad, Telangana, India.
Corresponding Author Email ID: msreddy33@klh.edu.in

Abstract. Because of the growing frequency sophistication, and severity of cyber security attacks, all businesses should ensure that cyber security risk is properly addressed in their enterprise risk management (ERM) programs. Risk Management, agreeing with the NIST Guide 8286 is the arrangement of "composed exercises to direct and control an association concerning risk". NISTIR 8286 characterizes a system and framework for hazard the executives. In any case, executing this norm without an in-depth plan can turn into a risk on associations. This paper provides a capability maturity model for risk assessment for threat intelligence using a risk register. This model helps the organization as reference and set clear path to survey risk assessments in accordance with latest threats.

Keywords: Risk Management, Risk Register, Threat Intelligence, Maturity model, NISTIR 8286

1. Introduction

Cyber-attacks are growing in scale, intensity, and ingenuity, as defense, enforcement, and risk management practitioners know. When someone in the industry tells us how important it is for our companies to pay attention to cyber security risk, most of us nod our heads in agreement. Nonetheless, several management teams and boards of directors continue to be perplexed by the magnitude of the problem.

NIST recently published a paper titled "Integrating Cyber security and Enterprise Risk Management" [1] to assist security practitioners in communicating the importance of preventative security to their management teams (NISTIR 8286). The use of a risk registry summarized as a "repository of risk information" – to effectively incorporate cyber security risk management is the focus of this guidance.
Threat Modeling is a systematic and organized protection methodology for identifying an application's security priorities, risks, and vulnerabilities, as well as determining where to focus efforts in designing, creating, and deploying safe applications. . Threat modeling [2] has the components like Threat knowledge, Asset Identification, Mitigation capabilities, Risk evaluation, and Threat planning.

The impact of uncertainty on the achievement of goals is known as risk. As a result, all organization's face threats and uncertainty, and they want to control risk in an organized manner are becoming increasingly apparent. Risk management (RM) is described as [3] "a set of organized activities designed to guide and monitor an organization's risk exposure". Various risk control practices are often used by organizations, and they do so in a variety of ways.

The risk assessment method is not always easy to execute, and certain companies give up before obtaining the expected results. This may be due to their inability to consistently and predictably carry out the risk assessment process over time. Maturity models are instruments that depict a journey toward a more streamlined and comprehensive way of doing business, which often involves a number of steps.

The goal of this paper is to feature the Risk Management (RM) frame of understanding via way of means of growing an artifact through the use of a well-mounted methodology. To outline a maturity version for RM, the version is primarily based totally on NISTIR 8286, which prescribes an RM shape and architecture.
We focus our attention in this paper on answering two research questions (RQ):

RQ1 - According to NISTIR 8286, what are the main tasks in a risk management process?

RQ2 – How will an RM maturity model be constructed to cope with the issues confronted with the aid of using diverse companies and industries?

To cope with those take a look at problems, this work is prepared into six sections. The most important terminology and meanings are defined in Section II, and its related work is given. Section III provides an overview of risk management and the risk register procedure. Section IV discusses hazard intelligence danger control. The RM Maturity Model is classified in Section V in phrases of the mapping among the RM Maturity Model Levels and the NISTIR 8286. Finally, in Section VI, the belief is discussed.

2. Background

After doing a literature review, the goal of this study is to evaluate cyber security maturity levels and make guidelines for improvement on the way to lower cyber dangers and vulnerabilities and grow cyber protection competencies. After evaluating many maturity models, the Cyber security Capability Maturity Model (C2M2) becomes selected to evaluate cyber protection competencies on the way to estimate an enterprise's risk.
To maintain a shared interpretation, this section defines the core terms and principles used in this article. Most organizations do not analyze or evaluate cyber security threats with the same rigor or reliable methods as other forms of risks within the organization, according to NIST's new risk paper, "Integrating Cyber security and Enterprise Risk Management. "As for the draft [2] of NIST, the key terms are defined as

Risk- Normally, it is calculated as a characteristic of the chance of a certain occasion happening and the possible effects of that event. Uncertainty is described as a condition of insufficient information for comprehending and knowing a specific occurrence, its consequences, or possibility.

Risk Register-"risk information repository" that includes an overview of a specific risk, the probability of it occurring, the possible cost effects, how it rates overall in importance relative to all other threats, the answer, and who owns the risk.

Risk Management –It is defined as "organized activities that lead and manage an organization's risk."

Risk tolerance- The willingness of the organization or related parties to accept residual risks in order to achieve their goals after responding to or evaluating risk. It is the acceptable degree of performance variation relative to the attainment of goals."

Watts Humphrey, the SEI group, and Mitre Corporation were hired to develop a tool to measure the capacities of the software firms with which the US Department of Defense collaborated in 1986 [4]. The first model of the CMM maturity model of capabilities was published in 1991. This model has had a lot of success and has developed into CMMI, which stands for Capability Maturity Model Integration.

This model [5] is very popular and is constantly updated and developed over time. It is today called the integrated version of the capability maturity model, hierarchical and continuous views can be used to connect to the maturity model of the system. There are maturity stages of step-by-step representations that correlate to the degree of process change over a series of predefined processes. Capacity thresholds are used in continuous representations to reflect the accomplishment of process change in particular process areas.

2.1. Examining and deciding on a cyber-security maturity model

The maturity models listed below were discovered in the literature offer a baseline against which an organization's present degree of maturity in its practices, processes, and procedures can be measured:

Researchers are currently working on a security maturity model. Several authors [6, 7] have examined the advantages and limitations of the current security maturity model. From 2012 to 2017, the comprehensive maturity model of information security [8] was strictly reviewed, and the current security maturity level [9, 10] was investigated in depth. Several of the most popular security models were selected for this study.

2.2 Identification of the model

The maximum important protection trendy, in step with Howe [11], is NIST 800-53.[12] It is a part of the National Institute of Standards and Technology (NIST) Cyber Security Framework, which has grown and been up to date on a everyday foundation over the preceding ten years. This framework defines protection measures [13] for Organizations like Federal Information Systems.

The Community Cyber Security Maturity Model (CCSMM), [14] which makes a specialty of a particular place of an organization, offers help on reacting to cyber threats on the network level.

The ISO/IEC 27000-series [15], a trendy for records control protection systems, presents pointers for growing one. ISA99/IEC 62443[16] is an association with present day computerization and manages frameworks. ISO/IEC 15408 suggests the concepts for a PC protection certificates. The Systems Security Engineering Capability Maturity Model (SSE-CMM), in any other case known as ISO/IEC 21827, is a protection designing improvement assessment that isn't always focused on on-line protection.

The Energy and Oil and Natural Gas Subsector cyber protection Capability Maturity Model (ES-C2M2)[17] and the Oil and Natural Gas Subsector cyber protection Capability Maturity Model (ONG-C2M2)[18] are each appropriate for the strength and oil and flammable fuel online areas.

The National Initiative for cyber protection Education Capability Maturity Model (NICE)[19] is a body of workers improvement, technique adulthood, and operational resilience version that doesn't encompass precise cyber protection pleasant practices. COBIT (Control Objectives for Information and Related Technology) is an IT management and manage layout that makes a specialty of statistics protection above cyber protection [20].
C2M2 is a cyber-protection Capability Maturity Model that makes a specialty of cyber protection and is meant for use in aggregate with the NIST framework; it's miles incredibly primary and is provided as a questionnaire.

One of the earliest adulthood models [21] used for records protection changed into Citigroup's Information Security Evaluation Maturity Model (ISEM). It makes a specialty of protection recognition and assessment, and it's been used as a version for growing opportunity cyber protection adulthood models, even though it isn't always presently applied with inside the industry.
The IBM Information Security Framework (IBMISF) [22] is a protection framework that makes a specialty of protection whole evaluation and assists agencies in figuring out their current protection posture. The (ISM3) [23] stands for Information Security Management Maturity Model is a commercial enterprise trendy that makes a specialty of statistics protection, hazard assessment, and technique integration.

3. Analysis of risk management with risk register process

Several authors and organization's studying RM have offered several frameworks to handle the topic, although they primarily apply to big businesses and not all industries; manufacturing and financial industries are favored [24, 25] RM is created in major companies and particular industries, with its own definition and principles.
RM is created in major companies and particular industries, with its own definition and principles. The level of hazards, according to prior studies, is dependent on the industrial sector [26, 27].Other studies show that risk management is impacted by experience, risk understanding, and the firm's ownership structure, but not by industry type or ownership structure.

The "NIST" [28, 29] contains essential RM principles and ideas, just as, more as of late, "risk-based thinking" as a necessary part of by and large quality administration in a wide range of organizations. At the point when applied in a firm, the plan and execution of RM should think about the association's specific evolving requests.

Dangers ought to be reported in a risk register with appropriate data and design. To help better network safety the board, digital protection hazards should be reported and followed in digital protection hazard registers. A threat register is described as "a store of threat information, comprising facts amassed over the years on hazards."
A threat register often entails a portrayal of the danger, the consequences if the threat happens, alleviation arrangements, hazard proprietors, and positioning to distinguish higher need chances". Cyber security risk registers are an important part of a company's cyber security risk management strategy. As further risk actions take undertaken, each register evolves and develops.

- The Risk Assessment process using the Risk Register is as follows.

- **Determine the context**- Context refers back to the surroundings wherein the enterprise operates, that's impacted via way of means of the dangers it faces.
- **Determine the dangers**- This involves spotting the entire variety of advantageous and bad dangers—that is, comparing which occurrences may assist or preclude objectives, which include the dangers related to passing up an opportunity.
- **Examine the dangers**- This involves calculating the opportunity of every indicated threat occasion occurring, in addition to the feasible impact of the consequences mentioned.
- **Rank the hazards**- In order of importance, every threat's publicity is evaluated primarily based totally on its chance and feasible impact, and dangers are then ranked in keeping with their publicity.
- **Prepare and put into effect threat Reduction** -An initiative for every threat, the right response is identified, with threat recommend from management guiding the technique.

- **Keep an eye fixed on things, assessment them, and make adjustments as needed-** As cyber protection dangers vary, non-stop tracking guarantees that organization threat occasions continue to be inside the set threat urge for risk thresholds.

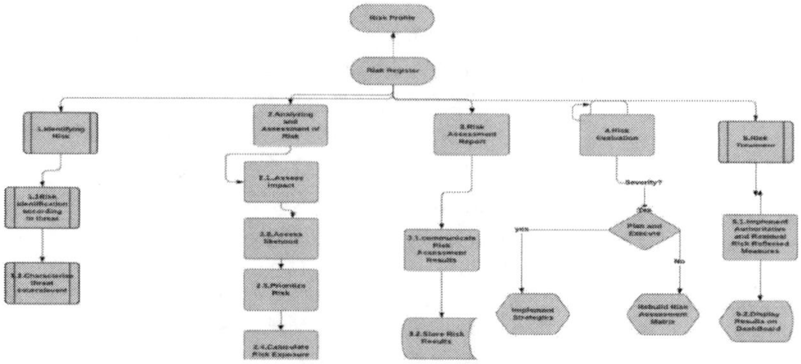

Figure1. FlowChart of Risk Management using Risk Register Process.

'Figure1' represents the flowchart of RiskManagement using Risk Register Process . The context evaluation section is when the RM manner' goals are first established, in step with the company's strategic goals, and then the resources.
The following are the important thing stages:[29,30] the detection of danger (relate the feasible risks), the risk assessment section (decide the probability and result of the recognized chance occurring), the risk remedy section (become aware of the way to lessen the incidence or effects of perfect risks, the way to lessen the effect of non-perfect risks, and pick the best responses to them), and the screen and evaluation section (have a look at the chance prevalence and if the remedy is powerful or not).
The conversation and interplay with stakeholders concerning the manner and the chance publicity is the procedure's very last action. Various methods and strategies for identifying, evaluating, and treating risks are used depending on the situation. These instruments might be qualitative, quantitative, objective, or subjective in nature.

4. Risk management for threat intelligence

The risk analysis process should enable an organization to assess its risk in terms of threats and vulnerabilities. The auditor will be able to classify the risk's severity and hence assign an overall relevance to each risk at this stage. It should be possible to develop a risk management strategy using this data [31, 32].

This should include the following:

• Preparing a dangerous treatment plan utilizing an assortment of control strategies.
• Analyzing singular dangers dependent on the effect of the dangers and weaknesses that have been distinguished from the dangers.
•Rate the individual dangers from most elevated to least significant.
•Create a dangerous treatment plan that classifies every one of the dangers and weaknesses arranged by its need to the association, along with certain potential controls.

Table 1 .Notational Cyber security Risk Register Template.

Notional Cybersecurity Risk Register												
ID	Priority	Risk Description	Risk Category	Current Assessment			Risk Response Type	Risk Response Cost	Risk Response Description	Risk Owner	Status
				Likelihood	Impact	Exposure Rating					
1											
2											
3											
4											
5											
Continually Communicate, Learn, and Update											

'Table 1' shows an example National Cyber Risk Register template, which includes risk priority, description, risk category and kind, who the owner is and what their current status is, as well as the cost and impact of the risk.

Quantitative and qualitative [33, 34] approaches can be used to assess the possible risk. The formulas may be used to calculate the Risk function.

$$Risk=Threat*vulnerability \quad (1)$$

$$Risk=Threat*vulnerability*impact \quad (2)$$

4.1. Cyber Threat Intelligence (CTI)

A threat is an undefined occasion that has the capacity to disrupt an organization's operational and practical operations. Threat intelligence is available in quite a few forms, which include strategic, operational, tactical, and technical danger intelligence.

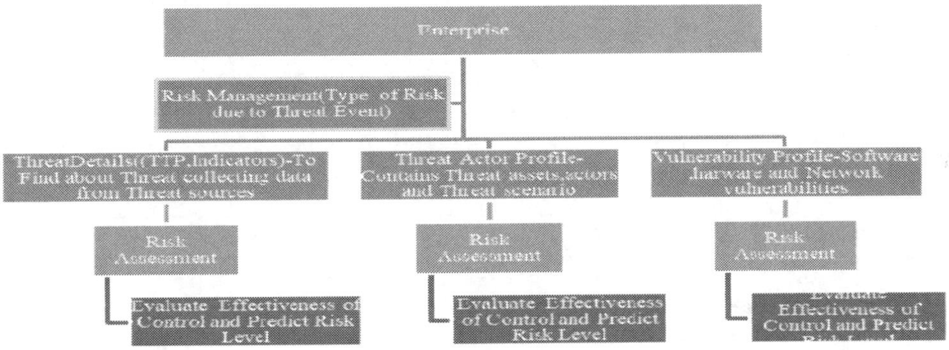

Figure 2. Integration of CTI (Cyber Threat Intelligence) with Risk Management

'Figure 2' depicts how risk management can be embedded in cyber threat intelligence in an enterprise/organization, where threat details can be collected from a risk register to identify external or internal threats and different preventive and deterrent control mechanisms can be used to mitigate the risk.

5. Maturity Model Design In Accordance With NIST Framework

The National Institute of Standards and Technology's (NIST) Cyber safety Framework [10, 13] is a hazard-primarily based totally framework that gives guidance for addressing cyber safety risks. Although the version outlines the implementation process, it gives little help to detect, analyze, and prioritizes gaps; set up a modern profile; carry out a hazard evaluation; set up a goal profile; and identify, analyze, and prioritize gaps.

The Risk Maturity Model (RMM) explains the primary metrics and activities that make up an enterprise risk management (ERM) software this is sustainable, repeatable, and mature. Organizations might also additionally use the hazard maturity self-evaluation to look at how properly their risk strategy control techniques align with the RMM indicators.

The RM Maturity Model's principal cause is to grow the effect of hazard control on an organization's business value. As visible in Figure, that have an impact will increase as you develop from a decrease to an extra adulthood level [10, 24]. Furthermore, the shortage of tactics and policies on the decrease tiers results in bad hazard control, which would possibly place the enterprise in threat without even knowing it.

Table 2. Levels in CMMI and RM-CMMI

SNO	CMMI (Capability Maturity Model Integration)	Purpose	RM-CMMI (Risk Management- Capability Maturity Model Integration)	Purpose
1.	-		Level 0 – No RM	-
2.	Level-1-Initial	Unpredictable, poorly conceived, and reactive process.	Level 1 – Initial RM	Risk Management is informal and Adhoc
3.	Level-2-Managed	Projects and reactive processes are characterized by this process.	Level 2 – Managed RM	Basic Risk Management
4.	Level-3-Defined	The process is characterized by its organization and proactive nature.	Level 3 – Defined RM	Standardized Risk Management Process
5.	Level-4-Quantitatively Managed	Control and measurement of the process.	Level 4: Quantitatively Managed Risk Management;	Quantitative Risk Management Process
6.	Level-5-Optimizing	Maintain a constant focus on process improvement.	Level 5: Risk Management Optimization	Continuous Risk Management Process

'Table 2' represents the proposed RM-CMMI (Risk Management- Capability Maturity Model Integration) comparison with CMMI (Capability Maturity Model Integration) at all stages.

As policies and procedures are adopted, specified, recorded, and assessed at greater maturity levels, this risk is reduced. Risk management is completely incorporated into the organizational structure at the greatest maturity level (level 5) and is used to give a competitive advantage.

We added level 0 to the previously described maturity levels 1–5, which signifies the company is not doing any RM processes or tasks at all. As a result, level 0 is not referenced directly in the RM maturity model.

5.1. Maturity stages emerge

We added level-0 to the formerly defined maturity tiers 1–5, which indicates the enterprise isn't always doing any RM tactics or obligations at all. As a result, level 0 isn't always referenced at once with inside the RM maturity model.
To develop from level zero to level 1, the enterprise has to understand that an RM system is needed as a crucial function. Additionally, essential danger control responsibilities are finished with the purpose of making sure that danger is treated at some point in the enterprise.

As a result, the enterprise has a view of the requirement for a danger control system at maturity level 1 (Initial). Some risk management actions are utilized by the enterprise; however, they're usually haphazard and disorganized. This kind of activity is extra reactive than proactive. The company lacks strong surroundings wherein to create a risk control approach.

The consequences of danger control are sudden and hard to repeat, and that they depend extra on the competencies of the human beings with inside the enterprise than on using a longtime procedure.
At maturity level 2 (Managed), the enterprise makes an attempt to plan and perform danger control system sports according to with the risk control coverage that the company has created with stakeholders.

Despite those efforts, danger control is suffering from the replication of preceding a hit acts in preference to a dependent procedure. Risk management tasks are delegated to persons with the skills, defined roles, and sufficient resources to generate results that are repeatable to some extent. The organization's uniformity is lacking at this level. Risk management might vary greatly amongst departments within a business.

The danger the executives cycle is described, perceived, and indicated in standard cycles, apparatuses, and techniques at maturity level 3 (Defined). The system is used to give consistency all through the organization, and there is a concentrated way to deal with hazard the executives.

The association utilizes quantitative and measurable techniques to screen, survey, and assess the danger the board cycle at maturity level 4 (Quantitatively Managed).

At long last, at maturity level 5 (Optimizing), the danger the board interaction is persistently upgraded dependent on the data acquired at before levels. Everybody is committed, and hazard the executives is seen as an essential arrangement.

To advance from level X to even out X+1, the association should meet the entirety of the prerequisites from level X, coming about in a "stages" approach in this development model. Because corporations at this maturity level are continuously inventing and improving the threat control process, they make contributions scientifically to the increase of threat control as a domain [5].

Table 3. NIST Risk Management Framework

Identify The Context	**SP 800-30-Rev.1**	**SP 800-30-Rev.1**	**SP 800-39**
Risk assessment (risk identification, risk analysis, risk level calculation)	3.2. Task 2-1: Identify and describe possible threat sources. Task 2-2. Identify possible dangerous situations and threat sources Task 2-3: Identify vulnerabilities and conditions	3.5.prepareTask P-14 , evaluate the Risk Assessment-System, Risk Assessment Report(PAR) Assess	3.2.Assessing Risk
Risk Assessment	3.2. Task 2-4: Calculate the negative consequences of threat occurrences. Task 2-5: Calculate the probability. Task 2-6: Assess the organization's risk. Report on Risk Assessment	Classify (3.2), Select (3.3), and Implement (3.4)	3.3. Risk Report
Risk Treatment	3.3.Task 3-1: Exchange of risk assessment results Task 3-2: Exchange of information on risk	3.4. Implementation, approval (3.6), POA, and M residual risks.	3.3. Respond To Risk
Monitoring and Review	3.4.Task 4-1: Maintain on going risk factor monitoring Task 4-2:Update the risk assessment	-	3.4.Monitoring Risk

'Table 3' shows the NIST risk management framework: SP 80030, 501 revision 1, risk assessment guide; SP 80037, revision 2, risk 502; and SP 80039

Table4. Proposed RM-CMMI Mapping With Maturity Model

Maturity level Assessment	Activity Point	Assessment criterion
Level 0 – No RM	-	No Criteria
Level 1 – Initial RM	Collecting Risk Data, Setting risk expectations and priorities	1.1. Ask the project manager for information. 1.2.Analyze the risk register 1.3.Continue to monitor the Risk Register for data
Level 2 – Managed RM	Explains risk appetite to define risk tolerance statements.	2.1 - Define the RM's scope and limits.
		2.2 - Establish RM policies.
		2.3 - Define the technique for threat evaluation.
		2.4 - Perform threat evaluation
		2.5 - Conduct a threat evaluation and evaluation
		2.6 - Define remedy options for people who are at threat.
		2.7 - Establish threat remedies and manipulate targets and procedures.
		2.8 - Obtain permission from control for any ultimate risks.
		2.9 - Obtain control approval to undertake and execute the Risk Management Plan.
		2.10.At the organizational level, risk tolerance statements (and measurements) are used to apply risk appetite direction. Direction on how to use RM (for example, centralized services, compliance / auditing techniques, and inherited and applied shared controls at the system level).
Level 3 – Defined RM	Risk tolerance statement is applied to reach the system level.	3.1 - Create a risk management strategy
		3.2 - Put the risk treatment plan into action.
		3.3 - Put controls in place to achieve control goals
		3.4 - Define the technique for figuring out the efficacy of the required controls.
		3.5 - Implement schooling and consciousness campaigns
		3.6 - Oversee the RM's operations
		3.7 - Manage the RM's resources
		3.8 - Implement risk detection and response processes and controls
Level 4: Quantitatively Managed Risk Management;	Assessing and Reporting system level risk responses	4.1 - Monitor and assess procedures as well as other controls.
		4.2 - Conduct frequent assessments of the risk type's efficacy.
		4.3 - Evaluate the efficacy of controls
		4.4 - Conduct a risk assessment review
		4.5 - Examine any remaining hazards
		4.6 - The review determined that the risk levels were acceptable.
		4.7 - Carry out internal risk assessments

		4.8 - Keep security plans up to date
		4.9 - Keep track of your actions and occurrences. Risk classification and metrics that enable continuing assessment, authorization, and continuous monitoring.
Level 5: Risk Management Optimization	Aggregating and integrate Business level Risk Management in enterprise	5.1 - Implement the RM enhancements that have been discovered.
		5.2 - Take the necessary corrective and preventative measures
		5.3 - Inform all interested parties about activities and improvements.
		5.4 - Ascertain that improvements meet their goals.
		5.5. An integrated Enterprise Risk Register with aggregated and normalized Enterprise Risk findings that are aligned with those of other risk categories.

'Table 4' represents Proposed RM-CMMI Mapping with Maturity Model, in which we compare risk parameters based on NIST Framework by considering latest Threats and vulnerabilities.

A maturity model is a tool that assists individuals in assessing a person's or group's present efficacy and determining what capabilities they need to learn next in order to enhance their performance while taking risk into consideration, as the number of potential cyber assaults grows every day.

Risk management maturity models are a great method for companies to assess where they are now, compare where they are now to where they want and need to go to get the most benefit, and talk about the value and cost of investing more in risk management. More precise defect detection in process development processes reduce software development or administration cost.

6. Conclusion

The creation of a maturity model for the RM system is the intention of this article. The latter can be used by the RM feature to observe and verify the present strengths and shortcomings of the RM system as a governance tool. The version, on the alternative hand, is not only for analytical purposes.
It can also be applied to create a roadmap for the RM feature's evolutionary development in phrases of capabilities, effectiveness, and efficiency. The first phase of the item describes the RM activities that had been used as a baseline to peer if current models are able to read an RM system comprehensively.

(RQ1)-The findings suggest that current maturity models do now no longer appropriately cover the complete reference baseline because they most effectively manage the activities selectively. As an end result, no current maturity version can cope with the said issue. Finally, according to the studies plan, we determined to create a brand new maturity version.

(RQ2)-To the solution, the second one studies question, "How should a maturity Model precise to RM be designed that objectives the demanding situations of various organization's and industries?" we portrayed the development of an improvement version for RM, which includes the real version simply as its evaluation with inside the second piece of the paper.

The version that has been constructed is primarily based totally on contemporary maturity model systems and consists of ideas and procedures from the NIST Risk parameters according to contemporary threats. Throughout the improvement system, we took an awesome attempt to make sure that usable observed end results become produced. The implemented methodology, of course, has its very own set of restrictions. In order to increase the scope of the maturity model's observation, we suggest that the RM maturity model be evaluated and subtle inside distinctive business sectors, ensuing in an extra popular RM maturity model that may be benchmarked throughout industries.

References

[1] Stine K, Quinn S, Witte G, Gardner RK "Integrating Cyber security and Enterprise Risk Management (ERM)" 2020.

[2] "National Institute of Standards and Technology" (Online), Available at https://www.nist.gov/cyberframework/informative-references 2020.

[3] ISO 31000:2009, "Risk Management – Principles and Guidelines," 2013.

[4] D. Ahern, A. Clouse, R. Turner, "CMMI Destilled: A Practical Introduction to Integrated Process Improvement, Third Edition," Addson Wesley Professional, 2008.

[5] CMMI Product Team, "CMMI for Development, Version 1.3," Carnegie Mellon Univ., no. November, p. 482, 2010.

[6] Le, Ngoc T. and D. Hoang. "Can maturity models support cyber security?" 2016 IEEE 35thInternational Performance Computing and Communications Conference (IPCCC) (2016): pp1-7.

[7] Lessing, M. "Best practices show the way to information security maturity", 2008.

[8] Karokola, Geoffrey & Kowalski, Stewart & Yngström, Louise," Secure e-government services: Towards a framework for integrating IT security services into e-government maturity models" P.1 – 9, Information security for South Africa, Johannesburg, 15–17 August 2011.

[9] Rea-Guaman AM, San Feliu T, Calvo-Manzano JA, et al. "Comparative study of cyber security capability maturity models" International Conference on Software Process Improvement and Capability Determination, pp.100–113, Springer, 4 October 2017.

[10] Rea-Guaman A, Sanchez-Garcia I, Feliu TS, et al. "Maturity models in cyber security: a systematic review" 12th Iberian conference on information systems and technologies (CISTI), Lisbon, Portugal, 14–17 June 2017.

[11] Howe N." Cyber security in railway signaling systems", Institution of Railway Signal Engineers News. 2017, p.1–4.

[12] Force JT and Initiative T," Security and privacy controls for federal information systems and organizations", NIST Spec Publ 2013; 800: 8–13.

[13] Barrett MP," Framework for improving critical infrastructure cyber security version 1.1", NIST cyber security framework, www.nist.gov/publications/frameworkimproving-critical-infrastructure-cybersecurity-version-11.

[14] White GB,"The community cyber security maturity model", IEEE international conference on technologies for homeland security (HST), Waltham, MA,USA, 15–17 November 2011.

[15] Disterer G," ISO/IEC 27000, 27001 and 27002 for information security management", Journal of Information Security, 2013; volume-4:92–100.

[16] ISA-62443-2-1: 2009," Security for industrial automation and control systems: establishing an industrial automation and control systems security program", 2010.

[17] ES-C2M2," Electricity subsector cyber security capability maturity model (ES-C2M2)", www.energy.gov/ceser/activities/cybersecurity-critical-energy-infrastructure/energy-sector- cyber security.

[18] ONG-C2M,"Oil and natural gas subsector cyber security capability maturity model (ONG- C2M2)", www.energy.gov/ceser/activities/cybersecurity-criticalenergy-infrastructure/energy- sector-cybersecurity-0/oil.

[19] Newhouse W, Keith S, Scribner B, et al." National initiative for cyber security education NICE) cyber security workforce framework", NIST Special Publication, vol.800. USA: NIST, 2017, p.181.

[20] ISACA. Cobit 5. USA: ISACA. "COBIT 5 for Risk "2013, Published by ISACA, COBIT 5, Risk IT ISBN: 9781604204582.

[21] Masinsin RQ," Secretary of defense corporate fellows program: final report", New York, NY: Time. 2008.

[22] Buecker A, Borrett M, Lorenz C, et al, "Introducing the IBM security framework and IBM security blueprint to realize business-driven security", IBM Red paper 2010;4528: 1–96.

[23] Aceituno V."ISM3-information security management maturity model", Version 2.1. San Francisco, CA, USA: ISM3 Consortium, the Open Group, 2007.

[24] Anass Bayaga," Information and Communication Technology (ICT) Risk Models and Performance of Small and Medium Enterprises (SMEs)", the Anthropologist, 15:2, 177- 183, DOI: 10.1080/09720073.2013.11891304, 2013.

[25] Teyssier, C," Risk Management: Lever for SME Development and Stakeholder Value Creation", United Kingdom: Wiley, 2017.

[26] S. Papastergiou, and N. Polemi, "MITIGATE: A dynamic supply chain cyber risk assessment methodology," In Smart Trends in Systems, Security and Sustainability, pp.19, Springer, Singapore, 2018.

[27] J. Meszaros, and A. Buchalcevova, "Introducing OSSF: A framework for online service cyber security risk management," Computers & Security, vol.65, pp.300–31, 2017.

[28] Steven Cockcroft, What is the NIST Framework? ITNOW, Volume 62, Issue 4, winter 2020, Pages 48–49, https://doi.org/10.1093/itnow/bwaa116

[29] National Institute of Standards and Technology, Gaithersburg, MD, NIST Interagency or Internal Report (IR) 8286. https://doi.org/10.6028/NIST.IR.8286

[30] D. Proenca, J. Estevens, R. Vieira, and J. Borbinha, "Risk Management: A Maturity Model Based on ISO 31000," 2017 IEEE 19th Conference on Business Informatics (CBI), 2017.

[31] Albakri, A., Boiten, E., De Lemos, R" Risks of sharing cyber incident information", Proceedings of the 13th International Conference on Availability, Reliability and Security - ARES 2018, pp. 1–10 (2018).

[32] Aksu, M.U., et al.: "A quantitative CVSS-based cyber security risk assessment methodology for IT system", Proceedings - International Carnahan Conference on Security Technology (2017).

[33] J.W. Candra, O.C. Briliyant, and S.R. Tamba, "ISMS planning based on ISO/IEC 27001:2013 using analytical hierarchy process at gap analysis phase (Case study : XYZ institute)," 2017 11th International Conference on Telecommunication Systems Services and Application (TSSA).IEEE, 2017.

[34] A. Yeboah-Ofori, and S. Islam, "Cyber Security Threat Modeling for Supply Chain Organizational Environments," Future Internet, vol.11, no.3, pp.1-25, 2019.

Efficient identification of node failure and recovery through end to end Probing techniques

Chitturi Sai Nikhil [1], Manjula Josephine Bollarapu [2], Kaja Sai Sree Neeraj [3], Alaparthi Praveen Kumar [4], Manchikalapudi Jaya Sesha Sai Raghuveer [5]

[1, 2, 3, 4,5] Department of Computer Science and Engineering, Koneru Lakshmaiah Education Foundation, Vaddeswaram, AP, India.

Corresponding Author: manjulajosephine@gmail.com

Abstract. Identification of Node failure detection and a localization is a very important challenge in a network community to get a quick recovery and avoid useless traffic in network. But it is very difficult to check the failure nodes or locations because of the large number of Screw ups in dense network. As finding the main source for failure of network is always challenging our proposed work will achieve that, it identifies the node failure by using probing measurement of binary state to end to end paths. Apart from identifying the network failure, it also quantifies the total failure nodes and the ip address or vicinity of failure nodes, Identification of node failure is done by monitoring nodes which are deployed in the netwok. Our Proposed word is divided majorly in two phases one is identifying the node failures by using Probing Packets and other is finding of the failure and its recovery.

Key words: Network probing, Node recovery, Measurement paths, probing mechanisms, Node Recovery, End to end path construction, localization.

1. INTRODUCTION

In order to get reliable and effective network performance, the network operators plays an important role. The main aim of network operators is to provide effective monitoring without disturbing any communication or any of the services. There are many factors which affect the performance of network such as loss of packet, delay in end to end delivery of packets, malfunctioning of nodes, nodes not reachable etc. Network monitors must continuously identify these issues. If these problems are identified in time then network administrator can take appropriate action. The admin can identify location of attacker. Admin can route the packets from one route to another where it is in high demand. Can specify alternate route so that the network performance is not degraded due to heavy packet loss. Tracing the location of malicious node is big challenge, there are many solutions proposed by many authors.

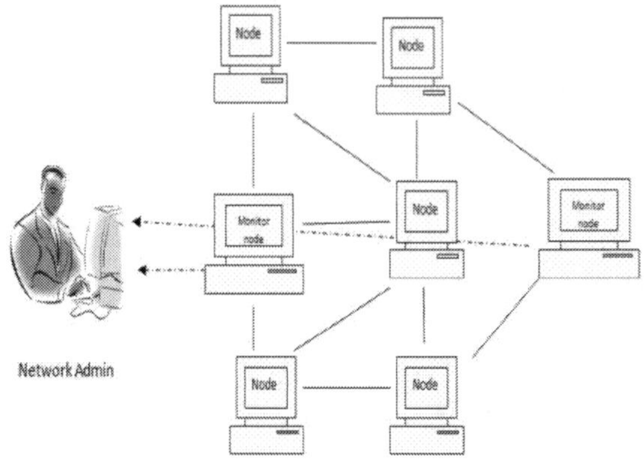

Figure: 1 Overview of Network Monitoring

There exists uncertainty in failure node detection and localization, there needed a novel technique, to identify both in network and sub-network area. Network properties highly helpful in identifying sub network area failures. To approach this problem, the solution provided are finding under which conditions node failure occurs, finding whether node failure nodes are in a subset of node[6], which is monitored in particular locations. The routing strategy of probe packets plays crucial role in network monitoring, which is unattended in existing systems. End to end path measurement approach can be implemented with routing probe packets satisfies scope of failure node localizing problem.

In the following sections some of the research related to network monitoring are discussed. In section 3, the details on implementation of proposed methodology is discussed. In Section 4, Results and discussed are given. Finally, Section 5, discussed work Conclusion and further enhancements are discussed

2. RELATED WORK

Network performance plays a crucial role for industries and business, there has been large amount of study in network failure detection. In Literature, many researched has worked on network failures and path measurements. Some them are discussed in detail in this section. Link delivery delays and link failures are discussed in work [7], The author proposed the greedy approximation algorithm. The Work in [8] about the virtual circuit routing for detecting node failure. The Work in [9] is all about the handling of the various resources in network failure conditions.

3. Methodology

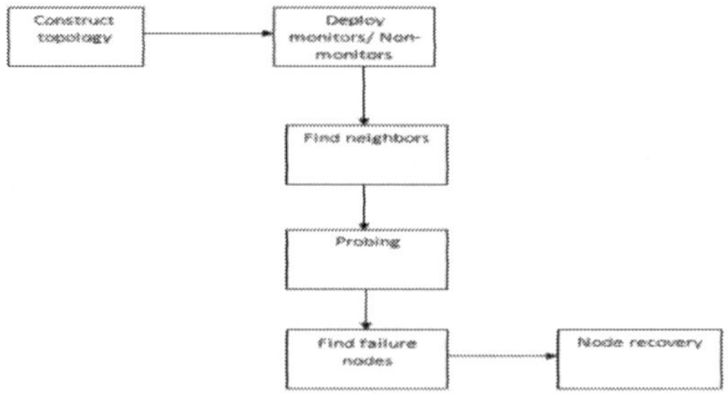

Figure 2: Work Flow

Journal of Physics: Conference Series **2040** (2021) 012006 doi:10.1088/1742-6596/2040/1/012006

The Implementation is done by using the below modules

3.1 Construction of Topology
3.2 Deploying Monitors/ Non Monitors
3.3 Finding Neighbors
3.4 Probing &Finding Failure Nodes
3.5 Node Recovery

3.1 Construction of Topology

Intially the network topology is constructed in this Phase with all the nodes, links and edges. The network topology can be undirected and can have sub network as well [14].

3.2 Deploying Monitors/ Non Monitors

By using the Binary State Process and unique rules for Measurement and Probing techniques Network will be monitored [15].

3.3 Finding Neighbors

The following figure represents the moderately controlled packet probing strategy, which is sent from monitoring node to normal node.

Figure: 3 Moderate control packet probing

The following figure represents the controlled packet probing strategy[17], which is sent from monitoring node to next monitoring node. The packed named CPP will be broadcast between one monitor and other monitor as shown below.

Figure: 4 Control packet probing

The following figure represents the uncontrolled packet probing strategy, which has no restriction on sending probe packet, it may take any route[18]. The route includes the all available neighbour nodes, or may be limited nodes based on a random selection done through this probing strategy. The packets names UPP is broadcast through the network.

Figure: 5 Un-control packet probing

3.4 Probing & Finding Failure Nodes

Here in this Phase by using binary type states and Probing Mechanisms Node failures will be detected and also there is a recovery mechanism to rectify the failure node.

4. Results and Discussion:

The proposed work is carried out in jdk 1.7 with mysql 5.5 as database. We initially Created dynamic topology, in which any range of nodes may be deployed.Nodes, linkas and edges are framed.

Nodes which are deployed in the network and its status can be monitored by monitioring nodes and as mentioned in following figure. For naming monitor nodes are named as M1, M2..etc Non-monitor nodes are named as A, B,C,D…etc. The proposed work considered deployed two monitor nodes M1 and M2 in the network topology. The number of normal nodes deployed in the network are four. Placement of monitor influences the performance of path measurements. If the monitoring nodes are available in between the normal nodes, the monitoring nodes can effective monitor the paths established through Moderate controllable probing packets of uncontrolled packet probing strategies. The following figure shows the monitoring of nodes or neighbour along the paths established.

Node Name	Status
A	Normal
B	Normal
M2	Normal
C	Faulty
D	Normal

Figure 6: Node Status Monitoring

Three types of Probing message are sent on particular button action. The nodes receive the neighbour and paths are arrived dynamically.

Node can be made as faulty node and the generated path will avoid faulty nodes in routing for message delivery[9]. The message is delivered on network without fail. The faulty node can be retrieved or recovered in the network, in such case, it again joins the routing path on message delivery.

In case of route failures, the proposed system identifies the all available route, to re-route the packets, this ensures the data delivery in network without any loss. The following screen shows the available path generated for sending packets from monitor node to non-monitoring Node 'C', the routing path creating, generated all available path, which has an alternate routes of more than 10 route to deliver. This ensures the packet delivery without any loss in network. The routing paths are generated whenever an event of data delivery on the network or probing packet deliver on the network. The routes are generated among monitor to non-monitor, monitor to monitor and monitor to all available nodes depending on the generated events like data or probe.

Figure: 7 All available paths for routing

5. Conclusion:

Our Proposed work has addressed various challenges in terms of network maintenance as it is the major part of networking and also it addressed various issues in terms of network performance through proper monitoring. The main idea of our work is to identify the failure node and its location through various probing strategies of network. We have achieved this by following various strategies like binary problem to identify the failure node and by using some distinct set of guide lines for calculating the paths and for finding and recovering the failure nodes. And our proposed research work experimentally proved the efficiency of node failure detection through probing strategies.

Future Scope

As the present work involved the detection of node failure as binary case only, in future this is can be extended to multiple class detection, along with identifying specific type of node failures.

REFERENCES

[1] O. J. Pandey, V. Gautam, H. H. Nguyen, M. K. Shukla and R. M. Hegde, "Fault-Resilient Distributed Detection and Estimation Over a SW-WSN Using LCMV Beam forming," in IEEE Transactions on Network and Service Management, vol. 17, no. 3, pp. 1758-1773, Sept. 2020, doi: 10.1109/TNSM.2020.2988994.

[2] M. Dhingra and N. Gupta, "Failure Node Reduction Algorithm to Enhance Fault Tolerance Capability of Cloud Nodes," 2020 12th International Conference on Computational Intelligence and Communication Networks (CICN), 2020, pp. 165-168, doi: 10.1109/CICN49253.2020.9242615.

[3] C. Pachajoa, C. Pacher and W. N. Gansterer, "Node-Failure-Resistant Preconditioned Conjugate Gradient Method without Replacement Nodes," 2019 IEEE/ACM 9th Workshop on Fault Tolerance for HPC at eXtreme Scale (FTXS), 2019, pp. 31-40, doi: 10.1109/FTXS49593.2019.00009.

[4] F. Barbosa, A. de Sousa, A. Agra, K. Walkowiak and R. Goścień, "A RMSA Algorithm Resilient to Multiple Node Failures on Elastic Optical Networks," 2019 11th International Workshop on Resilient Networks Design and Modeling (RNDM), 2019, pp. 1-8, doi: 10.1109/RNDM48015.2019.8949141.

[5] Y. Bejerano and Rajeev Rastogi, "Robust monitoring of link delays and faults in IP networks," IEEE INFOCOM 2003. Twenty-second Annual Joint Conference of the IEEE

Computer and Communications Societies (IEEE Cat. No.03CH37428), San Francisco, CA, USA, 2003, pp. 134-144 vol.1, doi: 10.1109/INFCOM.2003.1208666.

[6] A. A. Minhas, C. Steger, R. Weiss and S. Ehsan, "Node failure detection and path repairing scheme in virtual circuit routing algorithm for wireless ad hoc micro sensor networks," Proceedings of the IEEE Symposium on Emerging Technologies, 2005., Islamabad, Pakistan, 2005, pp. 86-91, doi: 10.1109/ICET.2005.1558860.

[7] I. Vardas, M. Ploumidis and M. Marazakis, "Towards Communication Profile, Topology and Node Failure Aware Process Placement," 2020 IEEE 32nd International Symposium on Computer Architecture and High Performance Computing (SBAC-PAD), Porto, Portugal, 2020, pp. 241-248, doi: 10.1109/SBAC-PAD49847.2020.00041.

[8] Zeng, Hongyi & Kazemian, Peyman & Varghese, George & McKeown, Nick. (2012). Automatic test packet generation. 241-252. 10.1145/2413176.2413205.

[9] H. X. Nguyen and P. Thiran, "The Boolean Solution to the Congested IP Link Location Problem: Theory and Practice," IEEE INFOCOM 2007 - 26th IEEE International Conference on Computer Communications, Anchorage, AK, USA, 2007, pp. 2117-2125, doi: 10.1109/INFCOM.2007.245.

[10] S. S. Ahuja, S. Ramasubramanian and M. Krunz, "Single-Link Failure Detection in All-Optical Networks Using Monitoring Cycles and Paths," in IEEE/ACM Transactions on Networking, vol. 17, no. 4, pp. 1080-1093, Aug. 2009, doi: 10.1109/TNET.2008.2008000.

[11] Lantz, Bob & Heller, Brandon & McKeown, Nick. (2010). A Network in a Laptop: Rapid Prototyping for Software-Defined Networks. Proceedings of the 9th ACM SIGCOMM Workshop on Hot Topics in Networks. 19. 10.1145/1868447.1868466.

[12] Mahimkar, Ajay & Ge, Zihui & Wang, Jia & Yates, Jennifer & Zhang, Yin & Emmons, Joanne & Huntley, Brian & Stockert, Mark. (2011). Rapid detection of maintenance induced changes in service performance. 10.1145/2079296.2079309.

[13] B. Schroeder and G. A. Gibson, "A large-scale study of failures in high-performance computing systems," International Conference on Dependable Systems and Networks (DSN'06), Philadelphia, PA, USA, 2006, pp. 249-258, doi: 10.1109/DSN.2006.5.

[14] E. Jeannot, G. Mercier and F. Tessier, "Process Placement in Multicore Clusters:Algorithmic Issues and Practical Techniques," in IEEE Transactions on Parallel and Distributed Systems, vol. 25, no. 4, pp. 993-1002, April 2014, doi: 10.1109/TPDS.2013.104.

[15] C. D. Sudheer and A. Srinivasan, "Optimization of the hop-byte metric for effective topology aware mapping," 2012 19th International Conference on High Performance Computing, Pune, India, 2012, pp. 1-9, doi: 10.1109/HiPC.2012.6507513.

[16] W. R. Heinzelman, A. Chandrakasan and H. Balakrishnan, "Energy-efficient communication protocol for wireless microsensor networks," Proceedings of the 33rd Annual Hawaii International Conference on System Sciences, Maui, HI, USA, 2000, pp. 10 pp. vol.2-, doi: 10.1109/HICSS.2000.926982.

[17] R. L. Carter and M. E. Crovella, "Server selection using dynamic path characterization in wide-area networks," Proceedings of INFOCOM '97, Kobe, Japan, 1997, pp. 1014-1021 vol.3, doi: 10.1109/INFCOM.1997.631117.

[18] Friedman, Timur & Towsley, Donald & Kurose, James. (2004). Scalable End-to-End Multicast Tree Fault Isolation. 1347-1358. 10.1007/978-3-540-27824-5_174.

[19] Y. Breitbart, Chee-Yong Chan, M. Garofalakis, R. Rastogi and A. Silberschatz, "Efficiently monitoring bandwidth and latency in IP networks," Proceedings IEEE INFOCOM 2001. Conference on Computer Communications. Twentieth Annual Joint Conference of the IEEE Computer and Communications Society (Cat. No.01CH37213), Anchorage, AK, USA, 2001, pp. 933-942 vol.2, doi: 10.1109/INFCOM.2001.916285.

[20] R. Bolliger and T. R. Gross, "Bandwidth monitoring for network-aware applications," Proceedings 10th IEEE International Symposium on High Performance Distributed Computing, San Francisco, CA, USA, 2001, pp. 241-251, doi: 10.1109/HPDC.2001.945193.

Smart Battery Management System Using LiFePO4 Battery for Offline UPS

BA Anandh[1]*, A Shankar Ganesh[1], R Sakthivel[1], D Mahesh Kumar[1], E Prem Kumar[2]

1Associate Professor, 2PG Student

Department of Electronics, PSG College of Arts & Science, Coimbatore, Tamil Nadu, India.

Corresponding Author E-mail: anandh.ba@gmail.com

Abstract. Power integrated circuits (ICs) have developed solutions for charging systems and power management system for battery powered devices. The main purpose of a battery management system comprises handling the charging cycle to limit the charging time without focusing on the battery. It also monitors the present status of the battery and reports the fault conditions if any malfunction or error occurs. This system is based on a smart LiFePO4 battery for offline UPS, which is used to power up a home AC application with 230VAC/50Hz. The main reason to select the LiFePO4 battery life cycle, it is made up of non-toxic material and this battery is not overheated or catches fire in case of overcharging or overload. In this system, the Arduino UNO microcontroller is used and it takes care of the parameters like battery charging voltage, charging current, and high voltage cutoff at backup mode. It also monitors the battery voltage, battery current, low voltage cutoff, and overload protection at inverter mode.

Key words: Uninterruptible Power Supply, LiFePO4 Battery, Direct Current, Alternating Current, Pulse Width Modulation.

1. Introduction

The main kinds of static UPS frameworks systems are on-line, off-line, and line-interactive arrangements. This system uses a LiFePO4 battery to store energy for backup power in offline UPS. Offline UPS works in two modes of operations. First "Inverter Mode" and another one is "Backup Mode". The microcontroller continuously monitors the grid power is present or not. When grid power is not present, then UPS goes to inverter mode. When grid power is present, then UPS goes to Backup Mode. In inverter mode, UPS converts the DC component in the battery into the AC component. Then UPS continuously monitors the grid power is present or not and it monitors battery voltage, battery current, and load capacity. This model consists of an oscillator circuit, a MOSFET driver circuit, step-up transformer, battery's low voltage cutoff, and overload protection circuit [1].

In the presence of grid power, then the micro-controller switches UPS into Backup mode. It converts the AC components from the grid into DC component for charging the battery. UPS monitor the battery charging voltage, charging current and switch grid power to load. Uninterruptible Power Supply is represented as UPS in short. UPS is an independent substitute power source that is used to source various sensitive electronic loads such as computers, medical assistance, communication exchanges, and many control and monitoring units in industries. Such type of application needs a power source that is always available in good quality.

A UPS is a solution for providing a good power source interface between the utility and the sensitive loads. The UPS will provide the voltage that is:

- Free from the disorders present in utility power and consistent with the required tolerances level by loads.
- Available in the event of utility blackout, within stated tolerances

UPS fulfill the above conditions as of power accessibility and quality by:

- Providing the voltage fulfilling with strict tolerances to the load, using an inverter
- Providing a self-governing alternate source, using a battery
- Change to alternate utility power with no handover time. i.e. without any interruption in the power to the load.

The above conditions make the UPS system an ideal power source for all sensitive loads since they guarantee the availability and good quality source, irrespective of the condition of the utility source power.

Basic Parts of a UPS System includes the following main components:

- Rectifier which converts AC to DC power to charge a battery.
- An Inverter that produces high-class electric power free from all the disturbances, particularly from micro-outages which within the tolerances level of sensitive electronic devices.
- A Battery, that delivers adequate backup to ensure the safety of the sensitive devices.
- A semiconductor-based static switch that switches the load back and forth from the inverter and utility power.

Different kinds of static UPS were characterized by standard IEC62040. The three operating modes for UPS are

- Offline - Passive standby
- Line-interactive
- Online - Double conversion

Leo Louis [2] presents Arduino boards can be a good tool in developing VLSI test bench, particularly for sensors. The main benefits are rapid processing and a simple interface. The author expressed the use of open source software and hardware devices in the present technological world. Creating and implementing new designs with Arduino becomes endless. Arduino can be used from wearable fashion to space research. Zhen Zhu et al [1] presented a technique for instantaneously increase the quality and efficiency of power for marine distributed generation. To limit the pollution treatment of external power, the self-regulatory capacity of the output voltage was enhanced by PI guidelines and the real-time feedback method. The efficiency of power conversion was improved by 10%. Pulse width modulation (PWM) is used in the control circuit for a single phase inverter by Omokere, E. S et al [3]. The 3525A controller is used to create the essential waveforms to control the frequency of the inverter using the switching pulse. The DC to AC inversion was effectively achieved by the switching signals; the inverter output has a frequency of almost 50Hz. Hence the conversion of DC to AC was successful. Michal Knapczykf. et al [4] analyzed the modulation schemes for the AC/DC line-side converters. The modulation techniques are mainly responsible for the process of the AC/DC converter. The problems in the over modulations were analyzed.

Journal of Physics: Conference Series **2040** (2021) 012007 doi:10.1088/1742-6596/2040/1/012007

Chosen modulation methods were examined by using Voltage Oriented Control of the AC/DC line-side converter. Switching frequency and line current distortion were also studied. Power inverters operate under dynamic loads were examined by F. Onoroh et al [5]; their aim is to make a thermal model to the IRF 3205 MOSFET chip switching operation and is realized in MATLAB R2013a setting to get the transient temperature response. The simulation demonstrates that exponential build up function is shown for transient device temperatures profile and rapidly increases till it attains the steady state during the switching operation. A set of equations were used for determining the switching losses of an IGBT device in an electromagnetic transient program and were integrated into the power electronic switching model. A.D. Rajapakse, et al [6]. Formulae were derived for estimating the switching losses during turn-on and turn-off by using predicted trajectories of the device between their pre and post switching-values.

Danijel Pavković et al [7] designed a cascade control system for a battery constant-current constant-voltage (CCCV) charger which has voltage PI controllers and dedicated battery current. Damping optimum criterion is used for designing the controller. LiFePO4 and VRLA battery cells were taken for testing. Superimposed voltage controller and inner current controller were implemented for the CCCV battery charging. Daniel Akinyele et al [8] analyzed the dependability of battery based on loss of power probability (LOPP) and the cost of energy (COE), by the life cycle cost study. The Hybrid Optimization Model was used to simulate battery banks for Electric Renewables (HOMER) environment. The results validate that the LOPP reduces with an increase in battery cells and the COE raises with the increase in the battery cells. Elie Ayoub et al [9] analyzed different charging techniques which are acceptable for commercial battery chargers and they have given the boost charging method as the appropriate charging method for recharging the battery. Christian Brañas et al designed a resonant converter based battery charger for high power $LiFePO_4$ Batteries. A model for charging a 48 V LiFePO4 battery with 50 Ah of capacity is designed and achieved a very good efficiency [10].

2. Implementation

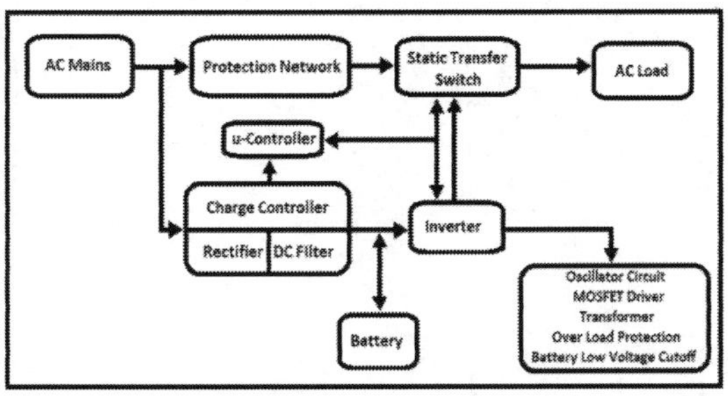

Figure 1. Block Diagram

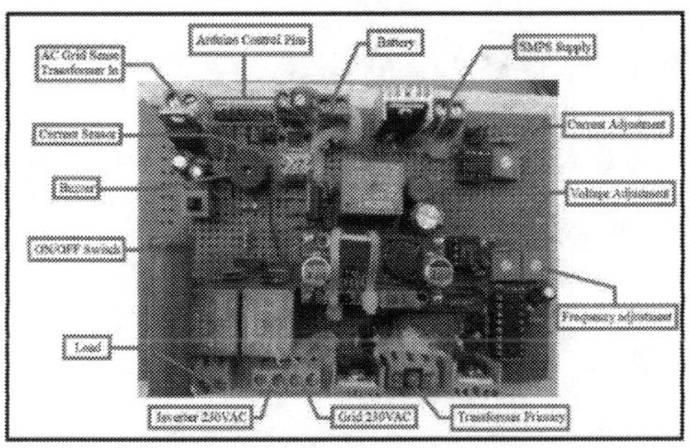

Figure 2. Top View

Figure 3. Practical View

3. Inverter Mode

When grid power is not present, then UPS goes to Inverter Mode. UPS converts DC power in the battery into AC power. UPS continuously monitors the grid power is present or not and UPS monitors battery voltage, battery current, and load capacity [11]. The inverter mode consists of an oscillator circuit, a MOSFET driver circuit, a step up transformer, a battery, a low voltage cutoff, and an overload protection circuit.

3.1 Oscillator

In India, the EB voltage is 220 volts with a 50 Hz frequency. Therefore, the inverter output is ~220VAC/50Hz. Transformer works only in AC [12]. Here SG3525 IC used to generate a 50Hz PWM signal for oscillates the MOSFET to convert the DC component of the battery into pulsated DC component.

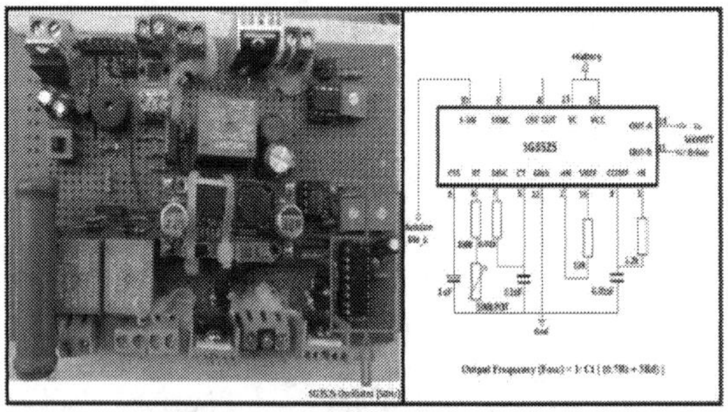

Figure 1. SG3525 Oscillator

3.2 MOSFET Driver

Here MOSFET act as a switch to drive the DC component of the battery into pulsated DC based on the oscillation frequency. IRF3205 MOSFET, which can drive continuous 110A of current. This MOSFET have very low R_{DS} resistance in the range of milliohms [13].

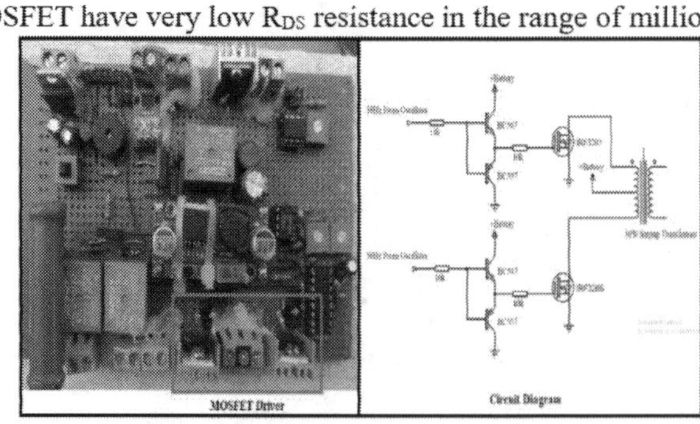

Figure 5. IRF3205 MOSFET Driver

3.3 StepUp Transformer
This transformer and MOSFET driver determine the load capacity. Here 12-0-12 EI core 3A transformer used to convert low voltage AC into high voltage AC. So it can load up to 36W (max). Due to the efficiency of the UPS, overload protection set to the maximum of 20W load at the output terminal.

Figure 6. Power Transformer (36W)

3.4 Output Voltage Adjustment
IC555 circuit is used to control the PWM signals applied to the gate of the MOSFETs driver circuit. This PWM adjustment can affect the energy transformation between the primary and secondary of the transformer [4] [14]. This way we can adjust the output voltage to control the ON-Time of the PWM signal applied to the gate of the MOSFETs.

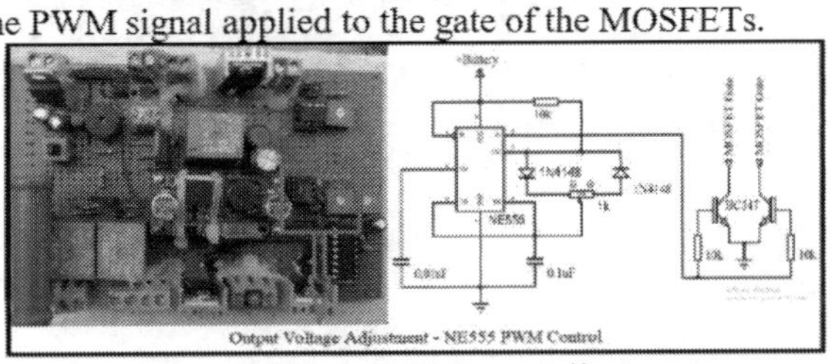

Figure 7. Output Voltage Adjustment

4. Operation

4.1 Inverter ON/OFF
Inverter ON/OFF state is controlled by ON/OFF state of an external switch. If the switch is ON, then the inverter is ON and vice versa. Microcontroller updates the status of inverter in the LCD display.

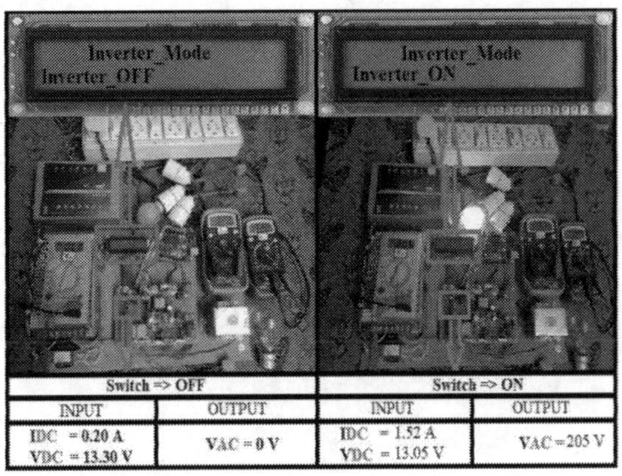

Figure 8. Inverter ON/OFF Control

4.2 Overload Protection
When connecting more loads to UPS the transformer drain more current from the battery.

The microcontroller monitors the battery current with the help of the ACS712 Hall-effect based current sensor. If the transformer drains more current, then the controller switches off the inverter as well as indicates overload in LCD display.

4.3 Battery Low Voltage Cutoff

If the power cut is too long, then the battery drains too much. This may cause damage to the battery permanently. So, cut off the battery at recommended voltage level. The microcontroller monitors the battery voltage and switches off the inverter if the voltage level of the battery is less than the recommended voltage level.

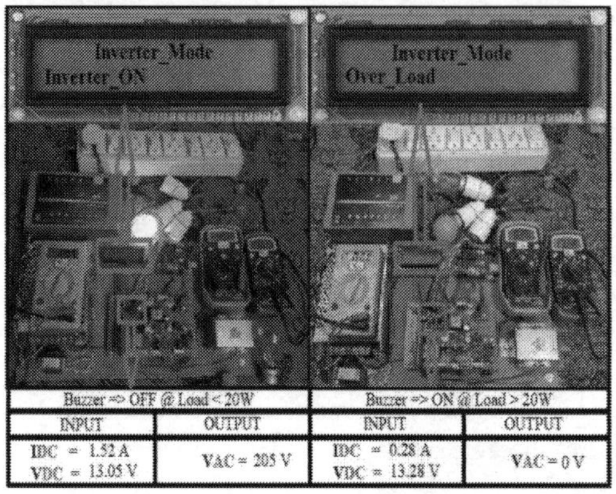

Figure 9. Overload Protection

5. Backup Mode

Grid power is present, then the UPS switch Backup mode. It converts the AC component from the grid into the DC components for charging the battery. UPS monitor the battery charging voltage, charging current and switch grid power to load.

5.1 Charging Voltage & Charging Current Control Circuit

Microcontroller monitors battery voltage and the current entering or leaving the battery. Charge control circuit controls the charging voltage, charging current, and battery overvoltage cutoff. Microcontroller takes care of all these things [15] [16].

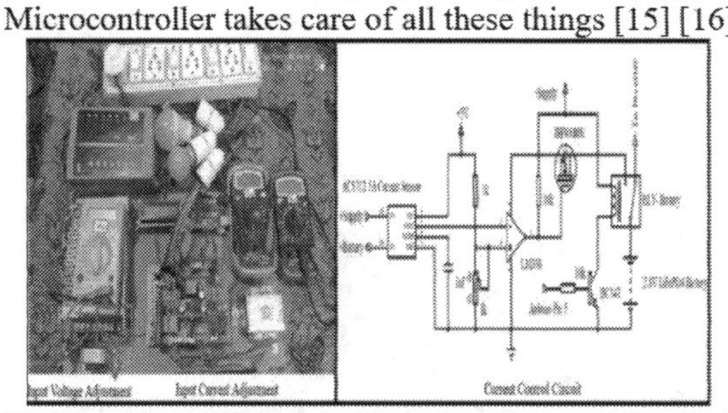

Figure 10. Charging Voltage & Charging Current Control Circuit

5.2 Power Switching

When the micro-controller sense grid power is present, then it monitors battery charging voltage and charging current as well as switches grid power to load with help of a relay circuit. The controller switches relay circuits to bypass grid power to load and at the same time battery charges with help of grid power [17].

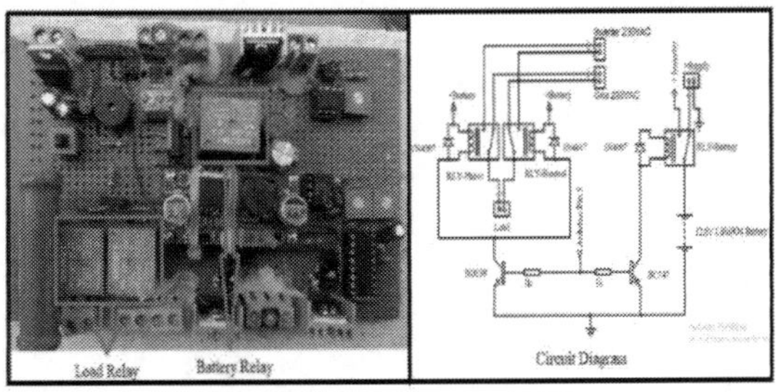

Figure 11. Power Switching

6. Microcontroller

Arduino UNO R3 micro-controller is a heart of this project. It switches the UPS into either backup mode or inverter mode based on the grid power. This monitors the output voltage, output current, and load capacity at inverter mode. In addition, it monitors charging voltage, charging current, and switch grid power to load at backup mode. Moreover, it controls the 16*2 LCD display via I2C communication for display UPS status.

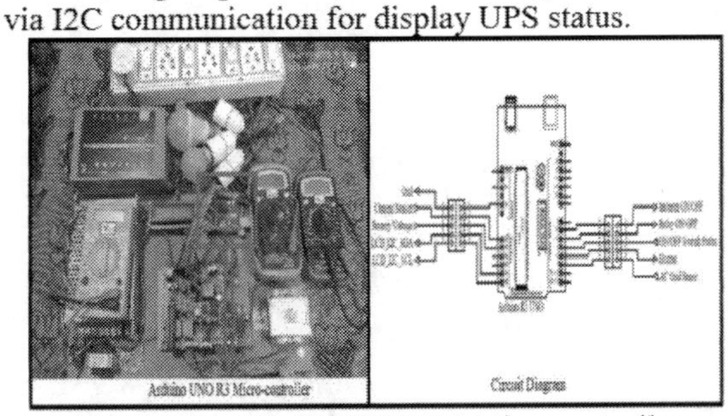

Figure 12. Arduino UNO R3 Microcontroller

7. Battery

A 12.8V 18Ah 0.5C LiFePO4 battery is used in this project. Table 1 shows the advantages of the LiFePO4 battery over the lead acid battery, SLA battery, and Li-ion battery [9]. LiFePO4 battery made up of non-toxic materials and battery not overheated or catches fire in case of overcharge or overload [14].

Table 1 Features of LiFePO4 Battery

Features	Lead Acid	Li-ion	LiFePO4
Efficiency	70%	80-90%	80-95%
Charging & Discharging Current Rate	0.1C	1C(max)	10C
Life Cycle	500-1000	1500-2000	2000-10,000

8. Features

8.1 Battery

- Battery Life cycle
- Non-toxic material
- Efficiency
- Very minimum space to store
- Less Charging and Discharging time when compared to lead acid and SLA batteries.
- Battery not overheated or catches fire in case of overcharge or overload.

8.2 Inverter

- Configurable output Voltage and Frequency
- Overload protection
- Over current protection

- Battery low voltage cut off

8.3 Charge Circuit
- Short circuit protection
- Battery over voltage protection
- Charging voltage control
- Charging current control

9. Application
- Industries
- Data Centers
- Banks and insurance
- Hospitals
- Telecommunications
- Some special projects (events)

10. Things to Improve

10.1 Inverter Output Voltage (V_{AC})
This project did not have a feedback circuit. Therefore, the output voltage changes with respect to load. In addition, battery voltage is near to transformer's operating voltage. After connecting load, output voltage decrease with respect to load. At that time, 100% duty cycle PWM pulses for MOSFET drives are not enough to increase the output voltage. Therefore, the selected transformer operating voltage is less than the battery voltage.
10.2 Inverter Output Waveform
Starting this project design was a pure sine wave inverter. But a square wave inverter was designed due to some practical issues. Therefore, the next improvement is converting this design into a pure sine wave inverter design.

11. Conclusion
Now a day, most UPS use lead acid or SLA batteries and few UPS use Li-ion batteries. But, another battery technology has much more advantage when compared to all the above batteries that are the LiFePO4 battery. The main objective of this project is a perfect battery that gives more reliability to the UPS. Because when compared to other batteries, LiFePO4 has a lot of advantages, especially battery life cycle, LiFePO4 battery is made up of non-toxic material and LiFePO4 battery is not overheated or catch fire in case of overcharging or overload. Other battery's UPS is not supported to LiFePO4 battery. Because the LiFePO4 battery's voltage rating and the current rating were different from each other.

Therefore, in this system, a UPS specially design for LiFePO4 battery's voltage rating and current rating. Other than that, features of UPS were added to this design. For example, in inverter mode overload protection, battery low voltage cutoff, and in backup mode overcharging cutoff are included in this design.

REFERENCES
[1] Zhen Zhu, Renda Wang, Yong Yin and Shengwei Xing 2019 Design of High Efficiency Single-Phase Bridge Passive Inverter Based on SG3525, IOP Conference Series: Materials Science and Engineering, Vol. 688, No 1, pp.1-6.
[2] Leo Louis 2016 Working Principle of Arduino and using It as a Tool for Study and Research, International Journal of Control, Automation, Communication and Systems (IJCACS), Vol. 1.
[3] Omokere E S and Nwokoye A O C 2012 Evaluating the Performance of a Single Phase PWM Inverter using 3525A PWM IC, International Journal of Engineering Research and Technology, Vol. 1, No 4, pp. 1-4.
[4] Elektrycznych, Pomiarów, K Pie and kowskif 2007 Analysis of Pulse Width Modulation Techniques for Ac/Dc Line-Side Converters.

[5] Onoroh F, Enibe S O and AdewumiO O 2018 Thermal Model and Experimental Validation of IRF 3205 MOSFET Switches for Inverter Application, FUW Trends in Science & Technology Journal, Vol.3, No 1, pp.73-78.

[6] Gayathri Monicka, J & Jamuna, V 2015, 'Hybrid Cascaded MLI topology using Ternary Voltage Progression Technique with Multicarrier Strategy', Journal of Electrical Engineering &Technology (JEET) vol.10,pp.1610- 1620,2015

[7] D Pavković, M. ć, M. Hrgetić, A Komljenović and V Smetko 2014 Battery current and voltage control system design with charging application, *IEEE Conference on Control Applications (CCA)*, pp.1133-1138, doi: 10.1109/CCA.2014.6981481

[8] Daniel Akinyele, Juri Belikov and Yoash Levron 2017 Battery Storage Technologies for Electrical applications: Impact in stand-alone photovoltaic systems, Energies, Vol. 10, No 11, pp. 1-39.

[9] Elie Ayoub and Nabil Karami, 2015 Review on the Charging Techniques of a li-ion Battery, *3rd International Conference on Technological Advances in Electrical, Electronics and Computer Engineering (TAEECE 2015)*, pp.50-55.

[10] Christian Brañas, Juan C. Viera, Francisco J. Azcondo, Rosario Casanueva, Manuela Gonzalez and Francisco J. Díaz 2021, Battery Charger Based on a Resonant Converter for High-Power LiFePO4 Batteries, Electronics, Vol. 10, No. 3, pp. 266. https://doi.org/10.3390/electronics10030266

[11] Gayathri Monicka, J.& V.Jamuna "Optimal Switching Strategy of Level shifted carrier based PWM technique for Asymmetric Multilevel Inverter" International Journal of Engineering Science ISSN: 0020-7225),vol.113.

[12] Ryder 2003 Diagnosing Transformer Faults using Frequency Response Analysis, IEEE Electrical Insulation Magazine, pp.16-22.

[13] Arvindan Sivasuriyan, D.S. Vijayan, Wojciech Górski, Magdalena Daria Vaverková, Eugeniusz Koda, "Practical Implementation of Structural Health Monitoring in Multi-Story Buildings", Buildings 2021, 11, 263. https://doi.org/10.3390/buildings11060263.

[14] Change CHEN, Chengxiong MAO, Dan WANG and Jiming LU 2009 The Research on Special Electronic Power Transformer, Istanbul University – Journal of Electrical & Electronics Engineering, Vol.9, No.1, pp.895-903.

[15] C. Amuthadevi, D. S. Vijayan, Varatharajan Ramachandran, "Development of air quality monitoring (AQM) models using different machine learning approaches", Journal of Ambient Intelligence and Humanized Computing, https://doi.org/10.1007/s12652-020-02724-2

[16] T Nguyen and L Bushnell 2003 Advanced Battery Charging Techniques: Pulse-Charging in Large-Scale Applications - Design of Divide and Conquer Technique for High Capacity Batteries, UWEE Technical Report Series, Seattle, USA.

[17] V Gupta 2010 Working and Analysis of the H - bridge Motor Driver Circuit Designed for Wheeled Mobile Robots, *2nd International Conference on Advanced Computer Control*, pp.441-444. doi: 10.1109/ICACC.2010.5486818

A study on vulnerable risks in security of cloud computing and proposal of its remedies

Rashi Saxena1, E. Gayathri2

[1]Research Scholar, [2]Associate Professor
Department of Computer Science & Engineering, K L Deemed to be University, Hyderabad, Telangana, India.

E-mail: rashisaxena.cse@klh.edu.in

Abstract. Data innovation worldview empowers pervasive access to shared pools of configurable framework assets and more significant level administrations that can be quickly provisioned with minimal administration exertion, regularly over the Internet. Cloud computing provides dynamic asset administrations for the client on request, and it is becoming increasingly popular. Cloud administrations are conveyed from server farms all through the world (as it is otherwise called Networks of systems). Outsider associations are additionally present there, because of which some significant security dangers came and make organize helplessly so the security of cloud information is imperative to keep over Cloud architecture to offer better support, ease administrations, and so forth. Despite that there are a few undermined hazards in cloud computing security. We will be concentrating majorly on cloud computing fraught with threats and how to overcome those threats. This paper is intended to introduce information about the most current attacks on cloud computing, just as safety efforts. The seriousness and impact of these attacks are talked about alongside genuine instances of these attacks. The paper likewise proposes alleviation strategies that can be utilized to decrease or dispose of the risks of the threats talked about. Moreover, the general cloud security proposals are given.

Key words: Cloud security, Cloud threats, Cyber-attacks, Security Tools.

1. Introduction

Cloud computing simplifies and adapts everything, but cloud architecture is complicated, and resolving any security or network-related issue is difficult. In brief, there are three sorts of clouds: public, private, and hybrid. In addition to Platform as a Service (PaaS), Software as a Service (SaaS), and Infrastructure as a Service (IaaS), cloud providers provided three categories of services [1] as seen in figure 1. A commercial name for "Cloud Computing" is Cloud [2].

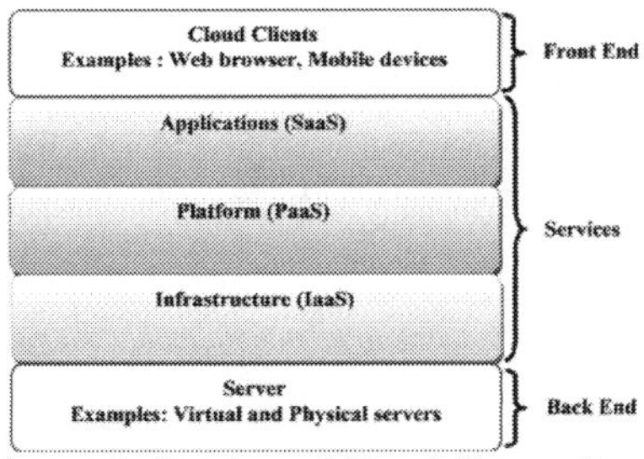

Figure 1: Service of Cloud Computing

Nonetheless, because cloud computing is not completely secure against specific attacks, it can cause difficulties for both end-users and cloud service providers. It is our primary goal to describe the vast majority of the threats while also attempting to outline some strategies for overcoming them.

2. Literature Survey

Author of [3] discussed about cloud security concerns. They sparked interest in security concerns that were centered on specialized security in cloud administrations, among other things. They discuss the hazards associated with cloud administration that have been overlooked, such as there are four types of assaults: VM-level attacks, compliance issues, isolation failure, and management interface compromise. They also used cloud security engineering in order to shield themselves from attacks and assaults on their infrastructure. Also covered are the critical components of this architecture: single sign-on, increased accessibility, a single administration dashboard, virtual machine protection, and a defense-in-depth cybersecurity strategy.

In [4], the author discussed several important risks that must be avoided to provide a secure cloud environment. Security elements such as those found inside threats, framework communication capabilities, and access control were built into the design (e.g., virtualization, data encryption, host Operating System, and visitor framework, backup, firewall, server location, etc.). The author shows how a major issue will shape cloud security in the future. Additionally, an in-depth exploration of such problems as vendor lock-in and contradictions is conducted. Security ideas might also be applied in cloud computing settings as cloud computing itself does not have any special security standards.

The author of the study article [5] investigated several risks and characteristics associated with cloud computing, as well as developing a strategy for its remediation. The researcher focused on the seven most major threats, as well as a few others, such as session riding and hijacking, weak cryptography, data protection, and portability. Also, customers will have better security because of the absence of uniformity. Struggled to demonstrate creativity while simultaneously implementing a very effective security system, which should be established in order to generate inventive innovation without going into detail.

3. Vulnerable threats/attacks on cloud computing with suggested moderations

Because a third party is engaged in cloud computing, security is the most important consideration. Consequently, certain tools and methods that are required to resolve this issue are presented. Several risks and assaults against the cloud computing environment are discussed, as well as some effective mitigations that may be implemented. Using the cloud computing platform as an example, [6] Figure 2 depicts the danger landscape –

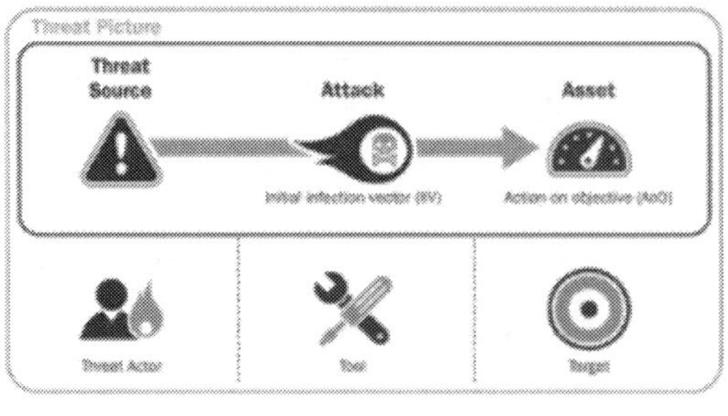

Figure 2: Threat Picture for a cloud computing platform

3.1 Nefarious usage of cloud computing

It has been designated as the most danger by the Cloud Security Alliance (CSA) [7]. Botnets are used to distribute viruses and spam over the internet. IaaS companies give their clients with a variety of benefits, including unlimited compute space, network bandwidth, and storage capacity. Then the attackers seize the opportunity to launch an assault against the seller.

Suggested moderation: Initial registration, credit card monitoring, and network traffic monitoring are just a few of the processes that can help to mitigate this issue.

3.2 Insecure Internet connections

Another hazard to cloud computing is the use of insecure interfaces. When developing APIs, it is critical to consider security factors such as authentication, access control, and a monitoring system.

Suggested moderation: Monitoring access control and authentication is permitted under tight conditions as specified.

3.3 Invading Saboteurs

When dealing with a hostile insider, it is critical to consider the sort of standard or employee who will be employed since there is no monitoring or reporting mechanism that would reveal their access data.

Suggested moderation: Through chain management, a security breach notification system and a reporting system are implemented.

3.4 Loss of data

Unreliable media, a weak password, no encrypted data, a lack of a secure backup system, hard drive failure, and other factors can all result in data loss and leaking, which is a serious concern.

Suggested moderation: It is necessary to implement an encryption and decryption mechanism, as well as monitoring, file backup, and a strong password.

3.5 Account, Service & Traffic Hijacking

Another concern that cloud clients should be aware of is account, service, and traffic hijacking, with the objective of preventing this from happening again. From a man-in-the-middle attack, phishing, and spam to denial-of-service assaults, re-used secret keys, extortion, and programming flaws, these dangers have evolved throughout the years. The hijacking of traffic is a recurring topic in cloud computing.

Suggested moderation: Understand a security strategy, put a few ideas into practice among clients and suppliers, implement a confirmation mechanism, and provide a means for ensuring the system's integrity.

3.6 DoSA (Denial of Service Attacks)

DoS attacks provide the impression that an authorized user is inaccessible; for example, when anybody visits a website, they may see an error page or a page that is not available for DoS attacks.

Suggested moderation: When dealing with this issue, an Intrusion Detection System (IDS) might be employed [8]. The defensive federation [9] is yet another type of detecting system.

3.7 Multi-tenancy natures

Since this threat targets service providers online that provide services to numerous customers on demand, the attack can occur regularly through VM (virtual machine), and the danger has an impact on IaaS (Internet of Things) services. A comparable program is used by several different customers who connect to the virtual machine over the internet.

Suggested moderation: This danger may be identified using strong authentication and access control.

3.8 Risk assessment

Programming variations, code upgrades, and disruption Endeavor security rehearsals are all things that clients or customers should be aware of. The user must be aware of the location in which the data or logs are stored. However, risk assessment never accomplishes this [7].

Suggested moderation: Clear information regarding logs and data, as well as monitoring and an alarm system, can help to avoid risk assessment from taking place.

3.9 SQL injection attacks

It is possible that malicious code has been embedded inside a carefully written SQL code, and this has resulted in a touchy assault and some unexpected results [10]. Access to a database without authorization is a type of assault that occurs from time to time.

Suggested moderation: SQL infusion can be stopped by filtering techniques; however, an intermediary engineering should be developed that gradually differentiates and eliminates contributions from clients that are suspected of overseeing SQL injection.

3.10 XSS (Cross-Site Scripting) attacks

In this sort of attack, a variety of malicious scripts are uploaded to the web and executed. It can occur in two ways: because of a stored XSS or because of a reflected XSS. One is utilized on a permanent basis, whereas the other is not [11].

Suggested moderation: Provide or implement some type of error-prone data disinfection mechanism, which protects components on both the consumer and server sides.

3.11 MIMA (Man in the Middle attacks)

It is an alluring danger in which an entity constantly seeks to remain between a sender and a receiver, implying that they constantly offer information between the sender and clients, resulting in the clients receiving erroneous information.

Suggested moderation: This challenge may be solved by implementing service security procedures that are distinct from endpoint and server security processes [12], the topology of the network can affect the setup of the network.

3.12 Theft of Identity

It is another danger to cloud computing. Sometimes identity alterations occur when a fraudster steals another person's identity and use it for all their advantages, and the victims suffer as a result in the long term.

Suggested moderation: Strong password is solution.

4. Tools for cloud computing security

These applications are designed to keep malware out of the system, as well as protect against unwanted system logins and external cloud applications like Salesforce and Google Docs. This is just the beginning. These concepts are discussed at the beginning of this article (shown in figure 3), we discussed hazard and remediation apparatus.

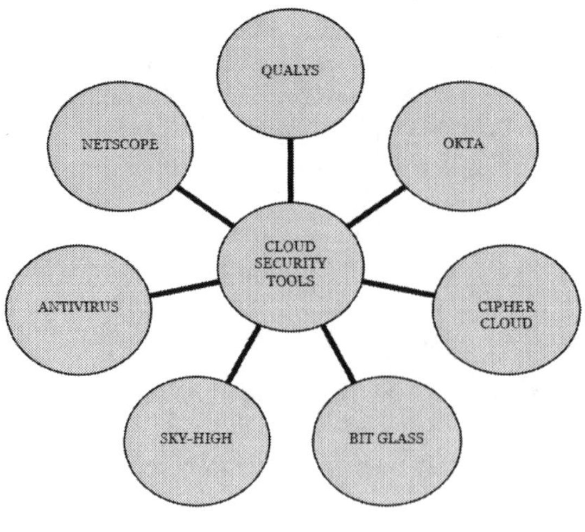

Figure 3: Cloud Computing Security Tools

4.1 Qualys

It secures devices, web applications, and web applications, but does not require any hardware or software. Malicious software is detectable, repair issues if found, scan apps, and save web apps in the cloud.

4.2 Okta

One of the most essential pieces of technology, it keeps a log of all the employees who use information. The first job is identity management. the overall objective of Okta is to make SSO (Single-Sign-on) available for all cloud-based applications.

4.3 Cipher Cloud

Cipher Cloud secures further, such as Salesforce, Chatterbox, Office 365, Gmail, Amazon Web Services, and various other products and services. It ensures that the administration model is encrypted, including for sending and receiving data, as well as to check traffic, which acts as a hostile to antivirus.

4.4 Bitglass

The ability to offer information and threat protection in any context, on any device, at any time, is what Bitglass delivers with aim to detect and reduce the risk of loss of data, while maintaining visibility of data even in the cloud.

4.5 Sky-high

It has the capability of discovering, analyzing, and securing applications that we use in the cloud. It has the capability of discovering or detecting which cloud applications are being used by the employees.

4.6 Antivirus

Security tools such as antivirus are not new to cloud computing. The virtual computer is protected by a Kaspersky Lab formula. E-mail security is provided by McAfee.

4.7 Netskope

Cloud apps are discovered and monitored by Netskope, which is a service that is utilized on your network to identify and monitor them. Based on the information collected by Netskope, it gives extensive statistics on users, sessions, shared and downloaded content as well as the contents of the shared content.

5. Conclusion

As far as we can tell, cloud computing is going to be with us for a long time to come. Cloud computing will be used in many ways soon. Even though it offers several features, the cloud is not reliable when it comes to security. A robust security model is required to manage all the layers of the cloud and it should be able to do so in a secure manner. This subject is covered in-depth in this article, and many strategies to counter these security concerns in cloud computing are laid forth. Threatened risk could not exist without all these protective and mitigative measures. Following this study, we can predict that a significant number of vulnerabilities would be solved in the upcoming years if these tools are employed. In this paper, a fundamental understanding of threatened risks is laid out, as well as various methods for dealing with them in cloud computing. All of these and similar resources are crucial for dealing with elevated risk. In this case, the tools can solve a lot of future security problems.

References

[1] Rabi Prasad Padhy, Manas Ranjan Patra, Suresh Chandra Satyapathy. "Cloud Computing: Security Issues and Research Challenges". IJCSITS Vol. 1, No. 2, December 2011.
[2] Qi Zhang, Lu Cheng, RaoufBoutaba. "Cloud Computing: State-of-the-art and research challenges". J Internet ServAppl (2010).
[3] Tripathi and A. Mishra, "Cloud computing security considerations" IEEE Int. conference on signal processing, communication, and computing (ICSPCC), 14-16 Sept., Xi'an, Shaanxi, China, 2011
[4] Amutha devi, Gayathri Monicka.J, "Emerging bio-medical applications and open research challenges in cognitive internet of things (CIOT)" International Journal of Pharmacy and Technology" Dec-2016 Vol. 8 Issue No.4 5049- 5054
[5] Mervat Adib Bamiah* et al. / (IJAEST) INTERNATIONAL JOURNAL OF ADVANCED ENGINEERING SCIENCES AND TECHNOLOGIES Vol No. 26, Issue No. 1, 012 – 015
[6] https://insights.sei.cmu.edu/blog/12-risks-threats-vulnerabilities-in-moving-to-the-cloud/
[7] CSA, "Security Guidance for Critical Areas of Focus in Cloud Computing V2.1" Cloud Security Alliance, 2009, [Online], Available: https://cloudsecurityalliance.org/topthreats/csathreats.v1.0.pdf,
[8] K. Vieira, A. Schulter, C. B. Westphall, and C. M. Westphall, "Intrusion detection techniques for Grid and Cloud Computing Environment", IT Professional, IEEE Computer Society, vol. 12, issue 4, pp. 38-43, 2010.
[9] Gayathri Monicka, J & Jamuna, V 2015, 'Hybrid Cascaded MLI topology using Ternary Voltage Progression Technique with Multicarrier Strategy', Journal of Electrical Engineering &Technology (JEET) vol.10,pp.1610- 1620,2015
[10] Liu, Y. Yuan, A Stavrou, "SQLProb: A Proxy-based Architecture towards Preventing SQL Injection Attacks", SAC March 8-12, 2009
[11] P. Vogt, F. Nentwich, N. Jovanovic, E. Kirda, C. Kruegel, and G. Vigna, "Cross-Site Scripting
Prevention with Dynamic Data Tainting and Static Analysis", Proceedings of the Network and Distributed System
[12] Eric Ogren, "Whitelists SaaS modify traditional security, tackle flaws", Sep. 17, 2009. [Eric Ogren is the founder and principal security analyst at Ogren Group]
[13] http://www.hongkiat.com/blog/cloud-security-tools/

Journal of Physics: Conference Series **2040** (2021) 012009 doi:10.1088/1742-6596/2040/1/012009

Face Recognition using CNN and Tracking Locations using Image Metadata

Ch.Sushma[1], P Lalitha Surya Kumari[2]

1Research Professor, 2Professor

Department of Computer Science and Engineering, Koneru Lakshmaiah Education Foundation, Hyderabad, Telangana , India.

Corresponding Author E-mail: chsushma22@gmail.com

Abstract. In This Paper Face Recognition has gain the more attention in the society of network multimedia for accessing information. Emerging Areas like network security, content indexing and extraction, and video compression benefits from face recognition technic because "people" are the center of attention in a lot of video footages. Face detection is advanced research area in the computer science vision. The research under this area is expanding in others areas like psychology. Localization of human face will be first and the initial stage as a study of face detection. In this paper the usage of Convolutional Neural Network (CNN) algorithm in detecting the criminal face and also the process of getting the location from image.

1. Introduction

Face Recognition System is a computer application used to identify humans from a video footage or else a image. It is used for security purposes to track the people who is coming into a location or to search someone in the location. It is not efficient for bio metric or iris scanner but it is easier for implementation purpose. This face recognition system will store the images of people in a database for tracking. The face recognition system will use the algorithms to verify that the person in camera or video footage matches with any image in the database or not. one algorithm the images of humans are stored in gray scale format and stores in a database .The Other algorithm searches for a face in the database for verifying purpose. Another algorithm compares whether the shape or area or size of the face matches within the database or not. The above algorithms will verify the image is matching in the stored database or not.

The Convolutional Neural Networks has brought the lot of drastic changes in developing of Face Recognition like a piece of cake[1].

The CNN is the game changers and used in many fields to provides the solutions for the problems related to computer vision. In computer technology the Face recognition process can be utilized in many applications and identifies human faces as digital images. It performs three basic steps which includes as follows detecting face, recognizing face, and extracting face process. The system will encapsulate the image and determines the location of the face where it is located[2].

Face recognition will take an image as input from a camera or video and output as the original image. Facial features may consider the changes in shapes, face, size, face regions, variations in the face structure, face cuts and angles styled and formatted designs. Face extraction process will include grabbing the features from camera. Face detection will include background process of removal and focusing on the eliminating foreground any other features apart from the face regions. The system still have few drawbacks it cannot able to detect the head count due to overlapping of faces or not proper recognizing the two faces having same facial features[3].

In this representation, the facial recognition is performed by using the trained CNN model. After finding the perfect match of the face from the input feed, the same image is given to the Plot function to use the metadata which will be helpful in tracking the location where the image was taken.

1.1. Related Works

The CNN model was built to detect the suspects by using the smart glasses, it was written by Khan et AL. [1]. Their model first detects the faces to make Haar – like features which will be given as feed to the CNN model. They have presented that their model of detection was most likely to be 98% accurate with 3099 features. They are used to transfer learning technique and can implement their proposed method. They have used AlexNet which is popular CNN architecture was used in transfer learning techniques. They have achieved 98.5% accuracy of recognising the suspect faces.

A Face Recognition model was developed by Musab et Al. [2] using CNN model. In their work they have more concentrated to normalize the model by adding the normalization to two of the layers and by using the batch normalization technique. Their model has reached a good accuracy.

The face detection model which was developed by Bogdan Kwolek [3], the gabor filters can be used in additional with the proposed CNN model to extract the face features and improving the accuracy of a model. The detection rate accuracy of the proposed model has reached around 87.5% on testing data and contains 1000 face samples and 10000 non-face samples.

A Varun et Al. [4], a CNN model was used to predict the animals emotions by transforming their voice into the sound wave images. The recognition and mapping of those images was performed with a trained accuracy 100% and the test accuracy 88%.

Boutell et Al. [5], had developed a model that will be differentiating the inside and outside images using the image content and the metadata.

2. Methodology

The development of this methodology for the purpose of the facial recognition along with metadata of the image for tracking the location. Due to the improvement of technology the tracking system has also been improved. The basic architecture of the proposed face recognition model is shown in the above figure. At first, the images from the social media or from any means are taken is pre-processed in which the size of the images are modified to the required size for the CNN model to be trained on. The other phase is the training of the model. In this phase the proposed CNN model is developed by using convolution and dense layers of artificial neurons.

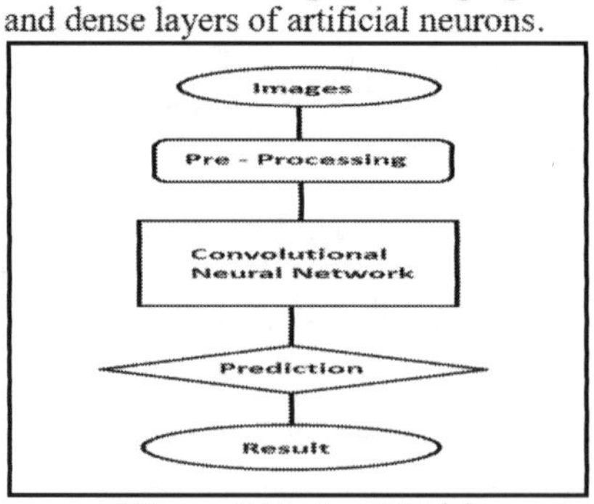

Figure 1. System Model

2.1 To fit the image into the model for training in the pre-processing as follows:

- Initially take all the images as a group and save them in a folder.
- The respective classes and images are stored in Blank multidimensional array.
- The images are can be read by computer vision library OpenCV, images will be converted into the numerical arrays.
- File is used for a set of 100 images after the combination of the [image_array, class] to save.
- To read all images along with their respective class the below file is used.
- The images and their respective class of the images which is stored in a file will be shuffled and saved in another file.

2.2 A Proposed CNN network is described below:

- The above created file is imported and adjustment of size from multidimensional array to compressed array to fit into the proposed CNN model.
- Firstly, 3 layers of Convolutional 2D layers are built and along with the respective softmax activating functions and pooling functions.
- The sequential network is designed with five layer is builded with three layers as hidden layers .
- For resizing the data and flattening layer is utilized which is known as initial layer.
- The neural network nodes with the initial hidden layer and rectified with the linear activation function as second layer.
- To construct 512 neural network nodes and relu activation function the third layer is used i.e second hidden layer.
- Taking 128 nodes relu as activation function and neural network as the fourth layer.
- Soft maximum to normalize the k real numbers into probability with a same number of nodes with available classes and activation function is a final layer.
- Compilation will done By using the Adam Optimizer
- With the help of sparse categorical cross entropy process and track the damage in the network.

After training the images with the above mentioned proposed model and it is ready for the testing process where the images are pulled from anywhere and can be pre-processed to find the accurate match or class. The next process is the node to node plotting in which the metadata of the images are considered.

Generally, in any image there will be metadata which is attached with the image that contains the details about the flash light, location, device name, camera quality, and more. But it is not visible to everyone as it is integrated or hidden inside the image. In this methodology, we have pulled out the information that is available I the metadata and then the we make use of location i.e., coordinates information for plotting in a map. The example of the plotting is as shown in the figure 2.

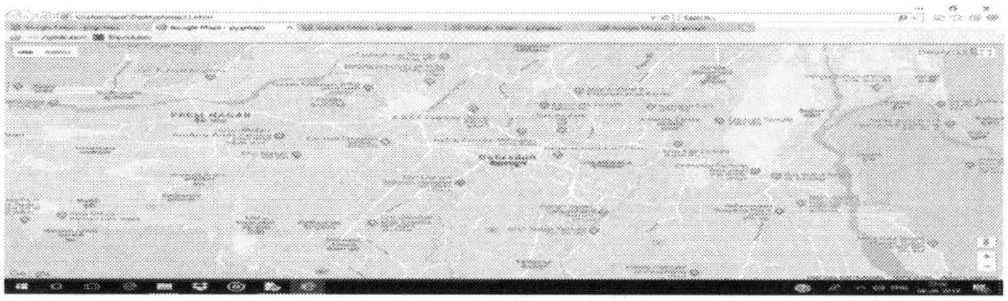

Figure 2: MapPlot Example

3. Results and Discussion

The accuracy of the Machine Learning or Deep Learning model is very much important in decision making. If there is any chance of getting the wrong decisions that will lead to a huge impact on the decision and the total process may become wrong. So, to achieve the good accuracy the It will be trained will good amount of data which will not under fit or over fit the model. The under fitting or over fitting will lead to false accuracy and the in the real time environment the model will not work well. In the training phase the proposed CNN model started its loss at 83% and reduced to almost 0% within 25 epochs. And the accuracy has started at 56% and has risen up to 100% at 29th epoch or iteration but the validation accuracy was shown as 99.86%.

4. Conclusion

The paper used in developing of the Convolutional Neural Networks model for the purpose of recognizing face. A matched face images are used to plot in the map using the metadata of the image. These models can be implemented in the criminal tracking or suspects tracking and even help in the form of gathering digital footprints. The image will provides the details like flash light, location, device name, camera quality. It will be helpful in tracking the criminals and provides security.

References

[1] S. Khan, M. H. Javed, E. Ahmed, S. A. A. Shah and S. U. Ali, "Facial Recognition using Convolutional Neural Networks and Implementation on Smart Glasses," 2019 International Conference on Information Science and Communication Technology (ICISCT), Karachi, Pakistan, 2019, pp. 1-6, doi: 10.1109/CISCT.2019.8777442.

[2] M. Coşkun, A. Uçar, Ö. Yildirim and Y. Demir, "Face recognition based on convolutional neural network," 2017 International Conference on Modern Electrical and Energy Systems (MEES), Kremenchuk, 2017, pp. 376-379, doi: 10.1109/MEES.2017.8248937.

[3] Kwolek, Bogdan. (2005). Face Detection Using Convolutional Neural Networks and Gabor Filters. 3696. 551-556. 10.1007/11550822_86.

[4] Totakura, Varun & Janmanchi, Mohana & Hussan, M.I.. (2020). Prediction Of Animal Vocal Emotions Using Convolutional Neural Network. International Journal of Scientific & Technology Research. VOLUME 9. 6007-6011.

[5] Boutell, Matthew & Luo, Jiebo. (2004). Photo Classification by Integrating Image Content and Camera Metadata.. 4. 901-904. 10.1109/ICPR.2004.1333918.

[6] Mullan, Patrick & Riess, Christian & Freiling, Felix. (2019). Forensic source identification using JPEG image headers: The case of smartphones. Digital Investigation. 28. S68-S76. 10.1016/j.diin.2019.01.016.

[7] "Facial Recognition System for Suspect Identification Using a Surveillance Camera" V. D .Ambeth kumar ISSN 1054-6618, Pattern Recognition and Image Analysis, 2018, Vol. 28, No. 3, pp. 410–420. Pleiades Publishing,

[8] Fahad P, M.d. Mahmudul, Md. Atiqur, Susan M, Moslehuddin M, Pandian VFace recognition based real time system for surveillance. Intelligent Decision Technologies, IOS Press (2017), pp. 79-9211(2017)

[9] Karthik HS, and Manikandan J. (2017) "Evaluation of relevance vector machine classifier for a real-time face recognition system." International Conference on Consumer Electronics – Asia, Bangalore, India, 5–7 October, IEEE Press, pp. 26–30.

[10] Shalmoly M, and Soumen B. (2016) "Face Recognition using PCA and Minimum Distance Classifier." Fifth International Conference on Frontiers in Intelligent Computing: Theory and Applications, 16–17 September, Bhubaneswar, India, Springer, pp. 397–405.

[11] Kukreja S, and Rekha G. (2011) "Comparative study of different face recognition techniques." International Conference on Computational Intelligence and Communication Networks, 7–9 October, Gwalior, India, pp. 271–273.

[12] Hu H, Shah A, Bennamoun M, and M Molton. (2017) "2D and 3D face recognition using convolutional neural network," IEEE Region 10 Conference, 5–8 November, Penang, Malaysia, pp. 133–132.

[13] Bansal Prachi, Mittal Saurabh, Gupta Maanak Using locality preserving projections in facial recognition. International Journal of Engineering Technology, Management and Applied Sciences, 2 (3) (2014), pp. 99-108

[14] Survey on Stress Emotion Recognition in Speech, Reddy, L.L., Kuchibhotla, S. Proceedings - 2019 International Conference on Computing, Communication, and Intelligent Systems, ICCCIS 2019 january

[15] Abdelmoghit Zaarane, Ibtissam Slimani, abdellatif Hamdoun, et al. Real-time vehicle detection using cross-correlation and 2D-DWT for feature extraction J. Electrical and Computer Eng, 2019 (2019), 10.1155/2019/6375176

[16] Ibtissam Slimani, Abdelmoghit Zaarane, Abdellatif Hamdoun, et al. Traffic surveillance system for vehicle detection using discrete wavelet transform J Theor Appl Inf Technol, 96 (17) (2018)

[17] Kenji Nakahira, Tetsuji Kodama, Shin Morita, et al. Distance measurement by an ultrasonic system based on a digital polarity correlator IEEE Trans on Instrumentation and Measurement, 50 (6) (2001), pp. 1748-1752

[18] Facial expression detection of all emotions and face recognition system Jahnavi, P.,

[19] Vamsidhar, E., Karthikeyan, C. International Journal of Emerging Trends in Engineering

[20] Research 2019 volume isuue12,778-783

[21] I. Slimani, A. Zaarane, A. et Hamdoun Convolution algorithm for implementing 2D discrete wavelet transform on the FPGA Computer systems and applications (AICCSA), 2016 IEEE/ACS 13th international conference of. IEEE (2016), pp. 1-3

AGRICULTURE CLOUD SYSTEM BASED EMPHATIC DATA ANALYSIS AND CROP YIELD PREDICTION USING HYBRID ARTIFICIAL INTELLIGENCE

D. Naga Swetha[1], Savadam Balaji[2]

[1]Research Scholar, Department of Computer Science and Engineering,

Koneru Lakshmaiah Education Foundation, Deemed to be University, Hyderabad,

Telangana-500075.

[2]Professor, Associate Dean, Department of Computer Science and Engineering,

Koneru Lakshmaiah Education Foundation, Deemed to be University, Hyderabad,

Telangana-500075.

Corresponding Author E-Mail: mailsforswetha@klh.edu.in,

ABSTRACT; In agricultural nations, such as India, where agriculture leads more to India's Economic growth, it plays a significant part. The prediction of the crop is one of the main tasks in agriculture. Crop prediction methods are employed by detecting different soil parameters and factors connected to the atmosphere for predicting the appropriate crop. The unstable climate exposes farmers to danger in the environment. Therefore the correct history data must be maintained is essential. The data stored may be evaluated to predict agricultural production. In a cloud server, experts analyze sensed data, land type, land, climate, and farmers' economies with a prediction effect. The method forecasts the use of artificial intelligence algorithms for appropriate crops and fertilizers. A crucial strategy for handling numerous challenges connected to agriculture is the domain of artificial intelligence with its high-quality learning capacity. Technologies to help farmers find better solutions around the world are being created. To benefit from the parallel computational and storage management of huge data sets, the agricultural community must establish an architectural design that would enable the identification of new statistical structures to extract valuable information from data structures. These processes assist to explore the field and different challenges and effectively respond to certain challenges. In the improved integration of diverse data collection types from multiple sources, artificial intelligence offers attractive computing and analytical methods. The main principle of AI and systemic approaches to understanding its use in agriculture are presented in this paper. It also addresses several algorithms for artificial intelligence which may be used to create models to deal with various agricultural problems.

Keywords: Sensors, Agriculture Cloud system, Artificial intelligence, SVM, NB, RF

1. INTRODUCTION

As one of India's most dominating fields, agriculture has been plagued by extraordinary tragedies in recent years. The system analyzes the farm, advises crops, proposes fertilizers, links with suppliers, and provides the farmer with more valuable information using sophisticated principles of artificial intelligence and predictive analysis. The supply of resources through the Internet is cloud computing [1]. The delivery of shared resources, software, programs, and services through the Internet, to fulfill the client's elastic demand with little effort or involvement. All necessary data must be stored in a centralized place [15]. It contains multiple, separate databases, the data relating to the soil, weather and research, crop, and farmers may be securely stored. End users like farmers, specialists, advisors, researchers, etc. may easily obtain this data from any location via cloud-based devices [2]. When AI is integrated with the cloud, processes get smarter and judgments are made intelligibly without human involvement. Various methods have been presented, spanning from databases to predictive analytics, for solving current difficulties in agriculture. Regarding accuracy and efficiency, systems that use AI have been proven to be the best performance. Agriculture is a dynamic field that cannot generalize conditions to provide a sensible solution [16]. The AI approaches allow us to collect the complex aspects of each circumstance and deliver the best response to that unique challenge. The evolution of diverse AI systems progressively solves exceedingly complicated issues. In agriculture, its different agricultural approaches are rapidly adapted to AI. The notion of intelligent systems is the one that gives farmers the ability to identify crop diseases, analyze soil, give expert advice and create a commercial opportunity [6,7]. This leads to chaotic AI technologies powering farming, which help understand, collect and respond to various conditions (based on the obtained knowledge) to increase performance. Farmers may give solutions via platforms like chatterbots to benefit from the field by keeping up with new advances in the agricultural sector [17].

2. LITERATURE SURVEY

Theyazn H. H Aldhyani et.al [2] Enhanced artificial intelligence (AI) techniques were discussed to predict water quality indices and water quality categorization. Deeper techniques for artificial neural network models, in particular the nonlinear autoregressive neural network, are used for the WQI forecasting (NARNET). The WQC Forecast, i.e. the WQC support vector machine (SVM), K-nearest (K-NN), and Naive Bayes, has also been followed by three machine learning approaches. The used database comprises 7 important elements, which were used for the analysis of the created models using different function vectors.

Rashmi Priya et.al [3] suggested the essential idea of machine learning and systemic procedures to understand its use in farming. Other learning machines may be used to create models to solve different agricultural concerns.

B.Ragavi et al. [4] designed seed sowing device IoT and AI-based "AGROBOT." Weather, fertilizers and pesticides need and water requirements tracking. Check and monitor the weather. The data is transmitted to the farmers for live surveillance of the field via cloud services. Reducing labor expenditures and increases agricultural returns.

T. Rajesh et al. [5] Discussed real-time soil and temperature data are captured through the use of many sensors for IoT and AI-based E-agricultural solutions. To priorities crops according to region, profitability, and soil and advice farmers, AI algorithms are deployed.

Pravin Kshirsagar et al [6][7] proposed mechanism is intended to measure track and accurately represent the state of the patient's health and well-being and vital biochemical knowledge. The competent community health may make vital medicinal requests by using the data found inside the material or e-mail address.[8][19] has Several nerve connections have been established in order to fulfill their objectives, such with one layer activating capabilities, multilayer experiences, Rbf, Ffnn, Pnn, Grnn, etc.. This conventional computer program was believed to be suitable for a restricted range of moderate run-times data and configurations. However, with digitized technology developments the number of data being processed and even the duration for execution of the design phase are reduced. Many academics have therefore modified conventional machine learning, which can accept a lot of input and then have less time to converge [20][21]. The impacts of complex and fake websites may be described or generated via biological systems.

3. DATA-DRIVEN MANAGEMENT FOR ADVANCED FARMING

Raw observations of critical crop characteristics must be treated efficiently so that values or pictures become usable data consistently. Crop management obtained from field data actually existed in precision farming years ago, but the digital information generation has undoubtedly changed [1,2]. In locations where technologies have not yet arrived, field management consists of visually evaluating crop progress, to get a diagnosis by farmers who decide and react to handle their crops differently. This strategy is based on knowledge in the areas and information that farmers observe. The affiliated farmers might also follow the rules of cooperative technicians or engineers engaged in the company to which they belong [3]. Field management changes depending upon the operational cycle as indicated in Figure 1 on farms with modern technologies.

- This management approach, based on empirical data on the field and intelligent decision-making, begins by managing the real crop, making use of its spatially and time-wise climatic variation.
- The platform consists of the physical mechanism by which information is collected and the sensors constitute the various components used to get detailed information.
- Data contains information obtained directly from the crop, soil, or ambient characteristics monitored. The information may be recovered from the sensors in several methods from plugging a pen drive into a USB port to recovering data through synchronized software programs onto the Internet [15]. The link between both the data and the decision phase comprises filtering procedures and AI algorithms to only provide the correct information and to enable the farmer to decide properly.
- Finally, the actuation refers to the physical performance of a decision-making activity that is generally done by modern technology, capable of receiving commands from a computerized control unit. Since each event comes over the crop [16] , the cycle begins and ends with crop level; specialized sensors register the reaction of the crop, and the loop methodically continues until harvest season marks the end of the crop life cycle.

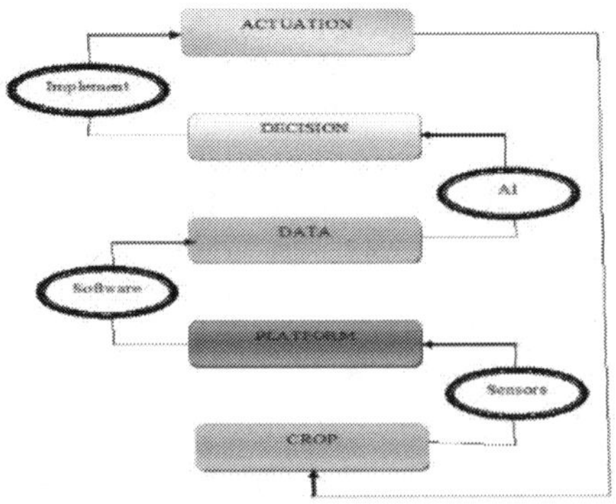

Fig 1:Information based management cycle for advanced agriculture

4. ROLE OF CLOUD COMPUTING IN AGRICULTURE

In the Indian agriculture industry, the notion of cloud computing aims to create a two-core cloud computing paradigm [10]. The first portion consists of monitoring and meeting customer requirements in a user-friendly and quicker way and the second part of saving all essential data in a centralized cloud area as illustrated in Figure 2.

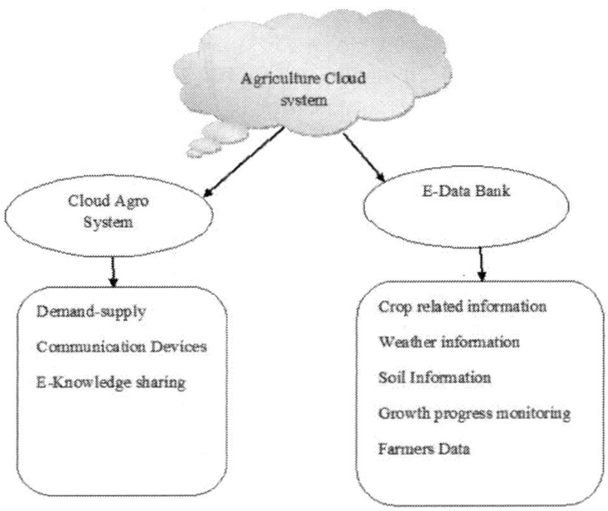

Fig 2: Agriculture cloud system

4.1 Cloud Agro System:

This feature of the system can be used to monitor the overall functioning of the system and deliver the services necessary [11]. The system will provide all users, from wherever in the country and at any time, with online services. To execute these activities, the agricultural system may contain the following services:

Demand supply: It can give up-to-date information about the current demand and supply of agricultural goods in various parts of the nation. It allows farmers to decide how crops are selected. It is also possible to analyze the demand and supply chain comparatively.

Communication Devices: Although most farmers in the surrounding area never heard of ICT, mobile services are used to it [10,11]. The system, therefore, includes mobile services and allows farmers to obtain data from an e-data bank from anywhere, anytime, through mobile telephones.

E-Knowledge sharing: The system also ensures that specialists are communicated online and attend online training programs that are used on a locally informed basis by community centers (CSC). Cloud Agriculture is a worldwide ICT strategy. It's not only about local information. The technology will therefore gather and distribute global information linked to agriculture to local farmers.

4.2 E-Data Bank:

It is a central database that may be used to store all information about agriculture in a central cloud which is available always and everywhere for all customers [12]. The fundamental premise behind an e-data bank is to distribute valuable information in decision-making to local farmers. The e-data bank has the following databases.

Crop-related information: It gathers information on all crops produced in different places in recent times [16]. This helps local farmers in making crop-related decisions in various sections of the country.

Weather information: The environmental data for the area and also the weather prediction is stored for a certain period. It will help farmers to make decisions on crop choices.

Soil Information: Soil knowledge also plays a significant role in the decision-making process for crops [17]. It can also contribute to earth trends in the prior and the prognosis of future soil trends.

Growth progress monitoring: It regularly monitors and collects agricultural growth data in various places. This would be particularly helpful in sensibly comparing the growing region and it will also provide a sharper view with previous data.
Farmers Data: This collects data connected to the region's wise farmers, to track and investigate local farms' participation in Indian farming. It will best serve to devise Indian farming strategies [10,11]. This will also assist understand the major agricultural areas of India to allow policy-makers to decide whether to encourage and develop agriculture. This can assist overcome difficulties like rural-urban migration and unemployment.

5. CHALLENGES IN AI ADOPTION IN AGRICULTURE

While AI offers huge prospects in agriculture, there is currently a lack of familiarity with state-of-the-art artificial intelligence at farms throughout the globe. Farming is very vulnerable to external influences such as weather, soil conditions, and insect attacks. A crop growing plan planned at the beginning of the season may not appear excellent at harvest, as it is modified by external circumstances [13]. To take exact guesses or projections, AI systems also require a lot of data for training equipment. Sometimes in the event of a really big agricultural region, geographic data might be easily acquired but it is challenging to gather time data. Only once in a year when the crops are produced may the diverse crop particular information be acquired. As the dataset requires time to grow, a solid AI machine learning model requires a lot of time. This is a key cause of the use of AI in agricultural goods such as crops [14], fertilizer, and chemicals, as demonstrated in Figure 3.

5.1 Intelligent spraying of chemicals

Farms create hundreds of data points per day on temperature, soil, water use, weather, etc. Using artificial intelligence and machine learning models [3], this data is used in real-time to get valuable insights such as selecting correct seed timing, deciding plant choice, hybrid seed options, and such like.

5.2 Improve the overall harvest quality and accuracy

The AI systems contribute to the overall quality of harvest and efficiency of so-called precision farming. Artificial intelligence helps to identify diseases in crops, pesticides, and inadequate farm nutrition. AI sensors can detect, target, and then determine which pesticide to use in the location [1,4]. This helps to minimize the use and cost reductions of pesticides. Many technology firms have created robots that accurately monitor and spray weeds using computer vision and artificial intelligence. These robots can reduce by 90percent the pesticide use of Eighty percent of the amount of the chemicals routinely sprayed on crops[6]. These smart AI sprayers can dramatically decrease the number of pesticides in the field, improving the quality and cost-effectiveness of agricultural products.

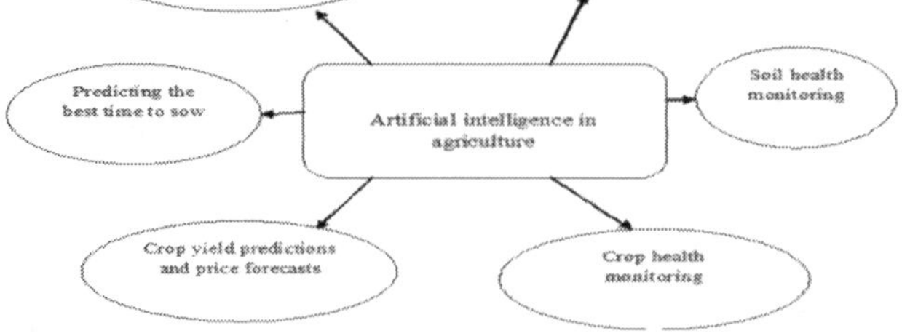

Fig 3: Use of Artificial intelligence in agriculture

5.3 Predicting the best time to sow

The variation between a highly profitable year and an unsuccessful harvest is only information on the timing of planting seed [17]. To overcome this problem, ICRISAT scientists devised a prospective analytical technique for seeding seeds on a particular date for optimal yield. In contrast to a 7-day weather forecast, it offers insight into soil conditions and fertilizer suggestions.

5.4 Crop yield predictions and price forecasts

The greatest concern for several farmers is the variation in crop price. Farmers can never predict a certain production pattern due to unpredictable pricing. This issue is extremely widespread in crops like tomatoes that have a very limited lifespan. Companies use satellite imaging and meteorological data for real-time area evaluation and monitoring of crop health [18,19]. Using technology such as AI and machine learning, enterprises can identify infections of pests and diseases, predict their tomato yield, and price forecasts. They would be able to assist farmers and administrations on future pricing trends, amount of demand, crop type, use of pesticides, etc. for best advantage[7].

5.5 Crop health monitoring

Evaluating a crop's health and detecting crop pests at an early stage is crucial in guaranteeing optimum crop yields. Climate stresses, nutritional deficits, weeds, insects, and infections of the fungus, must be discovered as early as possible so that farmers can counteract them [3,4]. Ai may be used to forecast seed advice, pest control, input controls that can assist the farming community in increasing its revenue and security. Using high-resolution weather, remote sensing data, AI technologies, and AI platforms, crop surveillance may take place in an integrated manner and give farmers extra insights as and when necessary.

5.6 Soil health monitoring

It helps them decide when, where, and what to seed, how to apply nutrients, and so on. Image recognition and profound supervised learning allowed widespread soil health monitoring without apparatus for laboratory testing [3]. AI technologies linked with remote satellite data signals and local picture capture on the field assist farmers to improve soil health immediately.

6. PROPOSED SYSTEM

This section discusses the proposed network sensor model [15]. The architectural diagram of the sensor-based data management is shown in Figure 4. Acquiring sensors needs three primary steps: the capture of information, communication, and processing of the data.Different sensors are employed to gather the different soil characteristics parameter values that are suited for agricultural development.

6.1 Sensor Unit

The data acquisition is made using various sensor devices such as

- PH Sensors
- Optical Sensors
- Temperature & Humidity Sensor
- Electro-Chemical Sensors
- Mechanical Sensors
- Dielectric Soil Moisture Sensors

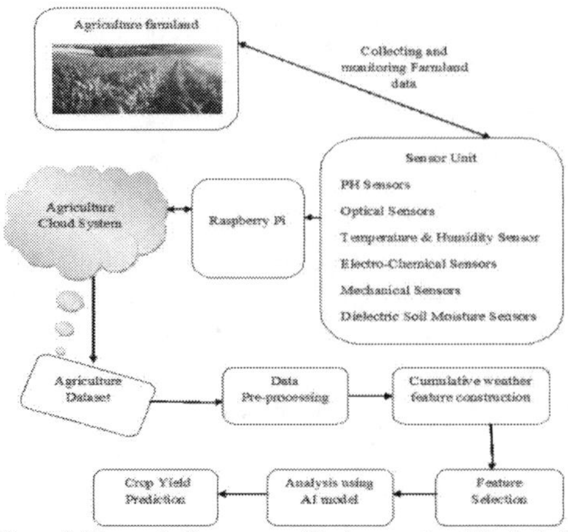

Fig 4: Architecture of Cloud-Based Sensors Driven Artificial Intelligence for Crop yield prediction

6.2 PH Sensors

PH sensors give effective information on the presence of undesirable compounds or soil nutrient deficits [16]. These sensors assist intelligent agriculture to track the soil pH and nutrient levels of agricultural industries daily, weekly, monthly, and annually.

Optical Sensors

Optical sensors in farming are used to comprehend the soil and cultivation features in real-time by analyzing the quantity of reflected light on cultivated sections. Optical sensors provide instruments for the assessment to improve nitrogen intake in weak and diseased plants and to manage the nitrogen intake in healthy plants [13]. Optical sensors also are utilized for the analysis of plant vigor by integrating soil biomass and nitrogen as parameters with other elements in the soil. This allows farmers to control air and soil moisture and prevent humidity.

Temperature & Humidity Sensor

In the agricultural growing area, the air temperature and humidity sensor can measure the air temperature and humidity fluctuations. The standard ranges of temperature and humidity controls are -40°C ~+80°C correspondingly and 0% of RH ~100% of RH.

Electro-Chemical Sensors

For sustainable and environmentally friendly agriculture, assessment of the pH level of the soil is crucial while increasing revenue [14]. Electrochemical systems are used to detect, analyze and implement pH-alteration procedures to maintain the present level of a crop for the future phases of the crop's life cycle. In outside farms and greenhouse farms, electrochemical sensors are employed. To track phosphorous levels, potassium, calcium, sodium, nitrogen, copper, and iron levels, the major application is electricity sensors.

Mechanical Sensors

Mechanical sensors are used to measure the resistance provided by the soil by assigning resistive forces [15,16]. This is commonly carried out with load cells or strain gauges. Mechanical sensors are utilized to assess the force exerted by roots for absorbing water and these data are employed to improve the tilling process and water input. In the initial phases of the life cycle of the crop, these data are used to deduce the appropriate amount of tiller pressure to achieve maximum output for a lengthy period.

Dielectric Soil Moisture Sensors

Dielectric soil humidity sensors are used to measure soil humidity to enhance watering management systems [17]. This enables the farmer to understand the composition of the soil, its hydration capability, the medium, and the mean rate of evaporation and also gives a complete overview of the crop's root zone. Some frequently utilized dielectric soil moisture sensors for smart agriculture are water potential sensors and calibrating sensors.

Air Flow Sensors

Airflow sensors are used to track the amount and perspective of the land to be grown before the seeding operation of the gaseous components existing in the soil in a given landscape, after irrigating it. The optimal pressure necessary to pump air to air the soil is determined and made more fruitful. It also helps to determine the soil's qualities, its compression, its moisture-keeping, and much more.

6.3 Raspberry PI

The system consists of Raspberry PI to collect data from several sensors. The GPIO pins of the Raspberry PI link the various sensors. The data is transmitted to the database cloud when data is gathered from a sensor via Raspberry PI through wireless networks [18]. Data collected from the cloud have been previously processed, analyzed, and interpreted by AI to anticipate the output of the specific crop.

6.4 Agriculture cloud system

The cloud facility used here is an Agriculture cloud system and the collected data is used for analytical purposes utilized for artificial intelligence.

Data collection on agriculture (crop, weather, soil, growth progress, farmer data, and expert consultation). Store all details about agriculture in a centralized cloud that is available wherever and anytime to all stakeholders [13,14].

Management through centralized decision-making support systems of all land, location, area, soil, and land properties data.

High interoperability and information exchange in agriculture.

The farmers' technical knowledge and resource restrictions can be eliminated.

Providing agricultural technology service and science

Improvement of the marketing of the agricultural products

Efficient use of agricultural resources

Promote the circulation of agricultural products and services at a wider level.

Data are accessible via a cloud device on the local PC. The technique is created on the device and the acquired data is checked to ensure that the results obtained are accurate.

6.5 Dataset

A series of data containing environmental information, administration and yields recorded was generated, consisting of 10,016 samples for 293 counties on the annual average grain production. Environment, genotype, and management are the primary elements that influence crop yields [18]. To this effect, weather and soil were included in the training dataset as environmental characteristics and as plant population, as well as planting progress. It is important to note that the processing of data was developed to handle the growing yield patterns as a result of technical and genotypic developments through time[19][20].

Plant population: One aspect which describes the annual population of plants and by state measured per square meter in plants.

6.6 Data Pre-processing

Several pre-processing works have been performed to guarantee that the data for artificial intelligence models are provided [19]. As some machine models are beneficial for balanced prediction methods with comparable ranges for input data, the initial pre-processing step was to scale input data from 0 to 1 using a min-max scale. Min-max scaling and normalization, which is the most popular technique of scaling, is chosen by Min-max scaling, as it retains input variables proportions.

6.7 Cumulative weather feature construction

Additional climatic information is provided for the Artificial Intelligence models, more weather characteristics have been developed [16], which include cumulative weather characteristics. The accumulation of weather characteristics calculates the aggregated precipitation, the increasing grade days, and the shortwave radiation features, whereas the aggregated minimum and maximum temperature characteristics are derived from the average of existing data.

6.8 Feature selection

Since the data set includes a large number of independent variables and is susceptible to generalization, it is vital to choose a feature to create generalizable AI models [14,15,16]. The selection of the most critical characteristics in the data set was accomplished in two stages and AI models were prevented from generalizing the high-dimensional training data. The choice of features was based on expert knowledge and the selection of permutation features using the support vector machine to execute the two phases.

- Feature selection based on expert knowledge: Expert knowledge has decreased the weather characteristics by reducing the characteristics during the interval between harvesting and planting. Furthermore, the number of progressive seeding characteristics was reduced since they did not offer helpful information by reducing the accumulated progress of the seeding during weeks before planting.
- Permutation feature selection with SVM: Significance for permutation assesses the priority of the input function by evaluating the decrease in the prediction error of the system when one feature is not provided.

6.9 AI model

In this section, some artificial intelligence algorithms, namely, Random forest, NaiveBayes, and support vector machine have been used to predict the yield [19,20,21].

Random forest

Bagging is another tree-based ensemble model based on a random forest concept. Through averaging predictions by sampling and substitution, Bagging seeks to minimize prediction variation [6]. Random forests provide a new element to the packaging, which randomly chooses and builds a tree with an alteration of characteristics and often repeats this method and ultimately measures all the forecasts provided by all trees. Random forest therefore handles and proves to be strong both bias and variance aspects of the problem.

Naive Bayes

This classifier has dynamically separate characteristics. Although most other classifiers forecast some link between the characteristics, the Naive Bayes model its distinct characteristics as independent characteristics given its class[7,8 9]. This restricts the data, although naïve bays utilize more advanced approaches to increase the efficiency and receive some theoretical support. With very little data training, Naive Bayes classifiers may have many high dimensional features and are also extremely highly scalable classifiers.

In Bayesian analysis, the probability of an event X given an event Y is not the same as the probability of Y given X as in Eq (1).

$$P(X|Y) \neq P(Y|X) \qquad 1$$

Assuming that X1,X2,.....Xn and Z are the features vectors and the class of the crop prediction dataset, respectively , the Bayes equation can be expressed as shown in Eq(2)

$$P(Z|X) = \frac{P(Z) \times P(X|Z)}{P(X)} \qquad 2$$

Where P(X) = prior probability representing the feature vectors of crop dataset.

P (X|Z) = prior probability representing the class of crop dataset.

Support Vector Machine (SVM)

An SVM support vector machine is a supervised learning model with several learning algorithms that examine the data required to solve both classification difficulties and regression difficulties. Some training examples are provided in which each sample is annotated such that it corresponds to one of the first two groups [20,21]. The Vector Model technique supports a model in which fresh samples are allocated for any of the categories specified. The SVM is a multitude of instances chosen as space points so that samples belonging to distinct groups are divided into two groups.
The input characteristics are transferred to a higher-dimensional space and can also be shown as Eq (3)

$$f(x,w) = \sum_{j=1}^{n} w_j g_j(x) + v \quad 3$$

Where $g_j(x)$ the set of nonlinear transformation and v is the bias

6.10 PERFORMANCE MEASUREMENT

To estimate the persistence of established models to prevent the yield, the statistic analysis, especially a mean square error (MSE) is utilized [19]. However, precise, specific, sensitive, accurate, and F-score assessment matrices were used to evaluate the classification model that has been created.

Table 1 specified the employed model measures:

Parameters	Formula's
Mean square error (MSE)	$MSE = \frac{1}{G}\sum_{j=1}^{G}(x_i - \hat{x}_i)^2$ Where x_i is predicted responses and $\hat{x}_i$ is observed responses respectively, G is the total number of variables
Root mean squared error (RMSE)	$RMSE = \sqrt{\frac{I}{k}\sum_{i=1}^{k}[[h(b_i) - y_i]]^2}$
Accuracy	$ACC = \frac{TP + TN}{TP + TN + FP + FN} \times 100\%$
Sensitivity	$Sensitivity = \frac{TP}{TP + TN} \times 100\%$
Specificity	$Specificity = \frac{TN}{TN + FP} \times 100\%$
Precision	$Precision = \frac{TP}{TP + FP} \times 100\%$
F-Score	$F = \frac{2 * Precision * Sensitivity}{Precision + Sensitivity} \times 100\%$
Where TP=True positive, TN=True negative, FP=False positive,FN=False negative respectively	

7. RESULT AND DISCUSSION

This paper uses the approach of crop prediction to anticipate the appropriate crop by sensing numerous ground parameters and atmospheric parameters. We applied artificial intelligence algorithms for this purpose. Figure 5 shows the MSE performance on different crops. Figure 6 shows the RMSE performance on different crops.

Journal of Physics: Conference Series **2040** (2021) 012010 doi:10.1088/1742-6596/2040/1/012010

Table 2: PH, Nitrogen, and temperature values of different crops

Crop	PH	Nitrogen (ppm)	Temp (°C)
Potato	5.6-6	40-50	24
Rice	5.5-8.5	50	22-25
Sugar cane	6.5-7.5	40	20-50
Wheat	6-8.5	80-150	16-22

Table 3: calculated Performance of an algorithm for different crop

Crop	**MSE**	**RMSE**
Potato crop	0.25678	0.56743
Rice crop	0.00493	0.06987
Sugarcane	0.03456	0.67430
wheat	0.39821	0.04583

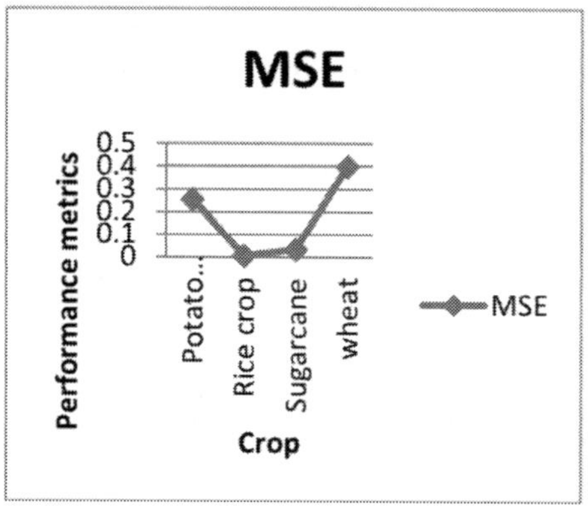

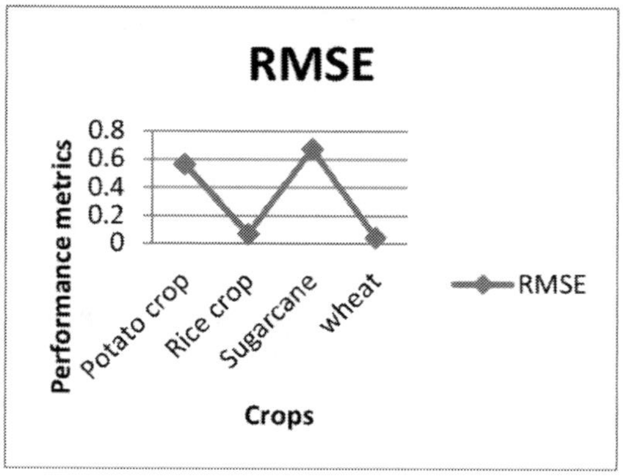

Fig 5: MSE performance on different crops **Fig 6:** RMSE performance on different crops

The dataset was separated into 70% training and 30% test sub-sets in order to validate the proposed model. For predicting the water quality of soils, RF, SVM and Naive Bayes were used. The results of the various AI classification models are given in Table 4. The performance metrics graph of multiple AI algorithms is shown in Figure 7.

Table 4: Performance of the used AI models to predict water quality prediction in soil

Models	Accuracy (%)	Sensitivity (%)	Specificity (%)	Precision (%)	F-score (%)
SVM	98.05	99.78	98.08	95.34	99.01
RF	84.98	82.67	94.56	88.89	86.76
NB	76.09	76.89	92.43	79.09	82.45

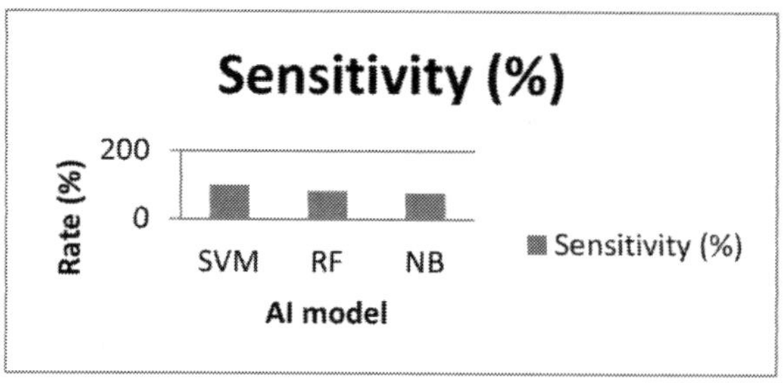

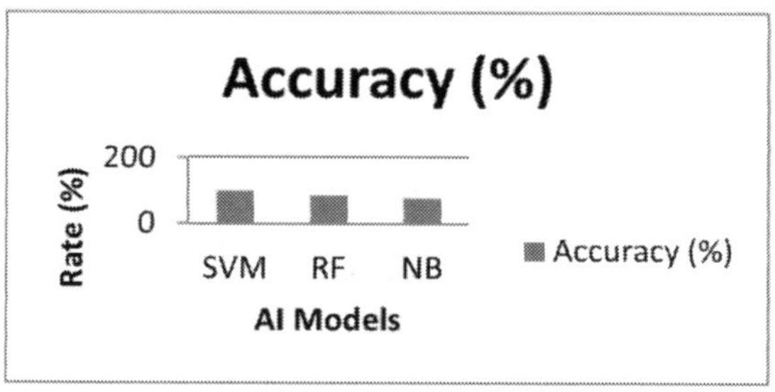

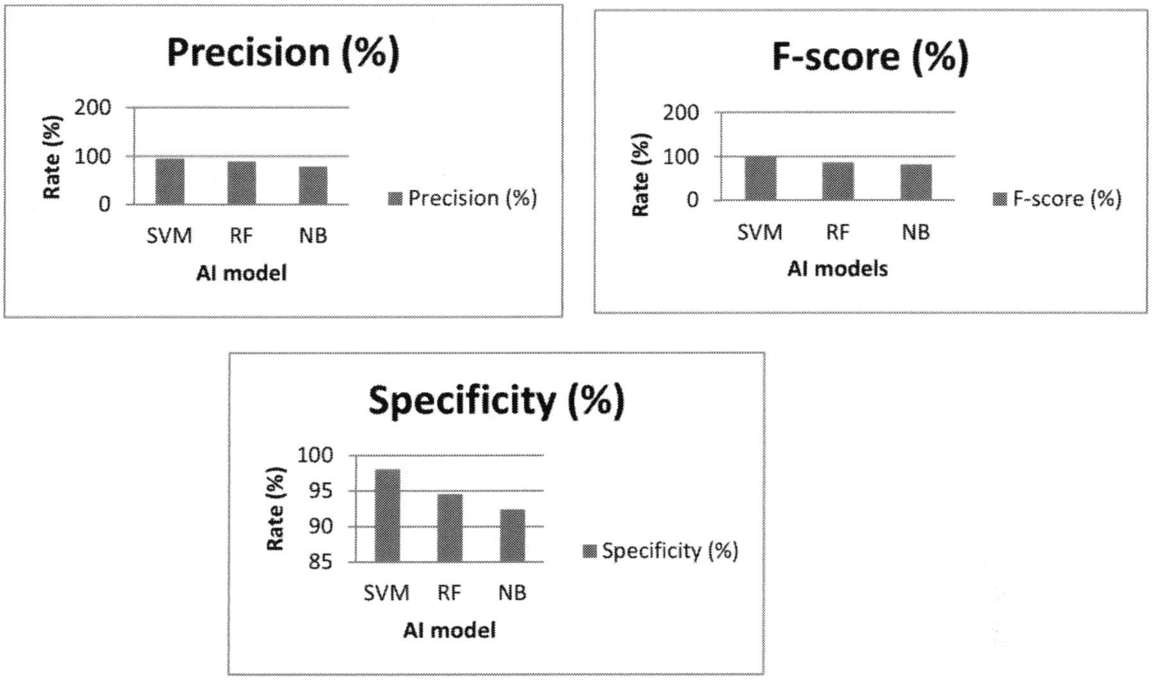

Fig 7: Performance metrics Vs AI algorithms

8. Conclusion

A cloud and AI-based approach were created to dynamically process, analyze, and display acquired data for customized monitoring and evaluation. Crop yield forecast remains to be an important issue to tackle by the farmers. The proposed system takes into account soil, temperature, and pH measurements and calculates which crops can best be cultivated in the appropriate soil conditions. Because all alternative cultures are listed in the database, helps the farmer to decide which crops should be cultivated in their region. Approaches for artificial intelligence give a wonderful technique to examine large-scale data acquired and to assemble the information obtained to produce a constructive and vital process. These methods are a fantastic technique for developing distinct deep models for looking at the relationship between distinct components and their effects. They may therefore be utilized for many predictions and forecasts according to the specific situations. This approach enables the farmer to decide on the most successful crop and helps to identify new crops that the farmer has not grown till now. For crops such as potato and rice, the stated crop production pattern predicts the maximum level. However, in comparison to RF and Naive Bayes methods, the SVM method has obtained the best accuracy of crop prediction.

References

[1] Vinayaka Manjunatha Purushan" Effective Data Analysis And Crop Yield Prediction Using Machine Learning Algorithm" Mukt Shabd Journal,Volume IX, Issue VII, JULY/2020, PP693-98.

[2] Theyazn H. H Aldhyani" Water Quality Prediction Using Artificial Intelligence Algorithms" Applied Bionics and Biomechanics" vol 2020, Published 30 December 2020, pp 1-12

[3] Rashmi Priya "ML based sustainable precision agriculture: A future generation perspective" Sustainable Computing: Informatics and Systems, Elsevier journal,2020

[4] Ragavi B, Pavithra L, Sandhiyadevi P, Mohanapriya GK, Harikirubha S, "Smart Agriculture with AI Sensor by Using Agrobot," in IEEE, 2020.

[5] Rajesh T, Kumar KT, Balan MM, Sidhartha MS, "E-Agriculture based on AI and IOT," IJRASET, vol. 7, no. II, pp. 343-347, 2019.

[6] P. Kshirsagar and S. Akojwar, "Classification & Detection of Neurological Disorders using ICA & AR as Feature Extractor", Int. J. Ser. Eng. Sci. IJSES, vol. 1, no. 1, Jan. 2015.

[7] Pravin Kshirsagar, Dr.SudhirAkojwar, "Classification and Prediction of Epilepsy using FFBPNN with PSO", IEEE International Conference on Communication Networks, 2015.

[8] P. Kshirsagar, S. Akojwar, Nidhi D. Bajaj , "A hybridised neural network and optimisation algorithms for prediction and classification of neurological disorders" International Journal of Biomedical Engineering and Technology ,vol. 28,Issue 4,2018.

[9] David S, Anand RS, Sagayam M, "Enhancing AI based evaluation for smart cultivation and crop testing using agro-datasets," JAIS, vol. 2, no. 1, pp. 149-167, 2020.

[10] Sukhpa Singh "Agri-Info: Cloud Based Autonomic System for Delivering Agriculture as

[11] Service" Elsevier journal,2020,PP.2991-99

[12] Deepak Indoria" Interrelation between Cloud Computing Technology and Agriculture Fields"

[13] International Journal of Current Microbiology and Applied Sciences,Vol 8 ,2019

[14] Rakesh Patel "Application of Cloud Computing in Agricultural Development of Rural India " International Journal of Computer Science and Information Technologies, Vol. 4 (6) , 2013, PP.922-926

[15] Diksha Manaware" Artificial Intelligence: A New Way to Improve Indian Agriculture" International Journal of Current Microbiology and Applied Sciences ISSN: 2319-7706 Volume 9 Number 3 (2020),PP.1095-1102

[16] Paras M. Khandelwal" Artificial Intelligence in Agriculture: An Emerging Era of Research"Research gate.

[17] Miss.Snehal" Agricultural Crop Yield Prediction Using Artificial Neural Network Approach" International journal of innovative research in electrical, electronics, instrumentation and control engineering vol. 2, issue 1, january 2014,PP.683-86

[18] Jaimaha prabhu A" Cloud Analytics based Farming with Predictive Analytics using Artificial Intelligence" Fifth International Conference on Science Technology Engineering and Mathematics (ICONSTEM),2019,PP.65-68

[19] Yiannis Ampatzidis" Agroview: Cloud-based application to process ,analyze and visualize UAV collected data for precision agriculture applications utilizing artificial intelligence" Elsevier journal in Computers and Electronics in Agriculture,2020

[20] Kevin Tom Thomas" Crop Prediction Using Machine Learning" International Journal of Future Generation Communication and Networking Vol. 13, No. 3, (2020), pp. 1896–1901

[21] P. Kshirsagar and S. Akojwar, "Novel approach for classification and prediction of non linear chaotic databases," 2016 International Conference on Electrical, Electronics, and Optimization Techniques (ICEEOT), 2016, pp. 514-518, doi: 10.1109/ICEEOT.2016.7755667,2016

[22] Pravin Kshirsagar, Nagaraj Balakrishnan & Arpit Deepak Yadav "Modelling of optimised neural network for classification and prediction of benchmark datasets" , Computer Methods in Biomechanics and Biomedical Engineering: Imaging & Visualization, 8:4, 426-435, DOI: 10.1080/21681163.2019.1711457,2020

[23] S. Akojwar and P. Kshirsagar, "A Novel Probabilistic-PSO Based Learning Algorithm for Optimization of Neural Networks for Benchmark Problems", Wseas Transactions on Electronics, Vol. 7, pp. 79-84, 2016.

ANT COLONY OPTIMIZATION ALGORITHM GPS CLUSTERING APPROACH

M.Grace Mercy [1], A.Kamala Kumari[2], A.Bhujangarao[3], V.Nooka Raju [4]

[1]Assistant Professor, Department of EECE, GITAM, Visakhapatnam, India
[2] Assistant Professor, Department of Instrument Technology, Andhra University, Visakhapatnam, India, Email: akkumari.inst@auvsp.edu.in
[3] Professor, Professor, Department of Instrument Technology, Andhra University, Visakhapatnam, India, Email: abhujangarao.inst@auvsp.edu.in
[4] Assistant Professor, Department of EECE, GITAM, Visakhapatnam, India, Email: nvaddadi@gitam.edu

Corresponding Author Email: gmatta@gitam.edu

Abstract: The geometry of the GPS satellite recipient (s), which reflects the recipient (s) of the satellites, has a major influence on the total positioning precision. The more precise the position, the stronger the geometry of the satellite. This article provides the geometry of satellite clustering for the selection of suitable satellite navigation subsets. This technique is based on the GDOP (Geometric Precision Dilution) satellite factor cluster with the Ant Colony Optimization (ACO) algorithm that has been created by simulating real and artificial ways to locate the quickest route between nesting resources and food. Pheromones are utilised in the suggested technique to assess the iterative outcome of single colonies. The ACO method can measure all subsets of satellites while reducing computer load by eliminating the need for a matrix inversion. Based on the simulation results, the GPS GDOP clustering technique is more efficient at achieving its optimum value.

Key words: ACO, GDOP, GPS, Clustering.

1. Introduction

The Global Positioning System (GPS) is a global navigation system created by the United States Department of Defence. It consists of 24 or more Earth-rounds in at least six aeroplanes, five control stations, and user receiving equipment. The GPS receiving signals are sent by these satellites. The distance (range) of each satellite may be calculated by timing when the signal arrives. The user's position may be calculated in three dimensions using the non-linear equation system below, with a range of at least four satellites [1-3].

$$p_i = \sqrt{(X - X_i)^2 + (Y - Y_i)^2 + (Z - Z_i)^2} + +b - c\delta t_i \qquad (1)$$

The ACO algorithm mimics a real group's collaborative route. Each ant individually searches the solution space for the solution and gives some information about the solution found. Ameisen leaves more information about the best-performing option, which has a higher probability of getting selected. All solutions have the same information in the early stages of the algorithm. The improved solution acquires more and more information with the development of computers. This ensures that the algorithm obtains the answer for optimization or almost optimization [4]. Compared to others, such as the Genetic Algorithm (GA), the ACO is able to look for optimal solutions both locally and globally by altering parameters. The list contains the best places. ACO may escape the optimal locally and reach the ideal globally. Moreover, the calculation procedure for ACO is faster since it does not use complex operators like aberration and intercross [5].

Conventional clustering techniques under supervision assume that all of the topic classes present in this dataset are known ahead of time. Unattended clustering, on the other hand, does not need any prior data knowledge. The main objective of clustering unsupervised learning is to split a data set into groupings (clusters) such that the data points in a cluster are more similar than those in other clusters [6-12]. To create an unsupervised clustering method, we used the Ant Colony Optimization (ACC) algorithm for GPS GDOP.

2. Ant Colony Optimization with Algorithm

The quickest route to food may be found by leaving a pheromone trail as travel. More ants follow the path to the summit. Ants that take the shorter route leave a stronger pheromone trail than those who take the longer route. Because the stronger the pheromone, the better it attracts the ants, the shortest route is increasingly chosen until all the ants choose the shortest route. Take three potential routes to the food supply, one shorter than the other. Ants have the same probability for each route. Ants that travel the shortest distance and return cause the most pheromones to be released as quickly as possible. New ants will then pick the track and a local optimiser will be linked to another algorithm. One issue is the early convergence towards an improper solution because too many virtual pheromones have been deposited quickly. Pheromone evaporation is carried out to prevent this standstill. In other words, over a period of time, the pheromone with a solution will disappear [13-15].

The objective of the GPS GDOP clustering is to achieve an optimum distribution of N satellites in one of the K clusters, where N is the satellite number and K is the cluster number. Artificial ants are referred to as software ants or agents in the algorithm and the number of agents is indicated by R. In the first iteration, bits begin with empty solution strings, and pheromone matrix components begin with identical values. The pheromone matrix is updated with a progression of iterations according to the quality of the products produced [16-20]. The solution strings stated in S of length N are specified by each algorithm agent and the solution string of each element is empty at the start of its algorithm. Each string element is equivalent to one of the test samples, and its value contains the cluster it is allocated. The agent utilises pheromones to assign each element of string S to a suitable class label for the purpose of building a solution. The pheromone matrix β is initialised at the beginning of the procedure at a modest value of $\mu 0$. Each solution string S element of each agent is thus allocated to one of the K clusters during the first iteration.

The optimal solution is the one that minimises the objective function value. If the best memory solution value is updated with the best solution value in the current iteration, if it is less objective than the best memory solution, the best memory solution is kept; otherwise, the best memory solution is kept. This method explains the completion of one algorithm iteration. The method iterates through these steps indefinitely until the optimal partitioning of subjects in a given dataset into different groups is reached by the required number of iterations and the solution with the lowest function value is obtained.

The ACO problem clustering method is summarised below [8]:

(1) New R solutions were produced employing software ants with information from earlier iterations of the pheromone trail modified.

(2) A local search for newly created solutions is conducted.

(3) Pheromone trail matrix update

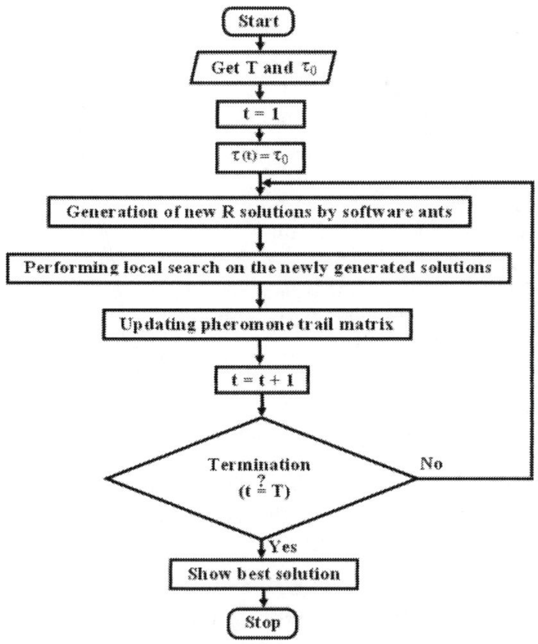

Figure1. The flow chart of ACO algorithm developed for solving data-clustering problem

3. Comparison with other algorithms

Advantages of ACO are inherent parallelism, positive feedback required for proper solutions, efficiency for problems similar to travelling sale man. It can be used in dynamic application by adapting to changes such as new distances. Further convergence is guaranteed though the time may me uncertain. ACO is more applicable to problems where source and destination are relatively specific and predefined whereas PSO is a clustering algorithm for multi objective optimization and constriction handling .It is more applicable for getting crisps results while PSO is applicable to problems which are fuzzy in nature . ACO as advantages over GA and simulated annealing as the ant colony algorithm and run continuously and adaptive to changes in real time. This feature makes it useful in network routing transport systems and GPS related problems.

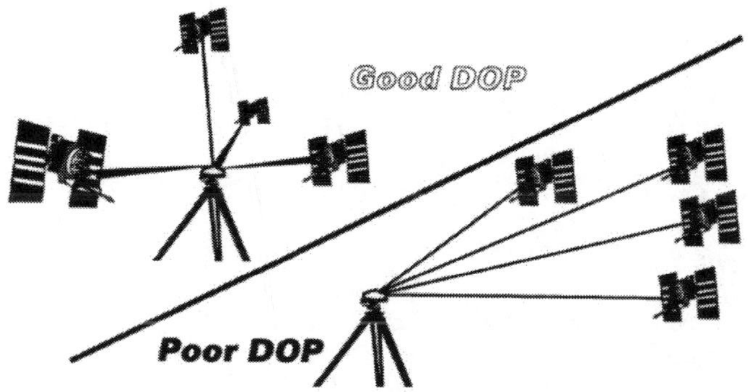

Figure 2. The visible GPS constellation's geometry

4. ACO algorithm for GPS GDOP clustering

Superb satellite geometry may be achieved at high elevations. Better geometry is provided with wider satellites. The DOP may be determined using precision dilution (DOP). As the DOP is lower, the geometric force is better. The DOP number is determined by relative satellite geometry (i.e., it requires the availability of both the receiver and the satellite coordinates). With early data, we can predict the DOP's value. The DOP will move due to the receiver and satellite movements (S). In today's world, many DOPs are being deployed. For example, consumers may wish to examine the geometry and the resultant quality of the three-dimensional location (latitude, longitude, and height). The value of the DOP position may be examined (PDOP). PDOP increases the three-dimensional positioning precision of the satellite geometry. Vertical and Horizontal DOP: (VDOP). Normally, GDP is larger than GDP. Low precision response to GPS elevation The VDOP may be enhanced by other sensors with GPS sensors. PDOP and TDOP are most often referenced in DOP (GDOP) variations.

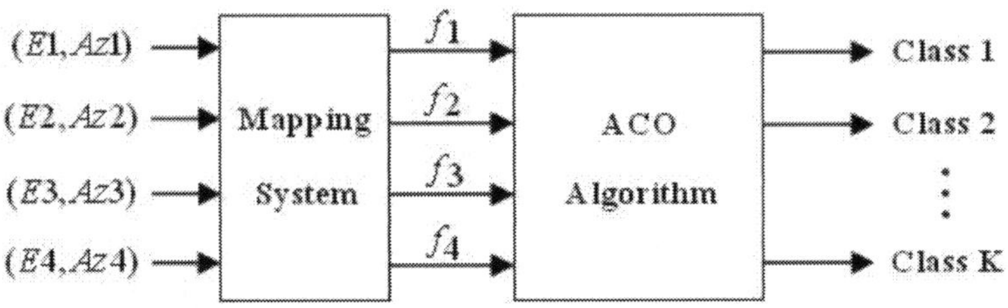

Figure 3. The GPS GDOP diagram using an ACO algorithm.

The GDOP determines the resultant solution's measurement noise magnification factor. The GDOP idea is a strong GPS utility. The receivers use an algorithm called GDOP to monitor the optimal set of satellites from a group of up to 11 satellites. After that, multiply the range precision by the dilution factor to get the precise location. The dilution factor (DOP) is solely dependent on geometry. Geometric differences are often considerably more significant than changes in the nominal satellite's constellation range. If the nominal satellites are not visible, the GDOP method calculates the effect. In order to provide position precision, the multiplier will receive accuracies from the GDOP calculation for the still-monitored satellites .Figure 2 depicts the 'Good DOP' and 'Poor DOP.

$$GDOP = \sqrt{\lambda_1^{-1} + \lambda_2^{-1} + \lambda_3^{-1} + \lambda_4^{-1}} \quad (2)$$

5. Simulations and Results

A DOP rating of less than 2 is considered outstanding, although it does not happen as often, and needs a clear vision of the sky to the horizon. The 2 to 3 DOP values are considered good. DOP values of 4 or less are typically shown if device accuracy capabilities are given. DOP ratings of 4 to 5 are usually considered good and appropriate for all but the highest accuracy requirements of the survey. A 6 DOP would only be suitable for low accuracy circumstances, such as rough positioning and navigation. The position data should not be recorded when the DOP value exceeds 6. The threshold dose selection affects the parameters and performance of the clustering algorithms minimally. One of the allowable subsets of satellites for navigation may be selected from the GDOP classification (e.g., one with a modest GDOP factor). In this study, the ACO classifier generates three types of output according to the GDOP parameters (the small, medium, and large GDOP). Similarly, it is feasible to organise into other categories based on the same idea. In "Microsoft Visual C#.Net," the application software was developed to provide the best solutions for the ACO algorithm for the problem of GPS GDOP clustering. Table 1 shows the number of clusters, agents, local search agents, pheromone iterations and beginning levels, evaporation rates of pheromones, and other variables needed for the technique.

Table 1 The parameters of ACO

Parameters	Value
K	3
R	10
L	2
T	1000
τ_0	0.01
ρ	0.01
q_0	0.98
p_{ls}	0.01

The algorithm parameters must be chosen carefully. The algorithm settings recommended in this article are based on experimental results. When a large number of repetitions are conducted, it is possible to get very good clustering. The results show that the recommended approach's correct percentage of clustering is higher than 98 percent.

6. Conclusions

GDOP must be reduced to improve the performance of a GPS receiver. It's also crucial to construct a GPS receiver that can receive signals from four or more GPS satellites at all times. Natural and biological systems have been the focus of recent advancements in heuristic optimization techniques. One of the most well-known swarm intelligence optimization techniques is the ACO. Ants' real power lies in their colony's brains and pheromone-driven communication. A technique based on GPS GDOP clustering with ACO algorithms was proposed in this article. The ACO technique can evaluate all satellite subsets without requiring a matrix initialization, reducing the computational burden. The simulation results in the programme indicated that the recommended technique is favored and successful, with an accurate clustering percentage of more than 98 percent for the method presented.

References

[1]. Fidanova, Stefka, Enrique Alba, and Guillermo Molina. "Hybrid ACO algorithm for the GPS surveying problem." In International Conference on Large-Scale Scientific Computing, pp. 318-325. Springer, Berlin, Heidelberg, 2009.

[2]. Mosavi, Mohammad-Reza, Amir-Ali Akhyani, and Abdolreza Rahmati. "A PMU placement optimal method in power systems using modified ACO algorithm and GPS timing." PRZEGLAD ELEKTROTECHNICZNY 88, no. 8 (2012): 346-349.

[3]. Mucherino, Antonio, Stefka Fidanova, and Maria Ganzha. "Ant colony optimization with environment changes: an application to GPS surveying." In 2015 Federated Conference on Computer Science and Information Systems (FedCSIS), pp. 495-500. IEEE, 2015.

[4]. Fidanova, Stefka. "Hybrid heuristic algorithm for GPS surveying problem." In International Conference on Numerical Methods and Applications, pp. 239-246. Springer, Berlin, Heidelberg, 2006.

[5]. Mosavi, M. R. "An effective method for GPS GDOP clustering using ant colony optimization algorithm." Asian Journal of Geoinformatics 10, no. 4 (2012).

[6]. Azarbad, Milad, Hamed Azami, Saeid Sanei, and Ataollah Ebrahimzadeh. "New neural network-based approaches for GPS GDOP classification based on neuro-fuzzy inference system, radial basis function, and improved bee algorithm." Applied Soft Computing 25 (2014): 285-292.

[7]. Mosavi, M. R[illegible], and M. Shiroie. "Efficient evolutionary algorithms for GPS satellites classification." Arabian Journal for Science and Engineering 37, no. 7 (2012): 2003-2015.

[8]. Cuevas, Erik, Primitivo Díaz, Omar Avalos, Daniel Zaldivar, and Marco Pérez-Cisneros. "Nonlinear system identification based on ANFIS-Hammerstein model using Gravitational search algorithm." Applied Intelligence 48, no. 1 (2018): 182-203.

[9]. Zhao, Jianna, Xinying Wang, and Zhuozheng Wu. "Forecasting GDP growth based on Ant Colony Clustering Algorithm and RBF neural network." In 2008 IEEE international conference on automation and logistics, pp. 1839-1843. IEEE, 2008.

[10]. Sivaram, M., M. Kaliappan, S. Jeya Shobana, M. Viju Prakash, V. Porkodi, K. Vijayalakshmi, S. Vimal, and A. Suresh. "Secure storage allocation scheme using fuzzy based heuristic algorithm for cloud." Journal of Ambient Intelligence and Humanized Computing (2020): 1-9.

[11]. Chase, Jeff, and Ron Doyle. "Energy management for server clusters." In Proceedings of 8th Workshop on Hot Topic in Operating Systems, pp. 0165-0165. IEEE Computer Society, 2001.

[12]. Azarbad, Milad, Hamed Azami, Saeid Sanei, and Ataollah Ebrahimzadeh. "New neural network-based approaches for GPS GDOP classification based on neuro-fuzzy inference system, radial basis function, and improved bee algorithm." Applied Soft Computing 25 (2014): 285-292.

[13]. Mosavi, M. R[illegible], and M. Shiroie. "Efficient evolutionary algorithms for GPS satellites classification." Arabian Journal for Science and Engineering 37, no. 7 (2012): 2003-2015.

[14]. Xu, Tu, Jian Li, and Xiao Liu. "A Fast Satellite Selection Algorithm for GPS/BD Integrated Navigation System Based on K-means++ Clustering Algorithm." In 2019 IEEE International Conference on Signal, Information and Data Processing (ICSIDP), pp. 1-6. IEEE, 2019.

[15]. Almuzaini, Khalid K., and T. Aaron Gulliver. "Range-based localization in wireless networks using the dbscan clustering algorithm." In 2011 IEEE 73rd Vehicular Technology Conference (VTC Spring), pp. 1-7. IEEE, 2011.

[16]. Jian-jun, S. U. N. "Fuzzy Processing on GPS GDOP [J]." Gnss World of China 2 (2006).

[17]. Yao, Jintao, Hongli Li, and Bin Li. "Deployment Location Algorithm of Navigation Base Station Based on GDOP Analysis." In China Satellite Navigation Conference (CSNC 2021) Proceedings, pp. 478-486. Springer, Singapore, 2021.

[18]. Wang, Ershen, Caimiao Sun, Chuanyun Wang, Pingping Qu, Yufeng Huang, Tao Pang, Song Xiang, and Song Xu. "A new satellite navigation receiver data processing method using improved PSO algorithm and GDOP." In Proceedings of the 2020 International Technical Meeting of The Institute of Navigation, pp. 726-735. 2020.

[19]. Refan, Mohammad Hossein, and Adel Dameshghi. "GDOP Classification and Approximation by Implementation of Time Delay Neural Network Method for Low-Cost GPS Receivers." Iranian Journal of Electrical and Electronic Engineering 16, no. 2 (2020): 192-200.

[20]. Wang, Ershen, Caimiao Sun, Qizhi Fang, Xuan Li, Pingping Qu, Yuxia Bie, and Tao Pang. "A Satellite Selection Algorithm Based on PSO for a Integrated Navigation Receiver." In China Satellite Navigation Conference, pp. 219-228. Springer, Singapore, 2020.

Runoff volume prediction in the megadrigadda reservoir catchment due to past land use/land cover trends- a case study.

P Shiva Kumar[1] G Venkata Rao[2] Eswara Veera Raghava Rao[3] and CH Kannamu Naidu[4]

[1]Assistant Professor, Vignan's Institute of Information Technology, Visakhapatnam, Andhra Pradesh, India. sivaplvls@gmail.com
[2]Professor, Anurag University, Hyderabad, Telangana, India. venkataraocbit@gmail.com
[3]Professor, Nalla Malla Reddy Engineering College, Hyderabad, Telangana, India. evrrao999@gmail.com
[4]Professor, Srinivasa Institute Of Engineering and Technology, Amalapuram, Andhra Pradesh, India. kanarp@rediffmail.com
Corresponding Author. sivaplvls@gmail.com

Abstract: One major source of water on the Earth is Surface water resources such as Rivers, Rain water sewers, Streams etc. With the rapid change in the mode of usage of land for various purposes like residential, agricultural, industrial, water storage basins, the volume of Surface runoff due to Rain & melting of Snow changes. Estimating such quantities of runoff is essential for proper managing of Water resources over the Earth surface. Runoff estimation of each Mini Water has been done by SCS-CN Curve number method. The required data for this method include Land Use/Land Cover trends of past years, Hydrological Soil Groups, Rainfall trends and AMC Type. Based on the analysis a considerable change in the Groundwater Recharge in the study area is found due to the variations in the Built-up & vegetative cover features ultimately resulting in increasing Surface Runoff there by excess inflow may enter into the Megadrigadda reservoir resulting in Flash floods during monsoons. In the coming years there are chances of the Dam being breached thereby affecting the Visakhapatnam Airport area in its Downstream.

Keywords: Land use, Land Cover, Runoff, Groundwater Recharge, GIS, Surface Features.

1. Introduction:

With the increasing human activities & Urbanization tends the surface properties of Earth changes continuously. Finding such changes of Earth surface is termed as "Change Detection". Various new technologies have been developed for identifying such properties like Remote Sensing. The changes on the earth surface are occurs due to disasters, deforestation, change in course of river, urbanization etc., as observed by Carlson and Traci Arthur (2000) **[1]**. Electromagnetic radiations are the key source detectors in remote sensing. Reflected rays either signals from satellites (Active Remote Sensing) or the sunrays by the object (passive Remote Sensing) are captured by the satellites to detect various objects on the surface or in the Oceans else in the atmosphere. As shown in Figure: 1 the Electromagnetic radiations are reflected &

captured. Sunlight reflected by the objects are the major sources of passing sensing, Olokeogun et al.(2014) **[2].**

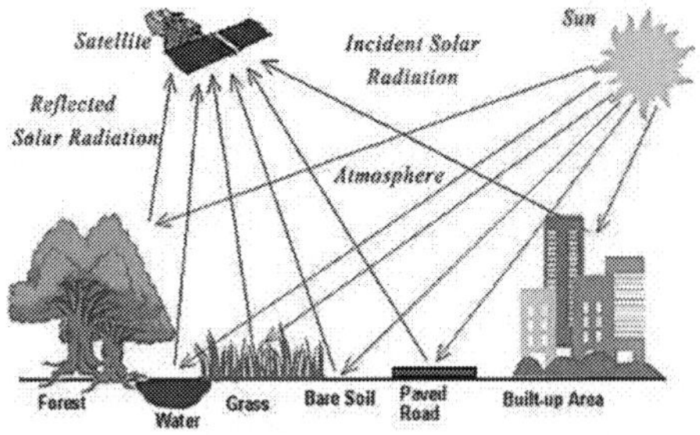

Figure 1- The Elements of Remote Sensing.

Land use and Land cover trends can be detected by using change detection comparisons and also the land cover changes analysed according to the topographic structure by using GIS functions according to Selcuk reis (2008) **[3]**. Hafez A. Afify (2011) **[4]** identified that proper estimation of the nature of LU, LC and their extent is vital in rapidly growing Cities. Multitemporal remotely sensed images were recommended to evaluate human impacts on long term land use changes as per Qiming zhou et al., (2011) **[5]**.

Land Use, Land Cover data is essential for land, water, and vegetation resource management. utilization, conservation and measurement. Land use includes vegetation types, rocks and buildings that exists on the surface. Land use, Land cover features to some extent reflect the geological/ geomorphologic controls sensing et al. (2017) **[6]**. As per Mercy Cheruto et al. (2016) **[7]** the land use & land cover map is developed in Arc GIS with a reference of Google earth map. The land use and land cover differentiates the land used by the people and land covered by vegetation, soil, etc. The land use and land cover map makes easy to help in determination of runoff. The runoff coefficients are to be noted according to the surface conditions provided in Codal provisions Shi et al. (2006) **[8]**.

1.1 Digital Elevation Model (DEM):

DEM is generated in Arc GIS with the help of elevation data as shown in Figure: 2. The DEM shows the rise and fall of ground of the study area which further helps in generation of slope map.

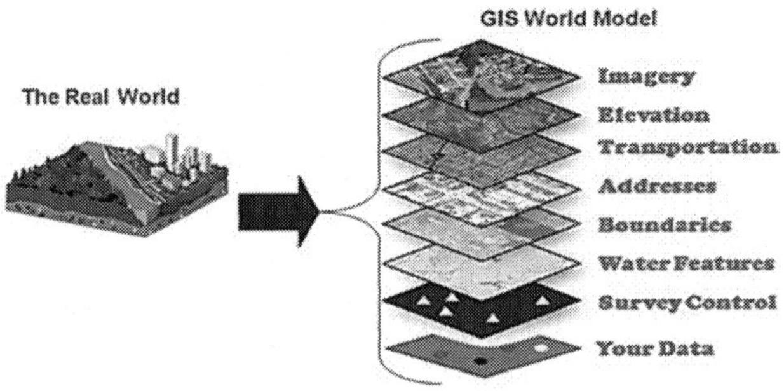

Figure: 2- GIS Model for generating DEM.

Slope maps are used to evaluate the runoff coefficients Carmelo Riccard O Fichera (2012) **[9]**. Slope map is obtained from the end result of the DEM. Based on steepness or gradient these are classified into several types among them Flat, Steep, Gentle, Moderate. Slope map gives the data required for the identification of coefficients.

2. Runoff Estimation:

For estimation of runoff the different methods are

1. Runoff coefficient Method
2. SCS CN Method

3. Unit hydrograph

4. Talbot's equation

5. Empirical Formulae

6. Flood frequency analysis.

2.1 Runoff Coefficient Method:
The Runoff coefficient Method is one of the simplest methods for runoff estimation. It is an approximate method. The direct runoff can be calculated using the following equation (1).

$$\textbf{Runoff} = \textbf{C*P} \quad \textbf{(1)}$$

R = Runoff in mm

C = Runoff Coefficient based on land use type

P = Rainfall in mm

Delineating the water shed and selecting the correct runoff coefficient according the land use type is very important as observed in Dhakal et al. (2012) **[10]**.

3. Methodology:

IMAGE SELECTION

↓

IMAGE REGISTRATION

↓

RADIOMETRIC AND ATMOSPHERIC CORRECTIONS

↓

MULTI- TEMPORAL ANALYSIS

↓

CHANGE DETECTION

Figure: 3- Methodology Flow chart. P. Shiva Kumar et al. (2020). [11]

4. Study Area:
Meghadrigadda is a major reservoir in the Visakhapatnam city of Andhra Pradesh. It is used as a key water resource for the entire city of Visakhapatnam and majorly in Southern Visakhapatnam. The GVMC monitord and maintains the water in this reservoir. Geographic coordinates of the reservoir are north latitude 17°45' to 18° 15' and east longitude of 83° to 83° 15'. The reservoir is formed by a DAM across the minor rivers of Meghadrigadda and Narava. A regulatory is constructed on the eastern side of the bund. Ground elevation is 129 meters. AMSL is at the bund level.

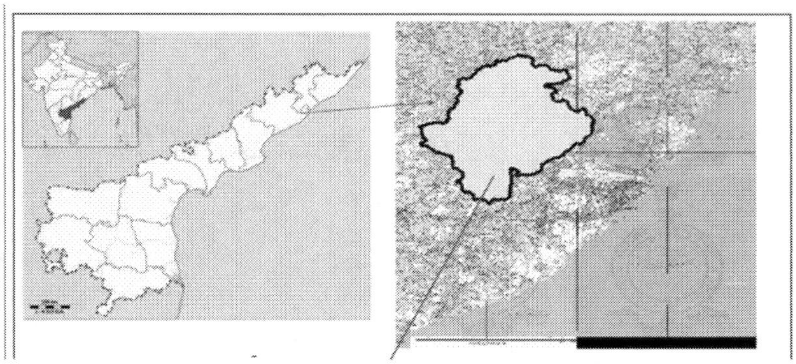

Figure: 4 Study Area under consideration

The following methodology has been adopted as derived from Figure: 3.:

1. The locations of boundary of Meghadrigadda reservoir have been created using the ArcMap 10.2.

2. The toposheets are collected from the website WWW.SOINAKSHI.uk.gov.in.

3. The collected toposheets are E44R1, E44R2&3, E44R4, E44R6.

4. The toposheets has been imported to ArcMAP 10.2 software to delineate the Land Use/ Land Cover features.

5. The DEM was composed from the United States Geological Survey Earth Explorer.

6. The Mini Watersheds have been delineated along with the stream line network using the DEM for Meghadrigedda reservoir.

7. The surface layers are estimated like Forest, Water bodies, Vegetation cover, Build up lands, Barren lands.

8. The daily rainfall data for 22 years has been collected from the nearest rain gauge station.

9. The total rainfall for each year has been calculated for the 22 years from 1997 to 2019.

10. The Runoff estimation is done by using the collected rainfall data.

5. Results:

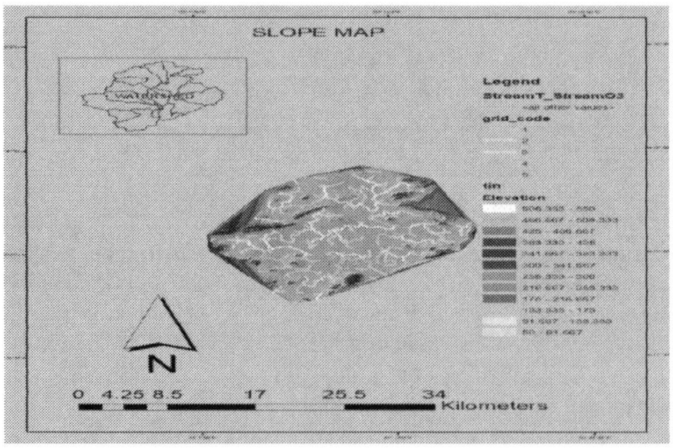

Figure: 5- Slope Map of The Meghadrigedda Reservoir.

 2040 (2021) 012012 doi:10.1088/1742-6596/2040/1/012012

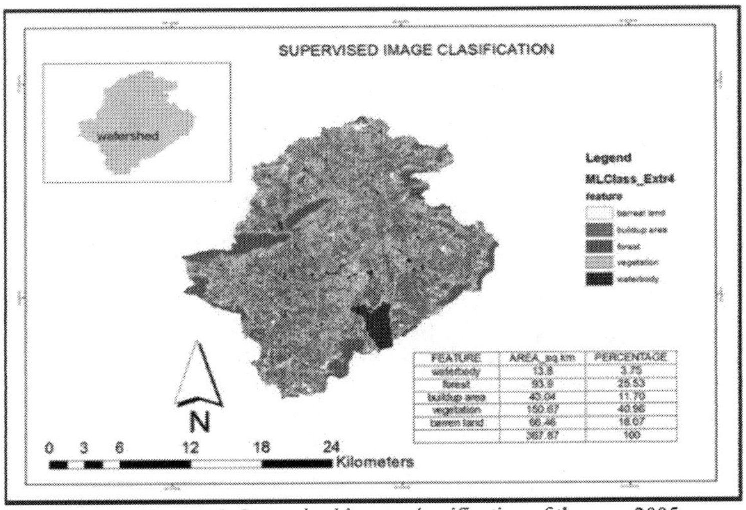

Figure 6- Supervised image classification of the year 2005.

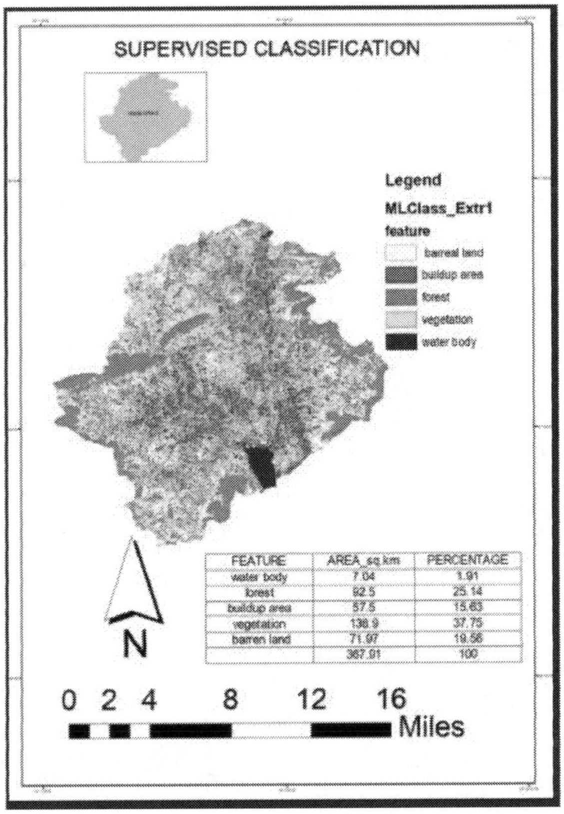

Figure 7 Supervised image classification of the year 2010.

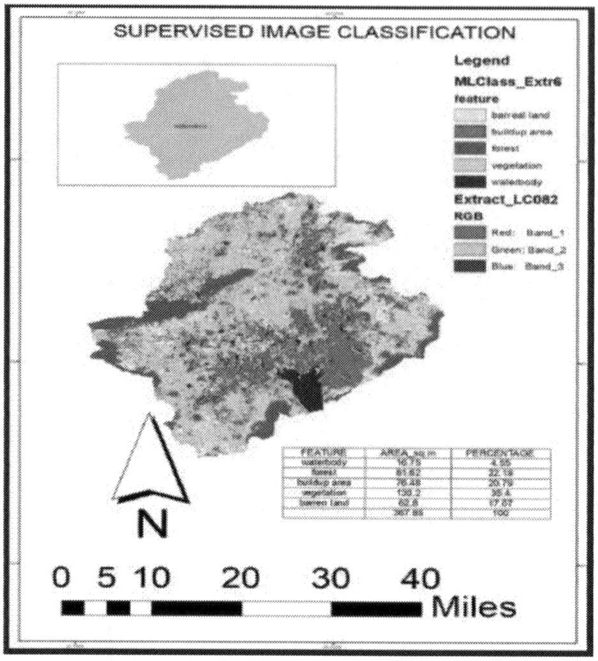

Figure 8- Supervised image classification of the year 2015.

As obtained from the ARC-GIS software, the amount of Area occupied by each of the Land Use, Land Cover Features as shown in Figure:6 for year 2005 satellite image, Figure:7 for year 2010 satellite image, Figure:8 for year 2015 satellite image are tabulated & percentage change corresponding to the overall Area under study are as below in table: 1.

Table: 1- % Change in LU/LC Areas from 2000 to 2015.

		Change in Area			Percentage Change		
S.NO	FEATURE	**2005**	**2010**	**2015**	**2005**	**2010**	**2015**
1	Water body	13.80	7.04	16.75	3.75	1.91	4.55
2	Forest	93.90	92.50	81.62	25.53	25.14	22.19
3	Buildup area	43.04	57.50	91.48	11.70	15.63	24.86
4	Vegetation	150.67	138.90	115.20	40.96	37.75	31.31
5	Barren land	66.46	71.97	62.80	18.07	19.56	17.07
	Total	367.87	367.91	367.85	100.00	100.00	100.00

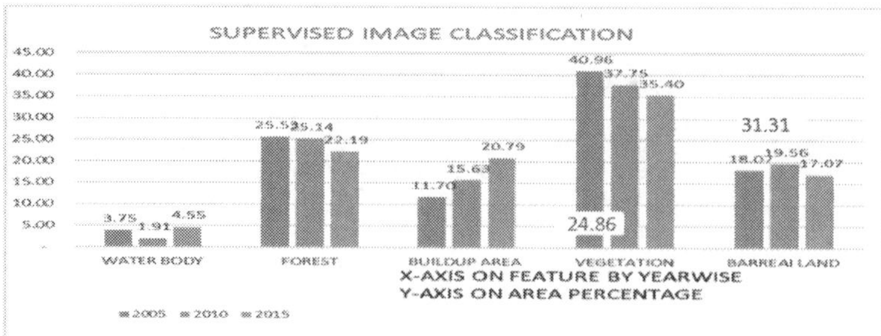

Figure: 9- % Change in LU/LC % Areas from 2000 to 2015.

Table: 2- Runoff depth corresponding to peak Rainfalls.

S.NO	LAND USE FEATURE	RUNOFF COEFFICIENT (C)	PEAK PRECIPITATION mm	RUNOFF DEPTH (R= C*P) in mm	RUNOFF DEPTH (R= C*P) in m
1	FOREST	0.2	208.2	41.64	0.04164
2	VEGETATION	0.35	208.2	72.87	0.07287
3	BARREN LAND	0.45	208.2	93.69	0.09369
4	BUILDUP AREA	0.9	208.2	187.38	0.18738
5	WATER BODY	1	208.2	208.2	0.2082

The Runoff corresponding to the peak precipitation in each of the Land Use/Land Cover Features are as per the values Tabulated in Table: 2 above. These Runoff depths will yield different runoff Volumes based on the relative Runoff Coefficients as shown in Table: 3.

Table: 3- Changes in Surface Runoff Volumes for each LU/LC feature.

S.NO	FEATURE	RUNOFF	AREA(sq.m)			VOLUME OF RUNOFF = AREA × RUNOFF (cu.m)		
			2005	**2010**	**2015**	**2005**	**2010**	**2015**
1	Water body	0.2082	13.8×10^6	7.04×10^6	16.75×10^6	2873160	1465728	3487350
2	Forest	0.04164	93.9×10^6	92.5×10^6	81.62×10^6	39099960	38517000	33986568
3	Buildup area	0.18738	43.04×10^6	57.5×10^6	91.48×10^6	8064835.2	10774350	17328902.4
4	Vegetation	0.07287	150.67×10^6	138.9×10^6	115.2×10^6	10979322.9	10121643	8394624
5	Barren land	0.09369	66.46×10^6	71.97×10^6	62.8×10^6	6226637.4	6742869.3	5883732

Table: 4- Changes in Rainfall Volumes for each LU/LC feature.

S.NO	FEATURE	RAINFALL DEPTH	AREA(sq.m)			VOLUME OF RAINFALL = AREA × RUNOFF (cu.m)		
			2005	**2010**	**2015**	**2005**	**2010**	**2015**
1	Water body	0.2082	13.8×10^6	7.04×10^6	16.75×10^6	2873160	1465728	3487350
2	Forest	0.2082	93.9×10^6	92.5×10^6	81.62×10^6	19549980	19258500	16993284
3	Buildup area	0.2082	43.04×10^6	57.5×10^6	91.48×10^6	8960928	11971500	15923136
4	Vegetation	0.2082	150.67×10^6	138.9×10^6	115.2×10^6	31369494	28918980	27107640
5	Barren land	0.2082	66.46×10^6	71.97×10^6	62.8×10^6	13836972	14984154	13074960

The net Amount of Surface Runoff is obtained after few of the Rainwater gets Infiltrated into the ground surface from the overall Volume of Rain occurred as in Table: 4. Hence the amount of Infiltrated water which contributed to the Groundwater recharge is as shown in Table: 5 after neglecting.

Table: 5- Change in Groundwater Recharge.

FEATURE	Runoff Volume (cu.m)			Rainfall Volume (cu.m)		
	2005	**2010**	**2015**	**2005**	**2010**	**2015**
Water body	2873160	1465728	3487350			
Forest	39099960	38517000	33986568			
Buildup area	8064835.2	10774350	17328902.4			
Vegetation	10979323	10121643	8394624			
Barren land	6226637.4	6742869	5883732		**76590534**	
Total in cubic meter	67243916	67621590	69081176.4	76590534	76590534	76590534
Infiltration (cu.m)	Rainfall Volume- Runoff Volume			9346618.5	8968943.7	7509357.6

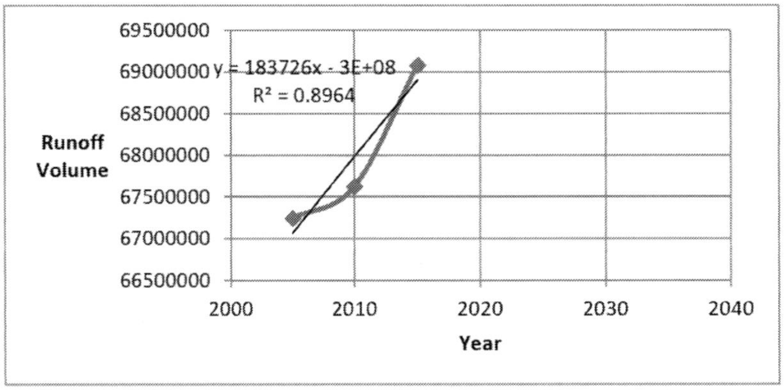

Plot: 1- Change in the Runoff volumes due to LU/LC Changes.

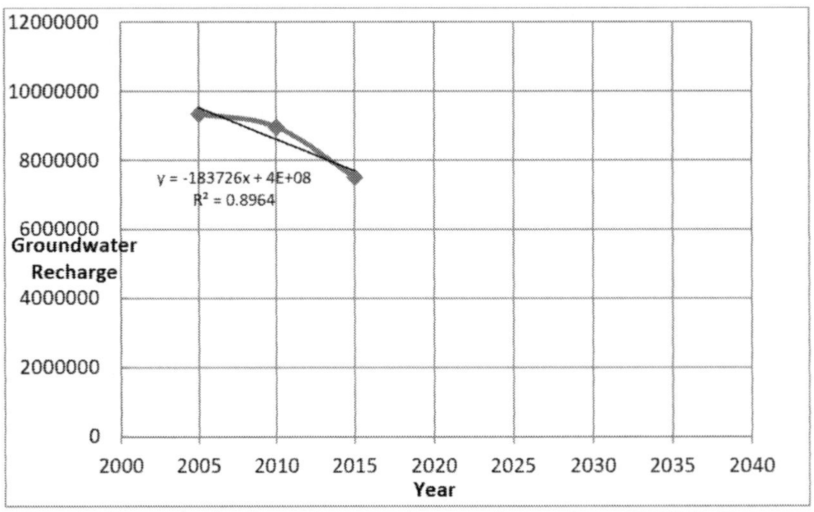

Plot: 2- Change in the Groundwater recharge volumes due to LU/LC Changes.

5.1 Future Predictions:

Based on the volume changes in both the ground water recharge & Runoff as shown in Plot:1 & Plot:2, we can develop the best fit predictions in order to estimate the changes in volumes for the next 20-30 years satisfactorily. The Volumes are as shown in Table: 6.

Table: 6- Predicted Runoff & groundwater volumes.

Year	Estimated Surface Runoff(cu.m)	Estimated groundwater Recharge
2020	71126520	5464014
2025	72045150	4545384
2030	72963780	3626754
2035	73882410	2708124
2040	74801040	1789494

6. Conclusions:

The forest area, water body and Vegetation gradually decreases from 2005 to 2015.The forest area decreases from 2005 to 2010 is 0.39% of the forst area and 2010 to 2015 is 0.05% of the forst area .The water body reduces from 2005 to 2010 is 1.84% of the water body and 2010 to 2015 is 2.64% of the water body area. The Vegetation area reduces from 2005 to 2010 is 1.49% of the Vegetation area and 2010 to 2015 is 2.49% of the Vegetation area The Build-up land area rapidly increases from 2005 to 2015. The Build-up land increases from 2005 to 2010 is 3.21% of the Build-up land area and 2010 to 2015 is 9.2 % of the Build-up land area.The Barren land area increases from 2005 to 2010 and decreases from 2010 to 2015 because the government have given instruction to people that to protect the water body's. The details are as shown in table 7.

Table: 7- Statistics of LU/LC in the Study Area.

Features	2005-2010	Increases	Decreases	2010-2015	Increases	Decreases
Forest	25.53%-25.14%	-	0.39%	25.14%-25.19%	0.05%	-
Water	3.75%-1.91%	-	1.84%	1.91%-4.55%	2.64%	-
Barren land	18.07%-19.56%	1.49%	-	19.56%-17.07%	-	2.49%
Vegetation	40.96%-37.75%	-	3.21%	37.75%-31.31%	-	6.42%
Build-up land	11.70%-15.63%	3.93%	-	15.63%-24.86%	9.16%	-

Even though the Megadrigadda reservoir Catchment is an Sub-urban area there is a considerable change in the Land use patterns within a small time thereby effecting the Groundwater Recharges & also increasing the surface runoff volumes which may result in Minor Flooding's if left unchecked.

References:

[1] Carlson, Toby N, and S Traci Arthur. "The Impact of Land Use — Land Cover Changes Due to Urbanization on Surface Microclimate and Hydrology: a Satellite

Perspective." *Global and Planetary Change*, **vol. 25, no. 1-2, 2000, pp. 49–65.**, doi:10.1016/s0921-8181(00)00021-7.

[2] Olokeogun, O. S., et al. "Application of Remote Sensing and GIS in Land Use/Land Cover Mapping and Change Detection in Shasha Forest Reserve, Nigeria." *The International Archives of the Photogrammetry, Remote Sensing and Spatial Information Sciences*, XL-8, 2014, **pp. 613–616.**, doi:10.5194/isprsarchives-xl-8-613-2014.

[3] Reis, Selçuk. "Analyzing Land Use/Land Cover Changes Using Remote Sensing and GIS in Rize, North-East Turkey." *Sensors*, **vol. 8, no. 10, 2008, pp. 6188–6202.**, doi:10.3390/s8106188.

[4] Afify, Hafez A. "Evaluation of Change Detection Techniques for Monitoring Land-Cover Changes: A Case Study in New Burg El-Arab Area." *Alexandria Engineering Journal*, **vol. 50, no. 2, 2011, pp. 187–195.**, doi:10.1016/j.aej.2011.06.001.

[5] Zhou, Qiming, and Yumin Chen. "Generalization of DEM for Terrain Analysis Using a Compound Method." *ISPRS Journal of Photogrammetry and Remote Sensing*, **vol. 66, no. 1, 2011, pp. 38–45.**, doi:10.1016/j.isprsjprs.2010.08.005.

[6] S. M. S. S, D. S. Vijayan, M. Anand, M. Ajona, and T. Jarin, " Biodegradation of P-nitro phenol using a novel bacterium Achromobacter denitrifacians isolated from industrial effluent water ," Water Sci. Technol., vol. 00, no. 0, pp. 1–12, 2021, doi: 10.2166/wst.2021.354.

[7] Cheruto, Mercy C, et al. "Assessment of Land Use and Land Cover Change Using GIS and Remote Sensing Techniques: A Case Study of Makueni County, Kenya." *Journal of Remote Sensing & GIS*, **vol. 05, no. 04, 2016**, doi:10.4172/2469-4134.1000175.

[8] Shi, Pei-Jun, et al. "The Effect of Land Use/Cover Change on Surface Runoff in Shenzhen Region, China." *CATENA*, **vol. 69, no. 1, 2007, pp. 31–35.**, doi:10.1016/j.catena.2006.04.015.

[9] Fichera, Carmelo Riccardo, et al. "Land Cover Classification and Change-Detection Analysis Using Multi-Temporal Remote Sensed Imagery and Landscape Metrics." *European Journal of Remote Sensing*, **vol. 45, no. 1, 2012, pp. 1–18.**, doi:10.5721/eujrs20124501.

[10] Dhakal, Nirajan, et al. "Estimation of Volumetric Runoff Coefficients for Texas Watersheds Using Land-Use and Rainfall-Runoff Data." *Journal of Irrigation and Drainage Engineering*, **vol. 138, no. 1, 2012, pp. 43–54.**, doi:10.1061/(asce)ir.1943-4774.0000368.

[11] P. Shiva Kumar, Dr. G. Venkata Rao, S. Nitish. (2020). A Study On Land Use Changes In Southern Part Of Visakhapatnam City Due To Urbanization-A Case Study. *PalArch's Journal of Archaeology of Egypt / Egyptology, 17*(7), **7318 - 7333**. Retrieved from https://archives.palarch.nl/index.php/jae/article/view/3149.

Design and Fabrication of Automatic Squeezing Machine for Food Industries

A.Imthiyas[1], R.Mahesh [2], Emin Eldho [3], Nevin Baby [3] and Hariprasath N S [3]

[1,2] Assistant Professor, Department of Mechanical Engineering, Aarupadai Veedu Institute of Technology, Vinayaka Mission's Research Foundation, Deemed to be University, Tamilnadu, India.

[3] UG Student, Department of Mechanical Engineering, Aarupadai Veedu Institute of Technology, Vinayaka Mission's Research Foundation, Deemed to be University, Tamilnadu, India.

E-Mail: imthiyas@avit.ac.in

Abstract: In the food sector, automation is increasingly a requirement to meet the requisite aspects of quality control, delivery performance, labour shortages, and revenue growth. Because of extremely competitive retailers always pulling to low prices, raw costs of raw material, and skyrocketing charges for utilization in an environment of labor-intensive production, food manufacturers are looking to automate their manufacturing operations. The capacity to mimic a product's aesthetic and quality with the least amount of materials not only increases line efficiency and profits, but it also has the potential to enhance sales. It also decreases the amount of work that has to be done. This food squeezing machine will squeeze food at a fast speed and in a more precise shape. In comparison to manual and traditional processes, the production rate is also high.

Keywords: productivity, quality, safety, adaptability, and precision.

1. Introduction

Computers technologies, robots, and information technology are used for automation system for handling various processes and machineries in an industry to replace a human being is known as industrial automation. Workers feel sophisticated by automation.[1] It is the next step in the industrialization process after mechanization. It increases the rate of production by improving production control. Freedom from microbial contaminants can be achieved by automation. . [2] It aids in the production of large quantities by reducing product processing time while improving quality. As a result, a given labour input yields a huge range of outcomes. Depending on the type and proportion of their quantity, the meal comes in a variety of flavors. In the food sector, automation can help to improve hygiene and food quality. It also has a high level of productivity, quality, safety, adaptability, and precision. Maintaining the nutritional value during production is primary task [3] Automating the process can result in high-quality meals with consistent shapes and sizes. It can also help with labour cost, time, and workload reduction. The advantages of using automated machines are less processing time, increases the efficiency of performance, increases the production capacity and reduces the waste and spillage. [4] Automatic machines can produce meals at a faster rate and with a more precise shape. It reduces the labour cost for production and improves profit. [5] There are numerous advantages to automating food production. The capacity to mimic a product's appearance and quality with the least amount of materials not only improves line efficiencies and profits, but it also has the potential to enhance sales. A fast food processing machine can able to deliver the food products within a short period of time mainly used for customer retention. . [6] At the same time, boosting raw ingredient traceability has the extra benefit of improving food safety. The issue that many in the food business are encountering is that line automation in the food sector is not the same like the industries of automobile and pharmaceutical ,in which automation is streamline and in huge level.. During the production of fast food, an automated fast food machine squeezes the duff mixture with the following category efficiency: time, human effort, safety, cleaning, and quality.7] Components or products in these industries are frequently homogeneous in shape, size, weight, and texture, and can be easily manipulated by artificial hands and placed precisely where they are required at high speeds and without damage. Because many basic products or components of a product – whether pasta, sausages, tomatoes, cheese, or potatoes – differ in length, structure, quality, mass, and appearance, fully automated robots find it very hard to modify them at the point of manufacture, primary pack-aging processes in the food industry, which require the exploitation of raw food into its initial carton or wrapper, are difficult. Raw food product automation requires high levels of machine sterilization. Food squeezing machines take less ability to process, increase capacity and efficiency, and reduce spoilage and wastage. [9] This needs easily accessible and dismantleable components for cleaning or chemical washing. Because the removal process does not always take place at the end of the shift. It's critical to go back into production as soon as possible, because components are frequently deluged throughout the clean-up procedure. In the food industry, waterproofing components and reassembly has proven to be difficult. In most food manufacturing situations.

2. Scope of the project

In general hand squeezing process induces work related muscular skeletal disorders , joint pain and muscle cramps for the workers[10] To reduce fatigue, a detailed design and development of an automated food squeezing machine for big food industry applications is required. During food preparation, an automated food squeezing machine squeezes a thick combination of food with the following category efficiency: time, human effort, safety, cleaning, and quality. The growth and development of the food product selling sectors employing modern selling methods is aided by automation. [11] Most manufacturing businesses seek automation engineers with a basic understanding of machines and the capacity to work on a variety of brand devices. This design primarily informs the user about the machine's cost as well as its efficiency in terms of time. As a result, the food squeezing machine has a high production rate when compared to alternative manual and commercially available options. Food manufacturing automation has enabled workers to complete a batch in almost half the time they were previously required, resulting in higher food unit output, lower dough and electric energy consumption, lower attendance requirements, and improved worker productivity and working conditions. Furthermore, hygiene is improved, and less labour is required, resulting in an increase in profit. Small-scale industries can benefit from this work. The Automatic Food Squeezing Machines are made of high-quality metal to provide maximum durability and smooth operation. The use of cutting-edge technology has helped to promote these automatic food squeezing machines in the food industry.

3. Purpose of the project

The ultimate purpose of the project is

- The number of labors is reduced.
- As the amount of labour required decreases, so does the cost of labour.
- In comparison to the old method, it reduces production time.
- In comparison to other machines on the market, it is less expensive.
- The linear actuator is critical in the automation of squeezing processes, allowing the

food sector to improve and maximize production.

4. Parts of food squeezing machine

This machine consist of various parts such as, Hydraulic cylinder ,Motor. Pressing plate, Relief valve Pressure gauge.

4.1. Hydraulic cylinder

A hydraulic is a mechanical actuator that delivers force along in an unique direction through a single-direction stroke. It is the base of most of the construction equipment, including engineering machines, automobiles and civil engineering equipment's. The fixed and regulated flow of oil to the hydraulic cylinder by the hydraulic pump is the "generator" side of the hydraulic system. The types of hydraulic pumps used in industrial applications are hand pump, electrical and pneumatic. The piston pushes the oil back from the chamber to the reservoir. The force F on the piston rod equals The pressure in the cylinder is equalized by the piston rod force times the area of piston A if the oil enters from the cap end during the extension stroke and the oil pressure in the rod end/head end is almost zero. Hydraulic and pneumatic cylinders are pneumatic and hydraulic devices that convert fluid pressure energy into useful mechanical energy. Actuators, as they are often known, are frequently utilized in a variety of control devices. Hydraulic or pneumatic cylinders for unique motion motors for rotational motion, reciprocating actuators for rotational motion. The pneumatic cylinder converts the gas's pressure energy into mechanical energy by using compressed air as a supply of gas. Hydraulic and pneumatic cylinders are classed based on how they move, as indicated in the diagram. Single acting cylinders and double acting cylinders can be classified into hydraulic or pneumatic cylinders

Figure 4.1 Hydraulic Cylinder

4.2. Pressing Plate

The primary function of pressing plates is to apply pressure to an object while maintaining a high level of strength. Mostly the work done by the machine is by pushing a plate or die against a work item, Depending on the type of machine the pressing action mechanics may vary. The machine press pushes the plate or die against the work piece's surface with hydraulic pressure, causing the work piece's shape to distort. Metal Forging, sizing, compression moulding, sheet metal blanking, die punching, for thin wire like structures deep drawing, and metal forming are all frequent uses for hydraulic presses. In manufacturing, and in food industries the hydraulic press is most commonly used as it allows for the creation of more complicated designs with material efficient. The pressing plate will revolve as a result of the pressure applied by the hydraulic motor. The food is

Journal of Physics: Conference Series **2040** (2021) 012013 doi:10.1088/1742-6596/2040/1/012013

then squeezed according to user input by the pressing plates, which are connected to the hydraulic press via a connecting rod. Only the pressing will move, while the rest of the parts will remain motionless.

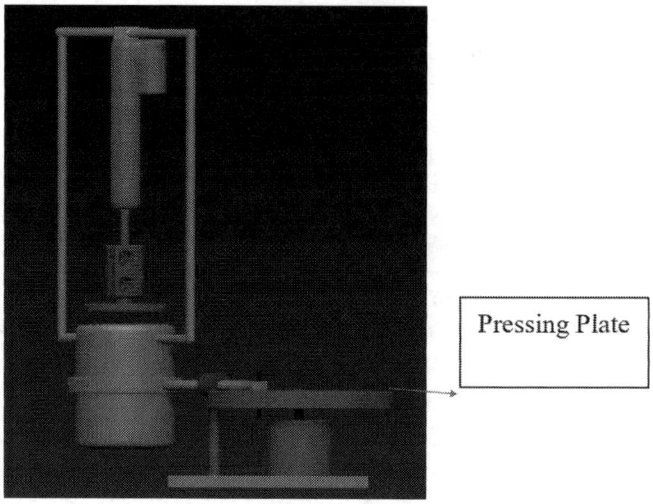

Figure 4.2 Graphical Illustration of Pressing Plate

4.3. Pressure Gauge

The type of material used to create the gauge is primarily determined by the pressure range at which it will be utilized. Gauges that operate at greater pressures are usually constructed of steel, whereas those that operate at lower pressures are usually composed of bronze. Gauges abound in pneumatic and compressed air systems, as pressure is recorded at numerous points throughout the system. , FRL or regulator alone in the system, are all monitored for pressure. Hydraulic actuators are monitored for pressure as it decides the desired shape of the squeezed object. Pneumatic pressure gauges are typically rated for not greater than the value of 300 psi, while most systems operate at around 100 psi. Depending on the type of gauge and the material it is composed of, the hydraulic gauge may sustain a wide range of pressures. As a result, two of the most essential selection factors for gauges are gauge style and material.

Figure 4.3 *Pressure Gauge*

4.4. Relief Valve

The fluid pressure in the hydraulic circuit is monitored and regulated by the relief valve. in which they are positioned. The valves seat are made in the form of ball shaped, poppet valve, or spool that is designed to mounted in a hollow or ported body and is resisted by a spring. A poppet is a disc or cone-shaped device that rests within an opposing machined seat and produces very minimal leakage when squeezed closed by spring pressure. A spool is a spring-loaded, cylindrical steel rod having measuring grooves or notches. The combination of pressure acting due to loading, pressure due to backpressure, and the energy required to flow through the valve causes a rise of pressure in relief valves. When the initial fluid force exceeds the seated force of the spring, cracking pressure occurs. The rate of pressure rise is stable as the valve flows additional fluid to the tank because the pressured fluid forces offset the spring's compression rate. As the valve bottoms out and is subjected to flow forces, pressure rises again as it approaches fully open. As the work or backpressure reduces, the valve begins to close at a different rate than when it first opened. The hysteresis of a curve is the difference between its opening and closing points, and it indicates the quality of its construction. Valves of higher quality with more advanced construction have a reduced pressure rise and better hysteresis. As hydraulic pressure in the circuit where the valve is situated rises, force against the spool or poppet begins to overcome the spring's opposing force, allowing a flow path to the tank to open. Energy is diverted (in the form of heat) as pressured fluid escapes the relief valve until downstream pressure meets the spring value force, which could be droplets of flow or the entire pump flow, depending on the application and state of the circuit

4.5. Hydraulic Motor

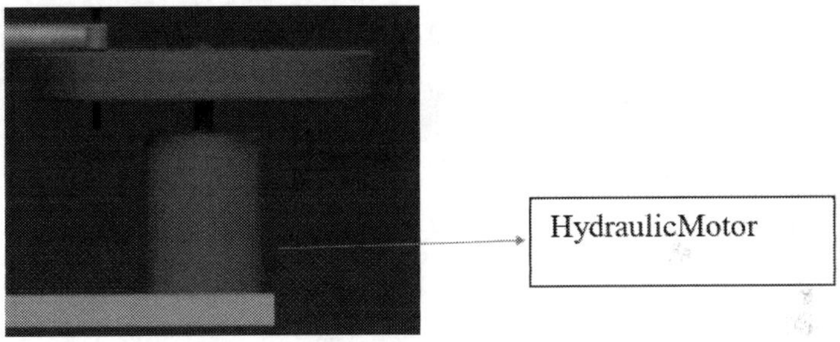

Figure 4.4 Graphical Illustration of Pressing Plate

In most cases the speed and torque required for squeezing operation at actual cases may sometimes be varied while keeping the required horsepower, the application of the hydraulic motor generally defines the required horsepower and motor speed range. The type of motor chosen is determined by the level of reliability, longevity, and performance required. Once the type of fluid has been decided, the actual size is chosen depending on the estimated life and the overall cost of the machine installation. A fluid motor running at less than rated capacity will have a service life extension that is more than proportionate to the lower operating capacity. When a motor is operating at maximum system pressure and maximum shaft speed, it produces its maximum horsepower. The motor's original cost will be the lowest if it is always used under these conditions. However, in cases where the output speed must be reduced, the entire cost of the motor with the reduced speed must be considered in order to reduce overall drive installation costs. The hydraulic motor will supply circular motion to the pressing plates attached to it, and the pressing plates attached to the hydraulic motor will rotate and squeeze the food to obtain the appropriate output.

5. Working Methodology

Figure 5.1 Fabrication of Squeezing machine

Figure 5.2 Assembly of Squeezing machine

This food squeezing machine uses a hydraulic press to squeeze the food fed into the machine and generate the desired output. The hydraulic press principle is based on Pascal's law, which states that the strength of pressure in a static fluid is distributed evenly in all directions. Two cylinders of different diameters make up the hydraulic press. One cylinder has a big diameter and houses the Ram, while the other has a smaller diameter and houses the plunger pipe runs between the two cylinders. A liquid is contained in the cylinders and pipe and is used to convey pressure. When a modest downward force f is given to the plunger, a pressure is created on the liquid in contact with the plunger. This pressure is distributed evenly in all directions and acts in an upward direction on the RAM. The ram is then hoisted up by the larger weight that has been placed on it.. Due to this the food is squeezed easily from the die with the desired shape without much effort

6. Advantages of squeezing machine

. Considering the proposal, it has numerous advantages, including

- Increased productivity and food manufacturing efficiency.
- Labor's workweeks are becoming shorter.
- Due to a reduction in manpower, labour costs are also lowered.
- Time Efficiency is a factor that contributes to reduce downtime.
- A higher standard of product
- Increased process control results in more efficient material usage and less scrap.
- Enhanced security

7. Conclusion

A low-cost food-preparation machine This project work has resulted in the development of a squeezing machine. All small-scale industrialists can use the built equipment as an affordable machine. Normally, the proposed system provides a cost-effective and long-term way for processing a variety of foods, as well as a hygienically safe approach for ingesting squeezed food for all developing countries. Automatic equipment have been a boon to the food processing industry, allowing for hygienic, high-quality food processing. A basic automatic machine can correctly and efficiently perform the work of several humans. Humans are being displaced by automated robots in every field, from agriculture to aerospace. With the introduction of this squeezing machine, both quality and quantity have improved.

Reference

[1] Shankar S, Maheswari C, Gowtham R, Kiruba P and Mohansrinivas K 2019 Design and fabrication of portable sugarcane harvesting machine International Journal of Scientific and Technology Research 8 2059-62

[2] Food pacaking and shelf life a practical guide " Laila Rfika" edited by Gordon L Robertson

[3] Dunford EK, Ni Mhurchu C, Huang L, Vandevijvere S, Swinburn B, Pravst I, et al. A comparison of the healthiness of packaged foods and beverages from 12 countries using the Health Star Rating nutrient profiling system, 2013–2018. Obes Rev [Internet]. 2019 Jul 22 [cited 2019 Oct 8];obr.12879.

[4] Tressler D K a J M A 1990 Fruit and Vegetable Juice Technology West Port Connectieut: AVI Publishing Company Inc 155-158

[5] Design and Fabrication of Chakali making machine " International journal of advanced resesrch in science and engineering , Volume No 7, special issue No 01, 2018

[6] Nish Talwar, Yogesh Kumar "Design & development of automatic fast food Machine"IJMERR,ISSN 2278- 0149, Vol. 3, No. 3, July, 2014

[7] Amit B Solankil, V R Solankil and D N Shah DESIGN & DEVELOPMENT OF AUTOMATIC FASTFOOD MACHINE," International journal of mechanical Engineering and Robotics research "ISSN 2278 – 0149

[8 Kousar A. Attar, Sammed A. Patil, Prafull D. Patil Development and Fabrication of Automatic Chakali Making Machine International Research Journal of Engineering and Technology (IRJET) Volume: 07 Issue: 05 | May 2020 Scientific and Technology Research 8 2059-62

[9] Tressler D K a J M A 1990 Fruit and Vegetable Juice Technology West Port Connectieut: AVI Publishing Company Inc 155-158 Connectieut: AVI Publishing Company Inc 155-158

[10] Shankar S, Maheswari C, Gowtham R, Kiruba P and Mohansrinivas K 2019 Design and fabrication of portable sugarcane harvesting machine International Journal of Scientific and Technology Research 8 2059-62 Scientific and Technology Research 8 2059-62

[11] Aviara N A, Lawa A, Nyam D S and Bamisaye J 2013 Development and performance evaluation of a multi-fruit juice extractor Global Journal of Engineering, Design and Technology 2 16-21

Comparative Study on Soil Stabilization using Industrial by Products and Coconut Coir

S.S.S. Nitish [1] ,Sangita De [2] , A.V.S.L Ramya [3] , G Sudheer Kumar [4]
[1]Assistant Professor, Department of Civil Engineering, Vignan's Institute of Information Technology, Visakhapatnam, India
[2]Assistant Professor, Department of Civil Engineering, Aditya College of Engineering, Surampalem, India, sangitade999@gmail.com
[3]Assistant Professor, Department of Civil Engineering, Vignan's Institute of Information Technology, Visakhapatnam, India ramya.attota@gmail.com,
[4]Assistant Professor, Department of Civil Engineering, Vignan's Institute of Information Technology, Visakhapatnamsudheercivil.viit@gmail.com
Corresponding author Email ID: nitish210795@gmail.com

ABSTRACT: Waste disposal has become a serious concern in developing countries like India. The substitution of these waste materials in the form of stabilizing agents in soil stabilization is a modern approach by which waste materials can be advantageously used. In many instances, the soil has proven to be problematic for the construction of various infrastructures like embankments, pavements, foundations, hydraulic barriers, etc. In the present study, a particular type of soil is stabilized to improve the physical properties by using multiple admixtures. In general, additives such as lime, cement, sawdust, stone dust, and other compounds are used for the stabilization of soil over years. This study is conducted to evaluate the viability of using Coconut Coir Fibre (CC) along with stone dust (SD)/pond ash (PA) as a stabilization material. A comparative analysis on the effect of CC with SD as well CC with PA on engineering characteristics of silty soil is presented in the present study. A sequence of laboratory experiments was conducted on silty soil blended with Coconut Coir Fibre along with proportions of Stone Dust/Pond Ash from 0.5% to 1.5% and 30% as constant respectively by mass of dehydrated soil. The experimental outcomes shown a significant change in properties of soil, which conclude that the coconut coir along with stone dust as a very potential additive to improve the characteristics of silty soil compared to that of pond ash.

KEYWORDS: Soil Stabilization, coconut coir, stone dust, pond ash, OMC, MDD, CBR, industrial waste, sustainability

1. INTRODUCTION

Soil Stabilization becomes necessary as there is a restriction on availability of decent sites for construction. The key objective of Stabilization is to recover the characteristics of soil as it is necessary where the available soil is weak. Soil stabilization is a process where natural or manufactured additives or binders are used to improve the index and engineering properties of soils. Additives such as lime, cement, saw dust, stone dust and other compounds have been used over the years with greater improvement to soil properties. In India, abundance in availability of waste material like coal ash, plastics, stone quarry, recycled aggregate, polythene etc. have been helpful in improving the soil strength over the recent years.

However, ash was proven to be less significant in soil stabilization [5][18]. Stabilization is mainly used to reduce the permeability and compressibility of the soil, while increasing the shear parameters. Stabilization is mainly used to condense the perviousness and compressibility characteristics of the soil, and increase the shear parameters. The improvement of soil properties is done by blending or mixing the materials. It is not ideal that with increase in the quantity of any additive to soil, the strength or properties of the soil are improved. Hence, an optimal proportion of additive shall be added for improvement of any soil is to be evaluated [5].

India is a country with a very huge population, in recent decades construction of roads and buildings are done even on weak soils. The throwing away of waste material is also an immense problem due to scarcity of land. The requirement for proper disposal techniques or dumping waste creates an impact on environment [4]. Also, upsurge in populace has increased the mandate for terrestrial acquisition. Industrial wastes have been the major concern considering this aspect of sustainable use of resources and waste materials. The use of these wastes in improving various characteristics of many other materials is efficient [17][18]. Use of industrial and agricultural waste to stabilize soil, improves the ground and also reduces the environmental pollution. As they are locally available, minimization of the construction cost and sustainable development are attained. Several studies were led around the world to know changes in various index and engineering properties of soil due to the addition of admixtures to the soil. The paper mainly focuses on the study of engineering properties i.e. O.M.C, M.D.D and C.B.R for a soil stabilised with coconut fibres and pond ash in comparison to the soil stabilised with coconut fibre and stone dust.

The crumpled stone or rock is a form of construction aggregate, formed by mining appropriate rock deposit and then breaking to the desired size using crushers. However, it is discrete from gravel, which is produced by natural processes of weathering and erosion. The use of crushed stone powder reduces the dry shrinkage and micro-cracks which help in high performance of stabilized materials [16]. Pond ash is a combination ash obtained due to the mixture of two by – products from the thermal power plants. Pond ash has been the greatest concern in environmental point of view [4]. Pond ash has been prevalent for its self-hardening characteristics [14].

Coir fibres are extracted from the fruit (coconut) of the coconut tree. The coconut fibre is extracted from the outer shell of a coconut [15]. The common name is coir, scientific name "cocos nucifera" and plant family of coconut fibre is Arecacea. Coconut fibres are categorized as tan fibre and white fibre. Tan fibre is extracted from the mature coconuts, where it is thick and strong with high abrasion resistance. And the white fibres are extracted from immature coconuts. These are smooth and fine in nature, However, these are considered weak. The main considerations regarding that, it would be a renewable resource and CO2 neutral material and biodegradable with low density and cheap availability. Coir fibre diameter is about 0.1 - 1.5mm and length is 3 - 15cm. The coconut coir is elastic in nature and an adequately used material as commercial product i.e., mattress [15].

2. METHODOLOGY:

The most common method of stabilization is to mix the soil with an admixture to form a product with improved properties. The stabilized soil material has relatively high strength, low permeability and compressibility to the actual virgin soil. Stabilization of soil does not alone indicate the improvement by additives but also indicates the advanced technology usage. Improved chemical and mechanical stabilization are necessary when considering waste materials like crushed asphalt pavement, copper – zinc slag and rubber tyre chips etc. Such materials need to be recycled and theses may contain potential hazardous materials [1]. Hence the use of industrial and agricultural waste can help in sustainable soil stabilization.

Coconut coir is collected from the locally available regions in the present study. The length of the fibre was considered 3 cm. Incremental percentages of coconut coir, say 0.5%, 1.0%, 1.5% and 2.0% when used in stabilization were identified to be optimal [13]. The coconut fibres, in proportions of 0.5%, 1.0% and 1.5% along with 30% stone dust is considered as Sample-I study and the same fibre proportions with 30% pond ash as the Sample-II study. The chemical properties of the stone dust and pond ash are presented in table-1 below:

S.no	Constituents	Stone dust (%by weight)	Constituents	Pond Ash (%by weight)
1	SiO_2	60.5	SiO_2	42.7 – 47.6
2	Al_2O_3	16.2	Al_2O_3	23.0 – 21.2
3	Fe_2O_3	3.27	Fe_2O_3	17.0 – 9.75
4	CaO	3.52	CaO	9.80 – 2.94
5	MgO	0.95	MgO	1.54 – 1.45
6	Na_2O	3.73	Na_2O	0.29 – 1.07
7	K_2O	4.84	K_2O	0.96 – 1.05
8	Loss on Ignition	1.7	Loss on Ignition	29.3

Table – 1: Chemical Composition of Stoneitu vhiii Dust [6] & Chemical Composition of Pond Ash [2] [6] [11].

3. FLOWCHART:

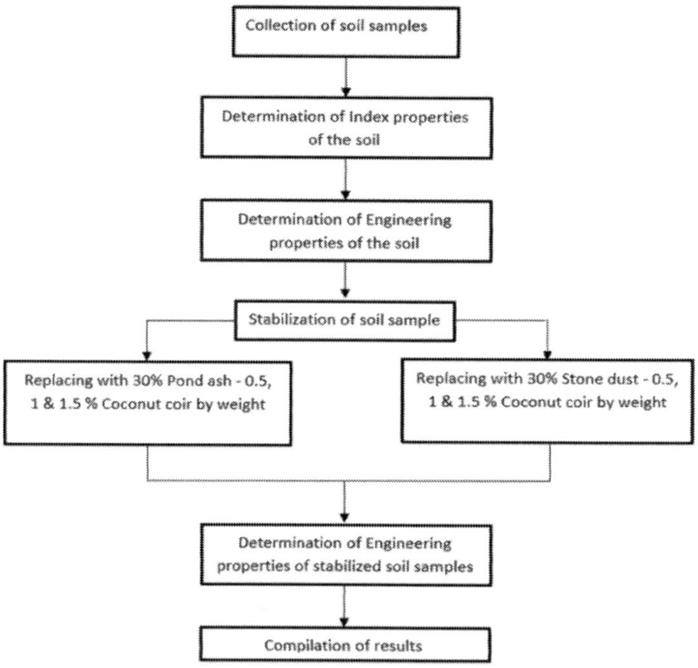

Chart – 1: Flowchart describing the work flow

4. Experimental Study:

Typical soil samples were collected at 1.5 m depth from two deposits, one in Aganampudi, Visakhapatnam, India and the other in Sheelanagar, Visakhapatnam, India. The properties of the soil samples designated as soil I and soil II. The depth at which the soil was collected it restricted to 1.5 m to avoid the inclusion of organic matter. Soil samples are classified as MI according to the Indian Soil Classification System. Silt has specific area with a characteristically non-sticky, plastic feel. [3]. The industrial by – products, stone dust and pond ash which are locally available along with coconut coir fibre are considered for soil I and soil II respectively.

The virgin soil properties have been understood by the experimental approach considering the index and engineering properties. Silty soils are very moisture subtle and their stability is being influenced significantly by the degree of compression attained during compaction.[13] Silty soils are aquaphobic. Hence, there would be difficulty in controlling moisture during active compaction. This could be problematic during pavements constructions as well as under regular vehicular traffic loads.[7].

S.no	Properties of the soil	Value
1	Specific gravity	2.65
2	Grain size distribution a) Gravel (%) b) Sand (%) c) Fines (%)	0.72 25.6 73.68
3	Atterberg limits a) Liquid limit (%) b) Plastic limit (%) c) Plasticity index (%)	39 32.6 6.4
4	Compaction characteristics (heavy compaction) a) Optimum moisture content (%) b) Maximum dry density(g/cc)	11.6 1.88
5	Strength parameters Un soaked CBR (%)	3.37
6	Differential free swell (%)	55%

Table – 2: Properties of Virgin Soil -I

S.no	Properties of the soil	Value
1	Specific gravity	2.65
2	Grain size distribution d) Gravel (%) e) Sand (%) f) Fines (%)	0.51 25.17 74.32
3	Atterberg limits d) Liquid limit (%) e) Plastic limit (%) f) Plasticity index (%)	38.4 32.1 6.3
4	Compaction characteristics (heavy compaction) c) Optimum moisture content (%) d) Maximum dry density(g/cc)	11.3 1.87
5	Strength parameters Un soaked CBR (%)	3.37
6	Differential free swell (%)	55%

Table – 3: Properties of Virgin Soil -II

Observing the similarity in index properties for both the soil samples as shown in table -2 and table -3, the soil sample from Aganampudi, Visakhapatnam, India i.e., soil sample – I is considered for the stabilization purpose. The present sample – I is experimentally studied for 0.5%, 1.0% and 1.5% proportions of coconut coir (CC) by weight of dry soil along with 30% stone dust (SD) initially and then studied for 0.5%, 1.0% and 1.5% proportions of coconut coir (CC) by weight of dry soil along with 30% pond ash (PA). These samples are mixed with utmost care to obtain a homogenous mix. Light weight compaction was performed for all proportions of mixed soil and the optimum moisture content (OMC) and maximum dry density (MDD) are found out. The samples are later tested for the California bearing ratio (CBR). These tests are performed based on the Indian Standard Code recommended for soil tests.

A comparative study based on the results obtained considering OMC, MDD and the CBR – value is made.

5. Results with discussions

The investigational results for optimum moisture content (OMC) and the maximum dry density (MDD) are obtained from the heavy weight compaction test. The results are presented in the table – 4 below for virgin soil, soil with varying proportions of coconut coir, stone dust and pond ash. Comparison of these results based on OMC and MDD are presented in Graph – 1.

COMPACTION							
S.no	**Sample**	**OMC**	**MDD**	**S.no**	**Sample**	**OMC**	**MDD**
1	Virgin soil	11.6	1.88	1	Virgin Soil	11.6	1.88
2	Soil + 0.5% CC + 30% SD	12.31	1.86	2	Soil + 0.5% CC + 30% PA	12.53	1.83
3	Soil + 1.0% CC + 30% SD	13.22	1.81	3	Soil + 1.0% CC + 30% PA	13.54	1.73
4	Soil + 1.5% CC + 30% SD	14.47	1.72	4	Soil + 1.5% CC + 30% PA	14.9	1.57

Table – 4: OMC and MDD values obtained for different considerations

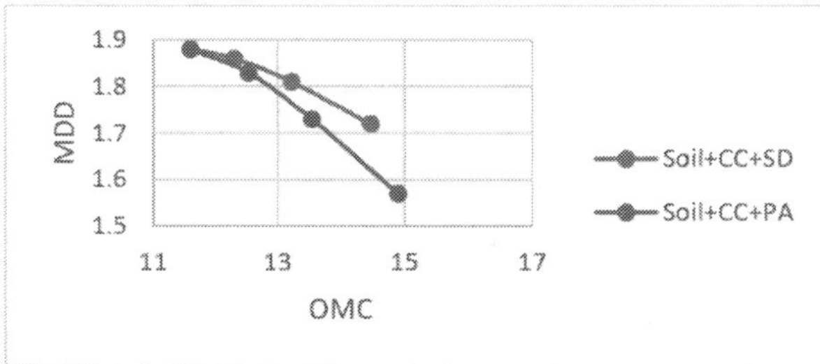

Graph – 1: Comparison of MDD & OMC for different considerations

These test samples are further tested for CBR value under unsoaked condition. The representative values obtained for all the considered proportions of admixtures are presented in the table – 5 below. A bar graph showing the comparation of CBR values for all the considered proportions are presented in the Graph – 2.

CBR					
S. No	**Sample**	**CBR (unsoaked)**	**S. No**	**Sample**	**CBR (unsoaked)**
1	Virgin soil	3.27	1	Virgin Soil	3.27
2	Soil + 0.5% CC + 30% SD	4.26	2	Soil + 0.5% CC + 30% PA	3.37
3	Soil + 1.0% CC + 30% SD	5.41	3	Soil + 1.0% CC + 30% PA	3.86
4	Soil + 1.5% CC + 30% SD	5.98	4	Soil + 1.5% CC + 30% PA	4.59

Table – 5: CBR values obtained for different considerations

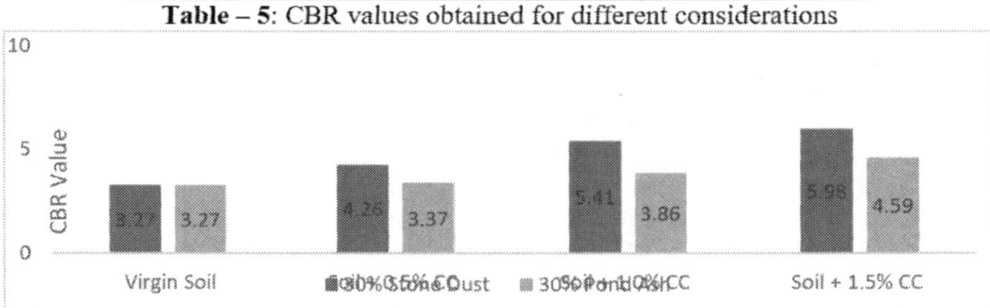

Graph – 2: Comparison of CBR for different considerations

For the above results, it can be clearly observed that, when 1.5% CC considered, there is 30% increase in CBR value for SD compared to PA in the soil sample considered. When 1.0% CC considered, there is 40% increase in CBR value for SD compared to PA in the soil sample

considered and when 0.5% CC is considered, there is 26.4% increase in CBR value for SD compared to PA in the soil sample considered.
For SD + CC Soil sample, with 1% increase in CC, there is an increase of 1.4 times the CBR value. For PA + CC Soil sample, with 1% increase in CC, there is an increase of 1.36 times the CBR value.

6. Conclusions

Based on the experimental test fallouts, conclusions are drawn:

- Slight increase in strength of stone dust considered soil sample with respect to coconut coir as compared to that of the pond ash considered soil.
- By addition of coconut coir and stone dust to the silty soil sample, is a great rise in CBR value.
- The effect of coconut coir in the improvement of this particular soil is predominant in both aspects as nearly equal proportion of improvement is observed i.e., 1.4 times in Soil + CC + SD and 1.36 times in Soil + CC + PA.
- At any considered proportions of coconut coir, the strength parameters of the CC + SD soil sample show better results than that of CC + PA soil sample.

Conflicts of Interest:
The authors state that they have no conflict of interest.

References:

[1] Afrin, H. (2017). A Review on Different Types Soil Stabilization Techniques. International Journal of Transportation Engineering and Technology. https://doi.org/10.11648/j.ijtet.20170302.12

[2] Awang, A., Marto, A. and Makhtar, A.M. (2011) "Geotechnical properties of Tanjung Bin coal ash mixtures for backfill materials in embankment construction" EJGE Vol. 16 pp. 1515 – 1531.

[3] Degirmenci, N., Okucu, A., & Turabi, A. (2007). Application of phosphogypsum in soil stabilization. Building and Environment, 42(9), 3393-3398. doi:10.1016/j.buildenv.2006.08.010

[4] Ghosh, A., & Bera, A. (2005). Bearing capacity of square footing on pond ash reinforced with jute-geotextile. Geotextiles and Geomembranes, 23(2), 144-173. doi:10.1016/j.geotexmem.2004.07.002

[5] Gomez, G. P., Martinez, J. G., & Cardenas, J. C. (2019). Soil stabilization with lime and fly ash Estabilización de suelos con cal y ceniza volante. 2019 Congreso Internacional De Innovación Y Tendencias En Ingenieria (CONIITI). doi:10.1109/coniiti48476.2019.8960697

[6] D.S. Vijayan, D. Parthiban," Effect of Solid waste based stabilizing material for strengthening of Expansive soil- A review," Environmental Technology & Innovation, Volume 20, 2020, 101108, ISSN 2352-1864, https://doi.org/10.1016/j.eti.2020.101108

[7] Heitor, Ana & Indraratna, B. & Rujikiatkamjorn, Cholachat. (2013). Laboratory study of small strain behavior of a compacted silty sand. Canadian Geotechnical Journal. 52. 179-188. 10.1139/cgj-2012-0037.

[8] IS 2720-8 (1983): Methods of test for soils, Part 8: Determination of water content-dry density relation using heavy compaction [CED 43: Soil and Foundation Engineering]

[9] IS 2720-16 (1987): Methods of test for soils, Part 16: Laboratory determination of CBR [CED 43: Soil and Foundation Engineering]

[10] JOYAL-Stationary and mobile crushing solution provider, Grinding mills manufacturer in China. (n.d.). Retrieved July 31, 2020, from http://www.crusherinc.com/

[11] Muhardi et al., (2010) "Engineering characteristics of Tanjung Bin coal ash" EJGE Vol. 15, Bund. K. pp. 1117 – 1129

[12] D.S. Vijayan, A. Mohan, J. Revathy, D. Parthiban , R. Varatharajan, "Evaluation of the impact of thermal performance on various building bricks and blocks: A review", Environmental Technology & Innovation 23 (2021) 101577, https://doi.org/10.1016/j.eti.2021.101577. Ramesh, A., & Kumar, M. (2009). An experimental investigation on coir fiber and fly ash in stabilized mechanistic pavements. In Proc. Indian Geotech. Conf (Vol. 1, pp. 201-204).

[13] Raymond, S., 1961. Pulverized fuel ash as embankment material. Proceedings of the Institute of Civil Engineers, vol. 19, pp 515–536.

[14] Singh S. K. (2014). Inclusion Of Coconut Coir Fiber In Soil Mixed With Coal Ash. International Journal of Research in Engineering and Technology, 03(11), 209-213. doi:10.15623/ijret.2014.0311035

[15] Song, T., Lee, S., & Kim, B. (2014). Recycling of crushed stone powder as a partial replacement for silica powder in extruded cement panels. Construction and Building Materials, 52, 105-115. doi:10.1016/j.conbuildmat.2013.10.060

[16] Zhang, T., Gao, P., Gao, P., Wei, J., & Yu, Q. (2013). Effectiveness of novel and traditional methods to incorporate industrial wastes in cementitious materials—An overview. Resources, Conservation and Recycling, 74, 134-143. doi:10.1016/j.resconrec.2013.03.003

[17] G.Sudheer Kumar, M.K.Sumanth, Madimi samuel. A REVIEW PAPER ON STABILIZATION OF SANDY SOIL BY USING CEMENT GROUTING TECHNIQUE. JCR. 2020; 7(14): 902-908. doi:10.31838/jcr.07.14.160

Simulation of Real-time Medicine suggestion box for COVID Screening

Vijay. J*, Sriram. D, Hemanth Sai. B, Kachana Santhosh Kumar R
Department of Electronics and Communication Engineering,
AarupadaiVeedu Institute of Technology,
Vinayaka Mission's Research Foundation, Paiyanoor, Tamilnadu, India
Email:vijay.ece@avit.ac.in

Abstract—Technology advancements have a rapid effect on every field of life, be it medical field or any other field. Artificial intelligence has shown the promising results in health care through its decision making by analysing the data. COVID-19 has affected more than 100 countries in a matter of no time. People all over the world are vulnerable to its consequences in future. It is imperative to develop a control system that will detect the coronavirus. One of the solutions to control the current havoc can be the diagnosis of disease with the help of various AI tools. The proposed system contains textual data analysis as well as real time physiological data analysis concept. The embedded platform reads the body temperature and heart rate of the patients. The patient is automatically induced to attend the pre-screening survey designed using the software GUI that collects most of the information on symptoms persists. A COVID-19 dataset is collected from publicly available websites. The read survey values and sensor values are pre-processed and extracted the unique features present in it. Those unique parameters are compared with the database to produce the output showing COVID positive status or Negative status and immediate medicine suggestions for them using the global collective medicine suggestions box.

Keywords— Machine learning, COVID-19, Medical analysis, artificial intelligence, medication suggestion system, Covid datasets, Neural networks

1. Introduction

1.1COVID-19 Virus

The novel corona virus is or the COVID -19 started spreading from the city of china, Wuhan. The virus directly hits the lungs of the peoples. The infection of Virus is identified with certain symptoms that starts from high fever, body pain, cough, head ache and cold. The virus transit from one person to another through air as medium.[1] The capacity of transit rate decreases if peoples started maintaining the social distance. On the other hand, many research works are undergoing taking COVID as serious infection, the earliest diagnosis procedure is under research. The global diagnostic procedures followed in various countries are considered for the research analysis. In the mean while precaution measure are taken by the government to find the earliest detection of Covid infection and to reduce the infection transit rate as soon as possible.

1.2 Input Data

[2]The ultimate input for the diagnosis of COVID infection starts from the physiological data collected from the patients in the form of blood pressure, temperature etc. Apart from physiological data, CT Scanning, X-rays and MRI scans are the next foremost options to find out the infection ratio and to find out the infected organs too. Fever act as the initial symptom, the pneumonia cases are initially screened through the X-rays only.

1.3 Datasets

[3] The Scientific computing team and Engineering research works focused on finding the methodology that predict the corona virus to the earliest. For the analysis of COVID research work, many publicly available datasets are considered. In the current scenario, publicly available open source datasets are helpful to them machine learning team to predict the transit and analysis of Corona virus. Data science act as peculiar role in finding out the most impacted factor for the fast transit of Virus.

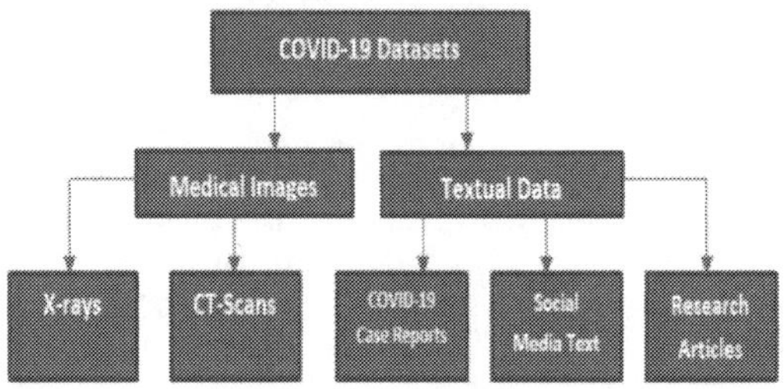

Figure 1. Various types of Data inputs available for COVID research

The presented work is focused on such kind of early prediction framework using machine learning algorithms. The screening model is developed with the oral survey work, and the physiological data measurement. Heart rate and temperature is considered as the physiological data. Based on the oral survey and real time health status, the maximum correlated COVID symptoms are predicted. Based on the classified results, the medicine suggestion is provided.

2. LITERATURE SURVEY

[1] Sohan et al., (2020) Research article cover various types of datasets accessible for COVID-19 forecast. The examination zeroed in on the assortment of top datasets that were considered for logical investigation of COVID-19. The test dataset covers blood tests, X-beams, Scan reports, the printed dataset of tweets, text report of the patients, and so forth The gathered examples are extensively named positive test tests, negative test tests, and both joined dataset. The people group joins are useful for scientists for reference to proper examination of the dataset.

[2]Vaid et al., (2020) led research work on COVID-19 location through the man-made consciousness strategy. During their work, (April 23, 2020), 2.7 million individuals were affirmed with the weak infection. Over 0.9 million individuals have passed on and 0.75 million patients recuperated totally from the boundless Coronavirus. They have used chest X-beam hints of patients and fostered a profound convolution neural organization (CNN) to recognize the irregularities present in the constructions. They moved toward the exchange learning strategy to identify profoundly covered up follows utilizing radiographs of the patients. The proposed CNN model scores 96.3% precision and deficiency of 0.151 through parallel cross-entropy.

Narinder et al., (2020) expressed in their exploration work on scourge investigation utilizing a dataset gathered from Johns-Hopkins. The datasets utilized for the examination are gathered as time-arrangement synopsis tables in CSV design containing the itemized construction of the refreshed status of COVID-19 affirmed cases, demise rate, and recuperation pace of the country. They fostered a novel model utilizing polynomial relapse to identify the situation with the quick spreading infection. The proposed procedure likewise contrasted and Support vector relapse (SVR), profound neural organizations (DNN), and Long-momentary memory in counterfeit neural organizations (LSTM). Reformist outcomes are showing least RMSE in the Polynomial relapse model.

M. J. Horry et al., (2020) assessed a VGG19 classifier model for expectation of COVID-19 utilizing CT Lung pictures of tainted patients. The solitary restriction past the identification interaction is the constraints persevering in the accessibility of ground truth information. The openly accessible pictures are poor in quality to deliver novel designs. The model created is constantly centered around diminishing the bogus positive conclusion. People and their contact follows are investigated for anticipating the feelings of anxiety and other clinical commitments produced by COVID-19.

Shuo Wang et al., (2020) assessed a Random Deep learning model utilizing a CT-EGFR dataset containing 5372 patients' unique records gathered from 7 urban communities. The preparation set comprises of 4106 patients with cellular breakdown in the lungs with EGFR (epidermal development factor receptor) transformation status. The assessed model uses DenseNet121 for the programmed division of the lungs. The prepared picture dataset is an example perceived and the proposed exactness scores around 80% for the Deep learning strategy in correlation with Random Deep learning techniques.

Zhang et al., (2020) led an examination and execution of computerized recognition on COVID-19 pneumonia utilizing profound learning instruments. Ultra man-made reasoning instruments are executed to precisely assess viral pneumonia that incorporates multi-center, respective GGOs where the normal contamination being the dorsal section of the correct lower flap of the lungs. The examination uncovers that confinement of irresistible districts that are needed for COVID-19 recognition as well as gives further help to the doctors in fostering the treatment plan.

Tahmina Z et al., (2020) Evaluated profound learning standards utilizing classifier models, for example, VGG16, ResNet50, and EfficientNetB with pre-prepared CT-Chest pictures of COVID-19 contaminated patients. The generative ill-disposed organization based framework is assessed. EfficientNetB0 accomplishes 96% precision in the forecast of the three sorted illnesses like pneumonia, COVID-19, and Normal. For planning, the expectation procedures that should be featured, inclination class enactment is utilized. Also, these procedures help envision the influenced districts of the lungs during infection movement and seriousness stages.

Ko H et al. (2020) developed an FCONet profound learning model to adequately analyze COVID-19 pneumonia in a brief period. The identification cycle with Res-Net 50 consolidated gives a superb analytic exhibition. They embraced an AI-based picture preparing grouping utilizing the Convolution neural organization system. The framework engineering is finished utilizing the openly accessible COVID-19 dataset set up by SIRM. Albeit the freely accessible datasets are useful for the forecast cycle in the underlying stages, the quickly refreshed experiments and the passing pace of patients with various physiological commitments made conflicting outcomes on the expectation interaction. The datasets considered are CT sweeps of individuals contaminated by cellular breakdown in the lungs, pneumonia, and lung pictures of ordinary patients.

3. SYSTEM DESIGN

The problem statement behind the system is considering the dataset alone making the decision in the previous works impact the accuracy. Hence to improve the accuracy a secondary decision is required. The proposed system focused on implementing hybrid decision model where the final classification depends on the previous decision model. The system design is implemented using MATLAB work environment. Signal processing toolbox is used to filter out the noise and handle the signal data in a proper way.

4. DESIGN METHODOLOGY

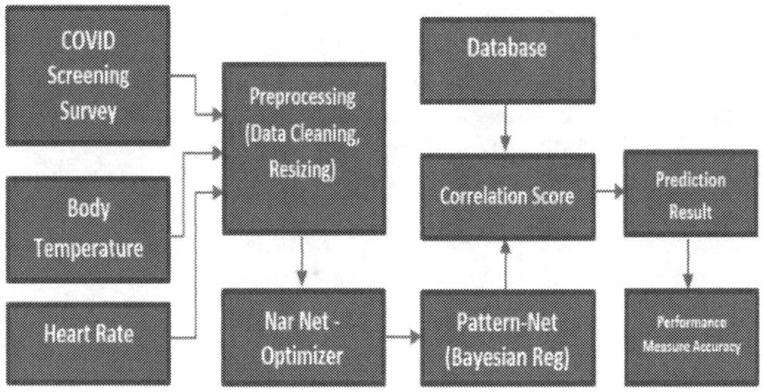

Figure 2. System Architecture of Proposed COVID screening model

4.1 Input Dataset

The input dataset is derived in the form of two methods.

- Through COVID screening survey
- Using Physiological data (Heart Rate, Temperature)

These two datasets are used as the primary data for the analysis. The preprocessing of the consists of applying the signal filter using Wavelet transform. The haar wavelet transform is used to determine the frequency of occurrence of the input data. The frequency constant is randomized before fetching it into the analysis module.

4.2 Analysis Module

The analysis module consists of two levels of algorithm that is used to produce the cascaded results. The first module is used to make a closed loop auto regression network that adopt the input data compare the same with the input as the feedback and have the optimized data. The MARX module is fetched with the COVID Survey data. The cross validation is done using the manual analysis of the Survey data.
The secondary analysis is done using the Pattern network designed using Bayesian regularization algorithm (BR). The proposed BR model is one of the effective optimized technique that adopts the back propagation model. The physiological data is split into training data of 70%, testing data of 15% and Validation data of 15%. The novel Hybrid COVIDNET is created using the cascaded decision model of NARX network and BR-pattern network. These two decisions are combined in final cascaded decision of COVID symptoms present or not. In case of symptoms the medications are updated in the warning dialog box as shown in SEC V.

5. RESULTS AND DISCUSSIONS

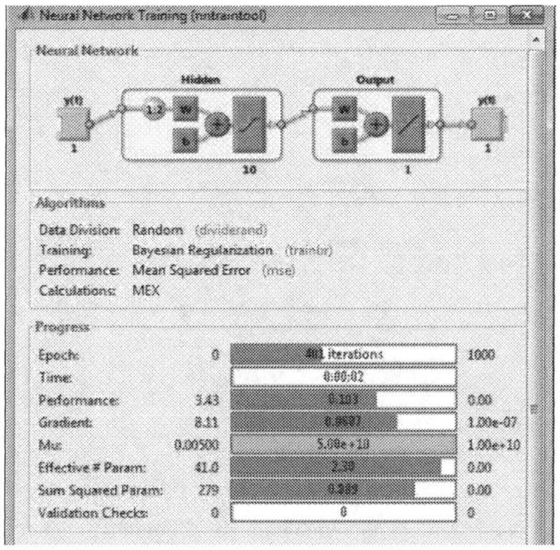

Figure 3. BR NN based neural network model

Fig 3. Shows the Configured neural network toolbox with training function as Bayesian regularization(trainbr). The performance measure is done using mean square error(MSE). The given input is compared with the training dataset using the 10 hidden layers defined in the nntraintool. The maximum number of epochs allowed are 1000. Depends on the complexity of the input, the iterations or epochs are limited.

(Fig 4.) Based on the given inputs and the expected targets the error rate is being calculated as Error = Target-output. Lesser the error value, then fairer the performance would be. The proposed model achieved the error rate of -0.00617 that act as the negligible value.

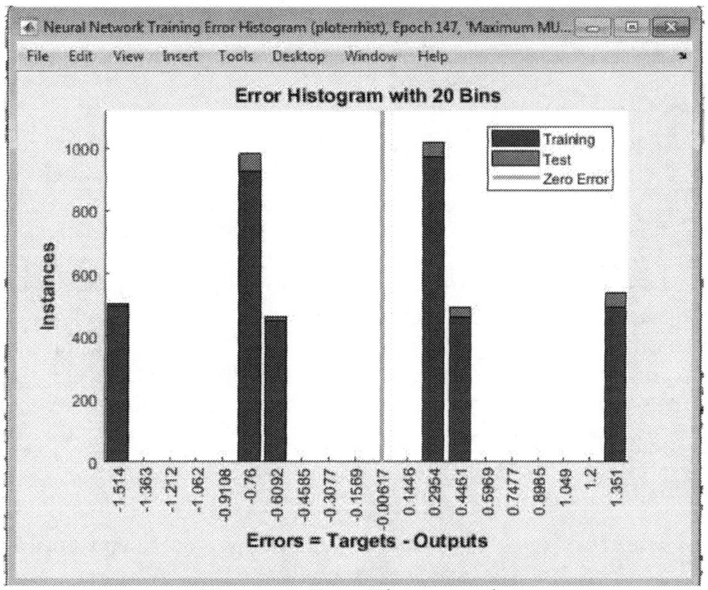

Figure 4. Error Histogram plot

Journal of Physics: Conference Series **2040** (2021) 012015 doi:10.1088/1742-6596/2040/1/012015

5.1 Notifications

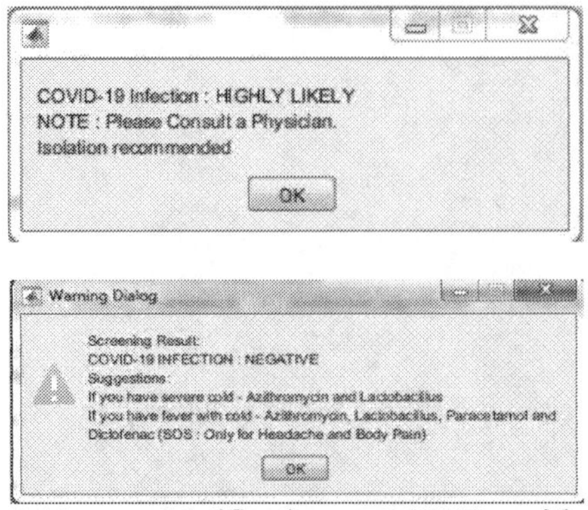

Figure 5. Message Notification on COVID positive & Negative

Fig 5. Shows the message box showing the notification on COVID positive status and negative status with the related suggestion provided by the screening algorithm.

5.2 Quantitative Measurements

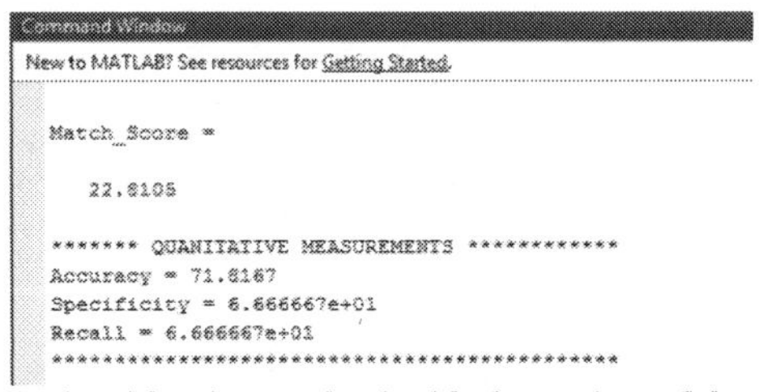

Figure 6 algorithm improvised with the tuning of data inputs

Since the proposed algorithm is derived with the help of two hybrid combination of the algorithms, the accuracy is likely to be 71.8%. The algorithm need to be further improvised with the tuning of data inputs too

6. CONCLUSION

COVID 19 has become one of the serious infection all over the world that leads to global pandemic. One of the foremost step required by the government is to control the transit of the disease and to detect the COVID infections at the earliest. Keeping those things as a motivation factor, the proposed system is developed. The system is focused on implementing a basic screening test of COVID infections and the real time physiological data analysis module using the Hybrid COVIDNET that might be helpful in detecting the infection by clinical screening and analysis. The system is further improved by implementing the same using more real time data and more likely screening questions using the word cloud from the web. The system achieved the error rate of -0.00617. Using deep learning algorithms the system could be further developed towards high accuracy.

REFERENCES

[1] Sohan, M.F. (2020). So You Need Datasets for Your COVID-19 Detection Research Using Machine Learning? ArXiv, abs/2008.05906.

[2] Vaid, S., Kalantar, R. & Bhandari, M. Deep learning COVID-19 detection bias: accuracy through artificial intelligence. International Orthopaedics (SICOT) 44, 1539–1542 (2020). https://doi.org/10.1007/s00264-020-04609-7

[3]Narinder Singh Punn, Sanjay Kumar Sonbhadra, Sonali Agarwal COVID-19 Epidemic Analysis using Machine Learning and Deep Learning Algorithms, , medRxiv 2020.04.08.20057679; doi: https://doi.org/10.1101/2020.04.08.20057679

[4] Amutha devi, Gayathri Monicka.J, "Emerging bio-medical applications and open research challenges in cognitive internet of things (CIOT)" International Journal of Pharmacy and Technology" Dec-2016 Vol. 8 Issue No.4 5049- 5054

[5] Shuo Wang, Yunfei Zha, Weimin Li, Qingxia Wu, Xiaohu Li, Meng Niu, Meiyun Wang, Xiaoming Qiu, Hongjun Li, He Yu, Wei Gong, Yan Bai, Li Li, Yongbei Zhu, Liusu Wang, Jie Tian A Fully Automatic Deep Learning System for COVID-19 Diagnostic and Prognostic Analysis
European Respiratory Journal 2020; DOI: 10.1183/13993003.00775-2020

[6] Zhang, Hai-tao et al 2020 Automated detection and quantification of COVID-19 pneumonia: CT imaging analysis by a deep learning-based software,European Journal of Nuclear Medicine and Molecular Imaging. 47. 10.1007/s00259-020-04953-1.

[7] Amutha devi, Gayathri Monicka.J, "Development of fuzzy approach to predict the fetus safety and growth using AFI" Journal of super computing" Dec-2019 DOI 10.1007/s11227-019-03099-8

[8] Tamina COVID-19 detection and disease progression visualization: Deep learning on chest X-rays for classification and coarse localization
Tahmina Zebin1 • Shahadate Rezvy2 (2020)

[9] Kumar, R., Nagpal, S., Kaushik, S. et al. COVID-19 diagnostic approaches: different roads to the same destination. VirusDis. 31, 97–105 (2020). https://doi.org/10.1007/s13337-020-00599-7

[10] Ghafoor, Kayhan (2020): COVID-19 Pneumonia Level Detection using Deep Learning Algorithm. TechRxiv. Preprint. https://doi.org/10.36227/techrxiv.12619193.v1

[11] D.Vijendra Babu et al 2020 Raspberry Pi based Real Time Vehicle Accident Intimation with Alert Message and Image Capture, *IOP Conference Series: Materials Science and Engineering, Vol.*993 012081

[12] D.Vijendra Babu et al 2020 Remote Monitoring of COMA Patients Using Zigbee, *IOP Conference Series: Materials Science and Engineering, Vol.*993 012082

[13] D.Vijendra Babu et al 2020 Line follower Robot & Obstacle detection using PID Controller,AIP Conference Proceedings,Vol. 2271,030005.

[14] D.Vijendra Babu et al 2020 Intelligent Street lightning using Traffic & ambient lightning, AIP Conference Proceedings,Vol. 2271,030008.

[15] V.Chandrasekar et al 2021 Designing of Tele-Health Smart Sensor Device to assist Home care Staff, Journal of Physics: Conference Series, *Vol.1964, 062106*

[16] R.Prabha et al 2021 A Novel Computational Rough Set based Feature Extraction for Heart Disease Analysis, First International Conference on Computing, Communication & Control System.

Design of Hybrid Deep Learning Approach for Covid-19 Infected Lung Image Segmentation

R.Prabha[1], M.Ramkumar Prabhu[2], SU.Suganthi[3], S.Sridevi[4], G.A.Senthil[5], D.Vijendra Babu[6]

[1,3]Associate Professor, [2,6] Professor, [4]Assistant Professor,[5] Research Scholar
[1,2,6] Department of Electronics & Communication, [3]Department of Computer & Communication, [4]Department of Computer Science & Engineering
[1,3]Sri Sairam Institute of Technology, Chennai, India.
[2]PERI Institute of Technology, Chennai, India.
[4,5]Vels Institute of Science, Technology & Advanced Studies, Chennai, India.
[6]Aarupadai Veedu Institute of Technology, Vinayaka Mission's Research Foundation, Deemed to be University, Tamil Nadu, India.

Corresponding author E-mail: r.praba05@gmail.com

Abstract. Lung infection or sickness is one of the most common acute ailments in humans. Pneumonia is one of the most common lung infections, and the annual global mortality rate from untreated pneumonia is increasing. Because of its rapid spread, pneumonia caused by the Coronavirus Disease (COVID-19) has emerged as a global danger as of December 2019. At the clinical level, the COVID-19 is frequently measured using a Computed Tomography Scan Slice (CTS) or a Chest X-ray. The goal of this study is to develop an image processing method for analyzing COVID-19 infection in CTS patients. The images in this study were preprocessed using the Hybrid Swarm Intelligence and Fuzzy DPSO algorithms. The findings suggest that the proposed method is more dependable, accurate, and simple than existing methods.

Key words: Fuzzy logic, Classifier, Swam Intelligence

1. Introduction

The COVID-19 pandemic, according to the World Health Organization, is an infectious disease that has infected millions of people worldwide and killed thousands of people since December 19, 2019[2]. Because of the pandemic's widespread impact, COVID-19 offers a significant challenge to medical professionals.COVID-19 preparation and response must include rapid diagnosis and contact tracing in order to prevent the virus from spreading further[4].As the number of new cases increases, particularly those requiring critical care, healthcare providers can use disease monitoring to make important treatment decisions. COVID-19 is a widespread disease that claims the lives of thousands of individuals every day. Early detection of this issue has proven to be one of the most effective strategies for infected tree cutting [8].

The rising number of COVID-19 patients is putting a strain on many countries' health-care systems. As a result, having a reliable automated approach for identifying and measuring infected lung regions would be invaluable. The creation of a system for linguistically segmenting medical lung scans of COVID-19 patients would help with the quantification of anomalies and research in this area [7]. It would aid front-line responders in better managing the situation of overwhelmed hospitals during the pandemic. While CT is a viable approach for diagnosing COVID-19, it has some disadvantages that make it impractical to utilise on a daily basis: CT scans are not widely available, take a long time to complete, and require patients to be transferred from their unit [1]. It's logistically challenging to use CT technology safely during a pandemic, and it can deplete available resources.

Even when adequately cleaned, CT scanners can be a source of infection for other patients who require imaging. CT scans are used in healthcare facilities to speed up the image acquisition and categorization process. However, a professional medical practitioner is required to verify the final results, which adds to the computation time. On the other hand, supervised learning models can be used to classify patients from CT images [6].On average, classification algorithms based on Machine Learning (ML) produce the highest reported accuracy rates. Machine learning models require a lot of processing power and high specifications to run because of their high categorization accuracy rates. Large hospitals in first-world countries may be able to afford this high-cost processing method, while hospitals in impoverished countries and rural areas may not. To reduce the computational cost, a machine learning model that consumes fewer resources while losing accuracy is necessary [8].

Machine learning researchers are seeking to battle the pandemic by putting together databases and developing algorithms that learn from them [10]. A vast number of research studies have been launched to try to improve diagnosis and predict the virus's propagation, number of fatalities, and genetic evolution. This could aid officials in determining how the virus spreads and the locations of quarantine zones. For COVID19 lung infections, a thoracic CT scan is used as a diagnostic technique in hospitals. By creating methods for CT image processing, the current hot study hopes to contribute to the automatic detection of the coronavirus. According to the researchers, algorithms that were used to identify lung cancer and lung collapse using X-ray images may also be useful for finding abnormal cases in COVID-19 patients [9]. As a result, stronger algorithms are still required.

1.1 Processing Techniques of Deep Learning

In the Medical profession, Machine Learning Technologies [13, 14] are gaining traction for diagnostics, predictive analytics, and general research. For the diagnosis of COVID-19, four distinct deep CNN architectures were built and tested on chest X-ray images. These models require fewer large training sets because the weights have already been pre-trained on the ImageNet database [3].Because data is frequently unlabeled or difficult to obtain, unsupervised learning networks are critical in the medical business. Unsupervised networks like Self-Organizing Feature Maps (SOFM) can be used to train unlabeled data. With a mean Euclidean distance of 1.1 between the first and second winning neurons in the testing set, the SOFM network was utilized to classify COVID-19 patients' chest x-ray photos and identified a strong differential between sick and healthy patients [5].

It can also show which properties in the input space influenced the classification the most, which can be used to evaluate the significance of features in an unsupervised network. As proven in this paper, unsupervised learning can extract features from medical data, such as COVID-19 patients' chest x-rays, while also successfully recognizing the image [6].The technique is given, which uses strong 2D and 3D deep learning models to adapt and alter current Artificial Intelligence (AI) models while incorporating clinical expertise. A 3D volume review and a Corona score are utilized to evaluate the system's efficacy in recognizing suspected COVID-19 thoracic CT characteristics and monitoring disease development in each patient over time.

The classification findings for Coronavirus versus Non-coronavirus infections per thoracic CT examination were 0.996 AUC on datasets of Chinese control and infected patients [10].This study presents a poorly supervised deep learning strategy for detecting and classifying COVID-19 infection from CT scans. Infection detection and discrimination between COVID-19 and non-COVID-19 patients is accurate, and manual CT image labelling is no longer required. Based on the positive qualitative and quantitative outcomes obtained, it envisioned a wide

deployment of our established methodology in large-scale clinical research [5].The use of artificial intelligence to analyze Chest X-Ray (CXR) photos for COVID-19 identification and clinical triage is becoming increasingly relevant in the aftermath of the global COVID-19 epidemic. Due to the pandemic's dynamic nature, systematic CXR data collecting for deep neural network training is problematic.
For COVID-19 detection, a patch-based convolutional neural network technique with a small number of trainable parameters has been developed. Our statistical analysis of the possible imaging indications of CXR radiographs inspired the proposed technique [4].This study focused primarily on the CT for the evaluation since it provides more clear information and more judgment accuracy than a chest X-ray. Because current methods for detecting the virus need the presence of a skilled radiologist, automating detection would be necessary to save radiologists' assessment time [5]. Machine Learning (ML) and Deep Learning (DL) algorithms have lately made major advancements in autonomously diagnosing diseases, lowering the cost and increasing the accessibility of diagnostics [6].This study focused primarily on the CT for the evaluation since it provides more clear information and more judgment accuracy than a chest X-ray. Because current methods for detecting the virus need the presence of a skilled radiologist, automating detection would be necessary to save radiologists' assessment time [2]. Machine Learning (ML) and Deep Learning (DL) algorithms have lately made major advancements in autonomously diagnosing diseases, lowering the cost and increasing the accessibility of diagnostics.

2. Methodology

The methodology of the paper is followed as which includes objective, explains the system model, describes the results and discussion followed by conclusion.

2.1 Objectives of the Paper

- The aim of this paper is CNN-based segmentation is proposed with the goal of employing more meaningful information to improve lung segmentation. This is prompted by the possibility of improved performance in general automatic lung segmentation systems, which are critical for a variety of medical and scientific applications.
- A new approach called Dropout CNN classification is applied to detect lung abnormalities automatically, which can improve the prediction accuracy of covid-19
- To improve the PSNR ratio and SSIM index, a Lee filter with equalization is utilized to enhance the pictures in order to reduce noise

2.2 System Model

Image segmentation is a critical step in the image analysis process. The purpose of developing a Convolutional Neural Network for segmentation is to use more meaningful information to improve lung segmentation. The promise of better performance in general automatic lung segmentation systems, which are crucial for a variety of medical and scientific applications, has encouraged this. Lung abnormalities are now only diagnosed by imaging after the development of neurological symptoms. The CT scans are interpreted by doctors to see whether there are any anomalies. In other circumstances, however, discrimination, decision-making, and diagnosis are extremely difficult for clinicians. Misdiagnosis and incorrect treatment approaches place a significant financial burden on the patient, diminish patient comfort, and result in conditions that are irreversible. As a result, we're going to show you how to use Dropout CNN classification for automatic covid-19 shown in Table 1.

2.3 Preprocessing

Due to preprocessing, automated COVID-19 lungs pollute area segmentation and measuring systems can better understand what's going on in the images. Bright areas in the original CT-scan images are used to introduce a lot of visible data. Some places, however, are overly bright, while others are overly black. To get a more exact segmentation [11, 12], it is required to improve local contrast before classifying details. Positive local contrast is seen in COVID-19 targets, indicating that the lesion areas are lighter in all directions than the surrounding backdrop. The proper brightness level is controlled using an exponential and logarithmic function. Selection of Reinforcement

2.4 Hybrid Swarm Intelligence and Fuzzy DPSO

Some of the assumptions in Discrete Particle Swarm Optimization (DPSO) are spawning a swarm, favorable adaptation, and negative adaptation. Simple ideas are executed in a way that resembles natural selection. When a swarm is exposed to a longer time of survival, it has a higher chance of producing progeny. The swarm's life span will be prolonged if it finds a more suitable state, such as favorable adaptation, but it will be reduced if it fails to find a suitable state, such as unfavorable adaptation. The fuzzy c-means clustering algorithm is effective, however the random selection of center points reduces the iterative process to the local best answer. Researchers have proposed numerous changes to the FCM in order to improve its convergence during the previous few years. The ideas of Fuzzy and DPSO are combined to create Fuzzy DPSO which is illustrated in Figure 1.The algorithm's junction rate is managed using fractional calculus, and swarms of Fuzzy DPSOs compete using Darwin's survival-of-the-fittest principles. By applying these concepts, the particles are spared the challenge of local minima. At the same time, several PSO algorithms are run.

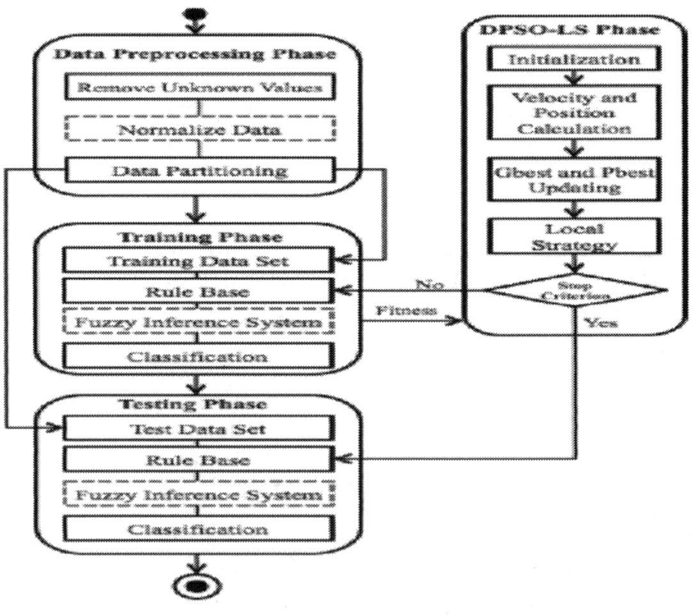

Figure 1. Fuzzy DPSO process

2.5 Swarm Intelligence Algorithm

- Initialize particle array (PA) and velocities (Vi).
- Allow the particle to traverse at Vi=Vmax.
- Find new global fitness.
- Check
- If µmin decreases, the neuron is deleted. µmin <m<=µmax
- if the maximum critical threshold Ct exceeds, a particle is detected from the input at scheduled time.
- Input collection is maintained.

2.6 Fuzzy DPSO Procedure

- In the beginning, a local optimum is sought in a certain area.
- If it can't be found, it's simply ignored, and local optima in other places are sought.
- Swarms that pass all of the tests for locating local optima sites have a longer life span, while particles that fail have a shorter one.

2.7 Dropout CNN Classifier

To segment lung images, the Dropout CNN classifier is employed, and the CLAHE algorithm is used to improve the lung picture. The CLAHE approach separates an original image into non-overlapping contextual sub-images, tiles, or blocks. Block Size (BS) and Clip Limit (CL) are the two most important elements of the CLAHE (CL). These two properties are significantly responsible for the improvement in image quality. Increased CL brightens the image and flattens the histogram because the input image has a low intensity. The image's dynamic range, as well as its contrast, increases as the BS increases. The two parameters determined at the location with the biggest entropy curvature, using the image's entropy, produce subjectively good image quality.

2.7.1 Procedure

- Create non-overlapping contextual zones using the original intensity image.
- Calculating the histogram of each contextual zone using the grey levels in the array image.
- The CL value is used to calculate the contextual region's contrast limited histogram.
- Re-distribute the remaining pixels till they're evenly distributed.
- The Rayleigh transform is used to improve the intensity values in each zone.
- Taking steps to mitigate the impact of an unexpected change.
- The new grey level assignment of pixels within a sub-matrix contextual region is produced using a bi-linear interpolation between four alternative mappings to eliminate border artefacts.

The CT image is converted from RGB to YIQ Colour space via linear transformation, and subsequently to HSI Colour space via nonlinear transformation in this method. In the YIQ and HSI Colour spaces, chromatic and brightness information are independent. Second, the brightness information is used to improve contrast while Rayleigh CLAHE is used to keep the chromatic information. Dropout is implemented per layer in a neural network. It works with a variety of layers, including dense fully connected layers, convolutional layers, and recurrent layers like the long short-term memory network layer. Dropout can be utilized on the visible or input layer, as well as any or all of the network's hidden layers. It isn't used on the output layer. The likelihood of the layer's outputs being dropped out or maintained is controlled by a new hyper parameter. We trained dropout neural networks for classification challenges on data sets from a variety of domains. We discovered that using dropout improved generalization p when compared to neural networks that did not use it.

3. Results and Discussion

The Kaggle dataset was provided by the Radiological Society of North America (RSNA). Hundreds of pictures of healthy and pneumonia-infected people remain in this collection, which are useful for this study. To generate a training dataset, images from those two datasets must first be downloaded. The desired photographs are then picked and saved to the appropriate location. Images are then loaded and preprocessed before being sent into the training process by being turned into numpy arrays of the necessary size. It is vital to set up the study's working environment before collecting data. Colab (Google Colaboratory) is a free cloud-based Jupyter notebook environment that requires no installation. Colaboratory allows users to build and execute code as well as access advanced computational resources directly from their browser. Colab, in particular, generously provides GPU, which significantly accelerates the computationally intensive training process. For these reasons, Colab has become highly popular among Deep Learning and Data Science enthusiasts who may not have access to a PC with expensive GPUs. To begin, we show that our method works on a COVID-CT-Dataset, which contains COVID-19-positive CT scans as well as ground-truth lesions that a radiology specialist manually recognizes. Figure 2 depicts the sample Covid-19 lung images. Second, we examined newly released CT images from a nearby hospital that contained individuals who had tested positive for the coronavirus. Finally, the COVID-19 lesion and its repercussions on the patient's lungs are depicted in three dimensions. The performance metrics compared in Table 1.

Performance Metrics	MIS method	Proposed method
Accuracy	0.843	0.99
Specificity	0.967	1
Sensitivity	0.854	0.99
FMeasure	0.812	0.12
Precision	0.914	1
MCC	0.917	0.97
DICE	0.923	0.981
JACCARD	0.891	0.960

Table 1: Performance Metrics comparison

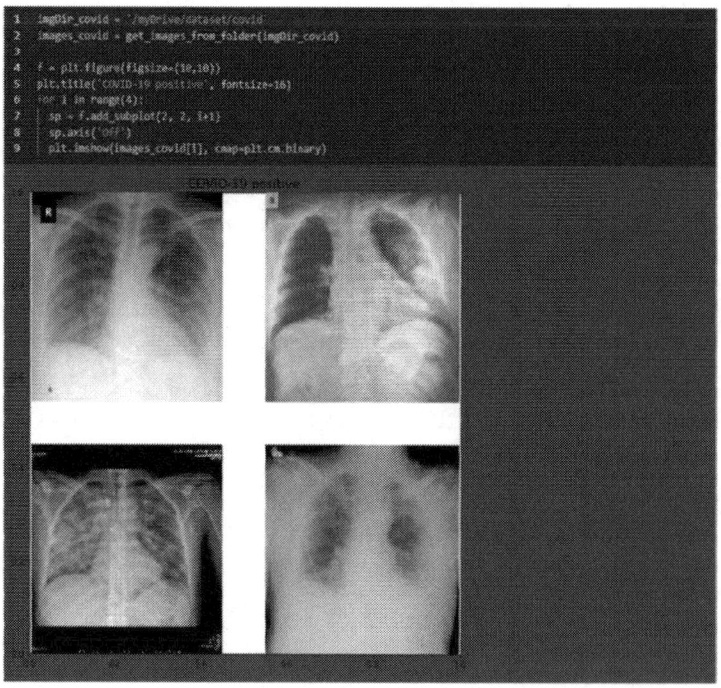

Figure 2. Covid 19-Chest X-Ray images

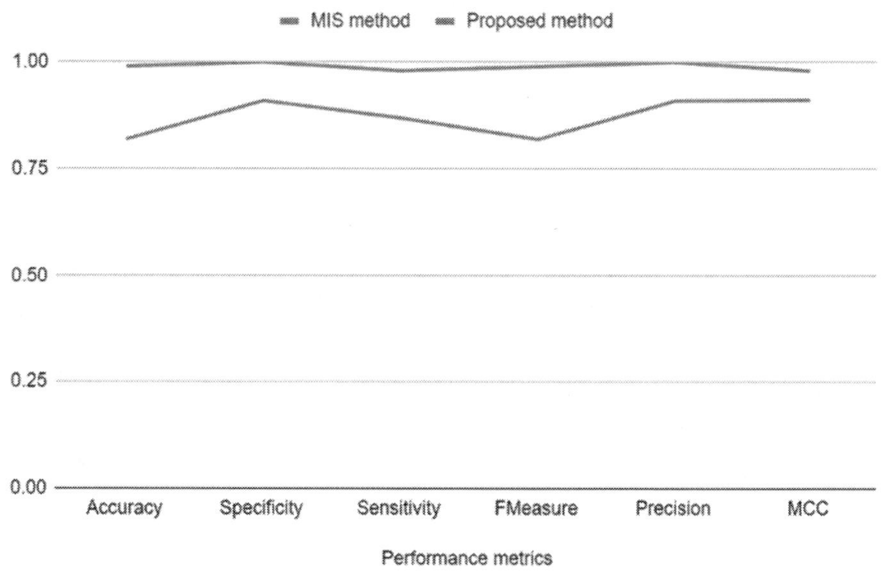

Figure 3. MIS Vs. HFDPSO method

Lung (Vx)	In cm	lesion	In(cm)	ratio
109530	2897	75149	1988.3	68.61
240827	6371.8	43752	1157.6	18.167
94919	2511.3	10578	279.87	11.14
38726	1024.6	20734	548.58	53.54
37637	995.8	30477	806.3	80.97
231634	6128.6	42215	1116.9	18.22
38372	1015.2	8920	236	23.24
151116	3998.2	7632	201.9	5.05
232166	6142.7	36725	971.6	15.81
44204	1169.5	7892	208.8	17.85

Table 2. Summary of Quantitative Segmentation Statistics

To analyze and determine the performance of the recommended segmentation approach, the statistical values of the segmented COVID-19 lesion are compared to the results of the Medical Image Segmentation (MIS) Methodology as shown in Table 2 and Figure 3. The performance of the provided algorithms was evaluated using generally established assessment scores. One of our work's primary characteristics is the capacity to assess the COVID-19 lesion, show the polluted region, and track disease changes in real time. Furthermore, the proposed method may detect aberrant regions despite the low-intensity contrast between lesions and healthy tissues. Even if our suggestions were effective, there are still disadvantages to be aware of. We also want to improve the algorithms so that lesions like ground-glass opacity, crazy paving, and consolidation can be identified. To improve COVID-19 detection, diagnosis, and assessment, we want to combine imaging data with clinical markers and laboratory testing results. COVID-19 continues to spread over the world in an uncontrollable and unpredictable manner.

4. Conclusion

Until date, CT-scan imaging has been a widely used, low-cost, comprehensive screening method that efficiently aids in the visualization and rapid assessment of COVID-19 lesion severity. We assessed the efficacy of an automated technique for COVID-19 lung infection segmentation and quantification using chest CT data. The goal of this project is to build and analyze automatic COVID-19 Lung Infection segmentation and quantification using Hybrid Swarm Intelligence and Fuzzy DPSO for chest CT images. According to extensive computer simulations, our end-to-end learning strategy for CT image segmentation with image enhancement is more efficient and adaptive than the Medical Image Segmentation (MIS) method.

5. References

[1] Abdullah Farid et al 2020 A Novel Approach of CT Images Feature Analysis and Prediction to Screen for CoronaVirus Disease (COVID-19), *International Journal of Scientific and Engineering Research.*

[2] O. Gozes et al. 2020 Rapid AI development cycle for the coronavirus (COVID-19) pandemic: Initial results for automated detection & patient monitoring using deep learning CT image analysis,*arXiv preprint.*

[3] Gozes et al 2020 Coronavirus Detection and Analysis on Chest CT with Deep Learning, *arXiv preprint.*

[4] Herath et al 2020 Deep Learning Approach to Recognition of Novel COVID-19 Using CT Scans and Digital Image Processing, *Research Square.*

[5] B. King et al 2020 Unsupervised Clustering of COVID-19 Chest X-Ray Images with a Self-Organizing Feature Map,*IEEE 63rd International Midwest Symposium on Circuits and Systems.*

[6] Purohit et al 2020 COVID-19 Detection on Chest X-Ray and CT Scan Images Using Multi-image Augmented Deep Learning Model. 10.1101/2020.07.15.205567.

[7] Shaoping Hu et al 2020 Weakly Supervised Deep Learning for COVID-19 Infection Detection and Classification from CT Images,*IEEE Access*,Vol. 8.

[8] Tello-Mijares et al 2021 Computed Tomography Image Processing Analysis in COVID-19 Patient Follow-Up Assessment, Journal of Healthcare Engineering,

[9] Thulasimani Thangamani et al 2021 IoT Defense Machine Learning: Emerging Solutions and Future Problems, Microprocessors and Microsystems.https://doi.org/10.1016/j.micpro.2021.104043

[10] Amutha devi, Gayathri Monicka.J, "Development of fuzzy approach to predict the fetus safety and growth using AFI" Journal of supercomputing" Dec-2019 DOI 10.1007/s11227-019-03099-8

[11] V.Deepak et al 2021 An Efficient Performance Analysis using Collaborative Recommendation System on Big Data,5th International Conference on Trends in Electronics and Informatics.

[12] R.Prabha et al 2021 A Novel Computational Rough Set based Feature Extraction for Heart Disease Analysis, First International Conference on Computing, Communication & Control System.

[13] C.Karthikeyan et al 2021 Stock Prices Prediction Using Machine Learning Technique, 6th International Conference on Inventive Computation Technologies.

[14] R.M.Balajee et al 2021 Requirements Identification on Automated Medical Care with Appropriate Machine Learning Techniques,6th International Conference on Inventive Computation Technologies.

Detection and prevention from different attacks in VANET: A Survey

Arun Singh Kaurav[1], Sushama Rani Dutta[2]

[1]Research Scholar, [2]Assistant Professor, Department of Computer Science & Engineering, KL Deemed to be University, Hyderabad, India

E-mail: aru.esoft678@gmail.com, sushamadutta@klh.edu.in

Abstract. Routing attacks will have distressing effects over the network and bequest a significant challenge once planning strong security mechanisms for vehicular communication. In this paper, we examine the effect and malicious activities of a number of the foremost common attacks and also mention some security schemes against some major attacks in VANET. The attacker's aim is only to modify the actual route or provides the false data about the route to the sender and also some attackers are only flooding unwanted packets to consume resources in available network. Various routing approaches are also mentioned in the paper because the routing of data is very important to deliver the traffic information to leading vehicles. It's advised that a number of the ways that to approach this made field of analysis issues in VANET might be to fastidiously design new secure routing protocols in which attacks are often rendered meaningless and because of the inherent constraints found in the network, there's a desire for light-weight and sturdy security mechanisms.

Keywords—VANET, Attack, Security, RSU, Nodes, Routing

1. Introduction

The wireless communication is the revolution in communication in between sender and receiver. The wireless network consists of towers, exchange offices, servers and mobile users [1]. The numbers of users are connected to tower/s and then the signals that are goes miles away [1,2]. The satellites also have very important role in communication [3]. In near living conditions associated with the location, the natural sensing process closes the sensor nodes and converts them into an electrical signal. Sensor's signal contains some properties about substance situated and events occurring within the vicinity of the mobile nodes. Wireless Mesh Network (WMN) is a very secure network since it links all the nodes within each other's range [4]. The main in WMN is extra overhead for establish link with all in the range nodes. In some applications involving unattended operations [2][4], an outsized range of those non-reusable sensors is often networked. A WSN includes many sensing nodes. These sensors are able to interact flexibly with each other or with the external base station (BS) [4].

In the area of wireless communication, VANET has been the important research area in wireless connection. The parent field of ad-hoc network is known as VANET. Attacks can also easily influence the open network and security restrictions must be established and stopped from attacking the network [4]. VANET is a part of MANET which without relying on the opposite infrastructure, organizes its own communication system [5]. The primary common application of MANET is a limited amount of knowledge exchange between various computers in the military due to its easy and basic communication mechanism. In distant and seldom hostile areas, it incorporates, by hand and gloves, space-dispersed autonomous small devices monitoring environmental or physical conditions. In recent advancements, development of micro -level and less costly sensors is feasible economically and technically. Before going into VANET, it is necessary to discuss its background related to the wireless communication i.e., shown in figure 1.

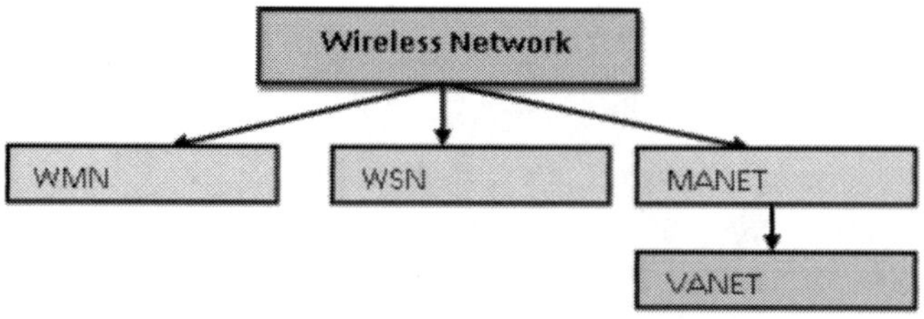

Fig. 1. Classification of Wireless Network

Along with certain improvements, VANET is analogous to MANET. VANET consists of handheld nodes, side-of-road units [6]. Remote nodes are devices embedded in automobiles called on-board data sharing networks between Road Side Systems. RSUs are fixed installed devices that links between the MN and the servers or the Internet for connectivity purposes. There are many services provided by VANET, but most notably, road safety services are distributed over the Internet to reduce road fatalities by data sharing.

Many applications are used by VANET to create contact between vehicles. It is possible to relay information about congested areas from one car to another vehicle, warning drivers of collisions or potential traffic jams. Drivers are told that they are heading to a place where an accident or traffic jams are occurring. When this information taken as an input for the system of vehicle navigation to escape a crowded area, the system often produces alternate routes. Real-time travel information regarding vehicle is obtained and analysed by the RSU [7]. First the RSU produces traffic information that includes data like average pace of the vehicle, vehicle density, and traffic jam incidents. Finally, it is transmitted to relatively distant vehicles by the RSU. After the traffic information is obtained, the vehicles evaluate it with road condition data.

2. Types of communication in VANET

Communication modes are of two types, which are differentiated as follows:
The first is vehicle to RSU or vehicle to Infrastructure (V-I) and second one is vehicle to vehicle (V-V) communication. In vehicle to infrastructure communication the road unit is needed for transmutation. The node or vehicle interacts with the RSU and collects information from the network. The V-RSU communication is more superior because it enhances the possibility reliable communication. The main target is to increase vehicle safety. Suppose, any vehicle detects an abnormal situation or condition, it can also inform other vehicles[8].

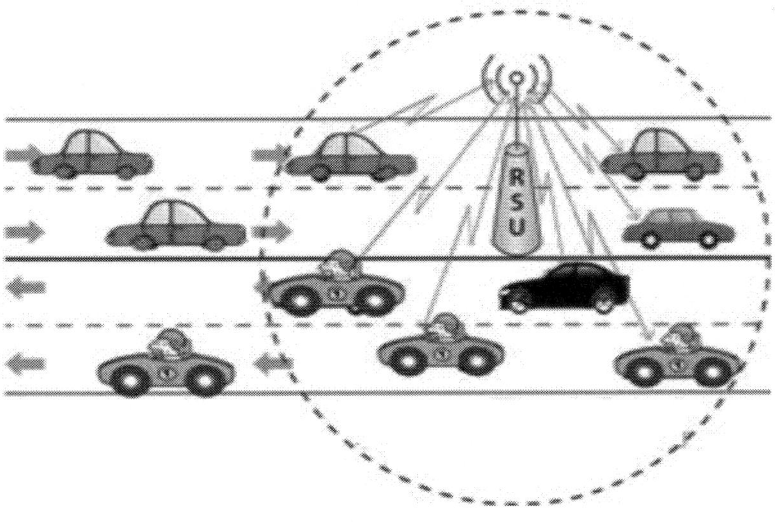

Fig. 2. Vehicle to RSU Communication

In vehicle-to-vehicle communication, as shown in figure 3, vehicles communicate between each another. While for Vehicle-to-RSU communication, vehicles form an Ad-Hoc network, called VANET mentioned in figure 2. VANET is a distinct kind of Mobile Ad-Hoc Network (MANET) instead of nodes which has vehicles. VANET is a self-organizing and both centralized and decentralized network composed of high-speed moving vehicle [7]. In vehicle-to-vehicle communication chances of malicious actions is more compared to Vehicle to RSU communication. The message or information is more accurate in V-RSU communication. The vehicles that are using the highway have no need to ensure traffic status because high mobility speed is there. It is also possible to gain traffic information at that time of also down the speed of vehicle/s. The communication is only in between the vehicles mentioned in figure 3.

Fig. 3. Example of Vehicle to Vehicle (V-V) Communication

Many applications are used by VANET to create contact between vehicles. Information on congested roads may be conveyed from one car to another vehicle, warning drivers about crashes or possible traffic delays. Drivers are advised they are going to a location where there is an accident or traffic congestion. The method also generates alternative routes as this information is used as an input for the system of vehicle navigation to avoid a crowded area. The real-time movement data of a vehicle is obtained and processed by the RSU [9]. First the RSU produces traffic information that involves data such as vehicle average speed, vehicle number, and traffic jam incidents. Finally, it is communicated by the RSU to comparatively distant cars. After receiving the traffic information, the cars evaluate it along with the road condition data.

3. Network Model

In VANET [10], network artefacts can be separated into three groups. These groups include servers for application and authorization, facilities on the road side, and nodes/vehicles.

3.1. Application and Authorization Servers

These are powerful workstations, responsible respectively for managing and providing service data. The authority knows all the keys and is accountable for maintenance planning. For cars, device servers provide operation details. The government or foreign operators will fund them. We assume there are powerful processing capabilities for authorization and application servers. So, here we ignored computation time.

3.2. Road Side Infrastructure

Road Infrastructure consists of power supplies located near roads and responsible for the collection and dissemination of data. Through wired networks, RSUs are connected to power and communicate via radio with vehicles.

3.3. Nodes/ Vehicles

Nodes or Vehicles are moves in the road and communication with the RSU or also their information exchange information is received by RSU in network. Every vehicle is presumed to be fitted with a differential GPS receiver with a meter-order accuracy and an on-board computer (OBU) [11] responsible for all communication and computing tasks

4. VANET Mobility Factor

Movement of vehicles are connected with cars, railways, bicycles, motorcycles and everything on the roads that moves with the wheels [12] [13]. For cars in the VANET, many factors influence their mobility, for example:

4.1. A. Street Construction

The highways, directions, traffic signals and signs on road determine traffic. Intersections and one or more roads, one-way or two-way, often affect it. In the simulation, intersections result in slower speeds, high knot density, and require to assign a probability value to predict each vehicle's rotation direction. Overtaking is banned on single-lane roads and speed is restricted and reliant on potential cars, whereas overtaking is permitted on multi-lane roads and driving is considered easier and safer.

4.2. Block Size

One of the small areas surrounded by streets can be considered. The size helps to assess intersections and how much the car decelerates and slows. Furthermore, streets with less intersections allow cars with more intersections to drive to higher speeds than city blocks. Vehicles are needed for slowing down more often and more accurate agility details will be considered.

4.3. Motion Control Mechanism

As mentioned in the first paragraph, streets differ in their designation. Traffic lights and stop signs are the major signs that consist predetermined location and help make any proposed mobility model more realistic [10]. They will summon the tail and slow down.

4.4. Interdependent Vehicle Traffic

Movement of vehicles effect the surrounding vehicles. Vehicle will be slow down, accelerate, change lanes, or even stop.

4.5. Average Speed

The higher the vehicle speed, the faster it will change position or location. In addition, speed limits affect the average vehicle speed. This can lead to changes in the connection to the network or network topology [11].

5. Routing Protocols in VANET

Routing protocols may be divided into several types based on their characteristics [14, 15]. Most popular method for differentiating VANET routing protocols depends on obtain and preserve routing information. Routing protocols are classed as follows:

5.1. Proactive Protocols

They monitor the networks topology all the times and the routes will be evaluated for all destinations continuously. Routes are stored by performing interchange of routing tables periodically in network, like the wired network. An advantage of this protocol is that routes must be decided and maintained in the buffer so whenever routing takes place these routes are used immediately so that the overall delay in the system must be reduced. To get the information of route and making a session is not very time consuming. This protocol has a disadvantage that whenever the topology changes, it reacts to those changes, though traffic is not affected, and results in the unwanted use of the bandwidth even there is not transfer of data. The protocols of this type are Optimized link state routing protocol and Destination sequenced distance vector routing protocol.

5.2. Reactive Routing

This protocol, often known as on demand routing protocols, only perform route determination on demand. In data transfer, reactive protocols select a route only at the start of a connection. It is done by launching the discovery of route by flooding it with packet known as route requests packets. once the route is created, the details are kept in the database of routing until the destination is unable to reach or the route expires. Because routing information is not updated on a regular basis, routing overhead is decreased dramatically during topology changes. The delay experienced during route finding is one of these systems' drawbacks. These protocols, however, must be particularly efficient for highly mobile networks. AODV i.e., Ad hoc On-demand Distance Vector routing protocol and DSR i.e., Dynamic Source Routing (DSR) are two routing methods that fall under this category.

5.3. Hybrid Protocols

The benefits of both proactive and reactive routing systems are combined in hybrid routing strategies. On-demand routing offers less routing costs than traditional routing, however it suffers from routing latency. Table-driven routing guarantees great quality in static topologies but cannot be used to mobile networks. Combining the best features of both, a few hybrids routing protocols have been devised, in which routing is launched with certain proactive routes and subsequently satisfies demand from other nodes via reactive floods. Hybrid protocols make use of hierarchical network designs. In a dynamic network, typical routing protocols of this category include Zone Routing Protocol (ZRP).

6. Types of attacks in MANET

It defines the intrusion as an action which tries to harm the main components of the security system [4]
1) The integrity
2) Availability of a resource
3) Confidentiality
In the same work, the intruder therefore was defined as a private or group of people who take the action in the intrusion. The plainness of the many routing protocols for wireless sensor networks makes them a simple target for the attacks [16] [17]. The routing attacks are classified into the subsequent categories are: -

6.1. Modified, or Replayed Routing Information
While sending the info, the data in transition could also be spoofed, altered, destroyed or resent. Due to short range communication of the nodes i.e., sensor nodes, an attacker having very huge power of the communication and the processing range can attack sensors parallelly and changes the sent information.

6.2. Selective Forwarding
In this reasonably attack the suspicious node may decline to forward every message it gets, acting as region or it could forward some messages to the incorrect receiver and simply drop others

6.3. Sinkhole Attack
The attacker aims to draw in all traffic. In case of a flooding-based protocol the compromised node may hear requests for routes, and then reply to the requesting node with messages restraining a bogus route with the shortest way to requested destination.

6.4. Sybil Attack
The malicious node will present itself as the multiple nodes. The attack of this kind aims to decrease the use and decrease the effectiveness of algorithms used. It is often performed against dispersed storage, routing, aggregation of data, correct resource allocation, voting and mischief detection [11].

6.5. Wormhole Attack
In this attack [12] the node that is malicious, tunnels messages from the one side of network on link, that normally not exists, to another side of network. The only sort of that kind of attack is to convince the nodes that they're neighbors. The attacks of this kind can be utilized with the select forwarding and eavesdropping

7. Literature Survey
The contribution of various researchers in field of VANET security is discussed in this section. These previous works are providing the idea of novel security scheme in VANET. Also found the research gad in recent work i.e. mention it in table 1.

Table 1 Related Work Overview

S.No.	Reference	Worked Done	Drawbacks
1	[18]	This research highlights key flaws for the protocols suggested for V-to-I and V-to-V communication and suggests an alternative protocol suite. Authentication of driver, key exchanges, data exchange, change of offline password, protocols for vehicle complain are part of our suite. V-to-I protocol for key exchange allows handoff capabilities to provide ongoing authentication when cars move from one RSU area to other.	In different groups, how they decide the direction of vehicles because no direction information not transfer with traffic request packets. Performance measure in average speed of vehicles but actually vehicles speed is random.
2	[19]	In this research focus is on hybrid device to device message authentication (HDMA) technique for the 5G-supported VANETs is suggested, which employs a unique group signature-based technique for the mutual authentication between V2V communication. In addition, to decrease the calculation overhead of the modular exponentiation process, an already calculated lookup database is used.	Storage or computation overhead for revocation is high. In city scenario security is important but on highways guidelines available.
3	[20]	This research developed a Swarm-based technique for detecting routing attacks on VANET networks. The method is based on IWD (intelligent water droplets), and the confidence model is applied. In dynamic moving networks, where nodes move at fast rates and the network architecture is unstable, IWD offers the following advantages: it is employed in systems that can adapt to shifts and have a highly crucial high processing speed. Trust is the most important aspect in establishing a trusted vehicle environment that promotes safety in car networks.	Trust based scheme is based on bandwidth and delay. But due to high packet dropping delay and bandwidth shows satisfactory values. How to reach to malicious node is not mentioned. The attacker identification is not possible because trust is calculated at receiver end and here the receiver is attacker.
4	[21]	This method outlines the suggested safety method for vehicle communications protection in ad hoc vehicle networks. The architecture of the security is called hybrid key cryptography, specially designed to defend VANET in a unique vehicle communications safeguard manner. The private key cryptography public key cryptography used to encode and decode facts distinguishes this security method. For security, AES and RSA are employed.	Both techniques are old and enhance overhead in dynamic network. Never identified dropping but performance of key exchange secure communication. The detection of attacker and prevention are not elaborated.
5	[22]	The author focuses largely on the hole formation assault in this research, in which malevolent or attacker drivers inside the network shatter the links by lowering or increasing their own speed to generate holes. The Stable Routing Protocol (RRP) protect messages from hole formation attacks between the source and destination. To protect against the vulnerability created by the VANET malicious drivers, RRP comprises of an attacker node or normal node security module and a recovery module.	From attacker nodes, loss is not mentioned and also number of attacker nodes which are flood the wrong information in traffic. The overhead of control also is not calculated which consist the request packets from attacker and reply information of packet.
6	[23]	The initial purpose of this research is to prevent erroneous information management concerns and the incompatible aims of protection and traceability. They suggest identity-based batch substantiation (IBV) techniques to provide anonymous verification, message integrity, anonymity, and traceability. The suggested IBV system may be utilized for both V2I and V2V communications. Second, the IBV system can handle our specified problems, such as forging, anti-traceability assaults, and loss of identity privacy.	The vehicles in network are always accelerating with random velocity but here the performance is evaluated in different fixed speed scenarios. The signature generation scheme is enhancing the overhead in communication then the better is to follow the accurate accepting message guidelines or standard communication guidelines.
7	[24]	In this research a hybrid architecture, consisting of automobiles, boulevard side units (RSUs), a server, and a trust authority (TA), is described, allowing for feedback storage and integrity ranking computation. Furthermore, vehicle communications serve as the foundation for platoon service requests and platoon management. The framework, in particular, is based on the Dirichlet distribution, which ensures great accuracy and dynamics.	Volume of traffic information packets is dropped because of presence of attacker or other issues. Then how to differentiate the attacker drop or other drop measurement. The dependency on platoon head vehicles is more. That means at initial level, if the platoon is fail then the further security will also not sure effectively work against attacks.

8. Optimal Route Selection in VANET

The optimized routing algorithm or route section in Vehicular ad-hoc network (VANET) is a critical issue, but some researchers have proposed some concepts of optimal routing [25][26]. There may fields in VANET on which they already authors doing hard work but without routing, it is challenging to transfer the traffic information to the far vehicles. so, optimal routing approaches also depends on the specific stream in the research field. This paper focuses on the authentication techniques in VANET, so optimal routing provides better performance but the security approach role is essential. The same is with other streams like congestion, location tracking and resource sharing.

9. Conclusion and Future work

In VANET, mobile nodes are equipped by the wireless technology and acting as computing nodes will move on roads. Network is self-organizing and provides better communication and coordination in vehicles by already maintained interaction among vehicles (V - V) and among vehicles (V - RSU). The vehicles are easily misguided by malicious vehicles because they only flood the incomplete data to leading vehicles. The procedure of VANET communication is completely different from MANET. In our paper, the focus is to how possible to deliver the complete traffic information in VANET. The overview is required to observe that what is needed to make the vehicle's communication more effective in a dynamic environment. The attacks and its role are completely different but the aim is only one i.e., to lost all valuable traffic information. The RSU is able to find the misbehaviour. Previous work promotes the new idea of modelling the security algorithm for secures vehicle communication. In previous work, the different security schemes were used against the attacker and almost all are very effective to prevent the network from attacks in VANET. IOT devices with Machine learning model can bring very big change in the VANET in future to make the process faster and more effective [27,28,29]. In the futuristic approaches, possible Artificial Intelligence and Machine learning techniques also may play a vital role in this work and for detecting the abnormal node based on the type of data sent with the packet [30,31,32]. More security features can be applied for accurate detection of attacks using some method [33]. The security schemes also have some drawbacks because it not possible to prove security effectiveness in all criteria. The classification of routing is discussed to know about the routing protocols techniques. The routing techniques performance measurement depends on the load in the network. The various researchers' contribution is useful for detection and secure communication.

References

[1] John R. Vacca, "Guide to Wireless Network Security," Springer, 2006.

[2] Jean Walrand, P. Varaiya, "High Performance Communication Networks," Science Direct, Second Edition, 2000.

[3] www.wisdomjobs.com/e-university/wireless-communication-tutorial-2262/wireless-communication-satellite-26541.html

[4] Pradeep Kumar Singh, Bharat K. Bhargava, Marcin Paprzycki· Narottam Chand Kaushal, Wei-Chiang Hong," Handbook of Wireless Sensor Networks: Issues and Challenges in Current Scenario's," Advances in Intelligent Systems and Computing 1132, Springer, 2020.

[5] Gaurav Soni, Mahendra Kumar Jhariya, Kamlesh Chandravanshi Deepak Tomar, "A Multipath Location based Hybrid DMR Protocol in MANET," 3rd International Conference on Emerging Technologies in Computer Engineering: Machine Learning and Internet of Things (ICETCE-2020), 07-08 February 2020.

[6] Sabih ur Rehman, M. Arif Khan, Tanveer A. Zia, Lihong Zheng, "Vehicular Ad-Hoc Networks (VANETs) – An Overview and Challenges", ResearchGate, Journal of Wireless Networking and Communications, pp 29 – 38, 2013.

[7] G. Guett, C. Bryce, "Using TPMs to Secure Vehicular Ad-Hoc Networks (VANETs)", IFIP International Federation for Information Processing 2008 pp.106-116, WISTP 2008, LNCS 5019.

[8] G. Guett, C. Bryce, "Using TPMs to Secure Vehicular Ad-Hoc Networks (VANETs)", IFIP International Federation for Information Processing 2008 pp.106-116, WISTP 2008, LNCS 5019.

[9] B. Rakesh Kumar and Mayank Dave " A Review of Various VANET Data Dissemination Protocols", International Journal of u- and e- Service, Science and Technology Vol. 5, No. 3, pp. 27-24 September, 2012.

[10] Y. Hao, Y. Cheng, C. Zhou, and W. Song, "A Distributed Key Management Framework with Cooperative Message Authentication in VANETs", IEEE Journal on Selected Areas of Communication, Vol. 29, No. 3, pp. 616–629, March 2011.

[11] G. Li, L. Boukhatem, and J. Wu, "Adaptive Quality-of-Service-based Routing for Vehicular Ad hoc Networks with Ant Colony Optimization," IEEE Transaction on Vehicular Technology, vol. 66, no. 4, pp. 3249–3264, 2017.

[12] A. Mahajan, N. Potnis, K. Gopalan and A.-I. A. Wang, "Urban Mobility Models for VANETs," in Proc. of 2nd Workshop on Next Generation Wireless Networks, 2006.

[13] D. S. Gaikwad and M. Zaveri, "A Novel mobility model for realistic behavior in Vehicular Ad Hoc Networks," in 11th IEEE International Conference on Computer and Information Technology, Cyprus, 2011.

[14] Kayhan Zrar Ghafoor and Marwan Aziz Mohammed, "Routing Protocols in Vehicular Ad hoc Networks: Survey and Research Challenges", Network Protocols and Algorithms, , Vol. 5, No. 4,pp. 39-83, 2013.

[15] Tarik Taleb, Ehssan Sakhaee, Abbas Jamalipour, Kazuo Hashimoto, Nei Kato, and Yoshiaki Nemoto, "A Stable Routing Protocol to Support ITS Services in VANET Networks", IEEE Transactions On Vehicular Technology, Vol. 56, No. 6, pp. 3337-3347, November 2007.

[16] Muhammad Sameer Sheikh and Jun Liang "A Comprehensive Survey on VANET Security Services," in Traffic Management System, Wireless Communications and Mobile Computing, Volume 2019, Hindawi, Article ID 2423915, 2019.

[17] Gaurav Soni, Kamlesh Chandravanshi, Mahendra Ku. Jhariya, Arjun Rajput, "An IPS Approach to Secure V-RSU Communication from Blackhole and Wormhole Attacks in VANET,"Contemporary Issues in Communication, Cloud and Big Data Analytics Proceedings of CCB, Springer 2020

[18] Basker Palaniswamya, Seyit Camtepeb ErnestFooa, Leonie Simpsona, Mir Ali Rezazadeh Baeea, Josef Pieprzykb, "Continuous authentication for VANET," Vehicular Communications, Elsevier, 2020.

[19] Peng Wang, Chien-Ming Chen, Senior Member, *Saru* Kumari, "HDMA: Hybrid D2D Message Authentication Scheme for 5G-Enabled VANETs," IEEE Transactions On Intelligent Transportation Systems, 2020.

[20] Vasiliy Krundyshev, Maxim Kalinin, Peter Zegzhda "Artificial swarm algorithm for VANET protection against routing attacks", IEEE Industrial Cyber-Physical Systems (ICPS), 2018.

[21] Hao Hu, Rongxing Lu, Zonghua Zhang and Jun Shao," REPLACE: A Reliable Trust-Based Platoon Service Recommendation Scheme in VANET," Transactions on Vehicular Technology, pp.1-11, 2016.

[22] Thanmayee Karimireddy, Ahmad Ghulam A Bakshi, "A Hybrid Security Framework for the Vehicular Communications in VANET", IEEE WiSPNET Conference, 2016.

[23] Sourav Kumar Bhoi, Rajendra Prasad Nayak, Debasis Dash and Jyoti Prakash Rout, "RRP: A Robust Routing Protocol for Vehicular Ad Hoc Network against Hole Generation Attack ", International conference on Communication and Signal Processing, pp. 1175-1179 April 3-5, 2013, India

[24] Shiang-Feng Tzeng, Shi-Jinn Horng, Tianrui Li, Xian Wang, Po-Hsian Huang, and Muhammad Khurram Khan Enhancing Security and Privacy for Identity-Based Batch Verification Scheme in VANETs IEEE Transactions on Vehicular Technology, Vol. 66, No. 4, April 2017.

[25] Tony Santhosh Gnanasekar, Dhandapani Samiappan "Optimal Routing in VANET using Improved Meta-heuristic Approach: A Variant of Jaya, IET Communication Journal, First published: 03 September 2020.

[26] Sumendra Yogarayan, Siti Fatimah, Abdul Razak Afizan Azman, Kirbana Jai Raman, "A Review of Routing Protocols for Vehicular Ad-Hoc Networks (VANETs)," IEEE 8th International Conference on Information and Communication Technology (ICoICT), 2020.

[27] Dutta, Sushama Rani, et al. "IoT in Autism Detection in Its Early Stages." Internet of Things: Enabling Technologies, Security and Social Implications. Springer, Singapore, 2021. 47-58.

[28] Dutta, Sushama Rani, Sujoy Datta, and Monideepa Roy. "Using cogency and machine learning for autism detection from a preliminary symptom." 2019 9th International Conference on Cloud Computing, Data Science & Engineering (Confluence). IEEE, 2019.

[29] Dutta, Sushama Rani, et al. "A Machine Learning-Based Method for Autism Diagnosis Assistance in Children." 2017 International Conference on Information Technology (ICIT). IEEE, 2017.

[30] Giri, Soumyajit, Sushama Rani, et al. "A real time epidemic alert generation system for rural areas using WBANs and kiosks." 2017 International Conference on Information Technology (ICIT). IEEE, 2017.

[31] Dutta, Sushama Rani, and Monideepa Roy. "EDOT: Context-aware tracking of similar data patterns of patients for faster diagnoses." 2017 Second International Conference on Electrical, Computer and Communication Technologies (ICECCT). IEEE, 2017.

[32] Dutta, Sushama Rani, and Monideepa Roy. "Providing context-aware healthcare services using circular geofencing technique." 2016 3rd International Conference on Computing for Sustainable Global Development (INDIACom). IEEE, 2016.

[33] Dutta, Sushama Rani, et al." Handling Scarcity of Data in Autism Identification using Binary Imputation Method" International Journal of Recent Technology and Engineering (IJRTE) ISSN: 2277-3878, Volume-8, Issue-1, May 2019

STUDY ON FLEXURAL BEHAVIOR OF BUBBLE DECK SLAB STRENGTHENED WITH FRP

V.ABISHEK [1], G.R.IYAPPAN[2]
[1]Student, [2]Associate Professor, Civil Engineering Department, SRM Valliammai Engineering College, Chennai, India.

Corresponding Author: abishekvel98@gmail.com

ABSTRACT : Bubble deck sheets are a technique to remove all the concrete from the center of the floor sheet, thus reducing structural dead weight. Inefficient concrete in the center of the slab replaces the high-density hollow spheres (HDPE). The weight decreases and increases efficiency in the plane. Because the bubble deck slab consumes less material, it reduces greenhouse gas emissions into the atmosphere, enabling us to achieve green construction. The benefits include reduced energy consumption - both in production and transportation - and emissions - from production and transportation, mainly CO2 - as well as material, load, and value reduction. It is also a green technology. Reduce the volume of concrete in the bubble deck technology by substituting locally available spherical bubbles. This implies that the monolithic slab element is used to determine load carrying, deformation (deformation), crack, and failure properties in the plant and is subjected to static gravity loads. The resulting conclusions will be used to define the failure mechanisms, and the benefits of the bubble deck slab will be highlighted.

Keywords: Bubble Deck slab, CFRP Strengthening, concrete technology, slab, CO2

1. INTRODUCTION

In building construction, the slab is a critical structural member that consumes the most concrete. Jorgen Bruenig invented the first biaxial hollow slab (now called a bubble deck slab) in Denmark in the 1990s. Bubble deck is a more sustainable construction method than traditional concrete floor systems because it uses less concrete and emits less CO2 during the manufacturing process. It complies with sustainability standards by utilizing recycled plastic spheres[1]

The spheres may be recycled even if the building is demolished or renovated in the future. By virtually eliminating the substantial portion of the conventional slab in the center, it results in a 30 to 50% lighter slab, which reduces the loads on the columns, walls, and foundations, and ultimately on the entire structure. The slab is a critical structural member in the construction of buildings because it forms the space [2].

A slab is a critical component of building construction. It is a significant structural member that consumes concrete. The primary impediment to concrete constructions, particularly horizontal slabs, is their high weight, restricting their span. Thus, it is necessary to concentrate on the span's weakness, reduce the load, or overcome concrete's inherent tension weakness [3].

In general, as the span increases, the deflection of the slab increases as well. As a result, the thickness of the slab should be increased. Increased slab thickness results in heavier slabs and can result in increased column and foundation size. As a result, buildings become uneconomical. To circumvent the disadvantages associated with increasing the self-weight of slabs, the voided slab system is used.

2. LITERATURE REVIEW

Based on the literature study following.
Shaimaa Tariq Sakin (2014) – Shaimaa Tariq Sakin is one of the leading European countries to participate. SCC has been shown to improve the punching shear power and reduce the deflection in the mid-span. We are increasing the box strength of the steel fiber in the crucial zone and reducing the angle of box failure. The 1% volume fiber ratio was more effective than the 0.8% volume fraction ratio. The crack pattern of the punching shear was reduced by the SCC use and the empty SCC use, but the empty SCC was increased by 1% by volume fraction, particularly by steel fiber [4]
Reshma Mathew, Binu (2016) – In this study GFRP strengthening system is used. Strengthened deck plates have a higher capacity of punch compared to a deck control. Increased load capacity carries up to 20% due to reinforcement of a GFRP bubble cover. A strengthened bubble deck has low deflection compared to an unstrengthen bubble deck slab. 8 % of global carbon dioxide is due to cement production. One ton of cement causes the emission of about 800 kg of carbon dioxide. One m3 of concrete causes carbon dioxide emission close to 300 kg. By inserting 180 mm diameter HDPE balls into a flat 230 mm diameter label, we save up to 23,62% on one ball [5]

Shikha Saini and Nishant Rajoria Muhammad Shafiq Mushfiq (2017), In this study (with less load capacity), the bubble deck platforms were not as efficient as the conventional plating, and in slab construction are very satisfactory, as the load-bearing capacity difference between them and the conventional is negligible. However, it is worth noting that the weight in bubble deck sheets is 10.5%–17% lower than in conventional decks, which benefits the bubble deck sheets, particularly in structures with a load problem. This is an added benefit [6].

S. Aswini (2017) – R.S. Dheepan, S.Saranya In this study, the bending strength of the plates increases when the spacing between the balls increases, regardless of the change in the thickness of the plates. The bending force of the plate in a diameter of 60mm balls are also observed to be higher than the bending strength of the plate in a diameter of 75mm balls for both 20mm and 30mm distance between balls. Finally, it is concluded that 60mm and 30mm can be the ideal diameter for the hollow balls used in the slab for a common purpose. The use of concrete is reduced by 35 – 50 percent, reducing the self-weight of the lath with equal bending strength compared to the conventional lath. It has an excellent thermal insulating property because of the vacuum in the plate. Lower total cost, lower building time, and green technology than conventional plates. The following inferences are noted. The load-bearing capacity of the bubble deck slab is mostly less than the conventional slab. The load-bearing capacity of the bubble deck slab depends on the arrangement of the slab, ball size, and ball diameter. The punching shear of the bubble deck slab is low. The self-compacting concrete increases the punching shear [7].

3. STAGES INVOLVED IN BUBBLE DECK SLAB

Bottom reinforcement placement - The first step is to lay the reinforcement on the underside. It puts the ball in a straight line and aligns it. Two directionally welded reinforcement meshes are bottom reinforcements.
Foot concrete laying - Concrete at the bottom of the assembly line is supplied. It acts as a material for ball connection because the ball is attached to the concrete. Hollow ball placing - the hollow ball is located rather than the concrete between the reinforcement.
Top enhancement - Top enhancement netting is supplied on the top of the sphere after placing the balls. It places the ball and also serves as the ball's cover. After placing the spheres in places, the two meshes are connected to make the shell rigid [8].

4. CFRP STRENGTHENING

Because of the material to enhance the cutting capacity of concrete plates, CFRP sheets were used. The CFRP increased bending rigidity and improved the punching shear strength of the sheets. It was observed. The increase in bending capacity has altered the failure from pure bending to a combined bending shaking mode or pure bending shaving mode [9].
The bending resistance of the two-way layer can be improved by binding the CFRP layers effectively to the tightening face of the laths. The flexural consolidation method offers several advantages: minimizing structural expansion and weight, easy on-site handling, and excellent corrosion resistance. The interface link between concrete and CFRP composite strengthening is of decisive importance for the overall performance of the strengthened framework since the CFRP sheets are linked to the concrete framework.

5. MATERIAL TESTING:

5.1 Cement
Cement is a binder material used in concrete shown in table 1. In this project work, cement of OPC 53 grade is used. The preliminary test is conducted based on IS codes [10].

Table 1 Properties of Cement

TEST	RESULT	As per IS standard
Consistency	29%	25-30%
Initial setting time	32min	Not less than 30min
Final setting time	259min	Not more than 600min
Specific gravity	3.06	2.5-3.15

5.2 Fine Aggregate:
Fine aggregates are usually sand or crushed stone. . The preliminary test is based on IS 2386 Part 3 (1963) observed in table 2

Table 2 Properties of Fine Aggregate

TEST	VALUE	As per IS standards
Specific gravity	2.75	2.6-2.8

5.3 Coarse Aggregate:
The ground compound is one of the essential elements of cement and occupies the most significant volume in the blend. The test is based on part 3 of the IS 2386 (1963). Properties of coarse aggregates shown in table 3 [11]

Table 3 Properties of coarse aggregates

Test	Value	As per IS standards
Specific gravity	2.60	2.6 – 2.85
Impact test	28.57%	not more than 30%

6. Experimental Description:

Cement: FA: CA was 1:1.5:2.8, and the M30 grade concrete was used in the design mix. As per the literature study and as per IS codal provision, the size of the slab is decided. The ball of size 75mm is placed between the top and bottom reinforcement. Since Flexural behavior is done, so two-point loading condition is adopted. The length of the slab is 700mm, and depth is 125 mm, and width is 300 mm. The mouth was placed on a smooth surface and arranged properly. The sides of the mold exposed to cement are oiled well, so the side walls of the mould do not absorb that water from the concrete, and the specimen can be easily removed. The cages were placed in the mould, and the cage and the shaped cover were kept to 20 mm. The materials in concrete, including concrete, cement, sand, aggregation, and water, have been precisely weighted and combined. Firstly, the bottom reinforcement is placed over the cover, and the hollow plastic balls were placed on the bottom reinforcement, and the top reinforcement was placed over the hollow balls. Immediately after mixing and well compacting, the concrete was placed into the mould [12]. At the end of 24 hours of casting, the test specimens were removed. On specimens, marks of identification were marked. For 28 days, they were healed in water. Three types of slabs are cast with variations in material properties along with cubes to find out the compressive strength. One specimen for the CFRP wrapping is a carbon fiber polymer laminate used to strengthen structural elements in load-bearing structures. CFRP sheet of bi-directional is used in the specimen. Surface preparation, including cleaning, one layer epoxy polyamine base, and a layer epoxy putty was made before the jacketing with CFRP sheets; then epoxy adhesive was used for binding CFRP sheets to the specimen [13]. Figure 1,2,3 and 4 shows casting of specimens. Table 4 depicts parameter of the specimens.

Table 4 Specimen Parameters

Sl.no	Specimen	Dimension (mm^3)
1.	S1 (Conventional Slab)	700X300X125
2.	BD2 (Bubble Deck Slab)	700X300X125
3.	BD3 (Bubble Deck Slab strengthened with cfrp)	700X300X125

Fig 1 Top and Bottom reinforcement

Fig 2 Mould for Specimen

Fig 3: casting of bubble deck slab

Fig 4 Casting of specimen

7. Testing of Specimen

The slabs are tested on a two-point load setup with support at the bottom. The distance between the end of the slab to support is 70mm, and the distance from support to the load point is 167mm the load is applied in the proportion of L/3. Jack used for the beam is of 50ton capacity. Flexural behavior beam is calculated after 28 days of curing, which consists of maximum load, deflection properties. The load was applied using a 50-ton jack in the total order of 2 kN, and deflection was observed using a digital dial gauge [14]. Load is applied till the ultimate point, and deflection is noted. Figure 5 and 6 shows testing of specimens.

Fig 5 Experimental setup for test specimen

Fig 6 Testing of Specimen

8. RESULT AND DISCUSSION

The three specimens' ultimate load and bending strength are shown in Table 5 for a dimension of 700mmx300mx125mm. The hollow balls used are 75mm in diameter.

Table 5 TEST RESULT OF SPECIMEN

SL.NO	Slab Specimen	Ultimate Load (kN)	Deflection (mm)	Flexural Strength (N/mm2)
1.	S1	55	4.59	8.66
2.	BD2	50	3.43	7.61
3.	BD3	84	9.32	12.5

8.1Load – Deflection Relationship

Figure 7,8 and 9 depicts relationship between load and deflections for conventional, bubble deck and bubble deck with CFRP [15].

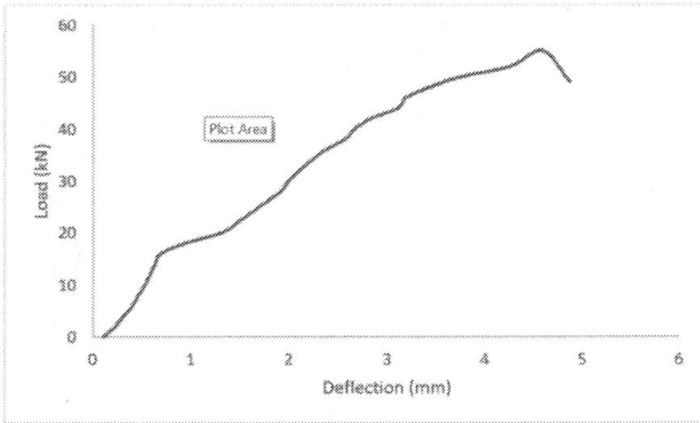

Fig 7 Load Vs. Deflection of Conventional Slab

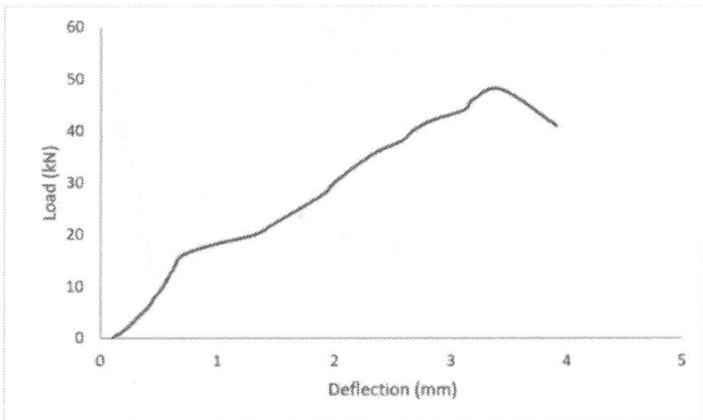

Fig 8 Load vs. Deflection for Bubble Deck slab

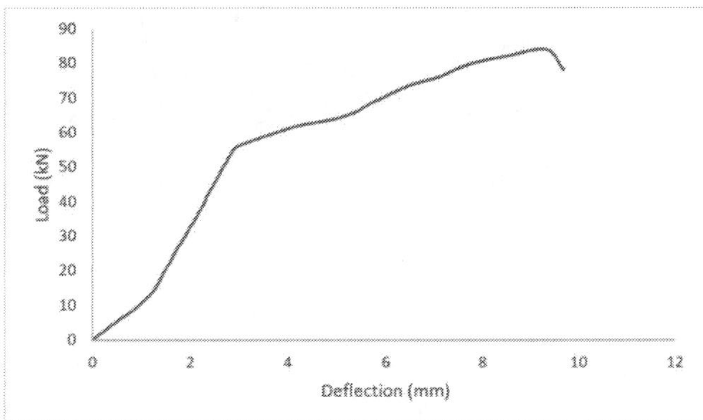

Fig 9 Load vs Deflection for bubble deck slab strengthened with CFRP

8.2 Crack Pattern:
Crack pattern observed in the three specimens are shown in the figure 10, 11 and 12.

Fig 10 Failure of Bubble Deck Slab

Fig 11 Failure of Conventional Slab

Fig 12 Failure of Bubble Deck Slab Strengthened with CFRP

9. CONCLUSION:

1. As a result of the preceding, it was determined through testing that while bubble deck slabs were not as efficient as conventional slabs.
2. The load-bearing capacity of a bubble deck slab reinforced with CFRP is 18% greater than that of a conventional slab and 32% greater than that of the bubble deck slab alone.
3. The experiment discovered that the bubble deck slab weighs less than the bubble deck slab.

REFERENCE

[1] R.Sagadevan, B.N. Rao (2019) -Experimental and analytical investigation of punching shear capacity of biaxial voided slabs, Elsevier.
[2] Thaar S. Al-Gasham, Jasim M. Mhalhal, Hussain A. Jabir (2019) - Improving punching behavior of interior voided slab-column connections using steel sheets, Elsevier.

[3] Reshma Mathew, Binu (2016) -Punching Shear Strength Development of Bubble deck Slab Using GFRP Stirrups,IOSR.

[4] Muhammad Shafiq Mushfiq, Shikha Saini and Nishant Rajoria (2017)- Experimental Study on bubble deck slab , IRJET .

[5] Jasna Jamal,Vishnu Vijayan (2019) -A Study on Strengthening of Bubble Deck Slab with Elliptical Balls by using GFRP Sheets , IJSRD.

[6] B. Gopi Raju, P. Poluraju (2019) - Numerical Study of Bubble Deck Slab using Ansys , IJRTE

[7] Nasr Z. Hassan , Mostafa A. Osman, Awad M. El-Hashimy, Heba K. Tantawy (2018) - Enhancement of punching shear strength of flat slabs using shear-band reinforcement, HBRC Journal.

[8] Saravanan, B., Vijayan, D.S. "Status review on experimental investigation on replacement of red-mud in cementitious concrete ", Materials Today: Proceedings, 2020, 33, pp. 593–598

[9] Vijayan, D.S., Nivetha, C., Dinesh Kumar, P., Saravanan, B., Gokulnath, V. "Green composite form of eco-friendly concrete by adding pva fiber", Journal of Green Engineering, 2020, 10(6), pp. 3084–3101

[10] Maha Habeeb , Adel A. Al-Azzawi, Faiq M.S. Al-Zwainy - Punching shear behavior of LWA bubble deck slab with different types of shear reinforcement -Science Direct

[11] Chung, JH., Jung, HS., Bae, Bi. et al. Two-Way Flexural Behavior of Donut-Type Voided Slabs. Int J Concr Struct Mater 12, 26 (2018).

[12] D S Vijayan, Dr. J Revathy, Flexural Behavior of Reinforced and Pre-stressed Concrete Beam Using Finite Element Method, International Journal of Applied Engineering Research, ISSN 0973-4562 Volume 10, Number 1 (2015) pp. 717-736

[13] BubbleDeck Technology. (2008). BubbleDeck voided flat slab solutions—Technical manual & documents. Delta: BubbleDeck UK.

[14] Naveen Kumar K.,Vijayan D.S.,Divahar R.,Abirami R.,Nivetha C.,"An experimental investigation on light-weight concrete blocks using vermiculite", Materials Today: Proceedings, Volume 22, Pages 987 – 991, 2020, 10.1016/j.matpr.2019.11.237

[15] D.S. Vijayan, A. Mohan, J. Revathy, D. Parthiban , R. Varatharajan, "Evaluation of the impact of thermal performance on various building bricks and blocks: A review", Environmental Technology & Innovation 23 (2021) 101577, https://doi.org/10.1016/j.eti.2021.101577.

STUDY ON FLEXURAL BEHAVIOUR OF RC BEAM STRENGTHENED WITH FRP

Dinesh Kumar J[1], **Sattainathan Sharma A**[2], **Suganya Devi** K[3]
[1]PG Student, [2&3]Assistant Professor, Department of Civil Engineering, SRM Valliammai Engineering College, Tamilnadu, India.
Corresponding Author Email: dineshkumarj1103@gmail.com

Abstract: This paper provides the analytical and experimental work of reinforced concrete beams strengthened with FRP. The main objective of this investigation is to study the flexural behaviour of reinforced concrete beams of different methods of strengthening methods using GFRP to find the flexural strength, failure modes, and ductility of the reinforced concrete beam. All beams were strengthened for flexure with external bonding to prevent flexural failure. The analytical and experimental results indicated that the externally bonded GFRP used for flexural strengthening of reinforced concrete beams increased the cracking load, increased the ultimate load-carrying capacity, and exhibited decreased ductility corresponding to the unstrengthen control specimen. The analytical work was carried out using ANSYS 17.0 Software and found the parameters of total deformation, stress strain curves, and Load deflection graph plotted.

Keywords: RC beam, Strengthening methods, Flexural Strength, ductility, Glass fiber reinforced Polymers, ANSYS.

1. Introduction

In reinforced concrete structures, the most common methods of strengthening are external bonding methods were used. In this paper, the GFRP were used on the external surface, which is wrapped on the different methods to increase the load carrying capacity of the existing concrete structures and bridges, which helps increase the life span of the structures and the service lives of the structures. In recent years, the primary material used for reinforcement has been fibres-reinforced polymer (FRP), which provides excellent tensile strength, corrosion resistance, high durability, and is readily available. Externally connected FRP reinforcement effectively strengthens several reinforced concrete structures, including beams, columns, plates, bridges, tunnels, chimneys, and silos, to increase reinforced concrete structures' flexibility and ductility. In this study, four specimens of the reinforced concrete beam were casted and strengthened with GFRP by different methods and tested for flexure and analyzed used ANSYS Software for better results. The results show the significant variation of reinforced concrete beam for the flexural capacity [1]

2. Literature review

Sharif, Al-Sulaimani et al., (1994) investigated the effects of various repair schemes using FRP plates on initially loaded RC beams. The plate thickness was varied to assess the initiation of premature failure at the plate curtailment zone. Different repair and anchoring schemes were conducted to eliminate such undesirable failures and to ensure ductile behavior. The authors discussed the behavior of repaired beams in terms of load-deflection curves and different failure modes. They concluded that I-jacket FRP

plates provided the best anchorage system; repaired beams provided enough ductility indicating the effectiveness of FRP plates; the shear and everyday stresses at the plate ends increased with increasing plate thickness leading to plating separation and concrete rip-off [8]
Arduini and Nanni et al., (1997) conducted a parametric study of beams with externally bonded FRP reinforcement. The parameters considered in the analysis were stiffness, strength, and adhesive stiffness. The authors reported that the bonded length of FRP should be as high as possible, and the adhesive should have high ultimate elongation [1]
Ross et al., (1999) presented experimental and analytical studies on the external bonding of FRP laminates to RC beams for flexural reinforcement. The authors used a non-linear finite element analysis and an inelastic section analysis to predict the retrofitted beams' load-displacement response. They concluded that significant strength enhancements in flexure can be obtained in lightly reinforced beams and that the bond strength between the composite plate and the concrete had a significant effect on the beam response [4]
Grace, Sayed et al., (2002) investigated the flexural behavior of reinforced concrete beams reinforced with various types of FRP laminates. They investigated the effect of varying the number of FRP layers, the type of epoxy, and the reinforcement pattern on the beam behavior. They discussed how strengthening affects deflection, failure load, mode of failure, strain, and ductility. The authors concluded that by combining vertical and horizontal sheets with a suitable epoxy, they could significantly increase the ultimate load-carrying capacity of the beams and that all strengthened beams experienced brittle failure, necessitating a higher factor of safety in design [3]
The load-deflection behavior of RC beams strengthened with GFRP sheets subject to various environmental conditions has been studied by Tarek Almusalam (2006). For this study, 84 samples of beams have been prepared. They were monitored in the laboratory environment, outside, wet-dry alkaline water, and second class covered with ultraviolet paint protection. The unreinforced and strengthened beams in each category. The specimens of various wet-dry environments were exposed within the solution to a two-week time cycle and outside the solution for two weeks. The tests were carried out in different environmental conditions following 6, 12, and 24 months of exposure. The author concluded that neither of the above-mentioned environmental conditions significantly impacts the beams' flexicurity [13]
Barbato (2008) conducted a study on efficient modeling with reinforced concrete beams with fiber-reinforced polymers. To forecast the capacity of the load carrier and the application of a three- and the four-point bending response of the RC beams, FRP-Force-based formulas (FB) were used. Due to the many literature tests available and published by different authors, numerical simulations and experimental results were compared. The numerically simulated reactions agree with the relevant experimental results remarkably well [2]

2.1. Summary of the Literature review
FRP platform-specific RC beams provide a wide range of information about the study's load deflection behaviour and parameters, including resilience, deflective, and ductility of RC beams. Many of the researchers suggested models for RC beams using FRP laminates. Some researchers have presented maths for predicting study parameters in which FRP enhances RC structural members. Few researchers have carried out reliability analyses for reinforced concrete beams from FRP. A vast number of studies have been carried out on FRP-stärker beams, and further research has thus been required to understand the interaction of GFRP-stärker reinforced concrete beams' load-bearing capacity, deflectors, and ductility. FRP wrapping is the most common method to strengthen the bending behavior and rigidity of the beam. With the GFRP enhancement methods, the load beam capacity is also increased. In strengthening degraded concrete structures, the efficiency of externally binding CFRP has been assessed. Decreasing crack widths was the main impact of the service loading. Further research has to be done for the easy understanding of FRP materials and their behavior [2].

3. Experimental Investigation
3.1. Preliminary Tests
To find the material property, the preliminary tests were taken for the cement, fine aggregates, coarse aggregate, and to increase the strength and workability, superplasticizer contrast SP 430 were used for

the concrete grade of M40 of concrete compressive strength were achieved, and results were given below. Table 1,2, 3 and 4 represents test results of the cement, coarse aggregate and fine aggregate along with specimen testing respectively [3]

Table 1: Test results on Cement

TEST ON CEMENT	RESULT OBTAINED	As Per IS Standard
CONSISTENCY	30%	25-35%
INITIAL SETTING TIME	32min	Not less than 30min
SPECIFIC GRAVITY	3.1	3.1 – 3.25

Table 2 : Test results on coarse aggregate

TEST ON COARSE AGGREGATE	RESULT OBTAINED	As Per IS Standard
Fineness Modulus	7.3	6.5- 8
Specific gravity	2.70	2.6-2.9

Table 3 : Test results on fine aggregate

TEST ON FINE AGGREGATE	RESULT OBTAINED	As Per IS Standard
Fineness Modulus	2.80	2.2-3.2
Specific gravity	2.60	2.6-2.8

3.2. Testing of cube and Cylinder

Totally six cubes and six cylinders were cast and tested on the 7th and 28th day to determine the characteristic compressive strength of concrete accordingly. The results obtained were shown below. The average compressive strength of 150mm3 cubes after 28days shows the better strength, and also the average split tensile strength of the cylinder also achieved the required strength. Test results were given below in the table. Fig 1 shows cube testing of cubes and cylinder [4].

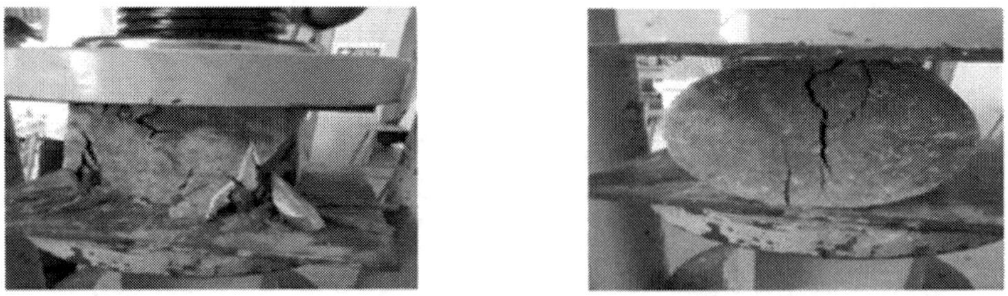

Fig. 1. Testing of cube and Cylinder

Table 4. Test results on Cube and Cylinder

S.NO	Compressive strength (N/mm^2)		Split tensile Strength (N/mm^2)	
	7th Day	28th Day	7th Day	28th Day
1	24.8	46.9	2.29	4.06
2	26.4	49.9	2.10	3.73
3	26.3	47.2	1.98	3.83

The average Compressive Strength of Cube (N/mm2) = 47.8 N/mm^2
The average Split tensile strength Of Cylinder (N/mm2) = 3.89 N/mm^2

3.3. Test Specimen and Strengthening methods

In the experimental investigation, six specimens of armored cement beams have been tested under a two-point load. The fig 2 shows dimensions and refurbishment details. The beam was 1000mm in length. The beams were 150mm and 200mm in width and depth, respectively. For all specimens, the cross-sectional geometries and the length reinforcements were identical. The length reinforcement tension consists of three high tensile bars with a diameter of 12mm. Two bears of high tensile steel deformed in diameter of 10 mm. The vertical stirred were spaced at 150mm center reinforcement with a diameter of 8mm, round, mild steel bars. GFRP reinforced all beams at different thicknesses. The beams are shown in the fig 2 for the refinement details.The Glass Fibre Reinforced Polymer type is Unidirectional Cloth fiber were used. The epoxy resin and the adhesive were applied and cured for 48hours at room temperature before loading. Fig 3 depicts casting and curing of specimens [4]

3.4. Specimen Details

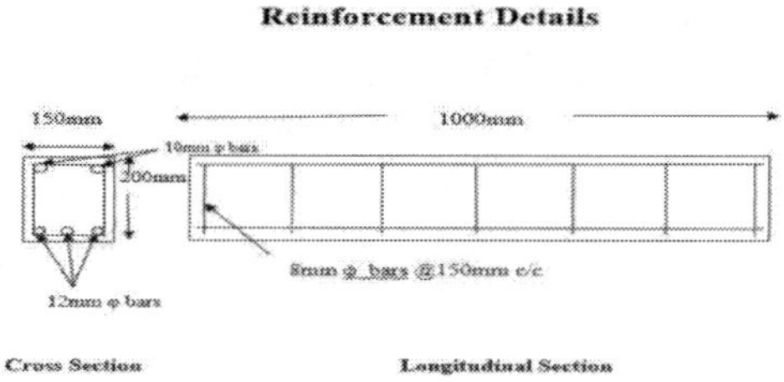

Fig. 2 Reinforcement details

The following specimen details were given below in the table.5

Table 5. Specimen details

SPECIMEN ID	STRENGTHENING METHODS	THICKNESS
BCN	Control specimen	-
BB01	Bottom wrap	1mm thickness
BB02	Bottom wrap	2mm thickness
BL01	Loading surface	1mm thickness
BL02	Loading surface	1mm thickness
BF01	Full wrap	1mm thickness

3.5. Casting and Curing of specimen

The specimen is cast and cured for 28 days in the curing tank, and after curing, the strengthening process has to be done on various methods.

Fig. 3. Casting and Curing of Specimen

3.6. Strengthening Process
For the strengthening process, the Glass Fibre Reinforced Polymer type is Unidirectional Cloth fiber were used. Both the epoxy resin and the adhesive have been used and cured 48 hours before loading at room temperature given in fig.4

Fig.4 Strengthening of specimen

4. Test Setup and Instrumentation
All beams with a capacity of four hundred kN have been tested in a two-point load. To get better results and read the measuring devices, every test started with several cycles of small loading (i.e., about 5-10kN). The load was applied until the failure happened afterward. The beams were fitted with the load control cell, the LVDT was fitted with midspan to measure deflection, and three electrical resistor gauges were attached on which the single pressure gauge is located. The fig 5 shows an experimental view of the test set-up. During the tests and the crack pattern, the visual inspection also observed the crack initiation, beam deflection, strains, and load values [5].

Fig. 5 Experimental view of test setup

5. Experimental results
The tests of all beams and the load causing the initial cracks, the load causing the yield of the voltage steel reinforcement are presented in the table 6. A baseline value for comparison with the strengthened beams is also calculated and presented on a table for the obtained deflection and the failure load for control specimens [6].

5.1. Enhancement of Load-carrying Capacities

The graph represents the load-deflection curves of the tested specimen, respectively. The result shows that the glass fiber reinforced polymer with full wrap gives high load carrying capacity, and the beam with the bottom wrap also gives a slight increase in the ultimate failure. Compared to the unstrengthen beam (i.e) Control specimen, the full wrap is achieved with a 28% higher load-carrying capacity. Table 6 depicts initial cracks and failure load comparison.

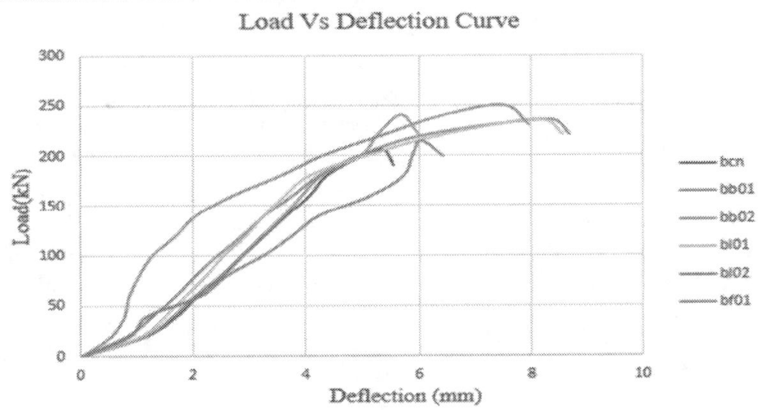

Fig.6 Load deflection Curve

Table.6. Initial Crack Load Vs Ultimate failure load

SPECIMEN ID	INITIAL CRACK LOAD (kN)	ULTIMATE FAILURE LOAD (kN)
BCN	90	205
BB01	70	230
BB02	80	240
BL01	105	225
BL02	60	215
BF01	110	250

5.2. Service Load Calculation

The maximum expected charge intensity during the structure's lifetime is called the service charge. The service load is calculated and given in the below table.7

Table.7. Service Load

SPECIMEN ID	SERVICE LOAD (kN)
BCN	136.67
BB01	153.33
BB02	160
BL01	166.67
BL02	156.67
BF01	167.33

5.3. Failure modes

The failure modes (i.e.) Crack Patterns for the control specimen, and strengthened beams are shown in fig 7. The given fig 7 show the failure modes of different specimens observed during the testing of beams and are described. The failure modes are briefly given below in the table 8 as follows[7].

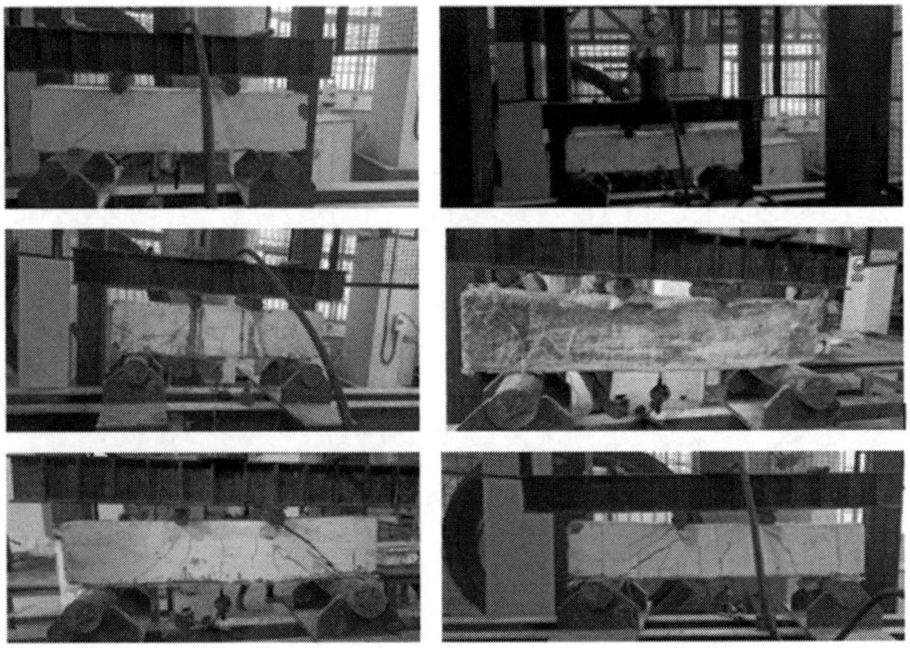

Fig. 7 Failure modes of the specimen

Table 8. Crack Pattern observed on the specimen

SPECIMEN ID	CRACK PATTERN OBSERVED ON THE SPECIMEN
BCN	The unstrengthened control specimen is failed by steel yielding followed by concrete crushing b/w loading points.
BB01	Initially, the flexural crack attains, and the crack propagated toward the one end of the support.
BB02	The beam strengthened at the bottom surface attains flexural cracks, and crack is initiated at the midspan and extended towards one support.
BL01	At a certain load level, the concrete started to break down, and a failure spread to the end.
BL02	In this, the crack occurred at two supports and extended up to the midspan.
BF01	The strengthened beam of complete wrapping is failed in the form of a horizontal crack propagated towards the midspan.

5.4. Beam deflection

The fig. 6 compares the load-deflection curves for a strengthened beam with the unstrengthen beam, a control specimen. The figure 6 indicates that the curves of beam B2 exhibit more load carrying capacity and attain high ductility compared to the unstrengthen beam. The table value shows that bottom wrap also has high tensile strength and load-carrying capacity. Comparing the beam B3 and B4 shows a slight increase in ultimate strength and shows a gradual increase in good ductile behaviour. Hence the maximum deflection curve could also be observed in these beams. The beam B2 strengthened with a full wrap of 1mm thickness achieved more strength, and the load -deflections observed that the beam is 39% higher and 28% higher load carrying capacity when compared to the unstrengthen beams. Table 9 shows deflection of beams [8].

Table 9. Beam Deflection

SPECIMEN ID	ULTIMATE LOAD (kN)	BEAM DEFLECTION (mm)
BCN	205	5.38
BB01	230	8.23
BB02	240	5.67
BL01	235	8.15
BL02	215	6.05
BF01	250	7.50

5.5. Beam Ductility (7)

The ductility of a ratio called the Ductility Index has usually been measured. The index of deflection ductility is used to measure the ductility of the beams, as shown in table 10.

$\mu\Delta = \Delta u/\Delta y$ where Δu = Midspan deflection at ultimate load

Δy = Midspan deflection at yielding of tension steel reinforcement

Table 10. Beam Ductility

SPECIMEN ID	DUCTILITY INDEX $\mu\Delta$	DUCTILITY RATIO
BCN	1.98	1
BB01	4.33	2.10
BB02	2.27	1.15
BL01	3.08	1.56
BL02	3.20	1.61
BF01	5.17	2.60

5.6. Initial Stiffness

Initial rigidity is the degree to which an object in response to applied force resists deformation. The initial stiffness is calculated and showed iin table 11.

Table 11. Initial Stiffness

SPECIMEN ID	INITIAL STIFFNESS (kN/mm)
BCN	33.21
BB01	36.08
BB02	32.12
BL01	39.77
BL02	28.03
BF01	75.86

5.7. Energy Absorption

It is defined as the maximum energy absorbed by a member within the elastic region. The energy absorption is found and given below table.12

Table 12. Energy absorption

SPECIMEN ID	ENERGY ABSORPTION (kN.mm)	
	ABSOLUTE	RELATIVE
BCN	4445	1
BB01	9710	2.18
BB02	11725	2.63
BL01	10275	2.31
BL02	662.5	0.14
BF01	144424	3.24

6. Software Analysis

An instant finite element analysis tool used in conjunction with CAD and Design Modeler is an intuitive, upfront workbench environmental system. ANSYS 17.0 is available. ANSYS 17.0 is a software environment for structural, thermal, and electromagnetic analyses. The course focuses on geometry development and optimization, attachment of the existing geometry, finite element modeling, resolution, and evaluation results. The class describes how you can use code and fundamental concepts for simulation and interpretation of finite element results. ANSYS 17.0 is software for the general-purpose, used to simulate the interaction among engineering engineers in all disciplines of physics, structural, vibration, fluid dynamics. CAD data can be imported, and geometry can also be built with its 'pre-processing' abilities. ANSYS 17.0. The finite element model (a.k.a. mesh) needed for the computer is generated similarly in the same preprocessor [9]. After loading definition and analysis, the results can be considered numerical and graphical. The material properties of the fibre GFRP UDC are given in table 13.

Table 13. Material Properties

Material	Density	Modulus of Elasticity	Poisson ratio
Concrete	2400 kg/m3	32 Mpa	0.2
Steel	7850 kg/m3	2x105 Mpa	0.3
GFRP (UDC)	1921 kg/m3	73 Gpa	0.26

6.1. Geometry creation and Mesh Generation

I have created the model with the primary reinforcement as 12mm, stirrups of 8mm, and a cross section of beam as 150x200x1000m, which is shown in the figure. After completing the geometry design, I have gone to the modeling phase and assign all the material properties. Then I have generated mesh that helps us solve the model in finite element method, and results can be obtained accurately, as shown below in fig.8 The Mesh generation is applied to convert larger size particles into several nodes. The process of generating the nodes is called Discretization [10]. This process helps to apply the loads in each node to be distributed throughout the structures shown in Figure.8

6.2. Loading pattern

The RC beam applied two-point loads to investigate the failure modes and improvement in strength, as shown in fig.8

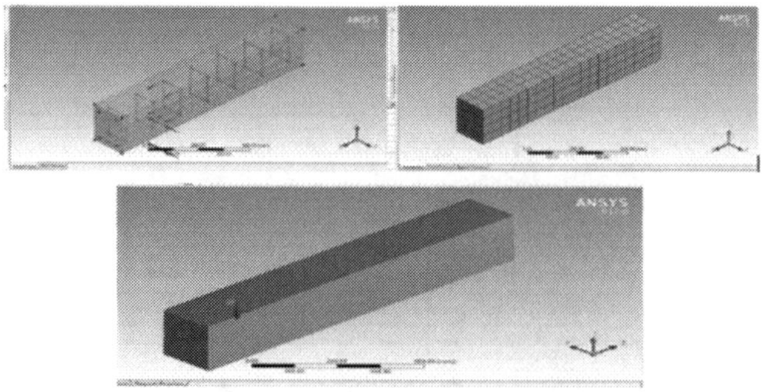

Fig.8 Meshing and Loading Pattern

6.3. Total Deformation

The total deformation of all the specimens, including the conventional beam and also the beam, which is strengthened with various methods are given in fig. and results are also furnished [11][12]. The deformation of BCN, BB01, BB02,BL01,BL02 and BF01 are given in fig.9.

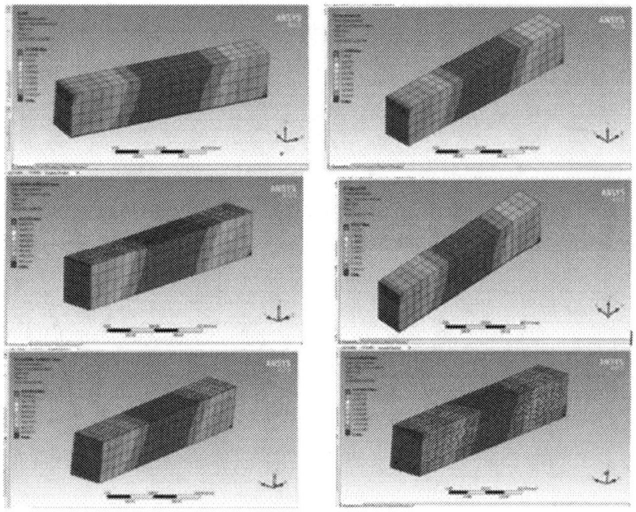

Fig. 9 Total deformation

6.4. Analytical Results and Experimental results

The analytical results show the results of various computations, and the results obtained are compared with experimental work is given in table. 14. The total deformation and load-deflection graph have been plotted from the analytical work, and the results obtained are related to the experimental work [13]. The comparison of results is given in fig.10.

Table.14. Comparison of Analytical and Experimental results

SPECIMEN ID	ANALYTICAL RESULTS		EXPERIMENTAL RESULTS	
	ULTIMATE LOAD (kN)	DEFLECTION (mm)	ULTIMATE LOAD (kN)	DEFLECTION (mm)
BCN	195	4.17	205	5.38
BB01	210	6.85	230	8.23
BB02	200	3.89	240	5.67
BL01	220	7.26	235	8.15
BL02	210	4.72	215	6.05
BF01	225	6.89	250	7.50

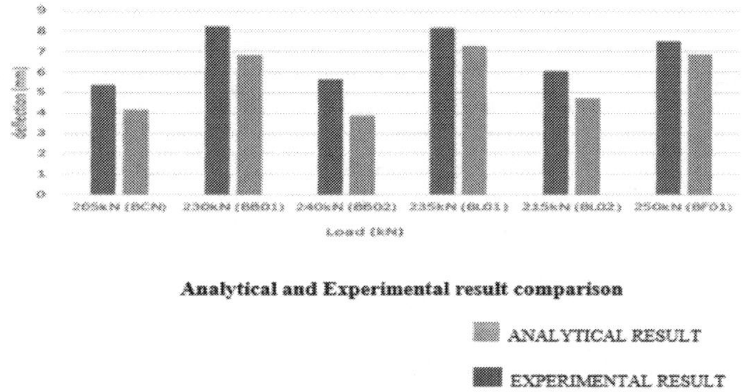

Fig.10 Comparison of results

7. Results and Discussion

- From the above experimental and analytical work, it is observed that GFRP strengthening exhibits higher load-carrying capacity. The control specimen has exhibited less load carrying capacity of 205kN while the beam strengthened with GFRP exhibits 250kN.
- The GFRP wrapped beam attains 28% of the ultimate failure load from the obtained results compared to the unstrengthen beam.
- There is no much difference in its ultimate failure load on the bottom wrap since the load increases slightly and is also considered the reasonable solution for strengthening the beams and the practical solution.
- The beam with full wrap indicated the additional flexural strength compared to the unstrengthen beam, which is the control beam.
- On the bottom wrap of thickness 1mm and 2mm, the ultimate load-carrying capacity increases up to 20.5% than the control specimen.
- The ductility of the increased when GFRP fully wraps the beam gives gradually increases with 2.7% of unstrengthen beam.

8. Conclusion

- The above experimental work shows that the wrapped beam has high deductibles and a wide bending to failure.
- From the conventional beam, 28% of load carrying capacity is achieved in complete wrapping, and 39% of the ductility is obtained on the complete wrapping
- The external wrapping on the bottom surface shows that it enhanced the cracking load and shows the significant variation in flexural load carrying capacity.
- Strengthening the RC beam with GFRP on the external surface with a double layer indicates the significant variation in increasing the strength and the ultimate load carrying capacity compared to the unstrengthen control beam.
- The analytical work was also done using ANSYS software, and the results were obtained for total deformation, equivalent stress, equivalent strain and load deflection curve, and stress strain graph.
- The experimental work concludes that reinforced concrete beams with GFRP will give an effective solution for premature failure.
- Hence it shows that the RC beam strengthened with GFRP will reduce the cracks compared to the unstrengthen reinforced concrete beam.
- Good results were obtained from the analytical and experimental work in terms of load-deflection curve and ductility behaviour of the beam.

References

[1] Arduini and Nanni (1997) "Behavior of Precracked RC Beams Strengthened with Carbon FRP Sheets" Vol.1 Issue 2 pp.63-70.

[2] D.S.Vijayan, S.Arvindan, K.Naveen Kumar, S Mohamed Javed, Seismic Performance of Flat Slab in Tall Buildings with and without a Shear wall, International Journal of Engineering and Advanced Technology (IJEAT) ISSN: 2249 – 8958, Volume-9 Issue-1, October 2019, Page no – 2672 – 2675.

[3] NF Grace, GA Sayed, AK Soliman, KR Saleh (1999) - Strengthening reinforced concrete beams using fiber reinforced polymer (FRP) laminates ACI Structural Journal-American Concrete Institute.Pg. 865-875.

[4] Ross, C. A. , Jerome, D. M. , Tedesco, J. W. and Hughes, M. L. (1999). "Strengthening of reinforced concrete beams with externally bonded composite laminates." ACI Struct. J., 96(2), 212-220.

[5] Rafid Saeed Atea (2017) "Torsional behavior of reinforced concrete T-beams strengthened with CFRP strips".Vol.03, No.02.pp.110-117

[6] Rami A. Hawileh, Hayder A. Rasheed, Jamal A. Abdalla, and Adil K. Al-Tamimi(2014) "Behavior of reinforced concrete beams strengthened with externally bonded hybrid fiber reinforced polymer systems" pp.972-982

[7] Sattainathan Sharma A , Sadhana , Shoban Prabhu , Solomon Sachin Samson Supriya. (2019) "Ductility Behaviour of Steel Fibre Reinforced Concrete Beam Strengthened with GFRP Laminates". Vol.06, No.03. pp.6530-6539.

[8] Sharif, A., Al-Sulaimani, G. J., Basunbul, I. A., Baluch, M. H. and Ghaleb, B. N., 'Strengthening of initially loaded reinforced concrete beams using FRP plates',ACI Structural Journal 91 (2) (1994) 160–168.

[9] R. Goplalakrishnan, A. Mohan, L. Ponrajsankar, D. S. Vijayan, "Characterisation On Toughness Property of Self- Compacting Fibre Reinforced Concrete", Journal of Environmental Protection and Ecology 21, No 6, 2153–2163 (2020).

[10] D S Vijayan, Dr. J Revathy, Flexural Behavior of Reinforced and Pre-stressed Concrete Beam Using Finite Element Method, International Journal of Applied Engineering Research, ISSN 0973-4562 Volume 10, Number 1 (2015) pp. 717-736

[11] D.S.Vijayan, J. Revathy, Flexural Response of Fibre Reinforced Polymer Laminated Pre-stressed Concrete Beams, Indian Journal of Science and Technology, Vol 9(42), DOI: 10.17485/ijst/2016/v9i42/101824, November 2016.

[12] V. Gokulnath, D. S. Vijayan, P. Kathirvel, et al., Behaviour of retrofitted beams using natural fibers, Materials Today: Proceedings, https://doi.org/10.1016/j.matpr.2020.06.505

[13] Yousef A Al-Salloum, Sherif El-Gamal, Tarek H Almusallam, Saleh H Alsayed, Mohammed Aqel (2013) "Effect of harsh environmental conditions on the tensile properties of GFRP bars. Volume 45, Issue 1, February 2013, Pages 835-844.

SUSTAINABLE DEVELOPMENT ON FERROCK MORTAR CUBES

S. KARTHIKA[1], A. LEEMA ROSE[2], G. PRIYADARSHINI[3]

[1&3]student, civil Engineering department, SRM valliammai Engineering College, Chennai.

[2]Associate professor, civil Engineering department, SRM valliammai Engineering College, Chennai.

Corresponding Author karthikasundar96@gmail.com,

Abstract. Concrete, after water across the world, the second most broadly utilized material involving 8-10% of all yields of CO2, is predominantly because of cement. This project ultimately aims to determine the potential use of Ferrock as an exceptional replacement for cement in concrete compared with other alternative alternatives. It is a steel-based restraining compound used to form a carbon-negative structure substance utilizing waste material absorbents. The iron residue (an iron business loss) that would end in sites somehow alongside small quantities of limestone, metakaolin, and fly ash is being used to make this an efficient substance. Our research focuses unexpectedly on their commitment to carbon dioxide contamination, energy use, water use, the ecologic impact of ordinary Portland cement and Ferrock (limestone 8%, metakaolin 12%, fly ash 20%, and iron residue 60%). By subtitling concrete with Ferrock in fluctuating proportions of 5%, 10%, 15%, and 20% in solid, we attempt to find the ideal proportion of substitution, which, along with sustainability, would boost wanted outcomes for both (compressive and divided tensile). In all this proportion, the test result shows 10% is more efficient than others.

KEY WORDS: Ferrock, mortar cubes, compressive strength of cubes, split tensile strength of cylinders.

1. INTRODUCTION

Concrete is perhaps the most commonly used material for structures on highways, structures, spans, and various foundations. Every year, approximately 1 tonne of cement is supplied for every human on the planet. This widespread worldwide use requires an accurate assessment of the natural impact of this material. Today, the effects of the material on ozone-hazardous substances and environmental change have been measured. From this perspective, a long-term inscrutable discrepancy has been generated in the supposedly solid green concept. The solid is responsible for the highest CO2 feelings by various materials frequently based on incompletely supplanting concrete.

Only 5-7% of all anthropogenic CO2 generated in concrete companies worldwide was assessed as responsible. Since no other GHGs are transferred in this sector, only about 3 percent of the complete GHG emanations generated by human exercises are responsible. The test of this century is to turn it into another system that supports the universal framework. This includes a fraught analysis of available services to provide the local area with secure buildings and structures. Maybe a concerted effort is needed to create innovative and optional new production materials. The use of modern waste to decrease daily assets and resources and environmental contamination is an excellent way to promote green concrete.

The use of these waste materials saves between 14 and 20 percent concrete measurement. On the other hand, the global goal of reducing ozone-related outflows of substances by 2010 was fixed in the Kyoto Protocol by 1997 to slow the rate of environmental change. Produced nations are very attentive to their needs, and they have put forward an environmental change charge. The UK Government has introduced the same sort of expenditure on 1 April 2001 to achieve a 12.5% reduction in ozone-harming material emissions, the home-grown goal of the public authority being a 20% decrease in CO2 outflows by 2010. It is also clear that the solid industry paradigm must be shifted towards sustainable development if its condition as a predominant commodity is later to be maintained.

2. FERROCK CONCRETE

David stone is brainy who patented a new concrete technology known as ferrock. As per the available literature, ferrock is a binder that consists of iron powder, fly ash, limestone, and metakaolin to make industrial waste materials as novel substances.

The mechanism which is involved in the ferrock as,

Fe+ CO2 + H2O

FeCO3 + H2

The iron residue among the steel reacts with water and CO2, which form an iron carbonate matrix; when it dries, it exhausts rock-like qualities called ferrock. It is also the best choice to substitute for cement. However, concrete has its own tensional limitation and poor ductility. To increase the tensile property of concrete with the help of industrial by-products, i.e., Ferrock. Ductility is an important characteristic of structure to resist seismic, impact, and blast loading.

3. APPLICATION

While using ferrock, 8 to 11% of co2 emission is reduced and economical in construction.

Ferrock actually five times stronger than OPC, and it can withstand more compression before breaking and is far more flexible, which means it resists the earth's movement caused by seismic activity or industrial purpose.

Ferrock is best suitable for marine-based construction.

Journal of Physics: Conference Series **2040** (2021) 012020 doi:10.1088/1742-6596/2040/1/012020

4. MATERIAL TESTING

4.1. CEMENT

In this project work,53 grade of ordinary Portland cement is used. Different tests were performed for the accuracy as per the Indian standard. It is everlastingly alluring to utilize the best concrete in constructions. In this manner, the properties of a concrete ought to be examined in Figs. 1 and 2. Even though captivating concrete properties could change contingent upon the sort of development, typically a respectable concrete has following properties appeared in Table1

Table 1: Physical Properties of cement

s.no.	Physical Properties	Readings
1	consistency	29
2	Initial setting time	35 mts
3	Final setting time	385 mts
4	Specific gravity	3.14

Fig 1: Vicat Apparatus

Fig 2: specific gravity test

4.2. Properties of Fine Aggregate

The total consists of granular material acclimated in cement or mortar, and if granular particles of the squares are thus fine to combine with a sieve of 4.75 mm, it is understood to be a thin total appearing in Fig 3. Fine total. The construction company is commonly used to expand the amount of cement. In this regard, the material saves the value, and you can consider the whole scale, thickness, and area to look for the least difficult material in Table 2.

Table 2: Properties of Fine Aggregate

s.no	Physical properties	Readings
1	Sieve analysis	Zone II
2	Fineness modulus	2.9
3	Specific gravity	2.67
4	Source	crusher

Fig 3: sieve analysis and specific gravity test on fine aggregate

4.3. Properties of Coarse Aggregate

Totals are translucent or grainy rocks that are extricated to fully utilized inside the development business. Totals are essential in the construction and fixing things like streets, railways, houses, etc., Fig. 4. Totals are used in cement to favorable to duce economy inside the cement expense and do not respond with concrete and water. These are physical properties or attributes of the total that impact the properties of the following solid blend shown in Table 3.

Table 3: Physical Properties of Coarse Aggregate

s.no	Physical properties	Readinds
1	Fineness modulus	3.6
2	Specific gravity	2.76

Fig 4: specific gravity test on CA

5. Experimental Description

In the design mix, Cement: FA: CA was 1:1.61:2.8, and the M30 grade concrete used experimental content testing estimates. 0 percent of the ferrock is applied to the primary blend with the weight of cement. 5% of ferrock is used as a replacement for the eventual blending by weight of concrete. In essence, the weight of concrete is 10%, 15%, and 20% of ferrock powder. The ingredients and their precise dimensions are shown in Tables 4–5 below. At first, various exploratory documents and other data should be cautiously focused. Each of the materials used in the formulation of cement, such as concrete, sand, coarse total, and ferrock powder, had been tried and marked with their behavior in the table structure. The M30 grade is intended for the solid blend plan with the defined qualities. The total weight of all materials for proper mixing concrete is necessary quantities for all examples is calculated with the help of blend proportions. In this instance, a split tensile cylinder size 300 mm X 150 mm was given as an example of the size of 3D square 70.6 mmx70.6 m mm according to select percentage replacements. The planned examples were saved in a dry place for 24 hours to consolidate the concrete and then removed from its shape. After the examples have been reshaped, it must save 7, 14, and 28 days in the water tank to be cured. After restoring the example, the water content will be removed from the relief tank and drained for not many hours. For compressive power, split strength, the dried example has to be attempted. Without replacing ordinary cement, the effects of replaced examples have been contrasted. At what level of a solid rack is reaching its most incredible power, it has provided an opportunity. From the test results, the ideal measurement was noted. The results and discussions demonstrate the characteristics and strength of the cement.

5.1. Compressive strength on mortar cubes

The mortar cubes are of size 70.6 mm x70.6 mm x70.6 mm for different extents of 0%, 5%, 10%, 15% and 20% has been casted. Then cubes are kept for curing periods of 7,14, and 28 days respectively. It was tried in the compressive testing machine of capacity 2000kN which appears in Fig. 5.

5.2. Split tensile test on cylinder

The split tensile strength of concrete is one of the essential characteristics that significantly affect the magnitude and cracking inside structure Fig.6. Besides, because of its porous composition, the concrete is highly fragile in stress. Direct stress is not supposed to stand. So, when tensile forces surpass their strength, concrete develops breaks. Therefore, the strength of the concrete must be developed to function the pressure to allow the concrete components to break. Besides, giving strength may be a technique for developing the strength of concrete. The analysis of concrete cylinders, ASTM C496, was endorsed by this protocol and almost as alternate codes as IS 5816 1999.

Fig 5: compression test on cubes

Fig 6: split tensile test

Concrete Grade	Number of mortar cubes	Ferrock (%)	Fine Aggregate (Kg)	Ferrock (Kg)	Cement (Kg)
M30	9	0	4.86	0	4.86
	9	5	4.86	0.24	4,617
	9	10	4.86	0.5	4.37
	9	15	4.86	0.73	4.131
	9	20	4.86	0.97	3.9

Table 4: Mix proportion of mortar cubes

Concrete Grade	Number of cylinders	Ferrock (%)	Ferrock (Kg)	Coarse Aggregate (Kg)	Fine Aggregate (Kg)	Cement (Kg)
M30	9	0	0	53.05	29.07	18.99
	9	5	0.95	53.05	29.07	18.041
	9	10	1.899	53.05	29.07	17.1
	9	15	2.855	53.05	29.07	16.14
	9	20	3.798	53.05	29.07	15.192

Table 5: Mix proportion of cylinders

Grade	Ferrock (%)	Compressive strength(N/mm^2)		
		7 days	14 days	28 days
M30	0	21.8	28.34	42.5
	5	23.8	30.94	43.7
	10	24.5	31.85	44.1
	15	19.75	25.7	27.5
	20	17.02	22.126	26.1

Table 6: Compressive strength of M30 mortar cubes

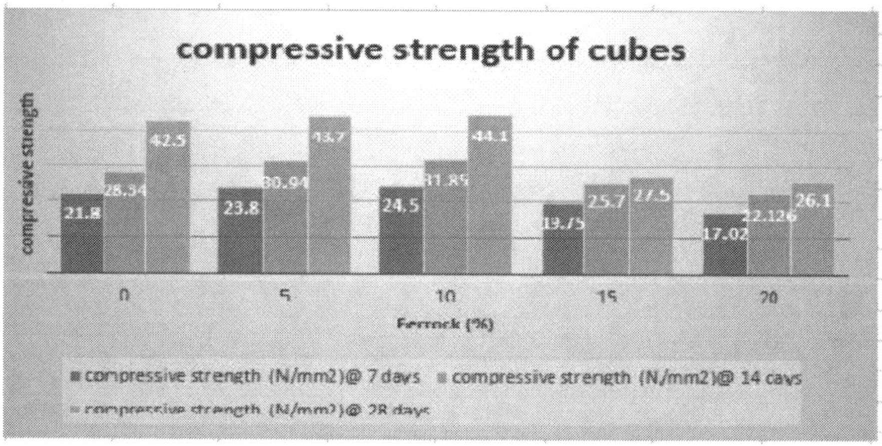

Fig 7: compressive strength for M30 grade

Grade	Ferrock (%)	Split tensile strength (N/mm^2)		
		7 days	14 days	28 days
M3O	0	2.56	2.91	3.35
	5	2.67	2.98	3.61
	10	2.82	3.05	3.75
	15	2.43	2.71	3.05
	20	2.32	2.65	2.91

Table 7: Split tensile test on cylinder

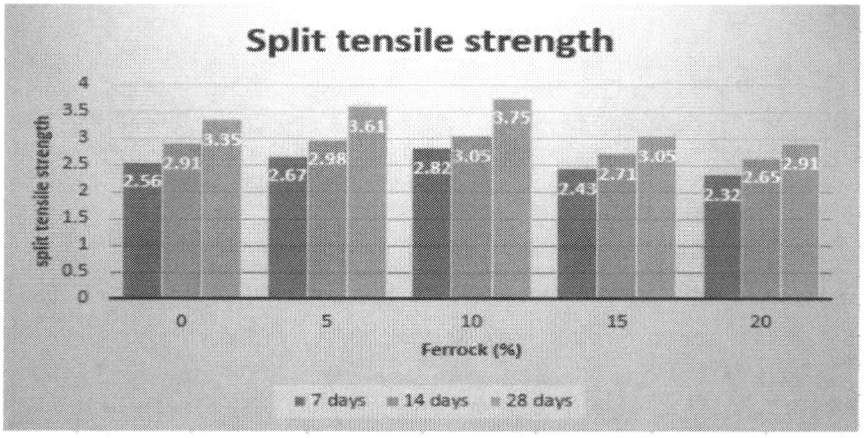

Fig 8: split tensile strength on cylinder

6. RESULT AND DISCUSSION

Compressive strengths are 21.8 N/mm2, 28.34 N/mm2, and 42,5 N/mm2, respectively, for the managed mix's cube for 7, 14, and 28 days. The compressed concrete capacities are 5 percent and 10 percent higher than standard concrete for 28 days. The size is 43.7 N/mm2 and 44.1 N/mm2. The cylinder split tensile test for the regulated mix is 2.56 N/mm2, 2.91 N/mm2, respectively, for 7, 14, and 28 days, 3.35 N/mm2. The 5% and 10% rack cement splitting tensile test for 28 days is higher than ordinary concrete. The figures are 3,61 N/mm2 and 3,75 N/mm2.

7. CONCLUSION

The Ferrock testing was performed in the current work to increase material strength when used as a solid block as an alternative for the reinforced concrete cement. The compressive strength test and the 7, 14, and 28-day Split Tensile tests were then completed. With 5 percent and 10 percent respectively, the compressive strength of the concrete in the ironwork has been increased to 43,7 N/mm2 and 44,1 N/mm2. Results show that the 10 percent share increased strength as compared to various concrete proportions. The divisional tensile measure on the rock concrete has been strengthened separately by 3.61 N/mm2 and 3.75 N/mm2, with a division of 5% and 10%. It is evident from the results that the 10% proportion increased strength in comparison to different concrete proportions. The findings of the testing of the compressive strength and the split tensile test have shown that the 10% of ferrock concrete results better in comparison to different proportions. The test results thus reveal that the 10% share is suitable for use in the production market.

8. Future scope

- With the obtained percentage of Ferrock, to conducting the experimental investigation on flexural behavior of composite beam.
- And it compared to do the analysis with the help of software.

REFERENCES

[1] Papayianni E Andtansion. "Production of high strength concrete using high volume of industrial by-products". Journal of construction materials, 2010, 1412-1417.

[2] Ammar A. Ali, Saad N. Sadik, Dr.wael S.Abdul sahib-"strength and ductility of concrete encased composite beams", Journal of engineering and technology, volume no 30, p.p no.2701-2714,2012.

[3] Xian li, Henglin Lv, Shuchun Zhou – "Flexural behavior of GFRP –reinforced encased steel composite beams", Journal of construction and building material, volume no 28, pp. no. 255-262,2012

[4] J. Rosales, M. Cabrera, F. Agrela. "Effect of stainless-steel slag waste as a replacement for cement in mortars. Mechanical and statistical study", journal of construction materials, 2017, 444-458.

[5] ChangzaiRen , Wenlong Wang , Guolin Li, " Preparation of high performance cementitious materials from industrial solid waste". journal of construction materials ,2017, 39-47.

[6] Alejandro Lanuza Gracia, Ashik Thitira Achaiach, John Bello, Thomas Donovan, "Ferrock: A life cycle comparision to ordinary Portland cement", ISE 576- Industrial ecology (April 24-2017)

[7] Naveen Kumar K.,Vijayan D.S.,Divahar R.,Abirami R.,Nivetha C.,"An experimental investigation on light-weight concrete blocks using vermiculite", Materials Today: Proceedings, Volume 22, Pages 987 – 991, 2020, 10.1016/j.matpr.2019.11.237

[8] Pradeep Jadhav, Vicky Jadhav, Giridhar Narule, shahid mulani, "co_2 absorbing concrete block", IJTEEER vol.2, issue-7,2347-4289

[9] D.S.Vijayan, C. Nivetha, P. Dinesh Kumar, B. Saravanan, V. Gogulnath, "Green Composite Form of Eco – Friendly Concrete by Adding PVA Fiber," Journal of Green Engineering (JGE), Volume-10, Issue-6, June 2020.

[10] D.S.Vijayan, Dinesh kumar, S Arvindan, Thattil shreelakshmi janarthanan, "Evaluation of Ferrock: A greener substitute to cement", journal of materials 2019.

[11] IS 383: 1970- Specification of Coarse Aggregate and Fine Aggregate Bureau of Indian standards, New Delhi.

[12] IS 10262 – Concrete Mix Design Bureau of Indian Standards 2019 New Delhi.

STUDY ON BEHAVIOUR OF DEEP BEAM STRENGTHENED WITH GFRP SHEETS

A. Meena[1], K. Suganya Devi [2], A. Sattainathan Sharma[3]

PG Student 1, Department of Civil Engineering, SRM Valliammai Engineering College, SRM Nagar, Kattankulathur, Chengalpattu, Tamil Nadu, India
Assistant Professor 2&3, Department of Civil Engineering, SRM Valliammai Engineering College, SRM Nagar, Kattankulathur, Chengalpattu, Tamil Nadu, India
Corresponding Author Email: meenamma.1698@gmail.com1

Abstract: This article discusses the behavior of a reinforced concrete deep beam that has been strengthened with GFRP sheets. The GFRP shows many essential and necessary properties such as low weight to strength ratio, non-corrosive, high fatigue strength, high tensile strength, etc. In this paper, four deep beams were cast, in which one beam is a control beam, and another one is with special confinement reinforcement. The rest of the two beams are strengthened with the GFRP sheets in standard control beam and special confinement reinforcement beam under static or monotonic loading under three points, and the corresponding behavior is determined. The behavior of the RC Deep beams, such as load-carrying capacity, first crack load, load-deflection behavior, crack patterns, and stress vs. strains, are determined. The analysis of the reinforced concrete deep beams was done using the finite element software called ANSYS. Additionally, this article provides preliminary support for proposing a novel shear strengthening technique during the member's design. In the finite element simulations, the proposed deep beams have been modeled and analyzed using the finite element software, ANSYS, and the results are compared with experimental work.

Keywords: RC Deep Beams, GFRP Sheets, special confinement reinforcement, shear behavior, ANSYS software.

I. INTRODUCTION

The Reinforced Concrete (RC) deep beam is a beam with a depth comparable to the span length or Beams with a significant depth to span length, or thickness quantitative relation is mainly referred to as deep beams. Different design codes propose different values for the span to depth quantitative relationship (Le/d) and the shear span to depth quantitative relationship (a/d) when defining deep beams. In step with the Indian Standard Code (IS 456:2000), the deep beam with depth/thickness or depth/span length is less than 2.5 for concentrated load and depth/thickness, or depth/span length is less than 5.0 Uniformly Distributed Load. [1].

Enhanced Concrete deep beams are used in various structural applications, including offshore structures, transfer girders, wall footings, foundation pile caps, tall buildings, floor diaphragms, and shear walls. Reinforced Concrete deep beams are usually used as load-distributing structural elements wherever the space between two columns or supports is ample to resist or withstand the bending moment that occurs due to heavy load.

The reinforced concrete deep beams have an excellent flexural strength when put next to the shear strength because of its shear span to depth quantitative relation. A more considerable depth resists the beam from the cause of the bending. The shear failure essentially happens in

RC deep beams instead of traditional RC beams because of the dimensional configuration. The defection could additionally happen at the diagonal portion and thus creates the diagonal shear cracks. If the depth of the profound beams increases, the shear strength decreases. Hence, the Shear carrying capability should be augmented within the RC deep beams while not increasing the depth of the deep beams [2].

1.1 STRENGTHENING TECHNIQUES

Structural strengthening is that the method of upgrading structures to boost their structural performance below existing loads or to enhance the strength of the structural members to resist the extra loads. The strengthening of the reinforced concrete structures is essential to increase the load carrying capability of the structural compenent, extend the flexural and the shear strength of the structures, decrease the propagation of the crack, and the development of the crack width and its dimensions [3]. There are several techniques in strengthening the reinforced concrete structures victimization Steel jacketing, Reinforced concrete jacketing, FRP confining or jacketing, the other replacement of various reusable materials or fibers with the partial quantity of cement or aggregates, etc.,

1.2 FIBER REINFORCED POLYMERS

Throughout the past years, strengthening or retrofitting structural members using Fiber Reinforced Polymers (FRP)is wide inflated for its various properties. Those fibers are generally made up of Natural fibers and Synthetic fibers. Some synthetic fibers are glass, carbon, polyolefin fibers, basalt fibers (volcanic rock fibers), or aramid, though different fibers like paper or amphibole are typically used. Some of the Natural fibers are jute, wood, coir, etc. The polymer is usually made out of epoxy, vinyl, or polyester plastic and resins of methanol are still in use. The regional, aerospace, car, marine, and construction industries typically use FRPs [5]. Fiber Reinforced Polymers (FRP) has become wide to extend or increase the high strength of the structural member to carry the heavy loads and increase the load-carrying capacity of the particular member. All FRPs have entirely different properties and have different strength parameters in keeping with the raw state of the fiber from wherever it is obtained. The price additionally varies in keeping with the supply and also the strength it gains. The FRPs are accessible in the various states as reinforcing bar, sheets, dry state, laminates, cloth, etc., each has its specific output when incorporated with the structural elements [6].

1.3 GLASS REINFORCED POLYMER SHEETS

The required strength for the structural components can be achieved by using some technologies with Glass Fiber Reinforced Polymers (FRP). The Glass Fiber Reinforced Polymers (GFRP) Sheets are used to enhance the shear carrying capacity of the deep beams. The GFRP sheets are most economical and readily available when compared to the other types of polymeric fibers such as Carbon Fiber Reinforced Polymers (CFRP), Basalt Fiber Reinforced Polymers (BFRP), Aramid Fiber Reinforced Polymers (AFRP), etc [7]. The characteristics properties of glass fibers are high strength, less weight, economy, low maintenance, good design flexibility, durability, chemical resistance, good tensile strength, etc., A key reason for using GFRP is their corrosion resistance, and one of the main reason is the fact that it does not conduct electricity. The glass fibers are available in many types, such as reinforcement rods, sheets, fabrics, the dry state in case of replacement, etc. The externally bonded glass fiber sheets with layers give additional stability and ductile property to the beam and decrease the reach of shear failure. The glass fiber sheets are also wrapped externally and bonded with the epoxy resin with its respective hardener to the RC deep beams. The GFRP sheets are available in different sizes and thicknesses, and the individual sizes show different efficiency [8].

1.4 SPECIAL CONFINEMENT REINFORCEMENT

The deep beams with special confining reinforcement with variation in the stirrups reinforcement give better ductility. Usually, to resist the shear failure, web reinforcements are provided to the RC beams. The capacity of the RC beam is generally the amount of Shear and moment that it can withstand. The spacing of the shear reinforcement at the center should be maximum, and at the support, it should be minimum. According to IS 13920: 1993, Cl: 6.3.5, Beam shall not exceed d/4 and eight times the diameter of the smallest longitudinal reinforcement bar on the shear side or zone. On either side, the beam shall not exceed d/2 of the longitudinal reinforcement on the flexure zone [9]. There is only limited literature in strengthening RC deep beams with Glass Fiber Reinforced Polymers (GFRP) and unique confining reinforcement methods. I have studied the various literature or review papers and described the research they have done in this paper [10].

As examined and described here, Extensive research work, Amr H. Zaher and et al (2019) inverstigated behaviour of RC beam by adopting glass fiber and carbon to enhance the shear strength. The beams have been divided in to five sections so that CFRP sheets shows better peformence by enhacning strength of 2,3, and 4th sections whereas GFRP provides better resistance 5th sections [1]. M. R. Islam and et al. (2004), This study explores a reinforcing technique using externally bonded FRP systems for structurally deficient deep beams. Three different availability of carbon fibers are used. They are fiber wrap called MBrace 130, in the form of strips called SIKA and girds called NEFMAC. The FRP grids placed in the normal orientation are concluded to be the most effective. An improvement of about 40% is achieved in this paper [8]. Rizwan Azam and et al. (2018), This literature explains the effectiveness, by contrast to epoxy-based carbon fiber-reinforced polymers, of cement-based composites for strengthening RC-deep beams. This paper conveys the concept because of the bi-directional fabric used in cement-based enhancement compared to the unidirectional sheets in the high-bonding epoxy-based enforcing system and its enhanced connection to the concrete substratum. The total thickness of the FRP used in this strengthening is about 6-8mm for Carbon Fiber Reinforced Cement Mortar and 8-12mm for CFRP grid in mortar. The epoxy-based system (CFRP) sheets are applied on the beams as per the manufacturer's requirements and specifications. The appearance and development of the cracks were readily visible and indicated a failure in the case of the cement-based mortar; on the other hand, the beams strengthened with CFRP sheets do not indicate failure and visible cracks [14].

From the research as mentioned above papers, it is concluded that using Fiber Reinforced Concrete (FRP) may be of any types will increase the structural behavior of the RC members, such as ultimate load-carrying capacity, ductility, shear strength, flexural behavior, and decrease in the failure modes, propagation of crack width, etc., The use of CFRP gives better results than any other types of frps. However, it is higher in cost when compared to other types of firms. The GFRP also gives additional strength in the case of flexure and Shear and the ultimate load carrying. Hence, while increasing the shear reinforcement in the shear zone than in the flexural zone, it can attain additional strength. When it is strengthened more with the GFRP sheets, it increases the load-carrying behavior, shear strength, etc. Three-point loading is given rather than two-point loading to determine the shear behavior of the beam. Analysis of the Control beam and GFRP strengthened with and without confinement reinforcement using (FEA) ANSYS software.

2. EXPERIMENTAL PROGRAM

To study the behavior of the deep beam strengthened with GFRP sheets, a full four beams were cast. Out of these four beams, two beams are the standard control beam, and the other two with special confinement reinforcement. Moreover, the one standard control beam and one special confined reinforced beam are wrapped with the GFRP sheets. The casted beams were tested under three-point loading conditions, and the respective load-deflection behavior, enhancement of shear capacity, failure modes are noticed. Details of the preliminary test carried out, specimen casting, GFRP sheets or fabric wrapping, experimental setup, testing and observation, results, and discussions are presented below [11].

2.1 PRELIMINARY TEST FOR MATERIALS

Cement could be a binding agent used to build it, hardening it and adhering to other materials to bind it. The cement used is OPC 53 grade cement. Various tests were conducted for preliminary tests for cement. The fineness value is 2, which is less than 10 as per is code (IS 4031 - Part 1 - 1996), Specific gravity is 3.15, which is within the range 3.1 – 3.16 (IS 2720 - Part 3 - 1980), Consistency ratio is 32%, which is within the range 25 – 35% (IS 4031 - Part 4 - 1988), Initial setting time is 32mins and Final setting time is 360 mins, which is within the range >30 mins and <600 mins as per code (IS 4031 - Part 5 - 1988) [12].

Fine aggregate usually comprises natural sand or crushed stone. The fine aggregate aims to fill the pores between the coarse aggregates in concrete, therefore reducing the porosity of the concrete paste and increasing the strength. M sand is employed during this project. The specific gravity value is 2.58, which is within the range of 2.6 – 2.8 as per code (IS 2386 Part – 3 – 1963), Fineness modulus corresponds to grading zone II (90 – 100% passed through 4.75 mm sieve), and the value is 2.8, within in the range (2.2 – 3.2) (IS 383 – 2016).

Coarse aggregates are defined as materials with a sieve size greater than 4.75 mm. Coarse aggregates are a type of rock quarried from underground deposits. Usage of coarse aggregates will reduce the shrinkage that occurs. Specific gravity obtained is 2.66, within the range between 2.6 – 2.85 (IS 2386 – Part – 3 – 1963). Fineness modulus passed 90% through 20 mm sieve, and the value obtained is 7.5, which is between the range 6 – 8.5 (IS 383 – 2016), Water Absorption is 0.5%, which is within the range of 0.5 % to 2 % (IS 2386 – Part – 3 – 1963) [13].

2.2 CASTING OF SPECIMENS

Concrete grade of M30 and Steel grade of Fe415 is adopted. Mix design was done as per IS:10262-2009, in which the proportion of cement, fine aggregate, coarse aggregate with a water-cement ratio of 0.45 is 1:1.5:2.7 per cubic meter. Four rectangular concrete deep beams with 1000 × 420 × 120 mm specifications are adopted and cast for the experimental program. Before the specimen casting, six cube specimens were cast with 150 × 150 mm dimensions and cured for 7 and 28 days. The average compressive strength of the cube for the 7th-day test obtained is 23.45 N/mm2 and the average compressive strength of the cube for the 28th day obtained is 35.76 N/mm2. The slump value obtained for the fresh concrete after three trials are 75 mm, a true slump [14]. The reinforcement detailing of the deep beam is shown in figs a and b. fig 1 and 2 displays plan of normal cocntrol beam and Special Confinement Reinforcement (BSCR).

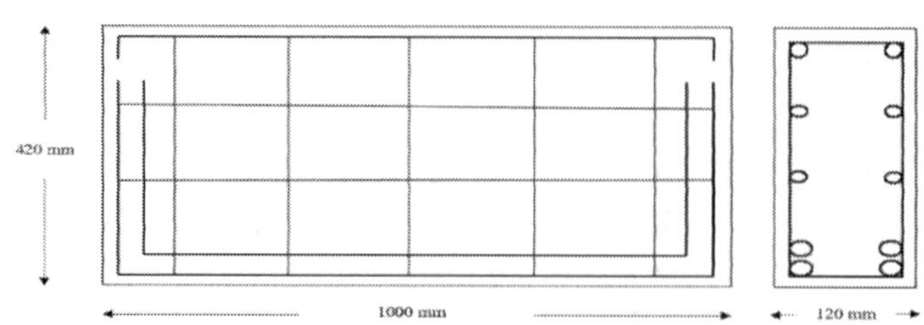

Figure. 1 Normal control beam (CB)

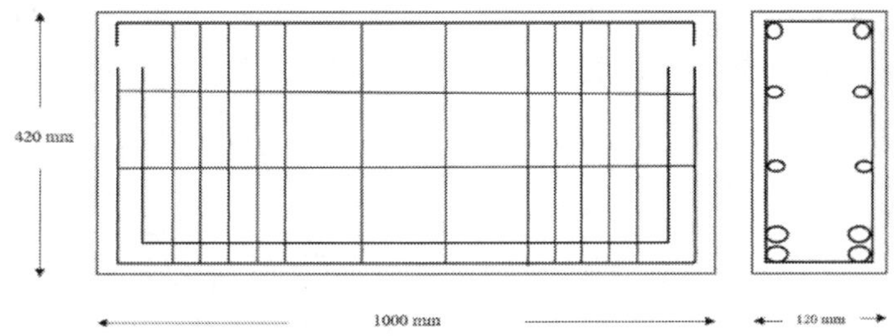

Figure. 2 Special Confinement Reinforcement (BSCR)

The tension reinforcement is four no. of 12 mm ϕ bars, compression reinforcement is two no. Of 10 mm ϕ bars, the skin reinforcement adopted is four no. of 8 mm ϕ bars, while the shear reinforcement is given as 8 mm, ϕ 2-legged stirrups @ 200 mm c/c spacing for the standard control beam (CB) shown in table 1 . The tension, compression, and skin reinforcement for the special confinement reinforcement (BSCR) are adopted as same as given for standard control beam while the shear reinforcement is given as 8 mm ϕ 2-legged stirrups @ 64 mm c/c spacing in shear zone and @ 110 mm c/c spacing in flexure zone [15]. Fig 3,4 and 5 depicts reinforcemnt details and catsting of beams.

Table 1 Designation for the beam specimens

SL.NO	BEAM DESIGNATION	GFRP LAYERS
1	CB	-
2	BSCR	-
3	SCB	3
4	SBSCR	3

Fig. 3 Reinforcement detailing

Fig. 4 CB specimens

Fig. 5 BSCR specimens

2.3 WRAPPING OF GFRP SHEETS ON TEST SPECIMENS

The GFRP sheets used for this project study are a Uni-directional cloth fiber with 0.3 mm thick and is usually made with cross glass fiber. These Uni-directional cloth fibers are wrapped all over the beam with the Epoxy resin with its hardener in a 1:9 ratio. First, the resin and hardener were mixed with proper proportion and then applied over the deep beam with the help of the roller after the beam surfaces are adequately cleaned. The GFRP sheet wrapped beams are cured for 3-4 days, and the tests are processed [16].

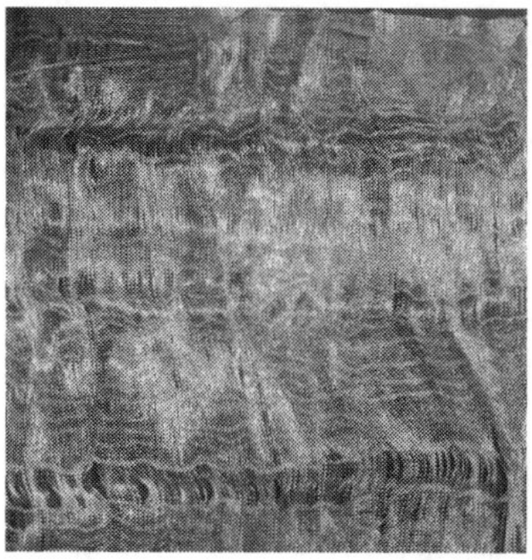

Fig. 6 Uni-directional glass cloth fiber

Fig. 7 GFRP wrapped beams

2.4 LOADING SETUP

The deep beams were tested under a three-point loading condition where the loadings point is given with the distance of l/4 on roller supports. The effective length is 800 mm. 100mm is taken for the support region. The loading frame used in this project has a capacity of about 1000 kN. The load from the hydraulic jack is transferred to the spreader beam, and it is split into three, where the load enters into the beam shown in fig 8. Here the load is applied through a pressure gauge. The deflection values are observed with the help of two dial gauges kept under the beam, each one at both of the supports. Loads were applied continuously with the constant rate, and the corresponding load and deflection value is noted [17]. Fig 6 and 7 represents GFRP and glass cloth fiber in specimen.

Fig. 8 Load set up in the loading frame

3. RESULTS AND DISCUSSIONS

During testing, the deflection for the corresponding applied load is detected using a dial gauge. The two dial gauges are kept at a point below the loading points, each at either support of the span. The load-deflection behavior is noticed. The deflection increases with the load's increment and increases after the reduction in the load after it reaches the ultimate load.

3.1 LOAD VS DEFLECTION

The load-deflection curve for the four specimens at the right side of the span and the left side of the span is shown in fig. 9 and 10, respectively.

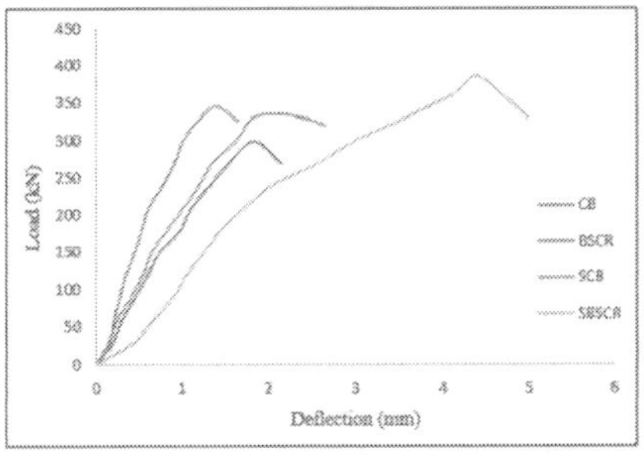

Fig. 9 Load vs. Deflection curve at the right side of the span

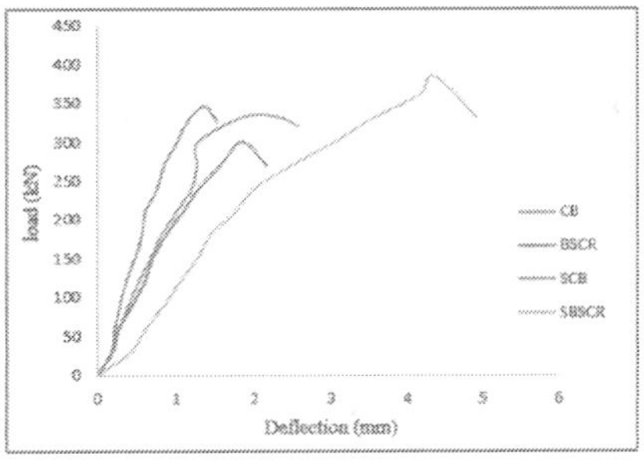

Fig. 10 Load vs. Deflection curve at left side of the span

From the above graphs, fig.9 shows the maximum deflection of CB is 1.83 mm at 300 kN, whereas the maximum deflection for BSCR is 1.4 mm at 345 kN. The maximum deflection of SCB is 2.11 mm at 335 kN, whereas the maximum deflection for SBSCR is 4.32 mm at 385 kN. Fig.10, the maximum deflection of the CB was 1.85mm at 300 kN, whereas the maximum deflection of BSCR was 1.36 mm at 345 KN. The GFRP strengthened beam SCB shows a maximum ultimate deflection of 2.13 mm at 335 kN, whereas the maximum deflection of SBSCR is 4.38 mm at 385 kN. Compared with the CB, the load-carrying capacity of the special confined reinforcement (BSCR) was increased. The deflection increases as the applied load increases, showing more significant load carrying capacity until the ultimate load. The load-deflection behavior shows slight variations in their curve, and finally, the right side shear region fails due to shear failure.

The initial stiffness of the CB was 197.53 kN/mm. The special confined reinforced beam (BSCR) shows a high stiffness value than CB, which shows that the resistance for deformation increases and exhibits high load-carrying capacity. The SCB also reaches a high stiffness value when they are strengthened with GFRP sheets, whereas SBSCR shows a low stiffness value due to the debonding of the GFRP sheets after the several increment of the applied load, the deflection value increases and also exhibits more significant load-carrying capacity. The energy absorption capacity of the CB, measured as the area under deflection, is the smallest as measured by the slope of the load-deflection curve, and thus the CB has a low ductility compared to the other specimens. The energy absorbed by the BSCR increases up to the ultimate deflection and shows high ductile behavior for the beam with additional shear reinforcements. The GFRP strengthened beams SCB and SBSCR also show high energy absorption capacity in which SBSCR exhibits more excellent absorption than SCB, and the ductile behavior also increases as per the slope of the load that it reaches [18].

3.2 CRACKING PATTERN

Testing is done, and the corresponding observations are recorded. The control beam (CB) reaches the ultimate load of 300 kN with the first initial diagonal crack in the proper side support of the beam, and the first crack load is at 160kN. After the increment of the load, the propagation of crack increases into the widened diagonal crack, and also the crack develops on the left side of the span, and finally, the beam leads to shear failure at the right side shear span region.

Fig. 11 Failure mode of CB Fig. 12 Failure mode of BSCR

In BSCR, the initiation of the crack was delayed after several increments of the load, and the first minor crack was observed at 185kN, and the ultimate load reaches 345 kN. The crack that occurred is a minor crack that starts on the right side shear span at the bottom face of the span, and the beam fails due to the shear failure. The formation of the first yield load cracks is also delayed in BSCR and hence carried a higher load when compared with CB [19]. Fig 11 and 12 shows failiure mode.

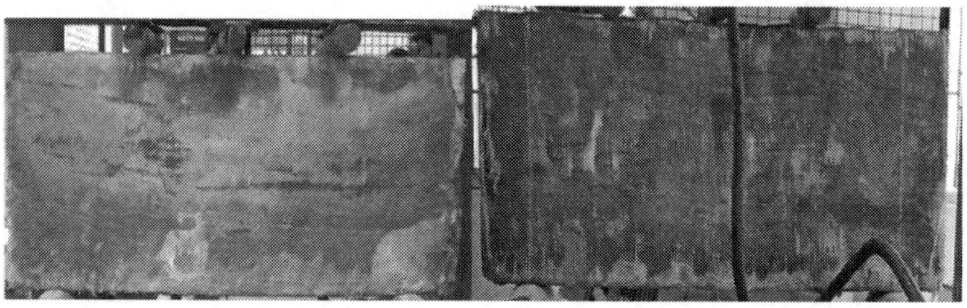

Fig. 13 Debonding of SCB Fig. 14 Debonding of SBSCR

The GFRP strengthened beams indicate the debonding of the strengthened GFRP sheets with debonding sound at the bottom face of the surrounding the whole beam at the tension face, and the shear cracks may be obtained at the shear region of either of the beams. The SCB beam reaches the ultimate load up to 335 kN with a maximum deflection of 2.12 mm with debonding sound, whereas the SBSCR beam reaches the ultimate of about 385 kN with a maximum deflection of about 4.35 mm[20]. The GFRP strengthened beams SCB and SBSCR show higher load carrying capacity and show high ductile behavior compared with CB and BSCR. The deflection increases gradually more than the CB and BSCR due to the debonding of the GFRP sheets at the bottom face of the beam at the shear zone [21]. Fig 12 and 13 depicts debonding of SCB and SBSCR. Table 2 provides experimental outcomes of the testied beams.

Table.2 Experimental results of the tested specimens

Test Specimens	First Crack Load (kN), P_y	First Crack deflection (mm), δ_y	Ultimate Load (kN), P_u	Avg. Max. Deflection (mm), δ_u
CB	160	0.83	300	1.84
BSCR	185	0.59	345	1.38
SCB	196	0.93	335	2.12
SBSCR	240	1.81	385	4.35

Table 3 Load deflection characteristics

Test Specimens	Initial Stiffness (kN/mm)	Energy Absorption (kN.mm)		Ducltility factor	
		Absolute	Relative	Absolute	Relative
CB	197.53	2755	1	2.22	1
BSCR	240.26	3597	1.31	2.34	1.05
SCB	210.75	5384	1.95	2.28	1.03
SBSCR	118.81	11405	4.14	2.40	1.09

4. ANALYTICAL WORK

The finite element method (FEM) for solving engineering and mathematical model issues is the most widely used. ANSYS WORKBENCH 19.2 version of finite element software is used for taking the analytical result for this study. The model of the beam is created as per the beam specifications, and the support conditions are applied. The meshing is done using excellent mesh in order to obtain an accurate result. The loadings are given in the experimental program, and the corresponding load-deflection behavior is obtained [22]. Fig 14,15,16,17,18,19 shows behaviour of beams thorugh ANSYS modelling.

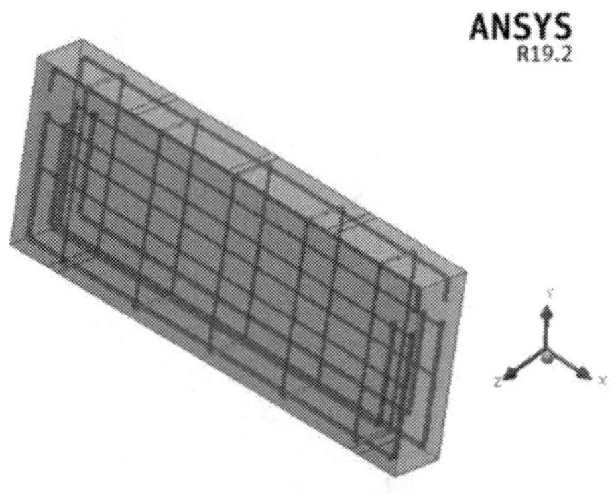

Fig. 15 Modelling of CB

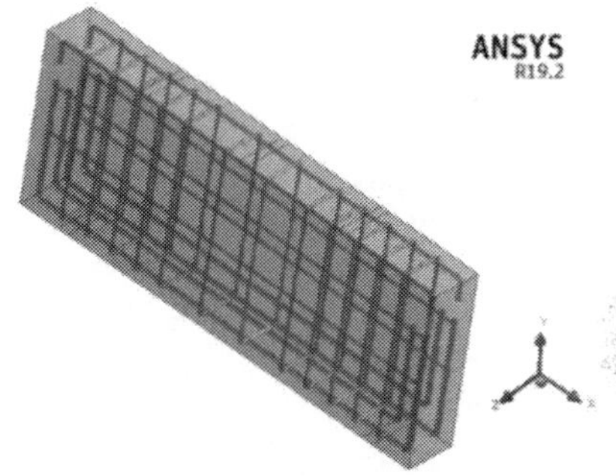

Fig. 16 Modelling of BSCR

Results obtained from the analytical study are the deformation behavior with the applied loads, which we observed from the experimental study. The beam specimens CB, BSCR, SCB, and SBSCR, are analyzed in ANSYS, and the corresponding deflection is shown in the below figures.

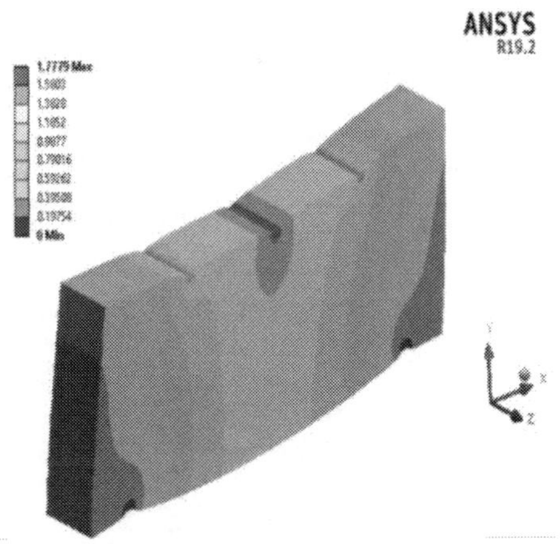

Fig. 17 Deformation of CB

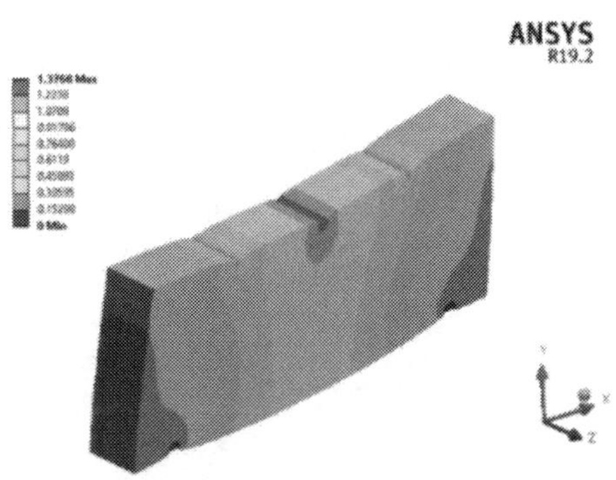

Fig. 18 Deformation of BSCR

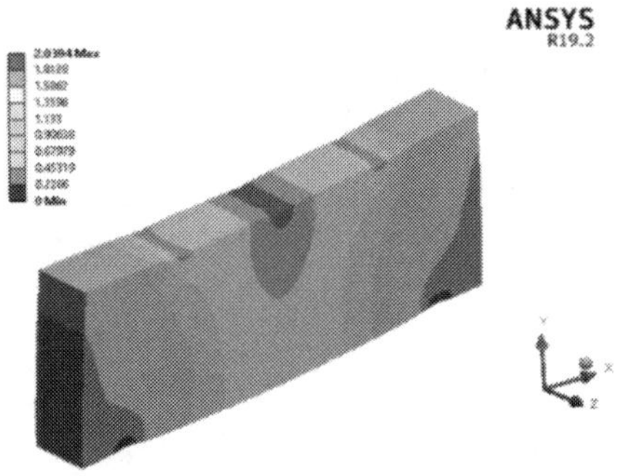

Fig. 19 Deformation of SCB

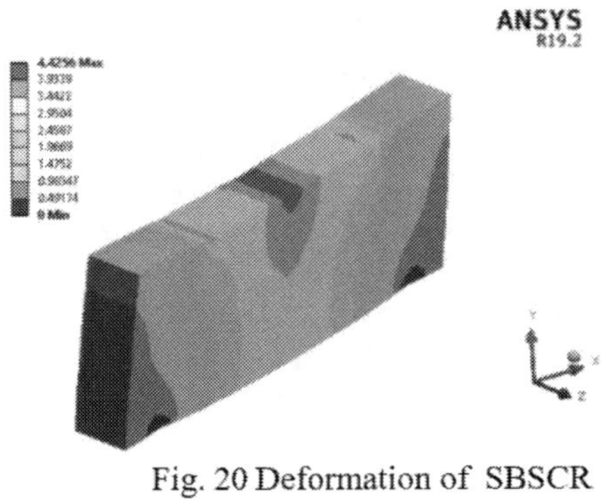

Fig. 20 Deformation of SBSCR

4.1 RESULTS AND DISCUSSIONS

In the analytical study, it is clearly shown that the deformation of the BSCR is reduced due to the implementation of the special confined reinforcement when compared with the control beam (CB). The GFRP sheets strengthened beams are also analyzed, and it shows the increase in the ultimate load-carrying capacity and the increase in the deflection as the GFRP gets debonding. Finally, the comparison between the experimental and the analytical results shows a good correlation in their percentage of variation. Nearly 96% of the experimental observations match the analytical results and are similar and are shown in table.4. The BSCR shows a 30% decrease in the deflection compared with the CB due to the special confining reinforcement [23].

Table. 4 Comparison of Analytical and Experimental results of the test specimens

Test Specimens	Ultimate Load (kN)	Maximum Deflection (mm)	
		Experimental Result	Analytical Result
CB	300	1.84	1.78
BSCR	345	1.38	1.37
SCB	335	2.12	2.04
SBSCR	385	4.35	4.43

5. CONCLUSION

- The experimental and the analytical study was conducted to investigate the effectiveness of the special confined reinforced deep beam and GFRP strengthened deep beams with the control beam, and the following conclusions are drawn.
- The special confinement reinforcement (BSCR) effectively increased the shear capacity of the deep beam by increasing its load-carrying capacity, which also delays its initial first crack load. It shows a 15% increase in ultimate load when the comparison is made with the control beam (CB).
- The GFRP sheets, when it is wrapped with the control beam and special confined beam, it shows an 11.6% increase in their ultimate load when both the test specimens SCB and SBSCR compared with CB.
- The Energy Absorption increases in BSCR with the area under deflection in the load-deflection slope and hence exhibits the high ductile behavior. The BSCR shows a 5.4% increment in their ductile behavior when compared with CB. The Energy Absorption capacity of the GFRP strengthened beams is high compared to the beams unstrengthen with GFRP sheets. SCB and SBSCR show a 3% and 8.1% increase in their ductile behavior than the control beam (CB).
- The Initial Stiffness value for the BSCR increases about 22% compared with the

control beam, which shows that the beam's resistance for deformation is enhanced. The initial stiffness of the SCB shows a 6.7% increase in their resist for deformation, whereas the SBSCR decrease in their stiffness due to the debonding of the GFRP sheets highly due to the high load-carrying capacity is reached.

- The analytical results also correlate with the experimental results in which the BSCR shows a 30% decrease in their deflection range to the CB. Both the experimental and the analytical results show 96% similar matches in their values. The analytical results show the BSCR is highest and it considerably increases the shear capacity of the deep beam.

REFERENCES

[1] Amr H. Zaher and et al. (2019), Strengthening and repairing RC deep beams using CFRP and GFRP, International Journal of Civil Engineering and Technology (IJCIET), Volume 11, Issue1.

[2] Giuseppe Campione Ph. D (2012), Flexural Behaviour of Steel Fibrous Reinforced Concrete Deep Beams, ASCE-Journal od Structural Engineering,138(2),235-246.

[3] Gouda Ghanem and et al. (2016), Behaviour of RC beams Retrofitted/Strengthened with external post-tensioning system, Research Gate- International Journal of Civil, Mechanical and Energy Science (IJCMES), Volume 2, Issue 1.

[4] Hussam El-Din. El-sheikh and et al. (2015), Application of Glass Fiber Reinforced Polymer Rods as an Alternative to Iron in concretes, Research Gate-Journal of Academy for Basic and Applied Sciences, Volume 14, no.2.

[5] Jiangfeng Dong and et al. (2012), Structural behaviour of RC beams with external flexural and flexural-shear strengthening by FRP sheets, Elsevier-Composites: part B, 604-612.

[6] Kumari et al. (2018), Shear strengthening of RC deep beam using externally bonded GFRP fabrics, The Institution of Engineers,99(2):341-350.

[7] Mohamed Said and et al. (2020), Experimental and Analytical Investigation of high-performance concrete beams reinforced with hybrid bars and polyvinyl alcohol fibers, Elsevier-Construction and Building Materials, 120-395.

[8] M. R. Islam and et al. (2004), Shear strengthening RC deep beams using externally bonded FRP systems, Elsevier-Cement &Concrete Composites (413-420).

[9] D.S.Vijayan, S.Arvindan, K.Naveen Kumar, S Mohamed Javed, Seismic Performance of Flat Slab in Tall Buildings with and without a Shear wall, International Journal of Engineering and Advanced Technology (IJEAT) ISSN: 2249 – 8958, Volume-9 Issue-1, October 2019, Page no – 2672 – 2675.

[10] Rajai Z Al-Rousan and et al. (2016), The effect of beam depth on the shear behaviour of reinforced concrete beams externally strengthened with carbon fiber-reinforced polymer composites, SAGE-Advances in Structural Engineering Volume 19(11) 1769-1779.

[11] Rami A. Hawileh and et al. (2013), Behaviour od reinforced concrete beams strengthened with externally bonded hybrid fiber-reinforced polymer systems, Elsevier-Materials and Design, 972-982.

[12] Ravikant Shrivastava and et al. (2013), Effect of Cyclic loading on flexural behaviour of FRP strengthened RC beams: A stability point approach, International Journal of Advanced Research in Engineering and Applied Sciences, volume 2, no.6.

[13] R. Balamuralikrishnan and et al. (2009), Flexural Behaviour of RC beams strengthened with Carbon Fiber Reinforced Polymer (CFRP) Fabrics, The Open Civil Engineering Journal, Volume 3, 102- 109.

[14] Rizwan Azam and et al. (2018), Shear strengthening RC deep beams with cement-based composites, Elsevier-Engineering Structures, 929-937.

[15] Weiwen Li and et al. (2015), Shear Span-Depth Ratio effect on the behavior of RC beam Shear Strengthened with Full-Wrapping FRP strip, ASCE-Journal of Composites for Construction, ISSN 1090-0268.

[16] Zhichao Zhang and et al. (2005), Shear strengthening of Reinforced Concrete Beams using Carbon-Fiber Reinforced Polymer Laminates, ASCE-Journal Composites for Construction (158-169).

[17] Indian Standard Code of 456: 2000 - Plain and Reinforced Concrete Code of Practice.

[18] Indian Standard Code of 13920: 1993 - Ductile detailing of reinforced concrete structures subjected to seismic forces.

[19] Indian Standard Code OF 10262: 2019 – Concrete mix proportioning – guidelines

[20] D.S.Vijayan, J. Revathy, Flexural Response of Fibre Reinforced Polymer Laminated Pre-stressed Concrete Beams, Indian Journal of Science and Technology, Vol 9(42), DOI: 10.17485/ijst/2016/v9i42/101824 , November 2016

[21] D. S. Vijayan, A. L. Rose, S. Arvindan, J. Revathy, and C. Amuthadevi, "Automation systems in smart buildings: a review," J. Ambient Intell. Humaniz. Comput., no. 0123456789, 2020, doi: 10.1007/s12652-020-02666-9.

[22] R. Goplalakrishnan, A. Mohan, L. Ponrajsankar, D. S. Vijayan, "Characterisation On Toughness Property of Self- Compacting Fibre Reinforced Concrete", Journal of Environmental Protection and Ecology 21, No 6, 2153–2163 (2020).

[23] E. Koda, A. Szymanski, and W. Wolski, "Field and laboratory experience with the use of strip drains in organic soils," Can. Geotech. J., vol. 30, no. 2, pp. 308–318, 1993, doi: 10.1139/t93-026.

Machine Learning With Health Care: A perspective

*Meenakshi Malik, Rekha Khatana , Aditi Kaushik

1,2&3Assistant professor, School of Computer Science and Engineering, Starex University, Gurugram, India,

Corresponding Author Email: meenakshimalik16@gmail.com

Abstract: ML "machine learning" is an ever-expanding research field with plenty of possibilities for study and implementation. Mr. James Collin stated at MIT that ML is the technology defining this decade, even though it has had a meagre effect on healthcare. Several fresh businesses in the ML industry are applying themselves earnestly on healthcare. Even google has jumped in to the race and it has designed a ML application for identification of cancer tumour on mammograms. To identify skin cancer, Stanford uses a Deep Learning algorithm. Around one trillion GB of data is getting generated per year by the USA health system. Various academic experts and scientists have worked out various characteristics and several factors of risk involved in of chronic illness. Additional data stands for more learning for machine, but for higher precision, these many features need a huge quantity samples. So if machines can harvest clinically greater risk oriented feature it would definitely be better. Precision is improved when the data in the form of exploratory data analysis and feature engineering is pre-processed. The multi-class classification might be capable of evaluating a patient's different disease risk levels. In health care, correct identification of the percentage of diseased individuals "sensitivity" is a primary concern rather than correct identification of the percentage of healthy individuals "specificity". Our paper introduces one of ML's super challenger and emergent application i.e. Healthcare. The careful and sympathetic relation with care providers will always be necessary for the patients. ML cannot remove this, but will become instrumental for health professionals in improving and strengthening on-going care. Our paper discusses the several established models of ML along with its applications in healthcare system. We also bring up directions for imparting more efficiency to ML model. In addition to this we have also discussed ML's case study for Brest cancer.

Key words: medical diagnosis, healthcare, Machine Learning, ML techniques

1. Introduction

AI "Artificial intelligence" can be simply described as a science and engineering stream that manages to integrate computer science and robust dataset to solve problems. It requires insights of computation by creating manmade tools and devices that perform demonstrate required type of behaviour [1]. ML and AI are integral part of computer science which are having close relation and connection with each other. Both of these are the trendiest technologies which are being utilised in the creation of smart systems, devices and gadgets. In clinical applications and diagnosis field also ML tools and techniques are being used over several years. The core aim was to enhance capability of healthcare field by improving the capabilities, performance and consistency of medical professionals. Enhanced precision and presence of huge ML data and techniques quickly ramped choice and shortlisting of ML for medical diagnostic troubleshooting. In further execution for medical usage, ML software contributed in great manner. This is why we are now in the fourth decade of AI having wide range of ML applications covering areas from medication to clinical diagnosis and surgery. The key benefits of ML are precision, cost-effectiveness and simulacrum based on constant and consistent learning. For instance; Various datasets from multiple online sources (UCI ML repository, Kaggle or data.gov.in) for the same ailment classification having distinct numbers and name for the functions esp. datasets of heart related diseases [2–5]. Our paper presents and attempts for providing a better understanding of the ML techniques in healthcare and clinical diagnostic applications continually evolving in recent years.

2. ML Techniques

ML is one of the main branch and most expanding sub-field of AI. Software of AI tracks a range of human brain energy features, including logic, pattern to learn, plan, understanding, sense, and the ability to transfer and use data and objects [6]. AI utilizes several methods in these areas to imitate human intelligence. ML methodologies collect scientific findings as inputs and predict data structures. Statistical evidence and scientific findings include sexuality, age, medical exam, imagery, expression of genes etc. Majorly following four ML techniques may be utilized to integrate huge data sets.

The same have been shown below as Fig. 1: (a) "Supervised Learning" (b) "Unsupervised Learning" (c) Semi Supervised Learning" (d) "Reinforcement Learning"

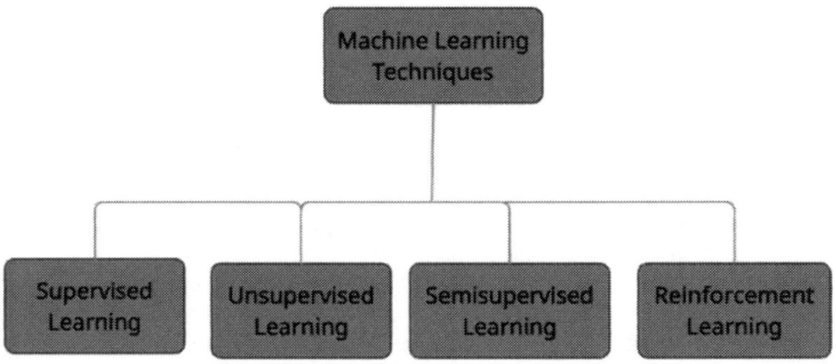

Fig. 1 Types of ML techniques

2.1. "Supervised machine learning":

The very first methodology is a supervised ML and it is related the latter known as unsupervised ML whereas the last one is semi supervised learning which enhances learning. Known classified dataset are always there in supervised ML. It is utilized mainly to predict the relationship between interest trait and result. We act as teachers interactively by training the data on the computer and feeding it. The data which is trained, contains the input means the predicator and the output of product, the results of which are used for testing the patterns. It builds relationships and dependencies with input and specific values of output. Classification and regression problems are the main category under supervised machine learning problems. Decision trees, the Naïve Bayes, Nearest Neighbour, Linear Regression, SVM "Support Vector Machines" [7] and neural networks are few of the most common algorithms of this type. Out of these SVM [8] may be termed as one of the simplest understandable ML algorithms. Supervised algorithm being used in clinical diagnosis more frequently and it produces more accurate results. SVM is devised for binary (two class) problems and may be used to classify data sets in multi-classes. Neural networks [7] are also one of the alternative ML methods which works best if the line between the two classes is not straight. In medical diagnosis, the SVM is widely employed. A neural network denotes a trinity of layers: the main is the input layer, covered with a second layers
made of plenty of layers, lastly showing an output layer (Fig. 2).

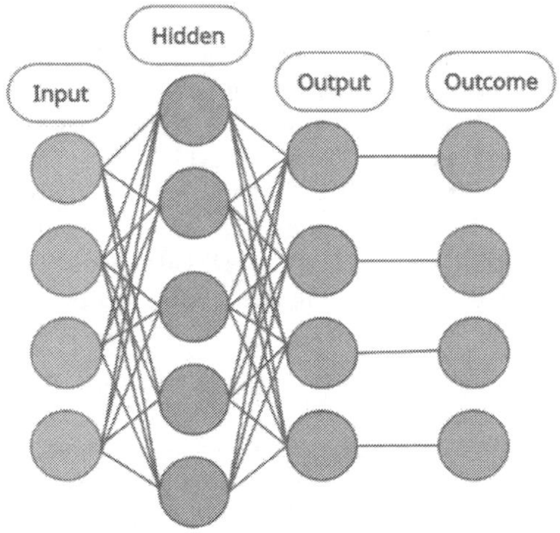

Fig. 2 Artificial Neural Network with different layers

The inputs of each node generate an output which is loaded to the next network node. The trait values are entered into the network as initial input. The result defines the traits classification. Ramesh et al [10] revealed usage of neural networks as the frequent diagnostics tools. Another study also showed that the most commonly used AI technique are "fuzzy logic-neural networks". It may be used rigorously in field of genetic, radiology, cardiology etc. The advanced version of the classical neural network technique is "Deep learning". In the past few years, revival of the same has emerged dramatically increased power of computation and availability of novel data sets.

Deeper learning can analyse further challenging data sets particularly non-linear ones. Due to the growing data size and its ramifications, deep learning has become very prominent in recent times [12]. It uses more hidden layers and has more complex data with different structures in comparison to classical neural network [13]. In clinical application, some of the prevalent and important networks are "network of neural recurrent", supported by the "network of neural deepness", "network of neural convolution (CNN)" and "network of deep belief" that has become the most common method of developing deep learning algorithms. Lately, with popular software such as Microdot's CNTK, Google's Tensor Flow, CNN has been efficient in diagnosing diseases.

2.2. "Unsupervised machine learning":

The chief categorisation of unsupervised learning technique consists of "Association Rule Learning Algorithms" and "Clustering Algorithms". In order to identify most significant data classification characteristics, popularly utilised unsupervised ML techniques are "Association Rules", "K-means clustering" and "Principal Component Analysis (PCA)" [14, 15]. Different clusters or groupings can be utilized in "K-means" to find the one where the points in each group are closest to one another. Whereas in the PCA procedure, the data is arranged into a centred matrix. Independent rows and columns portray the data model and the feature value respectively. We subtract the mean of the column from the values in each column in centred matrix. It is done to make mean of each column as zero.

2.3. *"Semi-supervised machine learning"*

This is a blend or fusion learning amongst supervised and unsupervised that is suited to data and situations in which the result is not complete for certain topics[16]. We have labels for all observations in datasets under supervised learning and we have no labels for them under unsupervised learning. Semi-supervised learning stands amidst both these learnings. The labelling costs are now high as they involve the knowledge of human experts. Some models have thus been developed, with presence of labels in few observations and unlabelled data in plenty of observations. These types of semi-supervised models are the perfect factors for building of model. Despite unlabelled data, the pattern is explored through hold of data info of group parameters.

2.4. *"Reinforcement learning"*

It is like a human child who learns to walk and speak. This technology has an agent that learns primarily from environment. Learning is carried out through the simulation of positive and negative rewards. If the actions are carried out by the agent as expected, they would be considered as positive reward else the results will be called negative reward. An example of reality is the formation of a robot to grab and stack the blocks at a site. The environment is therefore the blocks, grid and buckets. Robot's hand acts as agent. The robot is now programmed to collect the highest block, move it into a bucket and stack it. If robot slips the block in the middle of journey after completion of first simulation, it is concluded that robot does not do the job correctly. It sends out a false signal or alarm. Repeated picking and dropping actions are done until the robot hand does not learn to take action without a negative reward. Temporal difference, Q-learning, deep adversarial networks are the most famous reinforcement learning algorithms.

3. ML's case study in Brest Cancer

3.1. *Generalized Workflow*

ML support in Diagnosis for Breast Cancer is the most important and revolutionary step which is proving a support system in reducing the expiry rate of woman due to breast cancer, nowadays. A reliable futurist prototype for preliminary diagnosis can reduce women's weakening explicitly, the previous neutral estimate indicates whether the patients are owed to a cancerous/ malignant or non cancerous i.e. benevolent. In the area of malignant exploration, it is typically significant to retain trained data in an appropriate set of data which subsequently scrub up and revision of irrelevant or unacceptably inacceptable information aids to demonstrate additional cancer analysis interpreters.

Figure 3 depicts a ML system's general process flow. It begins with the user's raw data collection. Data features need to be extracted which can be added to the ML algorithm in order to make the data understandable. Thus, the data are purified and pre-processed in the next step. Irrelevant features are deleted in feature engineering and new features are created that will help train the ML model. This dataset with feature selection is split into a test and train datasets, where test dataset is reserved for the purpose of evaluation and prediction. For validation purposes, the train data set is further divided. ML model is trained on train datasets and validation is done with validation datasets. Once the model is finalized, the user can make prediction for the test datasets. Improvisation on model is done with further data or better ML algorithms by customer feedback and predictions.

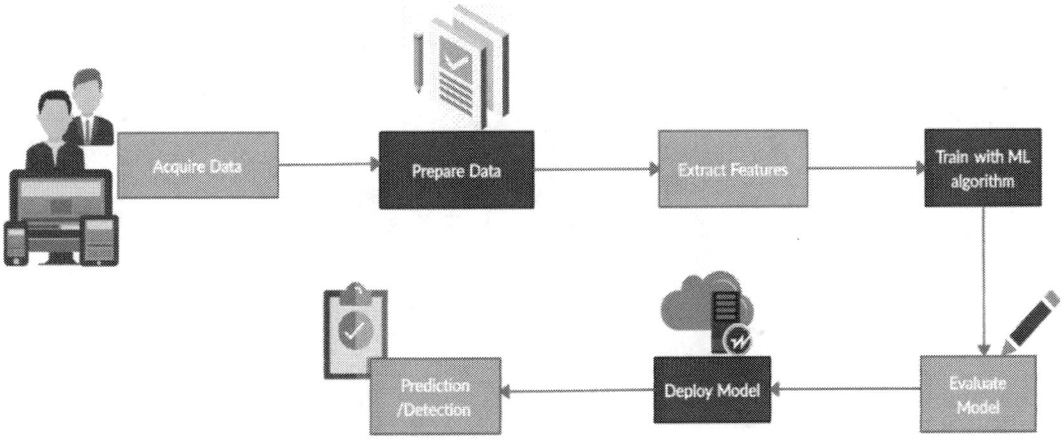

Fig 3 Generalized work flow of Machine learning

3.1.1. "Data Acquisition" or precise and early diagnosis of any illness and disorder, biomarkers are essential. Health technology progress has made it less complicated to obtain biomarkers today. The biomarker collected data becomes tools for driving ML healthcare applications. The data sets can be selected on the basis of factors like

- Data size and type
- Data acquisition cost and ease
- Continuation in data updation
- Amount of clinical attention needed
- Patient's physical visit to clinics

3.1.2. "Data Pre-processing" Data collected from whatever channel, such as sensor, must be in machine-understandable form. The first step in the removal of unnecessary data and outliers is data cleaning. Significant data for specific problems must be identified. For instance, it is possible to remove fraudulent or missed data field. Consequently, the data transformation, i.e. selection, scaling and normalisation of the data to be done. Seldom is the size of data reduced to save computing effort for huge datasets. (Fig. 3).

3.1.3. "Feature Engineering" Functional engineering can make or break an ML model the most important and crucial step. This is where knowledge of the domain can help as shown in figure 4. Feature engineering allows ML algorithms to interpret data. The processes of feature engineering include the creation of new features, removal of unused functionalities, correlation of features, categorical one-hot codification, detection of outliers, clustering operations, function transformation. The selection of features greatly influences model performance. The choice is done utilizing various combination of features, feature importance and model evaluations. In order to store a data subset for a final analysis of the model, feature engineered data is split into test and train dataset before training in the ML model. Training datasets are furthermore divided with Validation set for CV "cross validation" purpose. As selection of an apt ML model is not completely simple, general practise would be to train datasets through multiple ML algorithms.

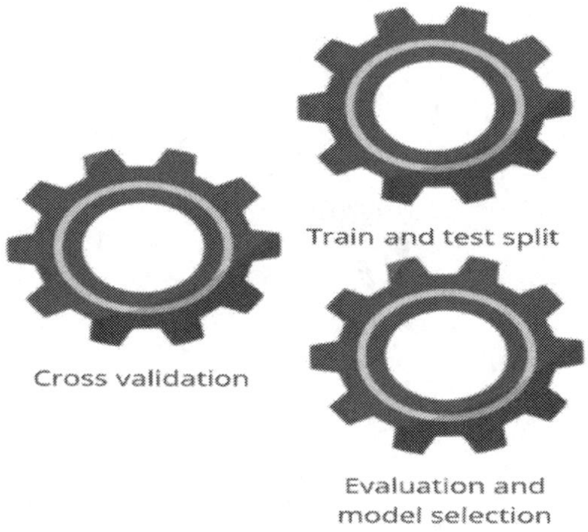

Fig 4: Feature engineering in ML

There are a lot of impressive open source ML libraries available with different algorithms. Linear and Logistic Regression, SVM "Support Vector Machine", KNN "K-Nearest Neighbours", Tree Regressors, Naïve Bayes, Neural Networks and Boosted Tree Recorder are among the most popular libraries and algorithms. The next step is to validation and comparison of few chosen algorithms with techniques like K-fold CV method. In case of insufficient accuracy or overfitting, every or all process is repeated and models are retrained for more data acquisition, data pre-processing and feature engineering.

3.1.4. "Evaluation and Prediction"

Only one metric like accuracy cannot support in evaluation of complete model efficiency effectively in a classification problem. Hence for every kind of skin lesion disease, the accuracy, precision, recall, F1 score and support are measured. The confusion matrix is also plotted to verify the performance of our model in each class.

4. ML's Applications in Medical Diagnosis

The increasing demands for precise and perfect knowledge in healthcare is forcing us to predict information, carry new tests and put novelty efforts hand in hand for the assistance of countless patients suffering from 'n' number of diseases. ML has made it possible which seemed tough or almost impossible earlier. As a specific area of AI, ML is the reflection of human being intelligence by computer system. It manages to combine both statistics and computer science; where computer science addresses identification of problems and their solutions and statistics takes care of hypothesis, data modelling and reliability measurement [17]. Machine Learning is a framework that learns from previous experiences for future performance improvements. Also development of programmes is majorly focused in which data is trained to learn by own selves. The following are few healthcare ML applications:

- Improved Radiotherapy
- Drug Discovery of drug and its Manufacturing
- Maintaining Smart Health Record
- Outbreak Predictions

- Identification of ailments
- Medical Imaging Diagnosis
- Medicine Personalization
- Clinical Trials and Researches
- Behavioural Modification based on ML
- Data Collections through crowdsourcing

The health domain is concerned with large data, like sensitive and detailed patient information, making it difficult for individuals to analyse and maintain these massive records. In order to overcome these issues, ML has shown the way forward by automatic discovery of patterns as well as providing predictions on data. This allows medical analysts to provide accurate medicine through personal care. We can handle large quantities of medical data efficiently by means of ML systems [18]. It provides numerous data classification and prediction algorithms, which is the main goal of the analysis of medical data. Hence it supports us to decide more accurately and more precisely.

5. Research Directions

All of the above-mentioned challenges offer several research possibilities to improve the accuracy, precision, and adaptability of computation models. Thus, some research directions are mentioned below by taking all of them into account:

- Integration: Various parameters from different aspects and sources should be integrated for better results in detection, analysis and measurement of biomedical information. Data from various sources could describe various aspects of the problem. Comprehensive insights into a particular disease can be obtained by means of integrative analysis.

- Security: The safety and privacy of patient data is very essential and crucial. Sun et al.[19] said that very slight changes to lab records can totally alter the prediction of a predictor pattern for the patient's EHR. In consideration of adverse attacks it is very important for researchers to develop defence mechanisms.
- Transparency: Computational models are very important for some areas of the medical field. Most of the models currently presented are black box models. In order to achieve a better precision, accuracy and adaptability, these Black Boxes should be explored more.

- Database creation: The database is not sufficiently available or does not take into consideration all the parameters and aspects, or any particular region or medical institute for many diseases like aids, dengue, mental ailments, flu, heart disease, etc. Hence many databases with various parameters are still required.

6. Conclusion

The roots of machine learning are firmly grown in history. If technology is to improve quality of care in future, the ability of analytics and machine learning must increase the electronic details available to doctors. Medicine machine learning has made headlines recently. The priority should be on how machine learning can be used to increase care for patients. The different models of ML and its applications in the health system are discussed in our paper. We also proposed guidelines to give the ML model greater efficiency. In addition to this we have also discussed the ML's case study for Brest cancer.

References

[1] S.C. Shapiro, Artificial intelligence, in Encyclopedia of Artificial Intelligence, vol. 1, 2nd edn., ed. by S.C. Shapiro (Wiley, New York, 1992)

[2] C.B. Gokulnath, S.P. Shantharajah,An Optimized Feature Selection Based on Genetic Approach and Support Vector Machine for Heart Disease (Springer Nature, Iran, 2018)

[3] E.R.Q. Fernanded, A.C.P.L.F. de Carvalho, X. Yao, Ensemble of classifiers based on multiobjective genetic sampling for imbalanced data. IEEE Trans. Knowl. Data Eng. 14(8) (2015)

[4] F. Babi˘c, J. Olejár, Z. Vantová, J. Parali˘c, Predictive and descriptive analysis for heart disease diagnosis, in FedCSIS, vol. 11 pp. 155–163, IEEE Catalog Number: CFP1785N-ART c 2017, Slovakia, https://doi.org/10.15439/2017f219. ISSN 2300-5963

[5] R. Pari, M. Sandhya, S. Sankar, A Multitier Stacked Ensemble Algorithm for Improving Classification Accuracy (IEEE, 2018)

[6] A. Agah, Introduction to medical applications of artificial intelligence, in Medical Applications of Artificial Intelligence (CRC Press, 2013), pp. 19–26

[7] Meenakshi, M. (2020, May). Machine learning algorithms and their real-life applications: A survey. In Proceedings of the International Conference on Innovative Computing & Communications (ICICC)

[8] C. Cortes, V. Vapnik, Support-vector networks. Mach. Learn. 20(3), 273–297 (1995)

[9] Meenakshi Malik, Rainu Nandal, Surjeet Dalal, Vivek Jalglan, Dac-Nhuong Le," Driving Pattern Profiling and Classification Using Deep Learning", Intelligent Automation & Soft Computing, Vol.28, No.3, 2021, pp.887-906

[10] J.H. John, Neural network and physical systems with emergent collective computational abilities. Proc. Natl. Acad. Sci. U.S.A. 79, 2554–2558 (1982)

[11] A.N. Ramesh, C. Kambhampati, J.R. Monson, P.J. Drew, Artificial intelligence in medicine. Ann. R. Coll. Surg. Engl. 86(5), 334 (2004)

[12] A. Yardimci, A survey on use of soft computing methods in medicine, in Proceedings of the 17th International Conference on Artificial Neural Networks, Porto, Portugal (2007), pp. 69–79

[13] D. Ravì, C. Wong, F. Deligianni, M. Berthelot, J. Andreu-Perez, B. Lo, G.Z. Yang, Deep learning for health informatics. IEEE J. Biomed. Health Inform. 21(1), 4–21 (2016)

[14] I. Goodfellow, Y. Bengio, A. Courville, Deep Learning (MIT Press, Cambridge, 2016)

[15] S. Lloyd, Least squares quantization in PCM. IEEE Trans. Inf. Theory 28(2), 129–137 (1982)

[16] C. Amuthadevi, D. S. Vijayan, Varatharajan Ramachandran, "Development of air quality monitoring (AQM) models using different machine learning approaches", Journal of Ambient Intelligence and Humanized Computing, https://doi.org/10.1007/s12652-020-02724-2

[17] F. Jiang, Y. Jiang, H. Zhi, Y. Dong, H. Li, S. Ma, Y. Wang, Q. Dong, H. Shen, Y. Wang, Artificial intelligence in healthcare: past, present and future. Stroke Vasc. Neurol. 2(4), 230–243 (2017)

[18] https://www.computerworld.com/article/3479833/machine-learning-essential-to-healthcaresays-orion-health.htm

[19] M. Sun, F. Tang, J. Yi, F. Wang, J. Zhou, Identify susceptible locations in medical records via adversarial attacks on deep predictive models, in Proceedings of the 24th ACM SIGKDD International Conference on Knowledge Discovery & Data Mining (2018), pp. 793–801

ANALYTICAL STUDY ON STATIC BEHAVIOUR OF A TRANSMISSION TOWER LINE SYSTEMS

B.MENAKA [1], VINOTH KUMAR[2]
[1] PG student, [2]Assistant Professor, Civil Engineering Department, SRM Valliammai Engineering College, Chennai, India.
Corresponding Author E-mail: monimenaka05@@gmail.com

Abstract: The need for significant transmission and distribution system has been raised in India due to the large population living worldwide and its huge need for electricity supply due to the technological boom, making it necessary to be available 24x7. As transmission line towers are critical lifeline structures that transport power from the source to various locations for various purposes and serve as lifelines for humans by illuminating our daily lives, it is critical to creating a damage-free transmission tower unthinkable life without such electric serving systems. The project here sheds on the type of connections that are said to be provided in the tower connection, and a design procedure is carried out to evaluate the tower member properties. A three-dimensional structure was performed using NEi-NASTRAN and tower designs. The members of the tower are then drawn up.

KEYWORDS: Transmission tower line, NEi-Nastran. Static analysis, 3D modeling

1. INTRODUCTION

A transmission turret is a large structure; usually, a steel lattice tower is used to support an overhead power line (alternatively electricity pylon). The phase conductors and earth wires of a transmission line support a transmission tower line with typical data on the transmission towers. Different designs, different sizes, different configurations, and materials are available. We see the axial, lateral, and deflection variation and a tower's weight between different bracing patterns. The standard cable structures are of these three main types: lattice, pole, and guyed tower construction [1].

India has a large population in every part of the country, which requires large transmission and distribution systems because of the need for power supplies. The disposition of the primary electricity resources, i.e., coal, hydro potential, is also quite uneven, thereby adding to the requirements for transmission. The transmission line consists of the conductor subsystem, ground wire subsystem, and one subsystem for each support category. A significant part of the cost, and they play an essential role in reliable power transmission, is the mechanical support of the transmission line [2].

The towers of the steel grid typically support high-voltage transmission lines. The tower costs are around a quarter to half of the transmission line cost, and therefore significant savings will be achieved through the optimized tower design. The choice of an optimal contour with the right type

of bracing system makes an essential contribution to developing a cost-saving transmission line tower design.
The user shall determine the tower's height, and the designer is responsible for the design of the general structure and the member and details. Various load formats, including normal conditions, are evaluated for the conductor and ground wire breakage status.

1.1. TOWER CONFIGURATION
The tower configurations are determined essentially by three factors:
- Tower height
- Base-width
- Top hamper-width

Depending upon the requirements of the transmission system, various line configurations have to be considered ranging from a single circuit to multi-circuit towers. Figure 1 indicates transmission tower.

$$H = H1 + H2 + H3 + H4$$

Height of Tower Structure
The factors affecting the height of a tower are as follows:
- Minimum allowable clearance of ground (H1)
- Maximum sag (H2)
- Vertical separation of conductors (H3)
- Clearance between the ground wire and the top conductor vertically (H4)

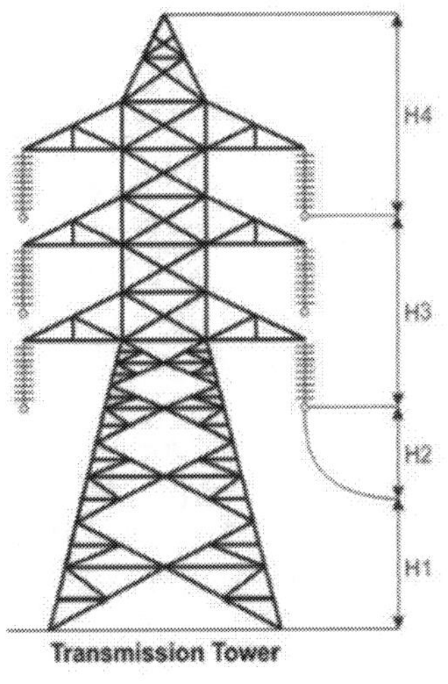

Figure 1: transmission tower

2. Literature study
James Robert Callaway (1983) A study was conducted to study the compressive capacity of a particular single-angle group of members. Compression testing in laboratories was conducted on 22 angle members, with four dimensions and two lengths. Further tests have been carried out to determine the material's yield strength. The results have been tabled and discussed, which have been normalized in respect of yield stress. To predict the test members' ultimate capacity, two existing analytical models were used. The first was based on the combined stress equation of the AISC elastic method. Mueller and Erzurumlu of Portland State University have developed the second method, an inelastic method. The results of the test program were compared. The result

shows that, when pinned ends are assumed, both analytical models provide conserve forecasts and unconservative results for fixed end conditions. For test members with ratios of L/r greater than 125, the results of the elastic method closely matched the test results, but the correlation was less consistent for those with ratios of L/r less than 125. All member tests have been complemented closely by results of the inelastic technology [3].

N. Prasad Rao, V. Kalyanaraman India, 2001 Microwave doorways and transmission towers are often made of angles that are directly or via gussets bolted together. Such towers are usually analyzed for design forces using linear static methods, provided only axial characteristics, and small deformations are subjected to the members. The effects of the end limits, eccentricities of the connections, and connections made via bracings on the secondary structure of the members will usually be taken into consideration in codal analysis recommendations. Non-linear analysis is carried out using MSCAN software to determine the angle compression grid members and 3D grinding frames in traditional grid towers. Test results will verify the performance of the analytical models. The redundant results are compared with codal provisions and specifications [4].

Linbo Hui – Canada, Yi Liu, Linbo Hui (2007) Twenty-eight single-angle specimens of rolled steel were tested to respond to axial loading at different end-excentricities. When required. The results indicated the critical eccentricity below which any consequent reduction of the final load is marginal when loaded accentuated concerning the principal axis. In contrast, since the loading eccentricities are increased relative to the minor principal axis, the load decrease is more pronounced, and there is no identical critical excentricity. Ascertained from test results: These two equations yield conservative compressive force and angle values for a single angle of ultimate compression, and instead, the equations proposed by Adluri and Madugula (1992) estimate a compressive capacity on the order of double. Results are more apparent in primary specimen orientation rather than in specimen orientation because of it. The third method offers an improvement of capacity estimates from a single angle, although it is intended for double-symmetric sections [5].

M. Samuel Knight – India, S. Vishnuvardhan, G.M. (2007) Cold-shaped steel single-angle, double-angled back-to-back welded and stellar angle compression tests were conducted on three different end connectors. The stub and short columns were evaluated in eight plain parts of varying sizes and thicknesses with varying strengths. The effect of a flat width-to-thickness ratio on load capacity and the effect of section size, material output strength, and cross-sectional symmetry will be investigated. The load compared with the behavior of axial shortening is also studied, the load against a lateral and the load vs. strain comportment. Initial rigidity and ductility coefficient are calculated and compared for various sections. The last loads of double angles with ball end connections are sold back-in and starred angles more than double, regardless of the slender ratio, compared to single angles. The initial rigidity of double corners welded back-to-back was found to be 1,25 times higher. In addition to 70% of the final load, the stress from compression to compound angle was reversed. The symmetric effect on the buckling mode in specimens is also reported in different slimming ratios [6].

Mizam Dogan – Turkey, Ruhi Aydin, Turkey; (2010) Experimentally and analytically, the behavior of equal legs of single angles in two-axial bending and compression is investigated using nonlinear techniques finite element analysis. Single-angles were tested with 100 to 100 to 10 mm equal-leg. The corner legs were placed flanging down, and their axes were laterally restrained as a supporting beam. The supported beams are loaded with an axial load parallel to the axis of the beam, which is applied at the middle of the length until a failure is achieved. ABAQUS software was used to generate the FEM model for the experimental system. Some code specifications and one design procedure were used for the testing and software results from ABAQUS [7].

Linbo Hui, Yi Liu; (2010) the answer to the axially eccentric loading of single steel angles are examined using numerical modeling based on the techniques of finite elements. Results show a critical eccentricity for every slender ratio studied in cases of excentric compression causing central axis bending. Below this eccentricity, any decrease of the final capacity due to excentricity is

marginal; with the slender angle ratio, the critical eccentricity increases. In contrast, the ultimate reduction of capacity is more pronounced, and no similar critical excentricities are identified for excentric compression, causing low axis bending. Compared to values from the design equations proposed in the AISC 2005, results show that, in general, the AISC equations provide the final capacity for axial compression and minor axis bending angles to be satisfactorily estimated. When it comes to excenter compression, the AISC equation is conservative in a wide range of parameters while overestimating capacity from a particular viewpoint.
Ke Cao, De Wei Zeng, Yao Jie Guo (2015) Although large-section and LHS columns have been increasingly used in steel structures in recent years, and existing design methods are based mainly on studies of mild carbon steels with normal-size sections. The humping behavior of LHS columns was investigated experimentally using 90 Q420 (420 MPa) specimens. The experimental buckling forces were compared to the corresponding design values in the current specifications. Nonlinear FE models were developed and validated to carry out extensive parametric studies in the context of residual stress, geometrical imperfects, and stress curves [8]. Effects of hardening progress were examined and discussed in Q420 stress-strain curves. Through FE analysis and then comparison with current design columns, slender theoretical strengths were obtained for slenderness from 30 to 120. It is concluded that the conservative degree of current codes has fallen with increased slenderness and that neither curve is well compatible with theoretical results, while design curves can be used conservatively in GB 50017-2013 and Eurocode 3. Finally, a new column curve to predict the buckling strengths of LHS columns was proposed based upon parametric studies.
In this present work, the failures of the steel angles are predicted by Finite Element analysis, then entirely created and analyzed NASTRAN workbench. In this study, the predictions are compared with real-time testing of the steel angles prototype. A solution will be conducted regarding using an analytical method of analysis of steel angles than a prototype test [9].

3. WIND LOAD CALCULATION

A sample wind load calculation is done below

- Basic wind speed (Vb) at tower location = 55 m/s from Fig.1 (200 kmph) North side
- Design wind speed Vd or Vz = Vb x k1 x k2 x k3
- Risk coefficients K1 = 1.08 from (Table 1)
- Terrain, height and structure size factor k2 = 1.05 @ 20m height of category 2 class B (Table 2)
- Topography factor K3 = 1.36 (refer 2.6.3)
- Design wind speed Vd or Vz = 55 x 1.08 x 1.05 x 1.36 = 84.82 m/s.
- Wind pressure Pd = 0.6Vz2
  = 0.6*84.82*84.82
  = 4316.65 N/m2
- Force coefficient Cf = for Cf, solidity ratio to be obtained
- Solidity ratio Φ = Exposed area / Total area

Exposed areas

- For leg members = 2*48.3*2000 = 193200mm2 = 0.1932m2
- For bracing members = 7*16*610 = 68320mm2 = 0.068m2
- For tie members = 7*12*540 = 45360mm2 = 0.045m2

Total exp. Area = 0.31m

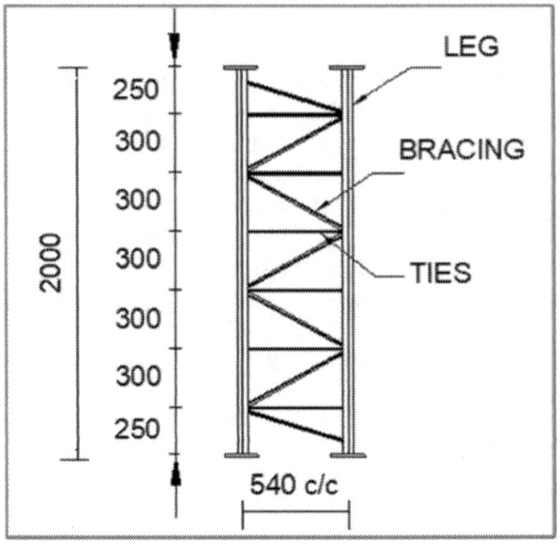

Figure 2: structural components of transmission tower

Total area
Height of the panel = 2000 mm
Width of the panel = 540 mm
= 2000*540
= 1.08m2
Hence Φ = 0.31m2 / 1.08m2
= 0.29
Check for Dvd, D= Dia of the members in meters

For leg, 0.0483*84.82 = 4.09m2
For brac., 0.016*84.82 = 1.36m2
For ties, 0.012*84.82 = 1.02m2
} < 6m2/s

The above values are < 6m2/s,
Hence the flow is subcritical
From table 2.3 (Table 32 of IS 875; 1987 part 3)
For Φ = 0.29, force coefficient taken from table 2.3 , Cf = 1.51
F = Cf*Ae*Pd
= 1.51*0.31*4316.65
= 2040.63N (2.04 kN) @ 20m height. Figure 3 depict wind load calculations

The load can be applied at any two points at 20m height. Figure 2 shows structural components of transmission tower Similarly, the loads can be calculated at each panel and applied at respective heights. The loads for all the panels are calculated using MS Excel is given below;

	A	B	C	D	E	F	G	H	I	J	K	L	M	N	O
1	Force coefficient method														
2	K1=	1.08	K3=	1.36		Vb=	55	55.56		Brac Leng	610.928				
3	Wind speed	200kmph					m/s								
4	Base width	540mm	Ht=	20	Leg=	0.0483	Bracing=	16	Ties=	12					
5					Leg Φ=	0.31	Brac.Φ=	0.31	Ties Φ=	0.31					
6															
7						Leg mem			Bracing mem			Ties mem			
8	PANELS	Ht	K2	Vz (or) Vd	leg exp	D*Vd	Cf	Brac.exp	D*Vd	Cf	Ties.exp	D*Vd	Cf	Pz	F(kN)
9	Penel 10	20	1.05	84.82	0.19	4.10	1.51	0.068	1.36	1.51	0.05184	1.01788	1.51	4317	2.04
10	Penel 9	18	1.038	83.85	0.19	4.05	1.51	0.068	1.34	1.51	0.05184	1.00625	1.51	4219	2.00
11	Penel 8	16	1.026	82.88	0.19	4.00	1.51	0.068	1.33	1.51	0.05184	0.99461	1.51	4122	1.95
12	Penel 7	14	1.012	81.75	0.19	3.95	1.51	0.068	1.31	1.51	0.05184	0.98104	1.51	4010	1.90
13	Penel 6	12	0.996	80.46	0.19	3.89	1.51	0.068	1.29	1.51	0.05184	0.96553	1.51	3884	1.84
14	Penel 5	10	0.98	79.17	0.19	3.82	1.51	0.068	1.27	1.51	0.05184	0.95002	1.51	3761	1.78
15	Penel 4	8	0.98	79.17	0.19	3.82	1.51	0.068	1.27	1.51	0.05184	0.95002	1.51	3761	1.78
16	Penel 3	6	0.98	79.17	0.19	3.82	1.51	0.068	1.27	1.51	0.05184	0.95002	1.51	3761	1.78
17	Penel 2	4	0.98	79.17	0.19	3.82	1.51	0.068	1.27	1.51	0.05184	0.95002	1.51	3761	1.78
18	Penel 1	2	0.98	79.17	0.19	3.82	1.51	0.068	1.27	1.51	0.05184	0.95002	1.51	3761	1.78

Figure 3: wind load calculation

4. MODELLING AND APPLICATION OF LOAD TO ANGLE SECTION IN NASTRAN

Angle member ISA 50 x 50 x 6 mm of different lengths varying from 250 to 2250 mm by the increment of 250 mm as plate shell elements are modeled in Fem NX NASTRAN. The material specification for the model is given as MS 250, whose details are shown [10][11][12]. The angle section is modeled as a plate shell element of 6 mm thickness. One flange of the angle section contains two bolt holes on top and bottom, respectively. An additional plate of 6mm thickness at a gap of 1mm from one flange of the angle section at the top is connected. The angle section and the plate are connected by a 16 mm diameter circular beam bar at the portion of the bolt holes. Various constraints are applied to the model based on the requirement. The self-weight of the model is assigned. The nodal load to the model is applied at specified nodes [13] [14][15]. A nonlinear static method of analysis is performed. The load at which the model fails is taken from the output [16][17][18]. The deformation and contour shapes for various cases can be observed. Figure 4 represents model of Non-linear analysis

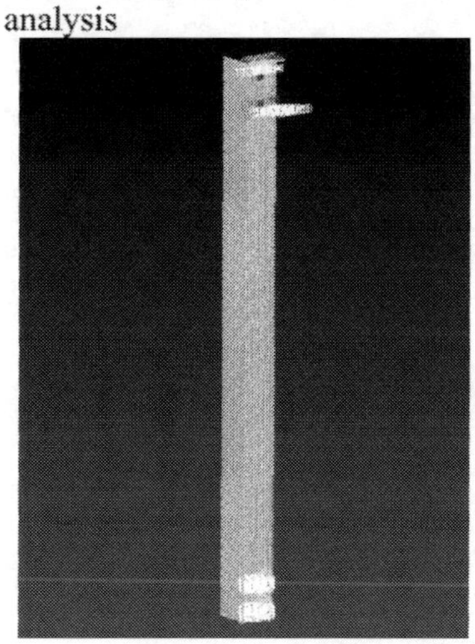

Figure 4: Model of Non-linear analysis

Model of angle section ISA 50 x 50 x 6 mm of 1000 mm length using plate shell elements after application of constraints and load (NASTRAN). Figure 5,6,7, 8 and 9 displays material properties for both general and Non-linear analysis. Figure 10,11,12 and 13 displays output of the analysis along with deformation in model

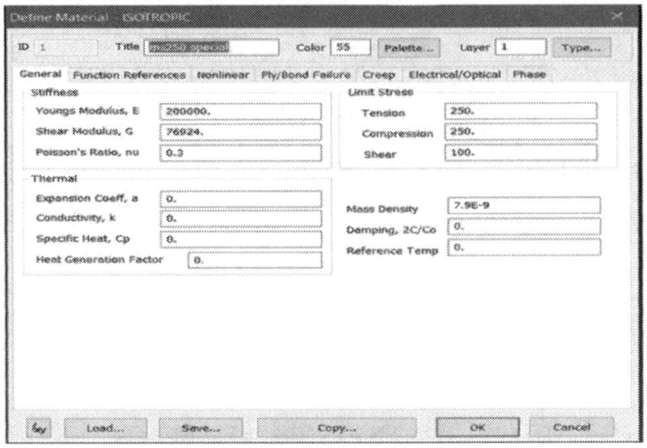

Figure 5: Material Model (General)

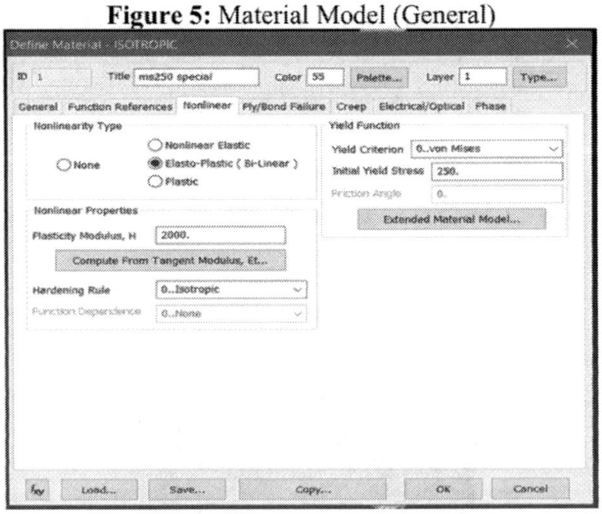

Figure 6: Material Model (Nonlinear)

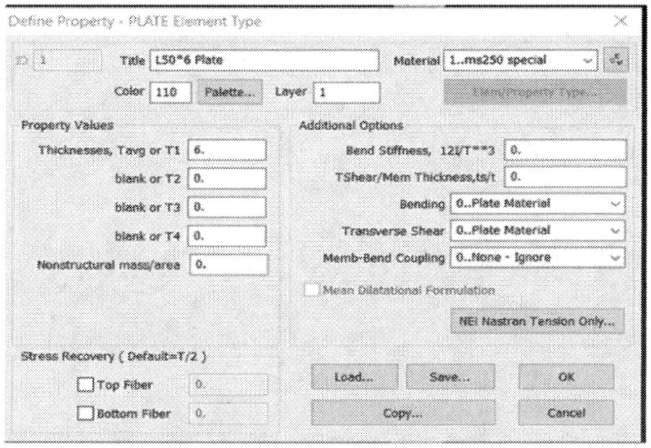

Figure 7: Property for plate element: for modelling angle flanges

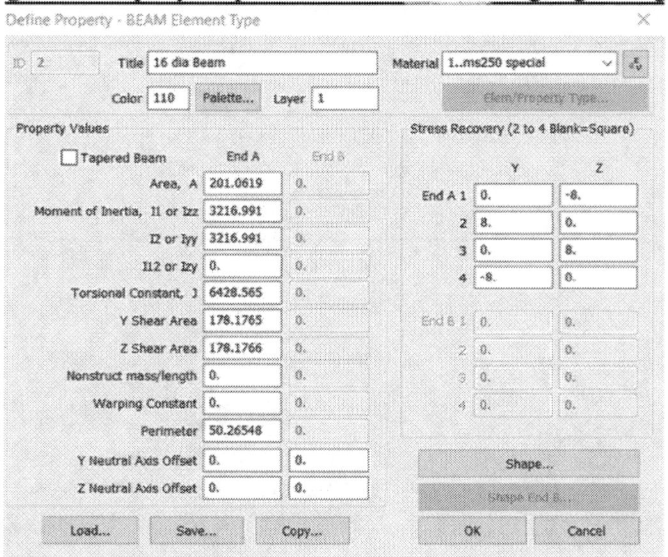

Figure 8: Property for beam model – Bolt

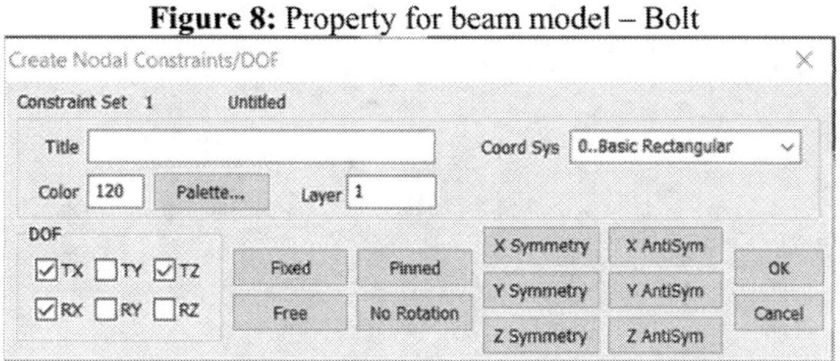

Figure 9: Constraints at Top of angle specimen

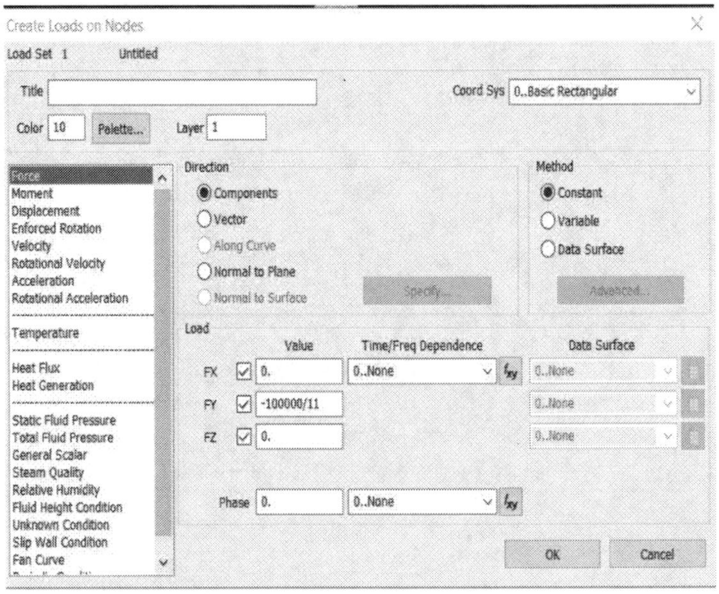

Figure 10: Application of Load

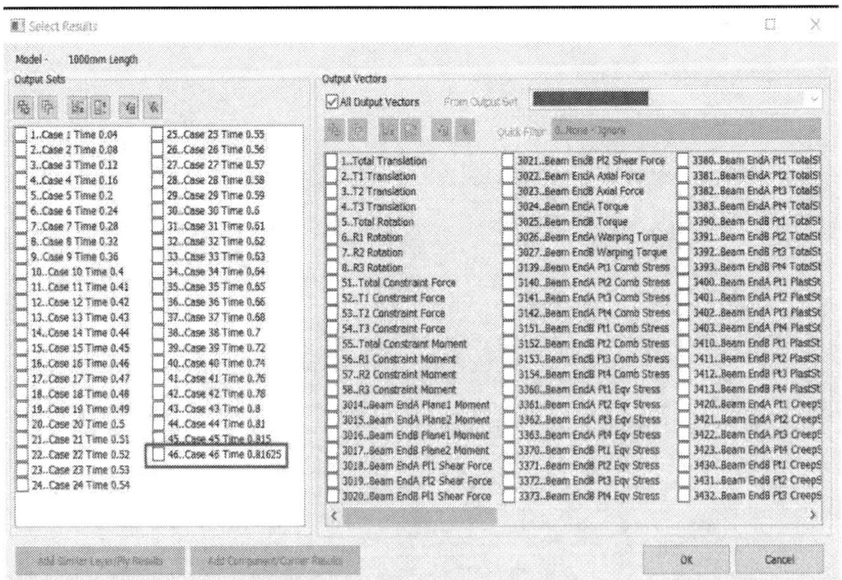

Figure 11: Output of the analysis

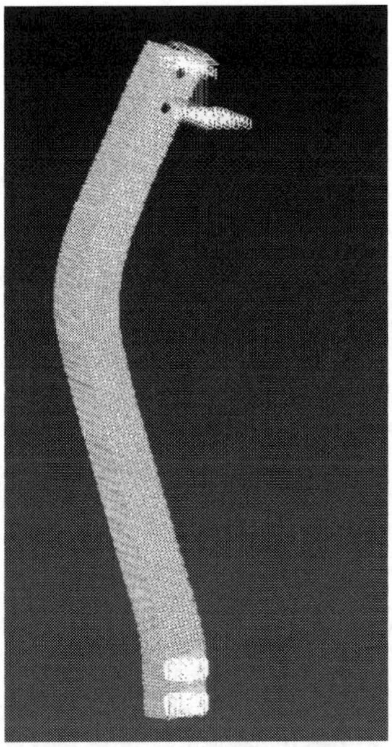

Figure 12: Deformed view for total Translation for case 38

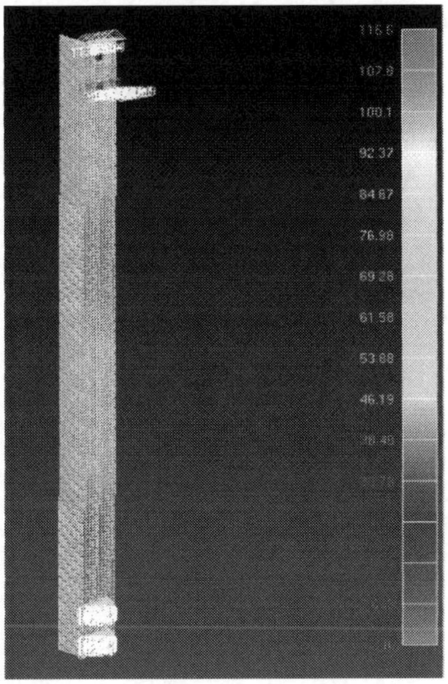

Figure 13: Contour View for total translation for case 38

5. CONCLUSION

The effective length should be reliable and efficient for a better design of compression strength of the tower. An indication of this statement that effective length is a significant criterion is provided in the chapter. The reason for the failure of NFD 30 given in the chapter is because the effective length was not appropriately considered. As many more tests have to be done to determine the effective length factors for eccentrically loaded steel angle compression members for IS 800, further studies have to be done.

REFERENCES

[1] IS 800 - 2007 - Code of practice for general construction in steel
[2] IS 875 - 1987 (Part 1) - Code of Practice for design dead loads other than Earthquake for buildings and structure.
[3] IS 875 - 1987(Part 2) - Code of Practice for design live loads other than Earthquake for buildings and structure.
[4] IS 875 - 1987 (Part 3) - Code of Practice for design wind loads other than Earthquake for buildings and structure.
[5] SP 6(1) - 1964 – Handbook for structural engineers (Structural steel sections)
[6] IS 802 (Part 1 / Sec 2) - 1992 – Code of practice for use of structural steel in overhead transmission line towers.
[7] Yi Liu, Linbo Hui – "Beam column behaviour of steel single angles" - Canada, 2007.
[8] Katalin Oszvald / Laszlo Dunai – "Behaviour of corroded steel angle compression members" – 2013.
[9] S.Vishnuvardhan, G.M.Samuel Knight – "Behaviour of cold – formed steel single and compound plain angles in compression" – India, 2007.
[10] Vijayan, D.S., Arvindan, S., Naveen Kumar, K., Mohamed Javed, S." Seismic performance of flat slab in tall buildings with and without shear wall ", International Journal of Engineering and Advanced Technology, 2019, 9(1), pp. 2672–2675, DOI: 10.35940/ijeat.A9732.109119
[11] Joseph Robert Yost, David W. Dinehart, Shawn P. Gross, Joseph J. Pote, and James Deeney – "Buckling strength of steel angle compression members"
[12] Yi Lui, Sabine Chantel – "Steel single unequal leg angles under eccentric compression" – Canada, 2011
[13] James Robert Callaway – "Single angle compression members" – 1983
[14] Ke Cao, Yao Jie Guo, De Wei Zeng – "Buckling behaviour of large section and 420mpa high strength angle steel columns" – China, 2015.
[15] Vijayan, D.S., Kumar, K.N., Aravindan, S." Analytical study on industrial steel building with braceing ", International Journal of Scientific and Technology Research, 2019, 8(10), pp. 1275–1278.
[16] Ruhi Aydin, Mizam Dogan – "Beam column behaviour of equal leg single angles" -Turkey, 2010.
[17] Yi Liu, Linbo Hui – "Behaviour of steel single angles subjected to eccentric axial loads" – Canada, 2010.
[18] N.Prasad Rao, V.Kalyanaraman – "Non-linear behaviour of lattice panel of angle towers" – India, 2001.

Prediction of undrained shear strength and correlation in between soil parameters

Yimam Mohammed Yimer[1], **A.Paul Makesh**[2]*, **SalihaShukri Muhammed**[3]
[1]Lecturer, Department of Civil Engineering, WolaitaSodo University, Sodo, Ethiopia
[2]Associate Professor, Department of Civil Engineering, WolaitaSodo University, Sodo, Ethiopia,
[3]Lecturer, Department of Construction Technology and Managment, Wolaita Sodo University, Sodo, Ethiopia
Corresponding Author Email: yimam2003@gmail.com

ABSTRACT: Correlation of soil parameters has undeniable benefit in the determination of engineering properties of soil to solve problems in geotechnical Engineering area. The tests were conducted within geotechnical laboratory.These tested soil parameters, used in the correlation analysis are unconfined compressive strength, bulk unit weight and dry unit weight.The aim of this study is proposing a relationship in between the strength parameter with some of the index properties of soils using statistical regression analysis. The linear regression analyses have been done for prediction of unconfined compressive strength (q_u) from bulk and dry unit weight as model-1 and model-2 respectively.And dry unit weight was predicted from bulk unit weight as model-3. Model-4 represents the multiple linear regression analysis to predictq_u. The health of developed models is measured by coefficient of determination (R^2) values. Though, model-1, model-2, model-3 and model-4 have R – squared values of 0.9112, 0.9333, 0.9109 and 0.9452 respectively. Therefore, they are correlated strongly and positively. The prediction of unconfined compressive strength of these soils correlated in linear regression, are fairly determined with Model-2 compared with model-1 and model-4 (MLR).

Key words: Regression analysis, unconfined compressive strength, unit weight

1. Introduction

The undrained shear strengths are determined in the laboratory using unconfined compressive strength apparatus. The undisturbed soil specimens were prepared and ready for tests. The results obtained from this test are unconfined compressive strength (q_u). If it is so, these samples are tested rapidly and it is a special case of the Unconsolidated undrained test in which cell pressure is zero. The unconfined compressive strength (q_u) is widely used for stability analysis of soil foundations under undrained conditions. This is primarily because the mean value of q u/2 accurately describes the undrained shear strength on the failure surface in a particular area, and additionally, the q u-value testing procedure is straightforward and inexpensive. The standard size specimen, 38mm in diameter and 76mm in height, is typically used in this study for unconfined compression tests [1]. Mechanical properties of clayey soils, particularly shear strength, are critical in engineering practise. One of the most important mechanical properties of soil, the undrained shear strength cu, was determined from unconfined and confined tests under undrained conditions [2]. Another very important soil properties investigated under soil exploration through laboratory tests are bulk unit weight and dry unit weight. The main purposes of testing unit weights are to compute the in-situ bulk and dry density of soils. Empirical correlations are widely used in geotechnical engineering practice as a tool to estimate the engineering properties of soils

2. METHODOLOGY

2.1. Sampling and Testing `

The soil samples were collected from Kemise town, which is located in Amhara region, Northern part of Ethiopia at latitude of 10°43′N and longitude of 39°52′E with elevation of 1450m above mean sea level. Soil samples were acquired from different places in the town. The soil samples collected from this area were prepared for correlating in between some of soil parameters with and unconfined compressive strength. The total numbers of soil samples used for this analysis were twelve. Phases I and II soil samples were subjected to an unconfined compressive test. The results included a stress-strain diagram. The stress-strain diagram was corrected for the initial concave portion of the curve, which is believed to be caused by sample preparation, irregular loading surfaces, and seating loads. The initial tangent modulus was determined by calculating the slope of the tangent to the initial straight segment of the corrected curve drawn through the origin. Finally, the conventional result of a standard UC test is noted as the failure strength. The sample preparation method, such as static versus impact compaction, had an effect on the stress-strain behaviour, and the corresponding data were considered and analysed separately [3]. The soils are classified in to silt, clay and silty sand according to Unified Soil Classification system. The soil properties tested in the geotechnical laboratory and used for this analysis are in-situ bulk unit weight, dry unit weight and unconfined compressive strength values. The methodologies followed under this study were presented in chart 1 below.

Stability analysis of a rock slope requires determination of the rock mass's cohesion (c) and angle of internal friction (). Typically, these parameters are estimated without extensive field testing. Su is the shear strength of an undrained soil when sheared at constant volume. A critical state is a state of stress in a soil that occurs when continuous shearing occurs at a constant shear stress to normal effective stress ratio and a constant volume. At the liquid limit, the undrained shear strength of a soil can be estimated to be approximately 1.7 kN/m2. As a result, the present work determined both the liquid and plastic limits using a single consistent method, namely the Swedish fall cone method. Composition of the soil: mineralogy, grain size and distribution, particle shape, pore fluid, and so on. Whether loose, dense, or overconsolidated in its initial state. Particle structure refers to the arrangement of particles, whether densely packed or randomly distributed. Conditions of loading: effective stress path (drained and undrained), and type of loading. The bulk density of the soil, its moisture content, texture, organic matter, and management all affect its resistance to metal penetrometers. The final two variables have a long-term but unknown effect on the soil strength of cultivated soil. Predicting the normalised, undrained shear strength of saturated fine-grained soils using plasticity-value correlations... because they quantify the interaction between the solid and liquid phases in soils in a straightforward manner, allowing for the classification of soils with similar mechanical properties.

Journal of Physics: Conference Series **2040** (2021) 012024 doi:10.1088/1742-6596/2040/1/012024

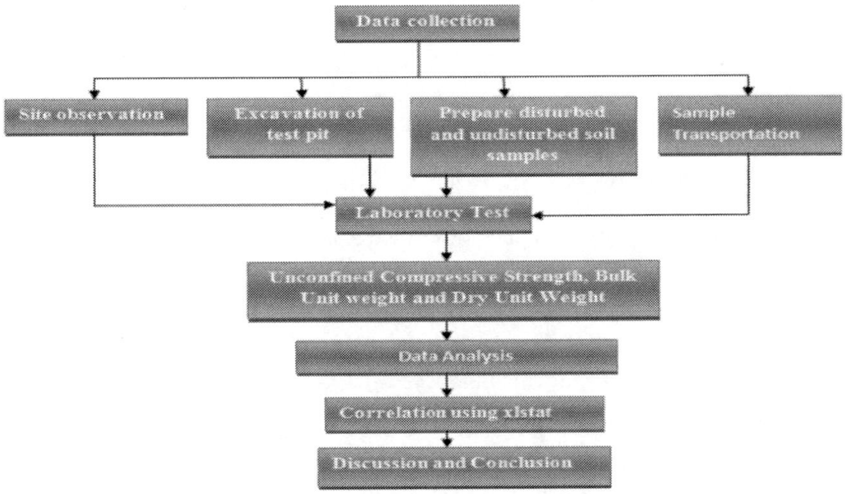

Fig .1 Research Methodology chart

3. DATA ANALYSISAND DISCUSSION

According to Unified Soil Classification system, the Soil types found in the study area are clay, silt and silty sand. Most of the soil samples are collected from 1.5m depth but three of them are from 1.6m, 1.7m and 1.9m having similar soil properties with the other samples. The results obtained from the laboratory test are unconfined compressive strength, bulk and dry unit weight. These values are tabulated in Table1 below.

Test Pit Location	Bulk unit weight (kN/m^3)	Dry unit weight (kN/m^3)	Unconfined Compression Strength (kN/m^2)
Zone Administration	14.6	10.59	114
GelmaAbageda	16	13.14	260
Kune	18	15.44	400
Green Area	20.2	16.32	494
Fewoz Factory	19.2	15.05	390
02 Elementary School	19.5	15.10	342
Amerach	16	11.56	120
Mickael Church	15.8	11.80	174
Segno Gebeya	13.25	10.35	72
High school	19.7	15.44	372
Preparatory school	17.4	13.78	248
Around TVET	14.35	11.72	140

Table 1: Presentation of laboratory test results.

Unit weight(γ_b)is the ratio of the total weight (w) to the total volume of the soil aggregate [9]:

$$\gamma_b = W/_V \quad \text{----------------} \quad (1)$$

The dry unit weight $\gamma_d$is defined as the ratio of the weight of soil solids (W_s)to the total volume (v) [9]:

$$\gamma_b = {W_s}/_V \quad \text{--------------} \quad (2)$$

3.1. Regression Analysis

Numerous studies have been conducted in recent years to predict the physical and engineering properties of various soils, which is generally beneficial for field engineers in terms of preliminary design, proper estimation, and higher quality control planning. Regression analysis is a statistical technique for examining, quantifying, and interpreting simple relationships between variables. The purpose of this article is to attempt to develop a mathematical relationship between two variables through the fitting of a linear equation to observed test data. Prior to fitting a linear model to observed test data, it is critical to ascertain whether any relationship exists between the variables of interest. The coefficient of determination (R2) is a measure of the relationship's strength. It is the ratio of the sum of squares in the regression to the sum of squares in the total. R2 is a value between 0 and 1. The closer the value is to 1, the more accurate the equation [4]. The true test of the quality of the resulting regression relationship is its ability to predict the dependent variable when no independent variables were used to estimate the regression coefficients [5]. The purpose of regression analysis is to determine the relationship between the values of Y and their corresponding values of X. The dependent variable or response is Y, whose value must be predicted, and the independent or regressor variable is X, which is used to predict the dependent variable's value [6].

3.1.1 Correlation

Correlation describes the degree of relationship between two variables. Product moment correlation efficient (r) can be measured to identify the correlation among different important geotechnical parameters. The equation is [5]:

$$r=\frac{S_{xy}}{\sqrt{S_{xx}S_{yy}}} \quad (3)$$

Where: S_{xx} = (Σxy) - nx_{avg}, S_{xx} =(Σx^2) - nx^2_{avg} and S_{yy} =(Σy^2) - ny^2_{avg}

To create valid prediction models, relationships between unconfined compression strength and soil index properties were established. These index properties, such as bulk unit weight and dry unit weight, can be determined more easily and quickly. With this in mind, relationships between unconfined compression strength and the previously mentioned index properties of soil were established. The health of the drawn relationships was determined using the R2 value, which is a more accurate indicator of the health of any correlation [7].

The simple linear regression developed models were determined by using xlstat2015 statistical analysis. The results are presented in Figure 1 to Figure 7.For model 1 and model 2 the linear regression has been performed by considering unconfined compressive strength as dependent variable and bulk and dry unit weight as independent variable. And for model 3 consider dry unit weight as dependent and bulk unit weight as independent variable [11].

3.2. Simple Linear Regression

3.2.1. Relationship between unconfined compressive strength and bulk unit weight

The correlation has been developed In between unconfined compressive strength (q_u) and bulk unit weight (γ_b)with scatter plot.

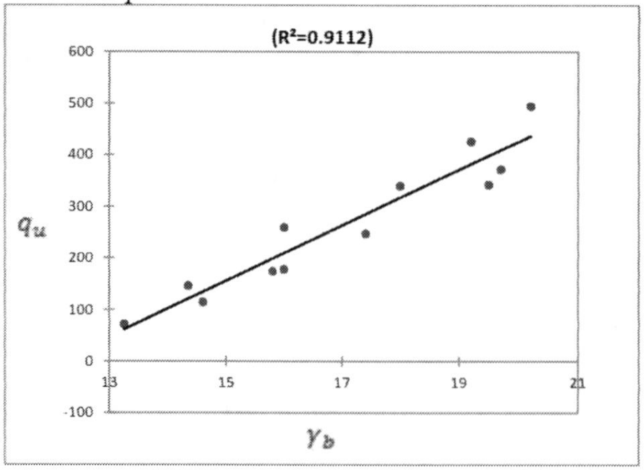

Fig 2: Relationship between unconfined compressive strength and bulk unit weight (Model - 1with correlation coefficient (R) of 0.95)

3.2.2. Relationship between unconfined compressive strength and dry unit weight

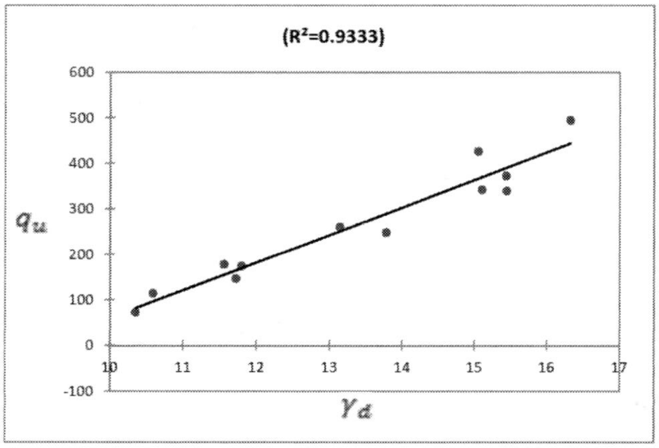

Fig 3: Relationship between unconfined compressive strength and dry unit weight (Model -21with correlation coefficient (R) of 0.97)

3.2.3 Relationship between dry unit weight and bulk unit weight

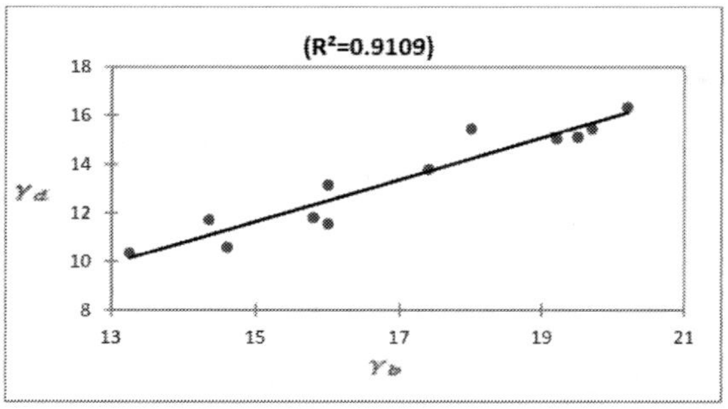

Fig 4: Relationship between dry unit weight and bulk unit weight (Model -31with correlation coefficient (R) of 0.95)

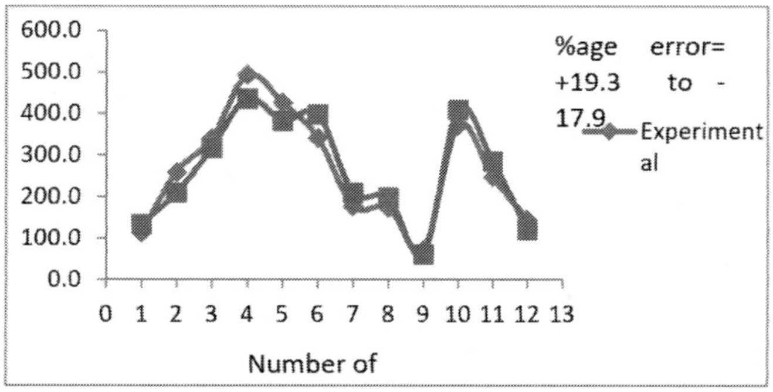

Fig 5: Comparison between predicted and experimental results (q_u) for Model-1

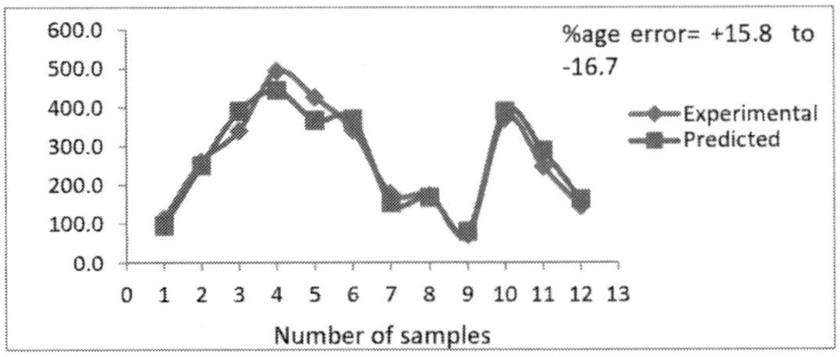

Fig 6: Comparison between predicted and experimental results (q_u) for Model-2

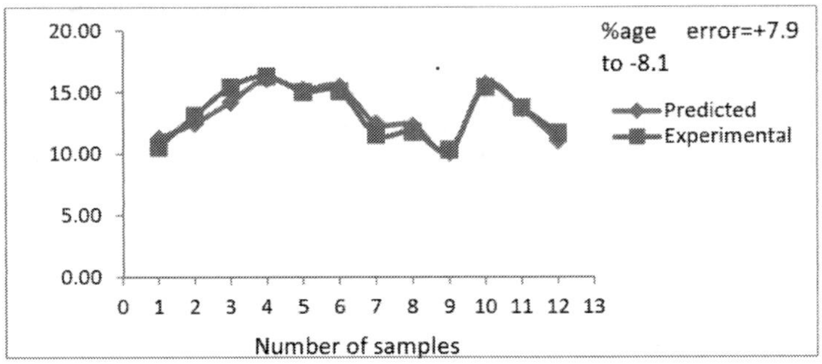

Fig 7: Comparison between predicted and experimental results (q_u) for Model-3

3.3. Multi-Linear Regressions

The most usually utilized correlation statistical analysis in the prediction of one soil parameter as dependent variable from other two or more soil parameters as independent variable is multi-linear regression analysis (MLR) [8]. Here, the soil strength is considered to be predicted, and bulk unit weight and dry unit weight are taken as independent variable. From the result obtained from MLR analysis, we can easily visualize the comparison in between predicted and experimental values shown in Fig. 7 below.

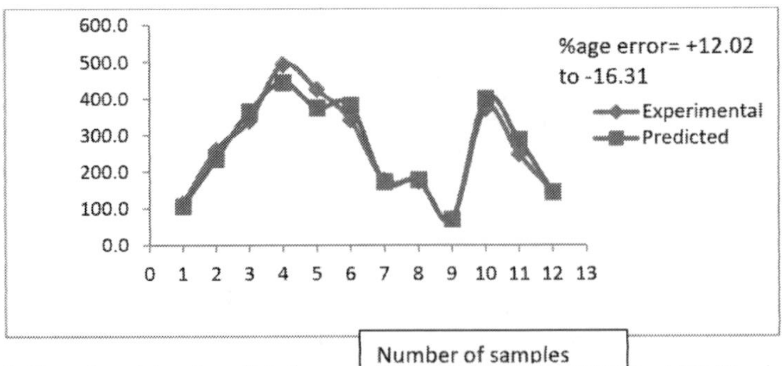

Fig.8: Experimental and predicted unconfined compressive strength (q_u) VS Number of samples.

All the soil properties tests conducted under this study were according to ASTM standards. The prediction of strength (kN/m^2) and dry unit weight are determined quickly from the developed

model equations.[9] For linear regression the predicted values for unconfined compressive strength are determined using Eq. (4) and Eq. (5), butEq. (6) is used to predict the dry unit weight (kN/m^3).
The percentage error for model-1 varies from 6.5 to 19.3 % and -10.1 to -17.9%. For Model-2, the errors vary from 2.7 to 15.8% and -8 to -16.7%. And for model-3 the percentage error is varying from 0.6 to 7.9 and -1.3 to -8.1%. The comparison of observed and predicted values are represented in these three modeled equations and presented in Fig.4, Fig.7 and Fig.8.

$$q_u = 53.9299*\gamma_b - 652.9751 \quad (4)$$
$$q_u = 60.6602*\gamma_d - 546.4603 \quad (5)$$
$$\gamma_d = 0.8587*\gamma_b - 1.2406 \quad (6)$$

Here, the developed equation for multiple linear regressions by using XLStat2015 statistical analysis software is listed below. Therefore, the errors within the developed models-4 for multiple linear regression analysis, is varies from +0.24 to +12.02% and -2.68 to -16.31%.

$$q_u = 20.6309*\gamma_b + 38.7767*\gamma_d - 604.8687 \quad (7)$$

The validations of the models are evaluated according to the results obtained from the equations and errors obtained from the variation in between observed (experimental) and predicted values.[10] For this correlation the developed models which gives least error for prediction of unconfined compressive strength (q_u) is Eq. (2). The prediction models-1, model-2 and model-3 have the coefficient of correlation (r) of 0.9546, 0.9661 and 0.9544.

4. Conclusion

The values of bulk unit weight and dry unit weight of soils are ranges from 14.35 to 20.2$\mathrm{kN/m^3}$ and 10.35 to 16.32$\mathrm{kN/m^3}$ respectively. The coefficient of determination (R^2) obtained from developed models are greater than 0.9. This shows that,the models have a strong relationship and correlated positively. The empirical relationship developed linearly for dependent variable ($q_{u)}$) as a function of bulk and dry unit weight with errors of predicted values verified with experimental values are varies from +19.3 to -17.9%. Model-1 and model-2 have coefficient of determination (R^2) value of 0.9112 and 0.9333 respectively. For the model-3 the errors vary from +7.9 to -8.1% with R^2 valueof 0.9109. From the equations developed under this linear regression statistical analysis good prediction of unconfined compressive strength is obtained from Eq. (2). In case of multiple linear regression analysis, the percentage errors in the prediction of unconfined compressive strength value verified with the experimental value is within the ranges of +12.02 to -16.31, and with the coefficient of determination (R^2) value of 0.9452.

Reference

[1] Associate Pro. Takaharu S. (2007), Effect of specimen size on unconfined compressive strength properties of natural deposits, national defense academy, Japan.
[2] Joanna S., Matylda T. (2014), the undrained shear strength of over consolidated clays, Wrockaw, Poland.
[3] Draper, N. R. and Smith, H. (1998) "Applied regression analysis," John Wiley and Sons, New York.
[4] Haan, C. T. (1994) "Statistical methods in hydrology," Affiliated East-West Press Pvt. Ltd., New Delhi, India.
[5] D.S. Vijayan, A. Mohan, J. Revathy, D. Parthiban , R. Varatharajan, "Evaluation of the impact of thermal performance on various building bricks and blocks: A review", Environmental Technology & Innovation 23 (2021) 101577, https://doi.org/10.1016/j.eti.2021.101577.
[6] Usama K., Zia-ur-R., Khalid F., Hassan M. (2015),Prediction of unconfined compressive strength from index properties of soils,Pakistan.
[7] Sorensen K.K., Okkels N. (2013), Correlation between drained shear strength and plasticity index of undisturbed overconsolidated clays, AarhusUniversity.
[8] Braja M. Das (2008), Advanced Soil Mechanics, third edition, Newyork, USA.

[9] M. ShabbirHossain, Ph.D., P.E., WanSooKim, Ph.D., P.E. (2014), Estimation of Subgrade Resilient Modulus Using the Unconfined Compression Test Virginia Department of Transportation, USA.

[10] Bruce, D. A., and Jewell, R. A. (1987). "Soil Nailing: The Second Decade." International Conference on Foundations and Tunnels, London, England, pp. 68–83.

[11] D.S. Vijayan, D. Parthiban," Effect of Solid waste based stabilizing material for strengthening of Expansive soil- A review," Environmental Technology & Innovation, Volume 20, 2020, 101108, ISSN 2352-1864, https://doi.org/10.1016/j.eti.2020.101108

Journal of Physics: Conference Series **2040** (2021) 012025 doi:10.1088/1742-6596/2040/1/012025

Numerical Analysis of Cold rolled structural steel angle sections subjected to tension members

A.Paulmakesh [1] , Gizachew Markos Makebo[2]

[1] Associate professor, Civil Engineering Department, Wolaita sodo university, Ethiopia.

[2]Associate professor, Civil Engineering Department, Wolaita sodo university, Ethiopia.

E-mail: paulmakesh1986@gmail.com

Abstract. Finite element method is a precise and flexible technology for predicting the performance, load conditions of a structure, mechanism or process. During the tests, the structural behaviour was used to build a FEM, especially for the nonlinear analysis. As the failure approaches, angles under tension exhibit non-linear behaviour. Everything is constructed, including the static structural component and all of its modules. Like ANSYS MAPDL modules (Mechanical Ansys Parametric Design Language). Build geometry, define materials, generate mesh, apply load, solving the model, and reviewing the results are all part of the static analyses of FEA. In order to determine the various structural investigations of a single or double section, a number of models have been developed. There were 72 specimens used in the analysis; they were modelled by bolt tensioners that were attached to the specimen. In order to calculate equivalent stress, equivalent elastic strain, and total deformation of single angle and double angle sections, equivalent stress, elastic strain, and total deformation of single angle and double angle sections were examined. In order to find out the ultimate load-carrying capacity of the single angle specimens and double angle specimens, a non-linear static analysis was done. The findings from the experiments, computational codes, and worldwide investigations into load capacity tension members are detailed in the paper. Failure modes are observed, too. In an ANSYS simulation using a numerical model, which used the ANSYS angle algorithm, researchers discovered that experimental and international codes values closely match each other.

Keywords: Cold formed steel, ultimate strength, lug Angle, single angle specimen, double angle specimens, non-linear, ANSYS 15.0 work bench and connection failure.

1. Introduction

Structural steel has been divided into two categories. Hot and other sections are cold and formed sections of steel. In recent years, sections of Cold Formed Steel gained prominence over sections of Hot Rolled Steel. The industry's standards for decades have been hot rolled and cold rolled steel processing. By applying high temperatures above 1 700 ° F, hot rolling steel is produced; we could see now on the market, like steel beam and column. High-speed, low temperature cold-forming steel [4] is formed. By pressing the steel turns into a sheet. The sheet is turned into different sections. The hardness, yield and tensile strengths have been increased. Hot rolled steel has 462 N/mm2 of final strength and yield strength is 310 N/mm2. The cold steel's ultimate resistance is 586 N/mm2 and yield resistance is 482 N/mm2. A wide range of buildings including industrial buildings, warehouses and health care are used for light-spread cold-shaped steel roof purlines. In many engineering areas the finite element method is a useful analysis tool. Finite element analysis is able to solve the problem prediction problem area in a material due to unknown stresses. Linear and nonlinear models can be used to carry out structural analysis. Linear models use simple parameters and suppose the material has not been deformed plastically. [2] It enables flexible design and very attractive, cost-effective and lasting solutions. The factor of safety due to block failure has been considered to predict the behavior of cold-shaped steel with bolted connections under tension capacity, load capacity [3]. Tests are presented, connected by bolts and tensed with loads in cold-shaped steel angles. The angle does not decrease evenly when the bound connection is taken into consideration, showing the phenomenon known as shear lags.

FEM of cold structures plays a growing role in engineering issues as it is time-efficient and expensive in comparison to physical experiments. Especially when the cross-sectional geometries are parametrically studied. FEA is an efficient analytical tool that predicts the behaviour accurately as long as the model developed is similar to the experimental prototype. [4] The results of the finite elements with the experimental results must therefore be verified. FEA is generally a strong tool for predicting cold formed steel members' ultimate loads and complex failure modes. In many fields of engineering the Finite element method is a highly valuable instrument of analysis. Finite element method is a precise and flexible technology for predicting the performance, load conditions of a structure, mechanism or process. The structural behaviour, especially during non-linear analysis, during the tests was employed to develop a FEM. In tension corners, as failure approaches, the behaviour is nonlinear. [5] [6] Static analysis is carried out with the ANSYS Workbench for Finite Element Analysis (FEA). [4]. Cold-formed steel functional properties The strength of the system, its durability, its sustainability, its efficiency, the cost-effectiveness of CFS, will not shrink or split, absorb moisture. Minimize the dead load because of light weight, fast building and minimum environmental impacts. The increased capacity of the cargo, moderately efficient, reduced cement consumption, reduces the impact on the environment

2. Review of Literature

According to the findings of Yu and Schafer (2003), the results of the investigation they did is based on cold-formed steel. We also limited distortional buckling and lateral torsional buckling to avoid excessive local buckling. Localised or distortional buckling is generally associated with the lateral bracing of cold-formed steel beams. Test results showed that both C and Z sections benefited from the improvement in elastic buckling and effective width calculation.

Shallow angles under tension are capable of displaying the shear lag effect (Prabha and Saravanan, 2011). IS: 801-1975 and BS: 5950-1998 believe the tensile strength of cold formed angle members will fall within a decent estimate. There is an increase in ultimate load capacity when the cross-sectional area is inversely proportional which indicates findings such as cross-sectional area, connection eccentricity will affect the tension members. Three bolted connection was considered in specimen with respect to AS/NZS 4600:2005 and IS: 800-2007. Findings shows enhancement in the outcomes of cold form angle as per IS: 801-1975 and BS: 5950-1998 (Part 5). The finite element analysis of a cold-formed steel bolted connection, which is loaded under shear, is investigated in the paper by Nirut Konkong (2017). As failure mode and ultimate strength of bolted connections are all dependent on each other, these characteristics are visually represented by this. While bolted connection tests detected the bearing failure modes, tensile coupon testing identified the material properties. Using the AISI: 2012 nominal strengths were calculated (manual). Comparing the predicted parameters with their actual measurements reveals that this specification is conservative when estimating nominal strength parameters. While finite element analysis used shell elements, solid elements were employed to model the cold-formed steel plate.

The actual behavior of the cold form steel angle members was examined by many technicians subjected under tension (2015). Also called an experimental load, this type of load is tested and compared to the most extreme load that meets the AS/NZS: 4600-2005 and BS: 5950-1998 standards (Part 5). It can be concluded from this that single and double angles most likely failed due to net section fracture failure, whereas the other angles failed due to local bend because of shear lag effect.

It has been shown by Abdul Bari and Easwaran (2017) that the CFS sections are used worldwide in construction as an industrial building component. In light of CFS adoption, we have advantages over the HRS. This work is concerned with the various CFS sections, including high fire resistance, minimum local buckling, high flexural strength, adequate seismic response, and bracing capabilities. The hollow sections with high loading rates and the possibility of infill material (geopolymer concrete) lead to the best joint results in building elements. Various building elements, including main walls, shear wall panels, tension members, compression members, and roof elements, are relevant to the various characters. Finally, finite element sections that were ready for finite element analysis were studied.

In their article "Cold-Formed Steel Structures: Structural and Functional Products", Bayan and Sariffuddin (2011) described cold-formed steel structures as products formed by forming flat sheets of steel at room temperature into various shapes that fulfil structural and functional requirements. While higher strength materials and a greater range of structural applications have increased the importance of CFS compared to traditional hot-rolled steel structural members, it has been over 15 years since this occurred. As a result, the understanding of CFS performance is crucial to study. This article contains three distinct works. The first step is to look at an introduction to CFS. Furthermore, it provides a general design guideline for CFS construction designs. Additionally, it contains a summary of the relevant research pertaining to joints and structures for cold-formed steel. As a result, it was concluded that conducting studies on the connections and portal frame for cold formed steel sections with both experimental and numerical approaches has emerged as a new field of study.

3. Finite Element Analysis

Finite element modelling of the experimental angle specimens is covered in this section. Finite Element Analysis (FEA) is a computational tool that can be used for performing engineering analysis. [Also,] [also] The 3D structural solid elements can represent large deformation geometric and material non-linear behaviour. For the current study, each of the angle specimens are individually examined. It is used to 3D model solids. Translations in the nodal of all the directions x, y, and z are part of what it means to have eight nodes with three degrees of freedom. Plasticity, hyper elasticity, stress stiffening, creep, large impact, and large strain capabilities are all characteristics of the element. It is also capable of producing deformations in elastoplastic materials with nearly incompressible properties, and in hyper elastic materials, both of which have incompressible properties. During their tenure in office, President George W. Bush and his Secretary of Defense Donald Rumsfeld faced a number of critical issues, such as the intelligence failures before the terrorist attacks of September 11, 2001, and Iraq's possession of weapons of mass destruction (WMD). Tension tests and measured cross-section dimensions were used to predict the static material properties of the angle specimens.

Can't be avoided social FEM is a set of ANSYS integrated pre-processing, solution, and post-processing applications that eases civil engineering issues for the user. The goal of Finite Element Analysis is to design models capable of simulating the behavior of single and double angle bolted cold formed steel components. The experimental outcomes will help to develop nonlinear analysis using several software's to predict the failure and effect of the materials.

4. About ANSYS Workbench

There are features exclusive to different types of materials and parts for steel structures in the ANSYS workbench. An entity within the ANSYS workbench library may be imported. section geometry and stress and strain listing for the plotting section are part of user-friendly beam and shell post processor software. Both of the worst loading designs for shell and solid components are used, along with the less than ideal combining module, part, and global setups [10]. This article gives a concise description of the various features found in the workbench project outline which are used to show the quality of the elements I in colour in the mesh branch (ii) using element quality bar charts (iii) and plotting the quality of the elements after post-processing distortion has occurred. Ansys capabilities include steel structural materials and sections that are unique and extensive. Additionally, users may add various shapes and materials to the corresponding ANSYS library. Ansys' workbench focuses on the design and optimization of geometric representations, with the addition of pre-existing geometries, the building of the finite element model, solution, and assessment. Figure 1,2,3,4,5 and 6 represents detailing of single and double angle specimens which is connected with gusset plate by with and without lip along with opposite and same side of the plate.

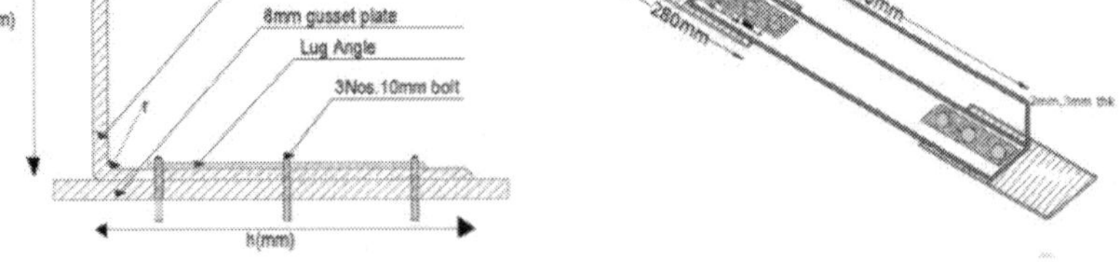

Fig 1: Details of single angles specimens without lip on connected with gusset plate

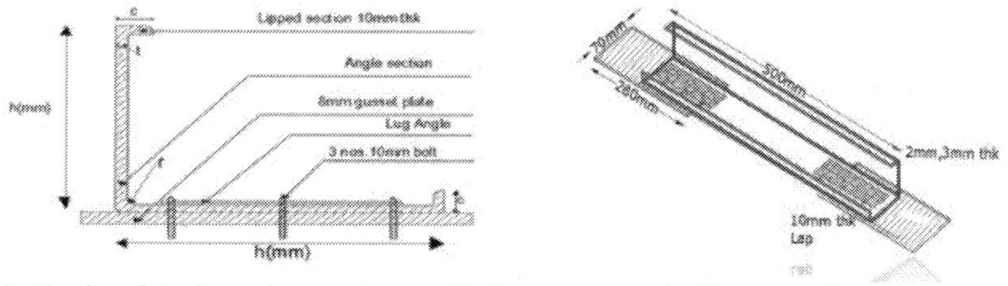

Fig 2: Details of single angles specimens with lip on connected with gusset plate

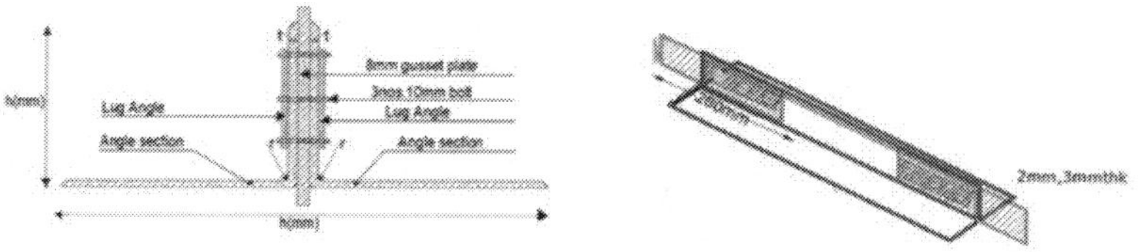

Fig 3: Details of double angles specimens without lip on connected to opposite side with gusset plate

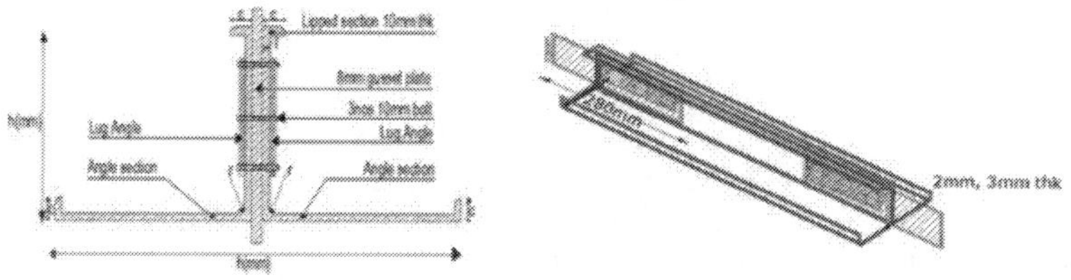

Fig 4: Details of double angles specimens with lip on connected to opposite side with gusset plate

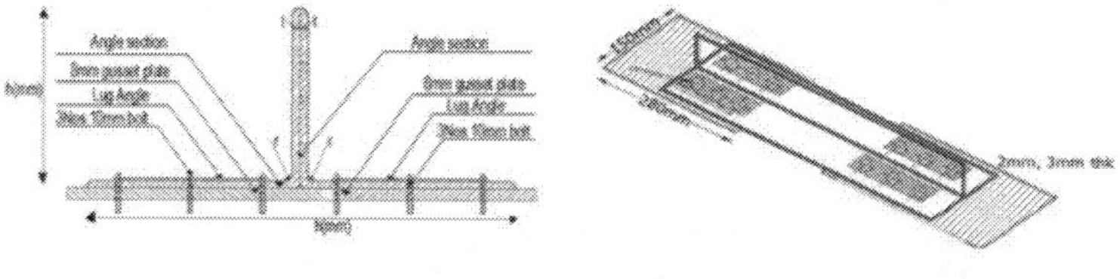

Fig 5: Double angles specimens without lip on connected to same side with gusset plate

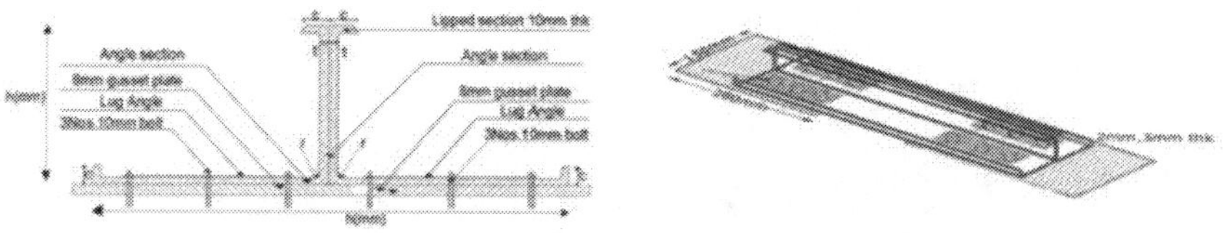

Fig 6: Double angles specimens with lip on connected to same side with gusset plate

5. Static Structural Analysis

Static structural analysis (also known as static analysis) may be used to identify structural displacements, stresses, strains, and forces as well as component inertias and damping in structures or components subjected to unidirectional loads. In other words, the loads and the structure's answer are assumed to adjust with respect to time slowly. Structural loads can be implemented in ANSYS using the software. A static structural analysis may or may not be linear or nonlinear, depending on the specifics of the problem. In order to assess the load of a static structure, a static structural load was estimated using the ANSYS 15.0. Additionally, the tension tests and the determined transverse cross-sectional dimensions were utilised to simulate the angle specimens, which have their own dynamic properties. Figure 7 shows modeling specifications for single angle without lip.

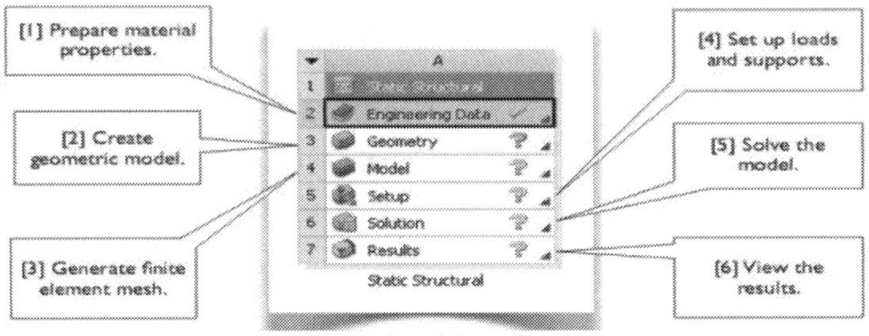

Fig 7: Single angle without lip 50x50 x2

5.1. Modelling

Modelling is one of the most critical parts of finite element analysis. The model created from cold-formed steel was used to define the material. Engineering specifications for the AISIES 1020 cold-formed steel can be input into the ANSYS workbench using the ANSYS input. These data inputs are important to the analysis of the structure's behaviour as the analysis is carried out. The information on these input is derived from both British and Indian standards. The linear elastic isotropic model is used to calculate the Elastic modulus, which is 2.05 x 105 N/mm2, Poisson's ratio is 0.290, shear modulus is 80 x 103 N/mm2, Density of mass is 7870 kg/m3, Tensile strength (ft) are 261-344 N/mm2, and yield strength (fy) are 216-244 N/mm2. Figure 8(a) and (b) displays ANSYS model of single angle with and without lip.

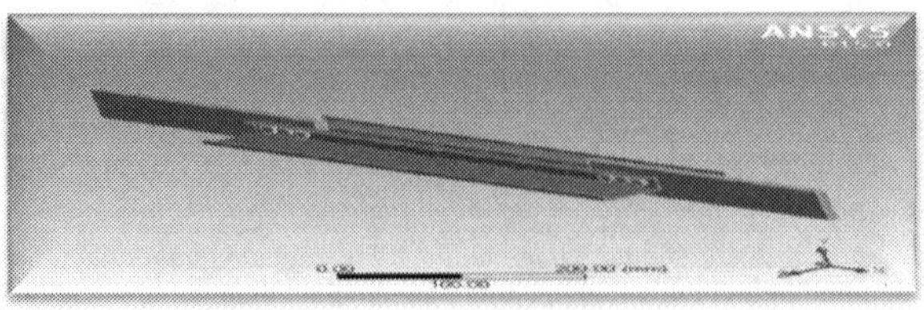

Fig 8a: Single angle without lip 50x50 x2

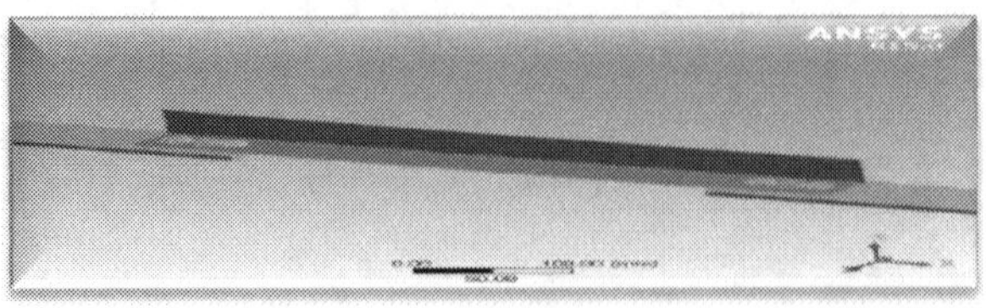

Fig 8b: Double angle connected to opposite side with lip 60x30x3

5.2. Element Type

As noted previously, the elements, Contra 174, Solid 186, Soild 187, and Surf 175, are the various element variant forms used to represent the single angle and double angle sections. The degrees of freedom of the element describe the discipline in which the element is used; in other words, it describes the context of the element. Plasticity is present, as well as hyper-elasticity, stress stiffening, creep, large deflection, and large strain [15]. Figure 9(a) and (b) depicts ANSYS modeling image of double angle connected to opposite as well as single angle with lip.

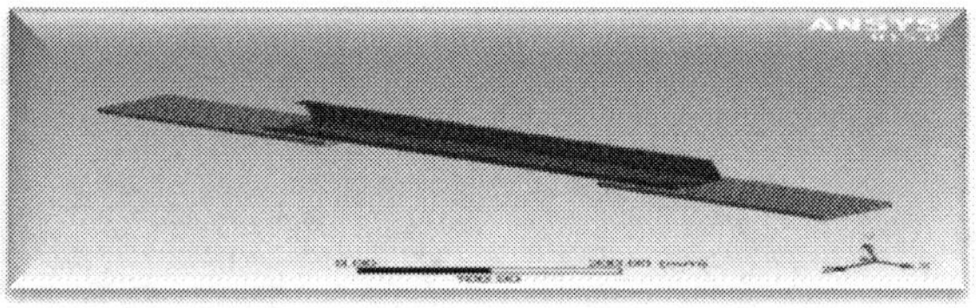

Fig 9a: Double angle connected to opposite side

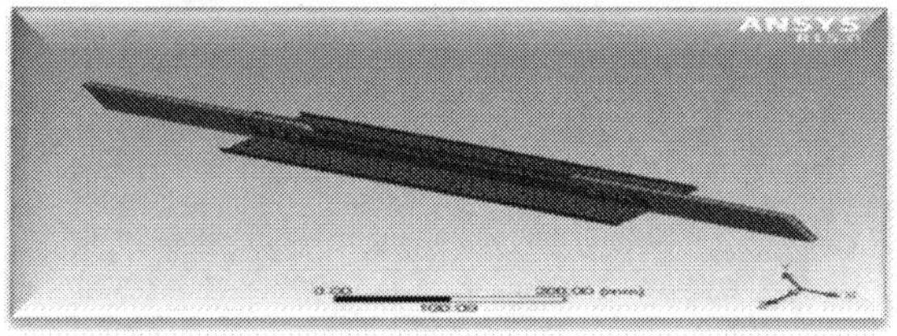

Fig 9b: Single angle with lip 50x50x10x3 with lip 50x50x10x2

5.3. Mesh Generation with Ansys Workbench

Meshing is defined as the process of dividing the whole component into a number of smaller units, so that it distributes the load uniformly. separation of the section into several smaller finite elemental fragments for efficient interaction.

5.4. Loading and boundary conditions

ANSYS Workbench version 15 is used to create the finite element model. A wide variety of elements are used to mesh the specimen geometry. The boundary conditions were set to mimic the actual experimental setting for all specimens. Specimens are 500 mm long. Fixed support conditions were considered to be fixed at the bolted end. Pitch and edge distance of the gusset plate were specified according to the code of practice in India. Both pitch and end distance are 30 mm. Figure 10 shows support condition (fixed) of double angle connected to opposite side without lip

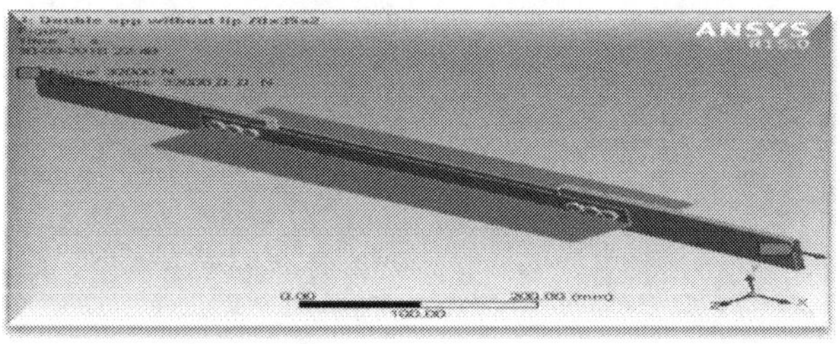

Fig 10: Fixed Support conditions double angle connected to opposite side without lip 70x35x3

Journal of Physics: Conference Series **2040** (2021) 012025 doi:10.1088/1742-6596/2040/1/012025

6. Experimental works

In accordance with theoretical design provisions outlined in practise codes, cold formed steel components and assemblies are typically designed. With the different thicknesses and dimensions, 72 specimens were tested for cold-formed single and double-angle sections. All specimens are equipped with tension grips using the UTM coordinate system with 40T capacity [12, 14]. The purpose of lug angle is to shorten the connection between the gusset plate and the bag, and to diminish shear lag effect. The lug angle to gusset plate is 20% greater than the force applied, while the lug angle to the main member is 40% greater than the force applied in the outstanding leg. Pitch and edge distance of the gusset plate were specified according to the code of practise in India [13]. The number of bolts that are required are calculated for all specimens and the design procedures were followed to supply those numbers. All of the specimens were created to measure 500mm long. Figure 11 (a)(b),(c),(d) ,(e), (f) displays fabrications, cutting , drilling of cold of steel angle and gusset plates.

11a: Fabrication of CFS

Fig 11b: Cutting sheet by Lazer machine

Fig 11c: Drilling Holes by Lazer Cutting machine

Fig 11d: Cold formed steel angles

Fig 11e: Marking the Specimen with dimension

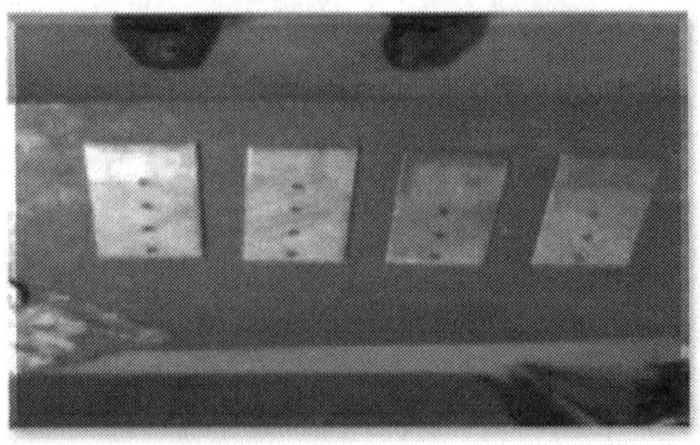

Fig 11f: Gusset plates

7. Various international codes

The relationship is a good one for all codal values [11]. To see if the recommended procedures found in various codes agree with the tested load carrying capacities of angle tested, comparison of observed load carrying capacities with predicted values suggested by the following codes is performed. Indian national standard IS: 801-1975 (R2001), as well as the Australia and New Zealand standards, AS/NZS: 4600-2005, and the American Iron and Steel Institute (AISI: 2010). (Part 1). Figure 12 (a),(b), (c) represents stress distribution and total deformation of angles. Figure 13 (a) and (b) displays graph of tension load applied on angle. Table 1 shows details of code books in terms of ultimate strength and table 2 provides Comparison of Experimental load and numerical value of thickness 2mm. Table 3 depicts Comparison of experimental load and predicted ultimate load obtained by various international codes of thickness 3mm thick.

Table 1: Ultimate strength for various international codes

IS: 801- 1975 (R2001)	**AS /NZS: 4600 – 2005**	**AISI: 2010**	**BS: 5950 - 2001 (Part 1)**
Tdg = Ag* fy *0.6 where, fy = yield strength of the material in MPa, Ag = gross area of cross section mm2 γm0= 1.10 partial safety factor for failure at yielding.	Nt = Ag fy (or) Nt = An 0.85 Kt An fu Ag = gross cross sectional area of the member, mm2 fy= yield stress of the material, N/mm2 Kt= correction factor 0.85 An= net area of the cross-section, fu= tensile strength N/mm2	Tn = Ag fy (or) Tn = An fu ϕt= 0.9 safety factor for yielding of cross section (LSD) ϕu= 0.9 safety factor for net section rupture (LSD) An= Net area fy = Design yield stress 0.8 fy (A2.2.2) fu = Tensile strength 0.8 fu	Pt = py * Ae Single angles Pt = py (Ae – 0.5 a2) Double angles Pt = py (Ae – 0.25 a2) Ae = effective area of the section. Ag = the gross cross-sectional area a1 = the gross area of the connected elements, a2 = (Ag – a1)

8. Analysis and Results

Table 2: Comparison of Experimental load and numerical value of thickness 2mm

S.No	Description	Size of specimen(mm)	Exp load (KN)	Ansys load (KN)	Ansys/Exp PAns /PExp	% increase in load
1	Equal angle Single angle without Lip	50x50	34.58	36.78	1.06	6.36
2		60x60	38.13	40.72	1.07	6.79
3		70x70	45.26	47.82	1.06	5.66
4	Single angle with Lip	50x50x10	41.25	43.72	1.06	5.99
5		60x60x10	51.47	54.28	1.05	5.46
6		70x70x10	62.47	65.21	1.04	4.39
7	Double angle opposite side without Lip	50x50	75.48	77.82	1.03	3.10
8		60x60	87.29	90.72	1.04	3.93
9		70x70	108.17	113.05	1.05	4.51
10	Double angle same side without Lip	50x50	80.47	83.78	1.04	4.11
11		60x60	92.58	95.74	1.03	3.41
12		70x70	112.28	115.83	1.03	3.16
13	Double angle opposite side with Lip	50x50x10	86.27	91.26	1.06	5.78
14		60x60x10	92.47	95.72	1.04	3.51
15		70x70x10	117.23	120.78	1.03	3.03
16	Double angle same side with Lip	50x50x10	87.46	90.78	1.04	3.80
17		60x60x10	98.75	103.48	1.05	4.79
18		70x70x10	112.17	115.89	1.03	3.32
19	Unequal size Single angle without Lip	50x25	28.17	29.46	1.05	4.58
20		60x30	32.78	34.72	1.06	5.92
21		70x35	37.85	39.72	1.05	4.94
22	Single angle with Lip	50x25x10	42.58	44.88	1.05	5.40
23		60x30x10	48.75	51.89	1.06	6.44
24		70x35x10	53.78	56.79	1.06	5.60
25		50x25	59.76	63.73	1.07	6.64

26	Double angle opposite side without Lip	60x30	73.7	77.89	1.06	5.69
27		70x35	87.81	92.18	1.05	4.98
28	Double angle same side without Lip	50x25	58.81	61.89	1.05	5.24
29		60x30	62.78	65.73	1.05	4.70
30		70x35	83.47	87.18	1.04	4.44
31	Double angle opposite side with Lip	50x25x10	72.37	76.83	1.06	6.16
32		60x30x10	89.34	93.73	1.05	4.91
33		70x35x10	106.35	111.59	1.05	4.93
34	Double angle same side with Lip	50x25x10	71.58	75.29	1.05	5.18
35		60x30x10	83.147	88.79	1.07	6.79
36		70x35x10	92.47	97.83	1.06	5.80
Mean					1.05	
Standard Deviation					0.05	

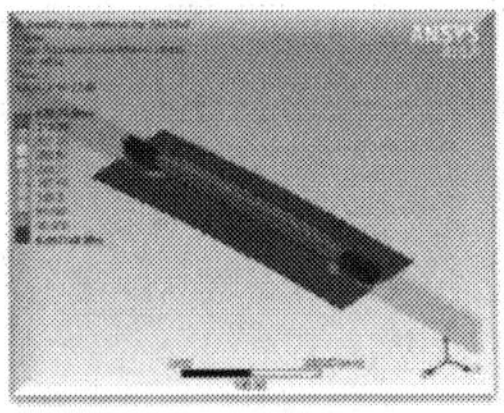

Fig 12a: Stress distribution for double angle connected to opposite side without lip 70x35x2

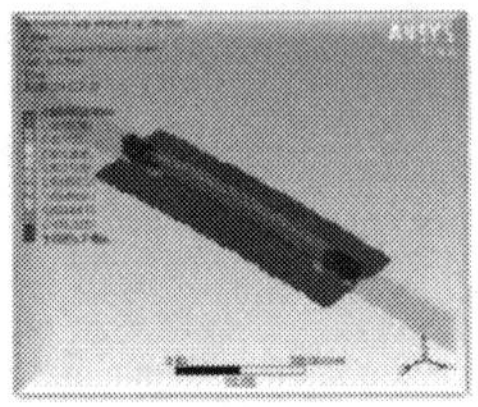

Fig 12b: Stress distribution for double angle connected to opposite side without lip 70x35x2

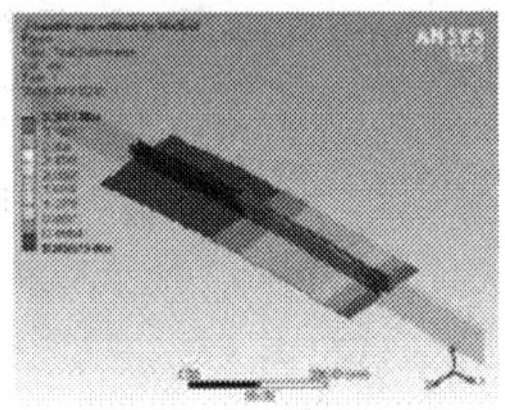

Fig 12c: Total deformation for double angle connected to opposite side without lip 70x35x2

9. Discussion of cold formed steel angle sections subjected to tension load.

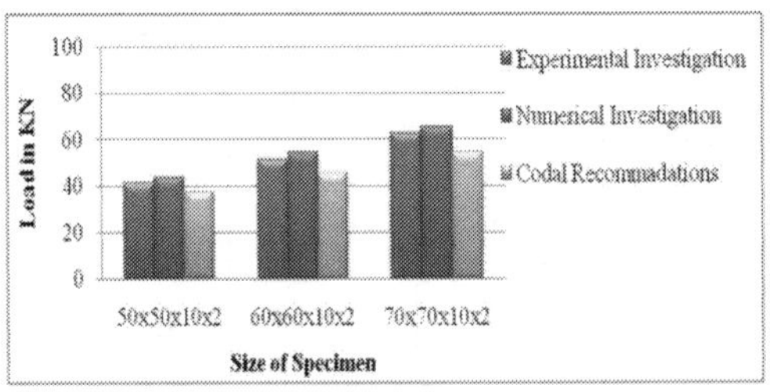

Fig 13a : Cold formed steel angle sections subjected to tension load for double angle connected to opposite side with lip 50x50x10x2

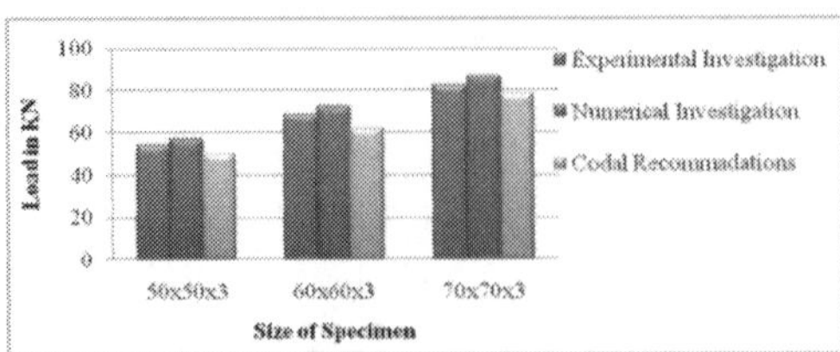

Fig 13b: Cold formed steel angle sections subjected to tension load for double angle connected to opposite side without lip 50x50x3

Table 3: Comparison of experimental load and predicted ultimate load obtained by various international codes of thickness 3mm thick.

S.No	Size of specimen (mm)	Exp valve (KN)	IS: 801-1975 (R2001) (KN)	IS : 801-1975/Exp	AS/NZS: 4600-2005 (KN)	AS/NZS/Exp	AISI:2010 (KN)	AISI 2010 /Exp	BS: 5950-2001 (Part 1) (KN)	BS : 5950-2001/Exp
1	50x50	54.28	50.12	0.92	51.91	0.96	61.07	1.13	56.01	1.03
2	60x60	68.45	61.85	0.90	63.98	0.93	75.27	1.10	69.90	1.02
3	70x70	82.45	78.43	0.95	76.05	0.92	89.47	1.09	83.76	1.02
4	50x50x10	78.25	72.58	0.93	61.57	0.79	72.43	0.93	60.15	0.77
5	60x60x10	94.28	89.75	0.95	73.64	0.78	86.63	0.92	74.36	0.79
6	70x70x10	102.5	96.13	0.94	85.71	0.84	100.8	0.98	88.69	0.86
7	50x50(o)	124.2	113.2	0.91	103.8	0.84	122.1	0.98	120.0	0.97
8	60x60(o)	154.2	123.8	0.80	127.9	0.83	150.5	0.98	125.1	0.81
9	70x70(o)	168.2	146.2	0.87	152.1	0.90	178.9	1.06	177.8	1.06
10	50x50(s)	113.2	112.0	0.99	103.8	0.92	122.1	1.08	120.0	1.06
11	60x60(s)	143.5	124.2	0.87	127.9	0.89	150.5	1.05	125.1	0.87
12	70x70 (s)	152.2	147.1	0.97	152.1	1.00	178.9	1.18	177.8	1.17
13	50x50x10(o)	105.2	101.2	0.96	123.1	1.17	144.8	1.38	129.9	1.23
14	60x60x10(o)	131.2	125.2	0.95	147.2	1.12	173.2	1.32	163.6	1.25
15	70x70x10(o)	168.2	161.2	0.96	171.4	1.02	201.6	1.20	189.3	1.12
16	50x50x10(s)	122.4	102.4	0.84	123.1	1.01	144.8	1.18	129.9	1.06
17	60x60x10(s)	142.5	127.8	0.90	147.2	1.03	173.2	1.22	163.6	1.15
18	70x70x10(s)	169.2	162.2	0.96	171.4	1.01	201.6	1.19	189.3	1.12
19	50x25	40.87	35.21	0.86	36.82	0.90	43.32	1.06	44.54	1.09
20	60x30	53.28	48.23	0.91	45.87	0.86	53.97	1.01	55.98	1.05
21	70x35	68.57	61.29	0.89	54.93	0.80	64.62	0.94	66.56	0.97
22	50x25x10	55.28	48.12	0.87	46.48	0.84	54.68	0.99	51.21	0.93
23	60x30x10	68.87	60.79	0.88	55.53	0.81	65.33	0.95	62.52	0.91
24	70x35x10	78.27	70.58	0.90	64.58	0.83	75.98	0.97	74.45	0.95
25	50x25	88.25	80.75	0.92	73.64	0.83	86.63	0.98	92.66	1.05
26	60x30	108.4	101.2	0.93	91.74	0.85	107.9	1.00	115.1	1.06
27	70x35	138.6	131.8	0.95	109.8	0.79	129.2	0.93	137.4	0.99
28	50x25	93.28	80.92	0.87	73.64	0.79	86.63	0.93	92.66	0.99
29	60x30	111.1	107.8	0.97	91.74	0.83	107.9	0.97	115.1	1.04
30	70x35	138.7	129.7	0.93	109.8	0.79	129.2	0.93	137.4	0.99
31	50x25x10(o)	104.3	98.42	0.94	92.95	0.89	109.3	1.05	110.0	1.05
32	60x30x10(o)	139.9	129.7	0.93	111.0	0.79	130.6	0.93	93.66	0.67
33	70x35x10(o)	151.0	146.7	0.97	129.1	0.86	151.9	1.01	156.7	1.04
34	50x25x10(s)	118.2	110.8	0.94	92.95	0.79	109.3	0.92	110.0	0.93
35	60x30x10(s)	122.8	116.2	0.95	111.0	0.90	130.6	1.06	93.66	0.76
36	70x35x10 (s)	147.2	138.5	0.94	129.1	0.88	151.9	1.03	156.7	1.06
Mean				0.89		0.88		1.03		0.98
Standard deviation				0.60		0.90		0.17		0.15

10. Conclusion

Experimental load versus deflection curves for cold formed steel angle sections revealed that the ultimate load-carrying capacity increases with the cross-sectional area and the number of bolts in the connection. The stiffness of the connection increases in direct proportion to the stiffness of the member. A member is loaded until strain hardenings are reached, whereupon it elongates excessively, leading to the members to fail at an angle. A failure mode study was conducted on the single angle and double angle sections. Failure occurs if the cross-sectional area and connection stiffness are both correct. Ansys software was used to calculate the tension strength behaviour of cold formed steel angle under load. The ANSYS-developed numerical model was found to be nearly identical to the experimental valves in terms of angles. In the finite element analysis, the stress contour denotes the stress that occurs in the innermost bolt holes, which is where the experimental failures were discovered.

References

[1] Bhavitha, E & Rosemol, K 2015.Comparative study on moment connections in cold formed steel sections with and without perforations. International journal of Innovative science, Engineering & Technology, vol. 2, no. 11, pp. 478-481

[2] Cam Tokay 2004. A finite element parametric study on block shear failure of steel tension members', Journal of Constructional Steel Research, vol. 60, no. 11, pp. 1615-1635

[3] Chakherlou & Oskouei 2008.Experimental and numerical investigation of the effect of clamping force on the fatigue behaviour of bolted plates', Engineering Failure Analysis, vol. 10, no. 5, pp. 563-574

[4] Cher Siang Tan & Mahmood, MD 2015. Experimental Investigation of Bolted Angle Joints for Cold-formed Steel Double Channel Sections. International Journal of structural analysis and Design, vol. 2, no. 2, pp. 14-17

[5] Chou, SM & Chai, GB 2000. Finite Element Technique for Design of stub Columns', Thin- Walled Structures, vol. 37, no. 2, pp. 97-112

[6] Chung, KR & Ip, KH 2000. Finite Element Modeling of Bolted Connections between cold formed steel strips and hot rolled steel plates under static shear loading', Engineering Structures, vol. 4, no. 22, pp. 1271-1284

[7] Chung, KF & Lau, L 1999. Experimental Investigation on bolted moment connections among cold formed steel members. Engineering Structures, vol. 2, no. 21, pp. 898-911

[8] Gotluru 2000. Behavior of cold formed steel beams having open sections which were subjected to torsion', Journal of Engineering structures, vol. 2, no. 3, pp. 1374-1975

[9] Gregory Hancock, J 2016. Cold-formed steel structures: Research review 2013-2014', Advances in Structural Engineering, vol. 19, no. 3, pp. 1-16

[10] Guangfan, LI & Yong Huang 2011. Experimental Investigation on Slip-In Connection for Cold-Formed Steel Double Channel Sections. Advanced Materials Research, vol. 25, no. 5, pp. 1038-1041

[11] Naveen Kumar K.,Vijayan D.S.,Divahar R.,Abirami R.,Nivetha C.,"An experimental investigation on light-weight concrete blocks using vermiculite", Materials Today: Proceedings, Volume 22, Pages 987 – 991, 2020, 10.1016/j.matpr.2019.11.237

[12] D.S.Vijayan ,J. Revathy,Flexural Response of Fibre Reinforced Polymer Laminated Pre-stressed Concrete Beams, Indian Journal of Science and Technology, Vol 9(42) , November 2016, DOI: 10.17485/ijst/2016/v9i42/101824

[13] D S Vijayan, Dr J Revathy, Flexural Behavior of Reinforced and Pre-Stressed Concrete Beam Using Finite Element Method, International Journal of Applied Engineering Research, ISSN 0973-4562 Volume 10, Number 1 (2015) pp. 717-736

[14] D.S.Vijayan, S.Arvindan, K.Naveen Kumar, S Mohamed Javed, Seismic Performance of Flat Slab in Tall Buildings with and without Shear wall, International Journal of Engineering and Advanced Technology (IJEAT) ISSN: 2249 – 8958, Volume-9 Issue-1, October 2019, Page no – 2672 – 2675.

[15] D.S.Vijayan, K.Naveen Kumar, S.Aravindan,Analytical Study on Industrial Steel Building with Braceing, INTERNATIONAL JOURNAL OF SCIENTIFIC TECHNOLOGY RESEARCH VOLUME 8, ISSUE 10, OCTOBER 2019, Page no 1275 – 1278.

ANALYTICAL STUDY ON FLEXURAL BEHAVIOUR OF LIGHT GAUGE STEEL CORRUGATED SECTION ENCASED IN CONCRETE

K. THILAGAR [1], S. SURESH BABU[2]

[1] PG student, [2]Assistant Professor, Civil Engineering Department, SRM Valliammai Engineering College, Chennai, India.

E-mail: thilakkannan449@gmail.com

Abstract. Light gauge elements for integrated beams were used, and thin webs are typically needed in constructed beams' economical design. However, the buckling problem may arise if the Web is fragile. This risk may be reduced by using thicker panels, web reinforcements, or web enhancements. The use of corrugated Web is a possible means of achieving adequate rigidity and shear resistance without hardeners. Analytical studies were conducted in this present work to study the bending behavior of the conventional beam and light-weight steel beam with corrugated, concrete enclosed Web. Analyzes of finite elements have been conducted using the ANSYS beam software. The results present the capacity for load carrying and the deformation of the concrete-covered corrugated web beams. This study's main objective is to acquire a better knowledge of a concrete embedded steel beam's behavior.

KEYWORDS: cold-formed steel section, triangular corrugation steel, trapezoidal corrugation steel.

1. INTRODUCTION

In the construction field, majorly two types of steel are used. They are hot-rolled steel and cold-form steel. Comparing with hot rolled steel section, light gauge steels are suitable to finish and dimensional accuracy. It is also suitable for the production of foils, plates, and sheets [1]. For the last ten years, India has been gaining popularity using cold steel for buildings and structures. Cold-shaped members are traditionally used for lightly loaded roofs as secondary structural components, like sprinklers [2]. Since 1990, cold-shaped steel members in low to medium-rise buildings have been increasingly used as the primary structural component. Usage of the light gauge steel section reduces the weight of the structural member [3].

Corrugated web beams may be used as beams, roof or slab beams, or components subjected to regular forces column or frame columns virtually without structural limitation [4-5]. The load-carrying for the beam encased with corrugated Web increased when compared to conventional beam. Corrugated Web is used to neglect the shear failure. With the introduction of corrugated profiles in I-sections, web instability and transverse reinforcements are consistently reduced. However, effects such as lateral-torsional buckling were observed even after webs provided [7]. Many researchers found that web corrugation significantly increases their buckling stability and can lead to very cost-effective designs [9]. The well-compatible web beams are therefore capable of eliminating many expensive web riggers.

Gopinath P, Manu S S (2016) Cold-formed steel sections have been used in place of conventional steel reinforcement bars in this project. The study incorporates reinforced concrete beams and reinforced concrete beams encased in lipped CFS channel sections with open and closed configurations. The flexural behavior of specimens is investigated experimentally using two-point loading. To increase the load-bearing capacity of the CFS-encased beam. Concrete encasing prevented local buckling of CFS sections, thereby increasing the section's flexural bending capacity. Local buckling, one of the primary disadvantages of CFS, is avoided due to the confinement provided by the encasing concrete. Beams encasing back-to-back arrangement (open section) suffered more significant vertical deflection but lesser cracks and only flexural failure. Beams encasing face-to-face arrangement showed lesser deflection but flexural, and shear cracks were formed at failure. Beams encasing open section exhibited equally spaced vertical flexural cracks of the lesser number. The ultimate load closed section encased beam undergoes 7 to 9% lesser deflection than the open section encased beam [8].

K.Elakkiya M.Anbarasan (2019) Beam of the I section with corrugated Web can be investigated by experimentally and analyzed by ANSYS software. Also, comparison between beams with flat Web to the corrugated Web typically the beam of I-section with flat Web can be used. Nevertheless, we aim to attain more strength with corrugated Web. Mainly the flexural strength can be studied and discussed. The experimental test should under two points loading in supported condition. Similarly, the numerical analysis could do with the same condition. As a result, the I-section with corrugated Web attains higher flexural strength than flat Web, and there was no failure in the shear zone. Also, the depth of the Web increases; thus, the strength increases[10].

Divahar RAVI, Aravind Raj PONSUBBIAH (2020) An analysis and numbers experimental study on the bending behavior of web-coated Web profiled cold-shaped steel beams, enclosed by trapezoidal corrugated concrete which were supported at both ends with the symmetric load of two points. 12 experiments were conducted in two different series to find the effect of concrete enclosure on the Web. Beams with three different 0°, 30°, and 45° web corrugation angles were tested with two different 60 and 80 web depth ratios. Experimental results are presented, such as load-deflection ratio, ultimate capacity, load-strain relation, moment-curvature curve, ductile, and fault mode indices improved bending strength by increasing the web depth. The specimens of ECWB30° offer more load carrying capacity, increased compressive longitudinal strain, and a higher bending ductility and obtain the elasticity of super elastic Web coated beams, increasing their usefulness in the earthquake-resistant design of building constructions [6].

A fully encased corrugated steel beam is compared with the conventional beam in this present work, increasing the load-carrying capacity and deflections. In Light gauge steel beam, Trapezoidal and Triangular corrugation is used to reduce the shear failure. The model was created and analyzed ANSYS workbench. In this study, the flexural behavior light gauge steel corrugated section encased in concrete and corrugation on the composite beam is analyzed.

2. DESCRIPTION OF MODEL

The concrete beam of size 1000x200x200 mm3.Conventional beam, Trapezoidal corrugated beam, Triangular corrugated beam is modeled and analyzed using ANSYS [8]. The primary reinforcement 12mm diameter bar is used, and the distribution reinforcement of the 8 mm diameter bar with the spacing of 150mm is provided. Similar material properties are used in the beam in order to avoid variation in the results. Specimens are shown in fig 1.

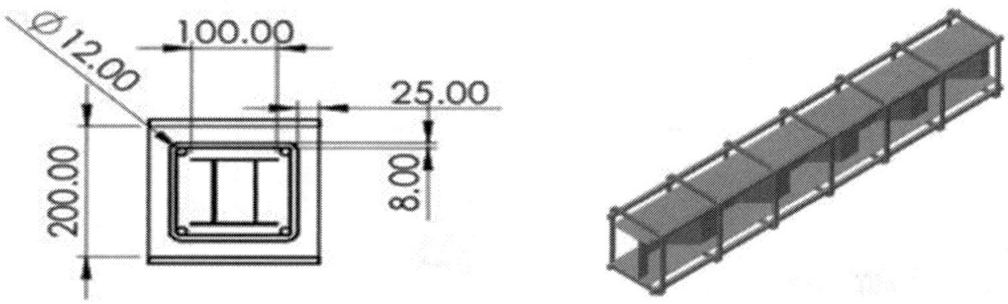

Figure 1: Encased corrugated beam

3. MATERIAL PROPERTIES

Properties of the material are fed into the ANSYS Software to analyze the beam. Table 1 shows the properties of the material.

Table1. Properties of material

Properties	Concrete	Steel
Density	2400 kg/m^3	7850 kg/m^3
Modulus of Elasticity (E_c)	31.2GPa	200 GPa
Reference temperature	200 ^{0}c	200 ^{0}c
Poisson's ratio (μ)	0.2	0.3

Span length of cold-formed steel 1 meter. The overall depth of the section is 100 mm. the width of the section is 100 mm.

Table 2: properties of cold-formed steel

DESIGNATION	PROPERTIES
THICKNESS OF FLANGE F_t (mm)	2.5
WIDTH OF FLANGE F_w (mm)	100
THICKNESS OF WEB W_t (mm)	2.5
DEPTH OF WEB W_d (mm)	95
OVERALL WIDTH W (mm)	100
OVERALL DEPTH D (mm)	100

4. FEM ANALYSIS

ANSYS Workbench is an FEM software which produced accurate output using experimental data furthermore it also can be used in structural engineering to analysis the structural elements even more effectively [11]. The class focuses on geometry creation and optimization, attachment of existing geometry, modeling, solution, and results review. The class discusses how to use the code and the basic concepts for simulation and interpretation of finite elements' results [12-13].

4.1 ENGINEERING DATA

Material properties have to be included, like the grade of steel, grade of concrete, admixtures, etc. there, I have to change the material properties as per my specifications. Assigning properties to the material is essential in this finite element software [14-15]. In this work, two different materials have been used. Cold-formed galvanized steel uses 2.5mm thick & Concrete material for concrete portions[16]. These materials were collected from various tests carried out on those materials.M30 grade concrete is used in this work. Figure 3.1 shows Engineering data of the beam.

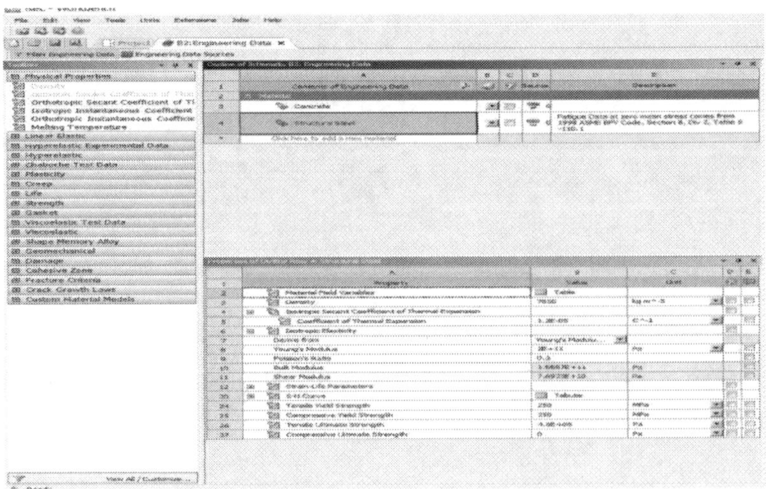

Figure 2. Engineering data of the beam

4.2 GEOMETRY CREATION

Geometry modeling in the ANSYS Workbench environment is highly automated and customizable to the type of analysis or application [17]. The parametric, feature-based ANSYS Design Modeler software enables creating parametric geometry from scratch or the preparation of an existing CAD geometry for analysis [18]. It features automated simplification, cleanup, repair, and defeaturing options [19]. I have created the model with the primary reinforcement as 12mm, stirrups of 8mm, and a cross section of beam as200x200x1000, shown in fig.3, fig 4.

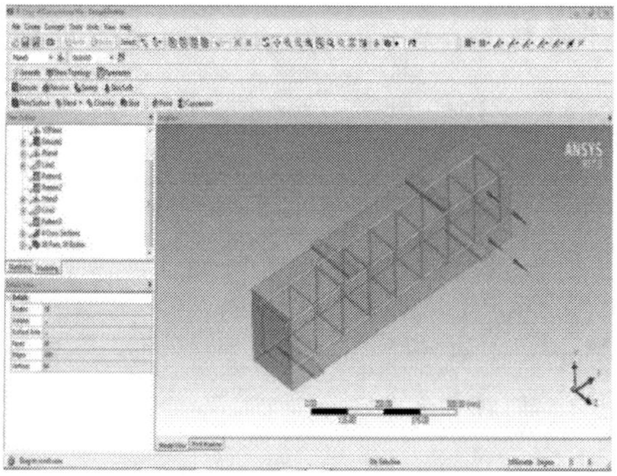

Figure 3. The geometry of conventional beam beam

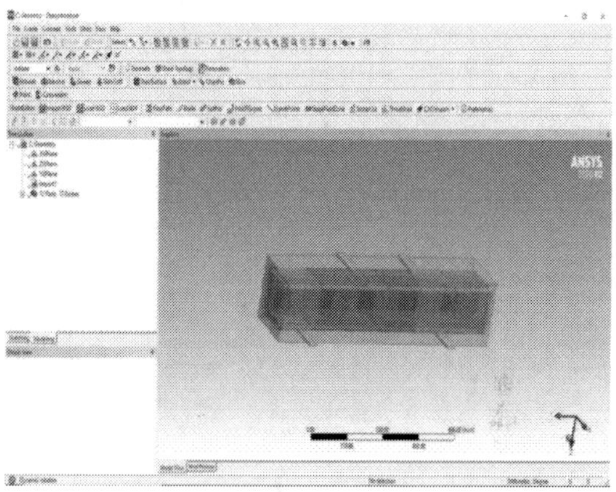

Figure 4. The geometry of the encased

4.3 MESHING OF THE BEAM

After completing the geometry design, I have gone to the modeling phase and assign all the material properties. Then I have generated a mesh that helps us solve the finite element method, and results can be obtained accurately, as shown below in fig 5.

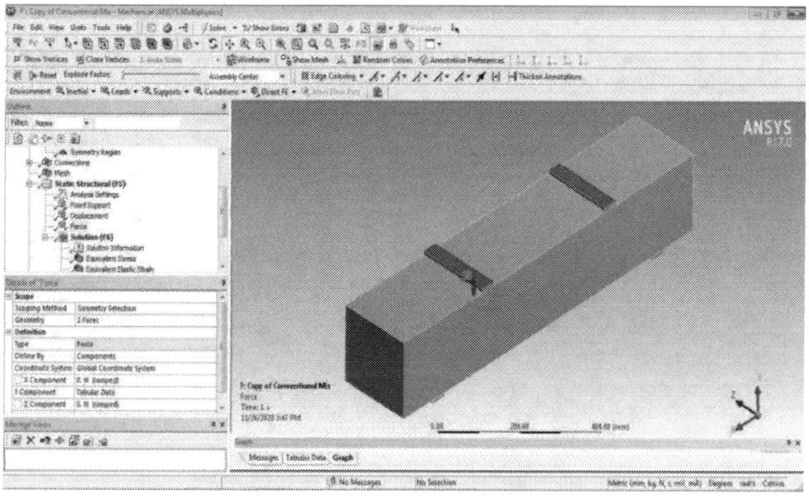

Figure 5. Generation of Mesh of the beam

I have adopted two-point loading in which load is applied at L/3 distance from both the ends and support conditions are given at 200 mm from the ends of the beam, as shown in fig 6. After the application of load, the model is solved for the required results, and the model's failure mode is obtained.

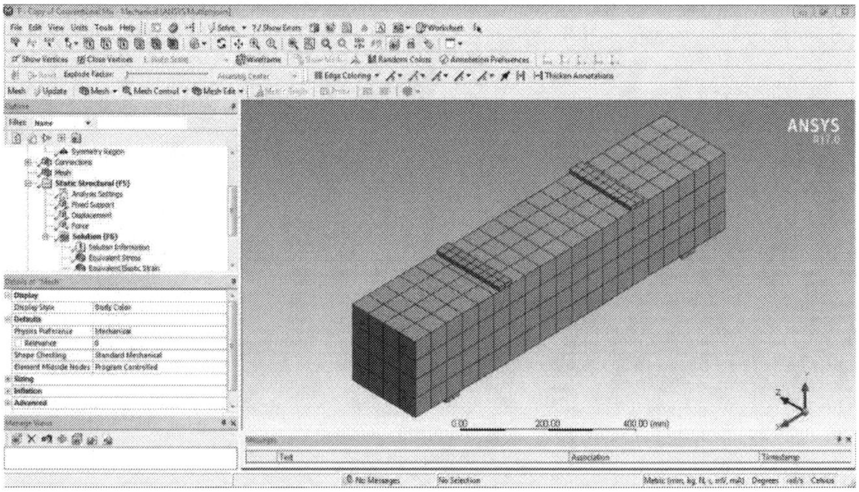

Figure 6. Application of load

5. RESULTS OF THE BEAM

Then I will go to the results phase to obtain the results as per our requirement. The evolved solution will be done using the finite element method, and this is the results for maximum deformation are shown in fig.7, fig.8, fig.9.

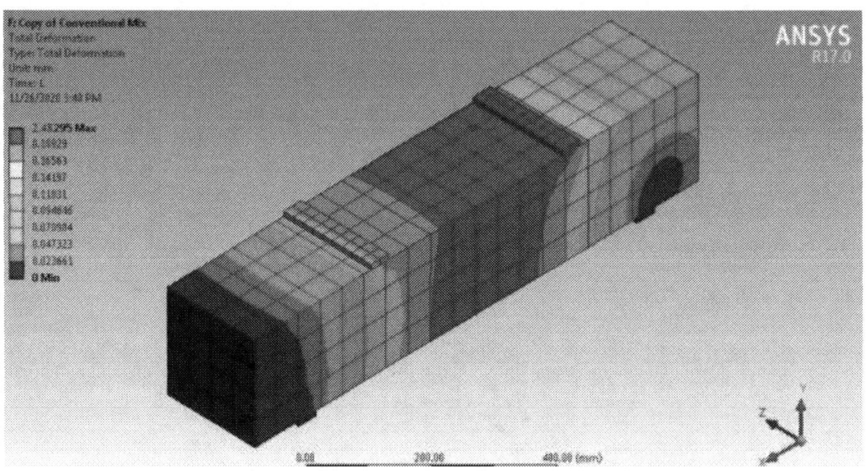

Figure.7. Total Deformation of conventional beam

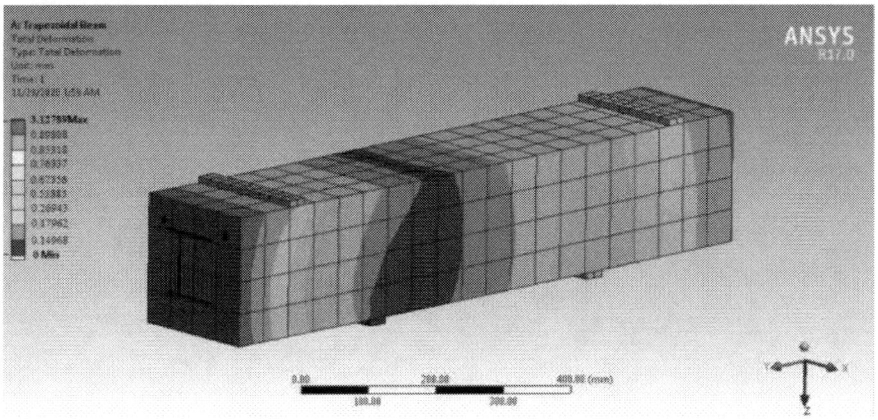

Figure 8. Total Deformation of encased trapezoidal beam

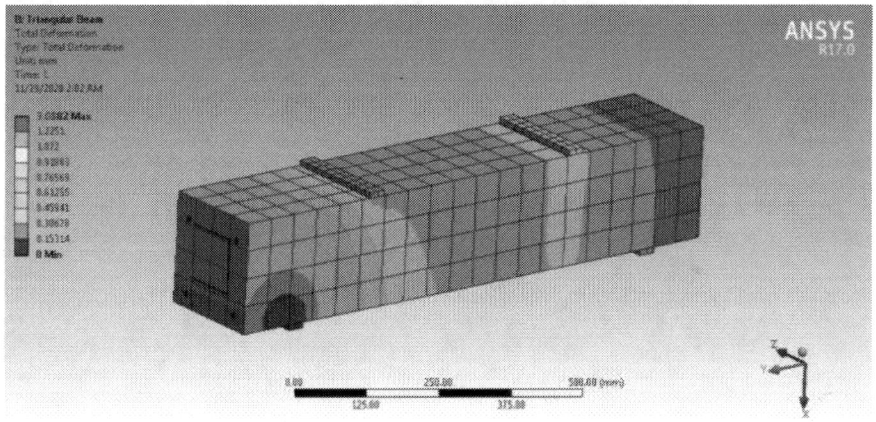

Figure.9 Total Deformation of encased triangular beam

Then I will go to the results phase to obtain the results as per our requirement. The evolved solution will be done using the finite element method, and this is the results for maximum stress are shown in fig.10, fig.11, fig.12.

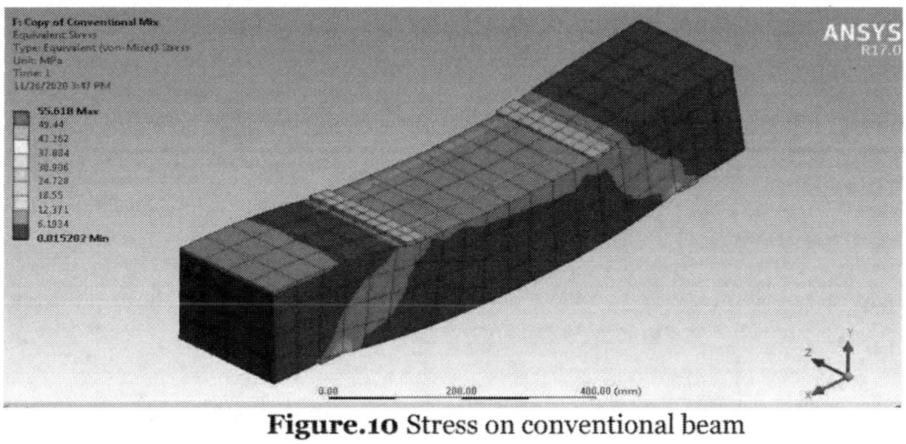

Figure.10 Stress on conventional beam

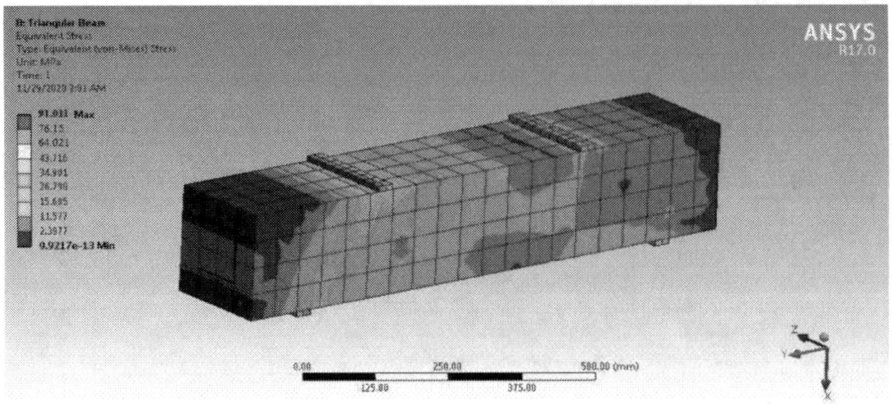

Figure.11 Stress on encased trapezoidal beam

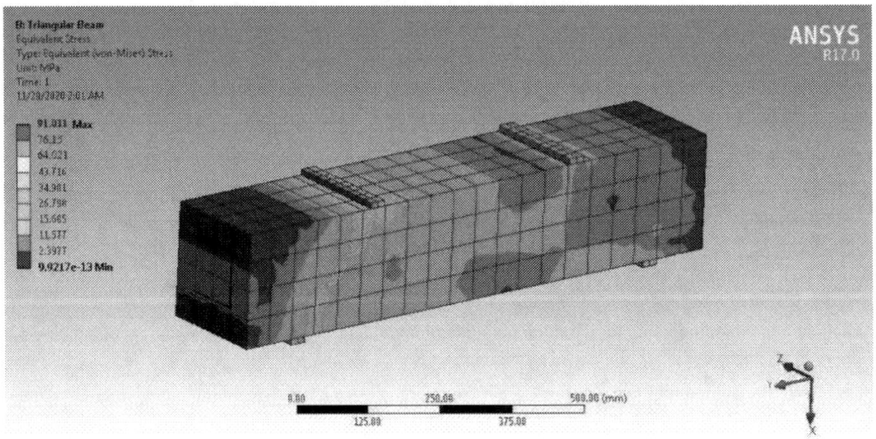

Figure.12 Stress on encased triangular beam

6. RESULTS AND DISCUSSION

The obtained results were discussed below.

6.1 DEFLECTION BEHAVIOUR

- The presence of encased trapezoidal corrugated beam reduces the deflection.
- At the ultimate load, compared to the encased trapezoidal corrugated beam, the deflection is less and shear failure.

The maximum deflection in an encased trapezoidal corrugated beam is 3.12 mm compared with a conventional beam; it has 2.48 mm. The ultimate load in a conventional beam is 170 KN but in an encased trapezoidal corrugated beam is 280 KN.

7. LOAD DEFORMATION CHARACTERISTICS

The load-deflection curve studied in the analytical study for both the conventional and encased corrugated beam is plotted below in the figure. From the results obtained, the presence of encased trapezoidal corrugated beam increases the load-carrying capacity.

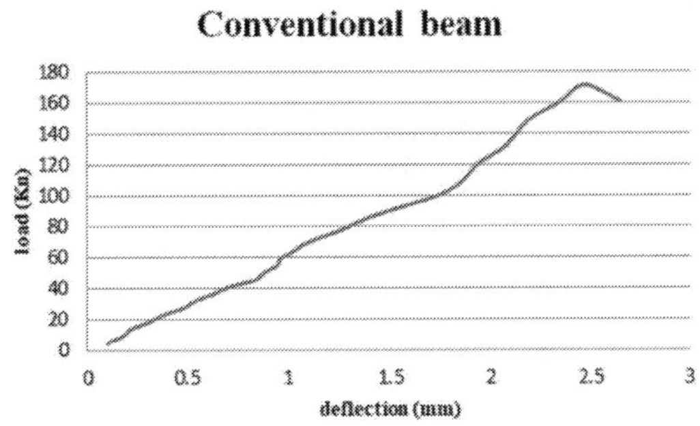

Figure.13 Load deflection For Conventional Beam

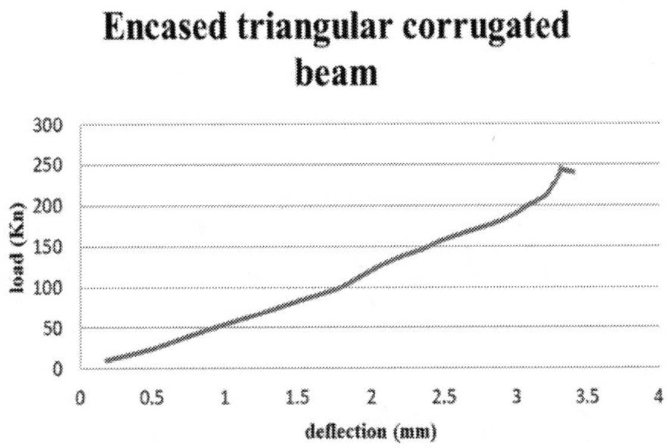

Figure.14 Load deflection For Encased trapezoidal Beam

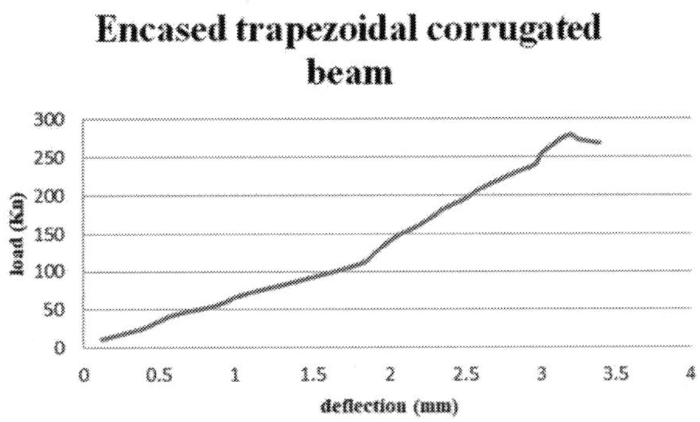

Figure.15 Load deflection For Encased triangular Beam

We can conclude from the above graph that the encased trapezoidal corrugated beam's load-carrying capacity is 40% higher than the conventional beam. It also resists more stress compared with the conventional beam.

Table 3. Results of Conventional Beam and Encased Beam

RESULTS	CONVENTIONAL BEAM	ENCASED TRAPEZOIDAL BEAM	ENCASED TRIANGULAR BEAM
TOTAL DEFORMATION (mm)	2.48	3.12	3.31
ULTIMATE LOAD (MPa)	170	280	240

From the above results, the load-carrying capacity for encased trapezoidal corrugated beam increases. In this, it is observed that the deformation gets decreased in the encased trapezoidal beam.

8. CONCLUSION

The analytical results show that the load-carrying capacity of encased beam is higher than the conventional beam. The above Analytical study shows that the load-carrying capacity of composite beam with trapezoidal corrugated Web is 10 % higher than composite beam with triangular corrugation web.

- This study was analyzing the effect of corrugation sections encased in concrete. Thus, it is to be concluded that the ultimate load-carrying capacity of composite beam with trapezoidal corrugation maximum value with minimum deflection than composite beams with triangular corrugation and conventional beam.
- The main parameter considered in the study is the ultimate load-carrying capacity of the section. An analytical study carried out using ANSYS, which will be compared with experimental results.
- With the analytical results, the study is further carried out in the second phase of the project. In the second phase of experimental testing, the specimen under two point loading ultimate load carrying capacity and deformation will find out and compared the results with the analytical result.

REFERENCES

1. Sumathi and M. Chandra Babu "FLEXURAL BEHAVIOR OF COLD FORMED STEEL I-SECTION BEAM WITH CORRUGATED WEB" ASIAN JOURNAL OF CIVIL ENGINEERING (BHRC) VOL. 18, NO. 1(2017) PAGES 31-38.
2. Aakash Karate, Ajay Ghude, Vipul Fulpagare, Kiran Jadhav, Vibhor Patil"Experimental Investigation of Steel Beam with Trapezoidal Corrugated Web Beam" International Research Journal of Engineering and Technology. Volume: 06 Issue: 04 | Apr 2019.
3. Alex. J, Iyyappan. G. R, "Experimental Study on Flexural Behaviour of Cold-Formed Steel Section" International Journal of Science Technology & Engineering. Volume 2, Issue 10, April 2016.

4. Ahmed S. Elamary, Amr B. Saddek and Mamdooh Alwetaishi, "Effect of corrugated web on flexural capacity of steel beams" International Journal of Applied Engineering Research ISSN 0973-4562 Volume 12, Number 4 (2017) pp. 470-481.
5. A Jayaraman, S Athibaranan , A Mohanraj , "FLEXURAL BEHAVIOUR OF LIGHT GAUGE COLD FORMED STEEL MEMBERS COMPARISON OF IS CODE AND EURO CODE" International Journal of Research in Engineering and Technology Volume: 04 Issue: 06 , June-2015.
6. Divahar RAVI, Aravind Raj PONSUBBIAH, "Experimental, analytical and numerical studies on concrete encased trapezoidally web profiled coldformed steel beams by varying depth-thickness ratio" Higher Education Press 2020 Front. Struct. Civ. Eng. 2020, 14(4): 930–946.
7. Fatimah De'nan , Nor Salwani Hashim, "The Effect of Web Corrugation Angle on Bending Performance of Triangular Web Profile Steel Beam Section" 978-1-61284-774-0/11 ©2011 IEEE.
8. Saravanan and D. S. Vijayan, Status review on the experimental investigation on replacement of red-mud in cementitious concrete, MaterialsToday:Proceedings, https://doi.org/10.1016/j.matpr.2020.05.500
9. Jun He, Yuqing Liu, Airong Chen, Dalei Wang, Teruhiko Yod"Bending behavior of concrete-encased composite I-girder with corrugated steel web" Elsevier Ltd. Thin-Walled Structures 74 (2014) 70–84.
10. K.Elakkiya M.Anbarasan, "Experimental and Analytical Investigation of Cold-Formed Steel ISection with Corrugated Web" International Journal for Scientific Research & Development| Vol. 7, Issue 01, 2019.
11. Krishnarani. K. K, Swathi Krishna, "Lateral Torsional Buckling of Corrugated I-Beam with Sinusoidal Openings" International Journal for Research in Applied Science & Engineering Technology (IJRASET) Volume 4 Issue IX.
12. Khalid, Y. A., Chan, C. L., Sahari, B. B. & Hamouda, A. M. S. (2004) , "Bending behaviour of corrugated web beams" Journal of Materials Processing Technology, Vol. 150, pp.242-254.
13. L. Krishnan, C. S. Dinesh raj, S. Prema "Experimental Investigation of Cold-Formed Steel Section Flexural Member with Triangular Web" IOSR Journal of Mechanical and Civil Engineering. volume 12, (Mar -Apr. 2015), PP 36-39.
14. Lincy P Abraham, Divahar R, "Behaviour of Encased Cold-Formed Trapezoidally Corrugated Web Beam" INTERNATIONAL JOURNAL OF ENGINEERING SCIENCES & RESEARCH TECHNOLOGY [2657-2663] October, 2013.
15. Lect.RashaAbd Al-RedhaGhani, "A Finite Element Analysis of Shear Connectors Impact on Shear Behavior of Concrete-Encased Composite Steel Beams" International Journal of Research in Engineering & Advanced Technology, Volume 5, Issue 6, Dec - Jan, 2018.
16. R. Goplalakrishnan, A. Mohan, L. Ponrajsankar, D. S. Vijayan, "Characterisation On Toughness Property of Self- Compacting Fibre Reinforced Concrete", Journal of Environmental Protection and Ecology 21, No 6, 2153–2163 (2020).
17. D S Vijayan, Dr. J Revathy, Flexural Behavior of Reinforced and Pre-stressed Concrete Beam Using Finite Element Method, International Journal of Applied Engineering Research, ISSN 0973-4562 Volume 10, Number 1 (2015) pp. 717-736
18. D.S.Vijayan, J. Revathy, Flexural Response of Fibre Reinforced Polymer Laminated Pre-stressed Concrete Beams, Indian Journal of Science and Technology, Vol 9(42), DOI: 10.17485/ijst/2016/v9i42/101824, November 2016.

19. D.S.Vijayan, S.Arvindan, K.Naveen Kumar, S Mohamed Javed, Seismic Performance of Flat Slab in Tall Buildings with and without a Shear wall, International Journal of Engineering and Advanced Technology (IJEAT) ISSN: 2249 – 8958, Volume-9 Issue-1, October 2019, Page no – 2672 – 2675.

The Effect of Glass Fiber on Concrete and Reinforced Concrete Beam under Elevated Temperature

A.Paulmakesh[1] Gizachew Markos Makebo[2]
[1]Associate professor, department of civil Engineering, Wolaita Sodo University, Ethiopia e-mail: paulmakesh1986@gmail.com
[2]Head,department of civil Engineering, Wolaita Sodo University, Ethiopia, email: gizemark2009@gmail.com
Correspondence author e-mail: paulmakesh1986@gmail.com

Abstract: Any service structures could suffer in unexpected time from one of the most serious accidents is fire accidents. A small source is sufficient for a fire accident to be ignited, but its shore is difficult to control. In Ethiopia, so many fire accidents took place that in the near past have taken lives and house damage. For example, Merkato area burnt down nearby shops at the Greater Anwar Mosque in Addis Ababa. The cause of the accident has been investigated and a great number of homes and lives have been demolished. A long fire period leads to structural failure, loss of valuable lives, etc. The use of advanced finite elements and structural analysis models to determine the fire growth within a compartment is currently undergoing good development to determine temperatures within the component through a heat transfer analysis. This study focussed more on the influence of fiberglass under high temperature on concrete and reinforced concrete beams. In this study, different fiber glass percentage, concrete temperature and fire exposure duration were discussed. Experimental tests on both conventional and fiber-powered concrete at high temperatures have been done and the compressive strength of the concrete cube is found to be reduced to 37.41% for two hours at a 3000c temperature relative to initial concrete compression strength. Concrete cylinder tensile strength decreases up to 20 percent at 3,000 cm for two hours, relatively with initial concrete tensile strength, and the reinforced concrete beam's load-resisting capacity decreases to 21 percent with 3,000 cm held for two hours. But by adding 0.75% of the fiber on concrete, the compressive strength increases to 21% compared to normal cements. The concrete with0.75% fiber increases the divided tensile strength by 29% over regular cement. The load resistance capacity for beams with 0.5% fiber at the same temperature exceeds 9.4%. Thus fiber glass concrete has a high resistance to load and a high fire resistance than conventional concrete.

Keywords: Fiber glass, Reinforced concrete, Elevated Temperature.

1. Introduction

Concrete is the most widely used man-made construction material in the world today. It is relatively inexpensive and simple to shape into any desired shape. Steel has been and continues to be the most common reinforcing material for concrete. Steel material has numerous advantages when used as reinforcing concrete; it is strong in tension and compression and has a high modulus of elasticity[1]. It has the same thermal expansion as concrete and works well with it under load. However, using steel as reinforcing concrete has some drawbacks; it corrodes over time and has a low fire resistance. However, the main challenge for civil and structural engineers is to provide society with sustainable, environmentally friendly, and financially viable structures. Finding new materials that can meet these requirements is essential. However, most fiberized materials on the market today have at least one major drawback that prevents them from being widely used in structural applications. Fibers made primarily of carbon fiber, for example, have exceptional structural properties such as a high elastic modulus and a relatively high tensile strength[2]. However, their performance in fire testing is less than desirable, and their cost makes them unsuitable for use in most applications. Fiber glass, a relatively new fiber, is generating interest in the construction industry as a raw material and fiber reinforcement polymer alternative to carbon fiber reinforced polymers. It is made of glass fibers that are extremely fine. It is a lightweight, extremely strong, and sturdy material. The most common synthetic fiber is glass fiber, which is chemically inert, hydrophobic, and lightweight. They are manufactured as continuous cylindrical monofilaments that can be cut to desired lengths. Glass fibers also help to reduce plastic shrinkage cracking and subsidence cracking when used in conjunction with steel reinforcement[3].

1.1. Statement of the Problems

The concrete structure under high temperatures (fire) results in a severe deterioration and a number of transformations and reactions, leading to gradual disintegration of the cement gel structure and reduced durability and increased drying shrinkage trend, structural cracking and related changes in colour. Betons lose strength at high temperatures from previous experiments, but the rate of reduction varies with the various construction materials used. Not only fire is now increasing rapidly (global warming) until the date of this research, but it also loses its strength at different altitudes, which require special attention in the construction industry[4]. In fact it also increases quickly (global warming). There are no detailed and adequate provisions on the performance of various types of construction materials for high temperatures also in our country building code (fire). Limits, infrastructure growth and increasing security interest in our country, repetitive and continuous experiments and fire-resistance tests must be given considerable value in determining the characteristics of the various construction materials, construction methods and structural components.

Adding fibers is one way to enhance the resistance to fire and reduce spillage. Fiber application can moderate damaging effects of expanded steam. The polypropylene melting point lies at 100-200 degrees Celsius. Fiber evaporates from concrete at such temperatures and builds small cavities into concrete. The concrete structure becomes more pore and expanded vapor can escape without significant damage to the concrete microstructure [5]. This effect limits the splitting and breakdown of the structure of the concrete. Fibers with a higher slimming ratio have been shown to be better suited. A porous concrete structure is then easier to escape from interconnected and expanded water steam without damaging the cement [6]. Furthermore, the glass fiber contains a mixture of oxides with high heat and fire resistance and therefore cannot burn in the air or oxygen but can smooth, melt and melt at high temperatures (8500c – 9000c). The reason for this research is this situation. The key agenda for this research is therefore the identification of these effects on concrete and beam with and without glass fibers under high temperature.

1.2. Objective of the Study

Generally, this thesis investigates the resistance of the effect of adding glass fibers in concrete and reinforced beam under elevated temperature. To determine whether the fiberized concrete and RC concrete beam is highly resistance as compared with conventional one[7]. To compare the performance of concrete and axially un-restrained RC beam element under elevated temperature with and without fiber in terms of strength. To identify the different parameters (Temperature and duration) which affect the performance of concrete and axially un-restrained RC beam under elevated temperature with and without fiber.

1.3. Significance of the Study

The construction of wider infrastructures including high rise buildings in our country Ethiopia is increasing now and will be expected to grow further. The design of such complex structures will need not only good software but also other inputs of researches, studies and review of foreign Codes used in our country. As previous practices tells important findings and research results of important value comes from such type of fulfilment papers and the results of this experimental research will be useful. Since this construction industry is huge it will need huge number of inputs. Based on the output of this research will be used[8]. To evaluate concrete structure with and without fiber after fire accident. To justify the recommendations in other literatures used. For the design of concrete structure with fiber glass for fire. As a reference for further study.

2. Literature Review

David et al, 2008 Even though there is nothing in the world up to now that will assure the control of fire ccident in service structures and buildings without causing significant damage, there exists variety of active firefighting mechanisms and other manual options[9]. There are different types of firefighting mechanisms some of them they are direct fighting systems others are alarming types for the people living inside prior to possible total damage or partial damage. Sprinkling system, fire detection and alarm system, fire Extinguishers, hose cabinet and stand pipes of water for firefighting vehicle could be mentioned as examples of firefighting mechanisms ACI216R-89, 1994 to compare the loss of life and resources during fire accident, whatever it costs the design of structures for fire safety and fire resistance of desired rating should be implemented[10]. The "safety factor" is contained in the fire resistance rating in the design for fire. Therefore, for a particular situation a member with a resistance of four hours should have a "safety factor" greater than one with a resistance rate of two hours

Beitragel & Wankiw, 2002. The impact of structural continuity on the global response of buildings at risk of fire has shown the real impact of large-scale experiments and real events such as the fire of Windsor Mtower in Madrid (Fletcher et al) and the World Trade Center collapse in New York (Beitel and Wankiv 2002). Since then much work, particularly steel and composite structures, has been done to understand the global fire response.. The global behavior is still not fully understood in concrete structures, mainly due to the lack of reliable information on material properties at high temperatures.

3. Materials and Methods

The conventional method of testing any building construction and materials samples for fire performance is based in burning furnaces, which is well equipped with temperature measuring sensors thermocouples, controlled fueling mechanism for burning and simultaneous loading way during burning (ASTM E119, 2000). This type of methodology for testing the sample is realistic but it is very expensive to build it for individual research purpose. The way of testing fire for this research commonly used an oven dry machine[11]. The samples are burned in an oven for different specified concrete temperature and duration then allowed to cool at room temperature and they are tested for failure.

3.1. Material Investigation

Cement: The cement used shall be Portland or Portland-Pozzolana cement from dangote is used satisfying the requirements of the latest Ethiopian Standards on such cements.(EBCS -2, 1995).For this research PPC cement from dangote is used.

3.2. Fine aggregate (sand)

River side from awash is used here for production of the test samples. Sieve analysis is carried out to determine its sizes and properties to confirm the ASTM standard and having specific gravity 2.5. Coarse aggregate: as per ASTM standards, sieve analysis carried out to find 20mm size confirming to having specific gravity 2.7 aggregate is selected for this mix design. Also its crushing value is checked. In this thesis, the crushed type of basaltic stone, available around Adam city which is classified as siliceous aggregate type. Actually the basaltic aggregate type is the most widely used coarse aggregate in our country for production of concrete.

3.3. Rebar

10mm deformed bar and 6mm bar is used. The reinforcement used for the sample production is made of Abyssinia steel factory, Ethiopia with steel grade S-520and its ultimate tensile strength is 634 (MPa) from test results.

3.4. Water

Mixing water used to be clean and free from harmful matter. Potable water is used.

3.5. Chopped strand mat (CSM 450) glass fiber

This form of glass fiber from world fiber glass company, Addis Ababa is used and it is easily detached in to single piece with length 5mm and it diameter of 0.5mm. Different proportions of fiber (0.5, 0.75 and 1.0%) are taken for making specimens. The following table shows the trial percentage of glass fiber by weight of concrete and its aspect ratio.

Percentage of glass fiber by weight of concrete	**Aspect Ratio (l/d)**
0.5	100
0.75	100
1.0	100

Table 1: Percentage of glass fiber and its aspect ratio

No.	Material	Quantity 25 MPa	
		Reinforced concrete beam	Concrete cylinder
1	Cement (PPC)	393.72 Kg	393.72 Kg
2	Fine Aggregate	677.33 Kg	677.33 Kg
3	Course Aggregate	1015.75Kg	1015.75Kg
4	Water	0.19686 m^3	0.216546 m^3

Table-2: Standard mixing for ordinary concrete structure per 50Kg bag of cement (EBCS -2, 1995).

Concrete grade (MPa)	**Description**	
	Nominal size of aggregate (mm)	20
C-25	Workability	High
	Limit of slump that may expect for normal (mm)	50-100
	Total aggregate(Kg)	215
	Fine Aggregate (%)	35 - 45
	Tot. vol. of concrete	0.127m^3
	Max. water cement ratioof beam for moderate exposure	0.55
	Max. Water cement ratio of concrete for moderate exposure	0.60

Table- 3: The mixing proportion for 1m^3

4. Mix Proportioning

The basic mix proportion for C-25 grade of concrete is cement, fine aggregate, coarse aggregate and water with mixing ratio of 1:1.67:3.12. Mix 1 and mix 2 contains 0% glass fiber. Mix 3 and 4, mix 5 and 6, mix 7 and 8 contains 0.5%, 0.75% and 1.0% glass fiber by weight of concrete respectively. A total of 8 mixes were studied with water-cement ratio of 0.50 for both reinforced concrete beam and concrete cylinder to maintained C-25 MPa for all the concrete mixes. The workability of the concrete with and without fiber is also checked within the limit slump value according to the EBCS standard in using slump . The slump value lay under normal range of 50mm- 100 mm. But the concrete with fiber affects the workability of the concrete; it needs another recommended water cement ratio or special adding techniques. Increasing the percentage beyond 0.75%, it reduce the workability of the concrete mix and it gives the formation of mat and balling which is difficult to separate by vibration. When as we increases the percentage of fiber beyond 1.0%, it may affects the workability of concrete mix.

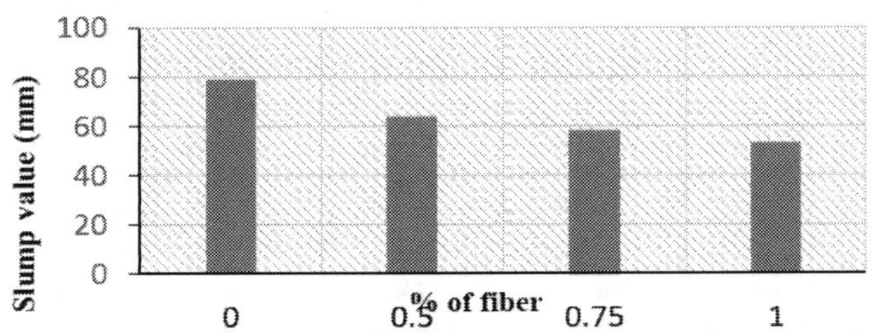

Figure 1: Slump test results for concrete with and without fiber

4.1. Concrete Specimen for Compressive Strength

Compressive concrete specimen Three concrete cube samples were tested for strength testing to achieve a representative compressive strength value for various burning tempers and fire exposure times. First design and make the mixing ratio with and without glass fiber as mentioned above. The concrete sample is cast correctly with the L*W*D aspect ratio (150X150X150) and the Concrete compaction. The next day it was demolished and placed in portable water for 28 days after casting the concrete cube. The cube sample is subjected to room temperature (200c) of up to 3000c after 28 days. Then the cube sample is placed so that the load is applied without a shock and continuously increase at a rate. The load is then increased until the specimen is broken and the maximum load on the specimen is recorded during the test. Finally, the specimen shall record the maximum load and notice failure.

4.2. Testing Specimens

The cylindrical specimens of 150mm*300mm concrete were the specimen used in this test. Two groups of samples would be produced to study the performance of concrete tensile strength under fire. The first group consists of 3 and 18 fibre-free samples at normal and elevated temperatures to assume their tensile resistance. The second group includes 9 and 54 fiber samples at normal and high temperatures, to assume its tensile strength, respectively. Preparation of the concrete cylinder samples for the divided tensile strengths and loading conditions of conventional and fiberized concrete. Three beam samples are tested to achieve a representative force value for different burning temperatures and fire exposure duration. Premiere design and make the mixing ratio with and without glass fiber as mentioned above. Then cast beam samples with the aspect ratio of W by D da L (150 Tod150 by 750) and compact concrete with a lateral vibrator correctly. After casting the sample on the cement beam, remove the mold and put it in a 28-day water reservoir[12]. The fiber effect is shown in the oven at a different heat level after 28 days, by the room temperature up to a higher temperature from 1000 c - 3000 cm in the furnace. Then take the test from the oven and expose it 24 hours before testing at room temperature[13]. The beam specimen is placed in a bending machine so that the charge is applied without shock and continuously increased until the specimen is unsuccessful[14]. Finally, a specimen records the maximum load and notices the error.

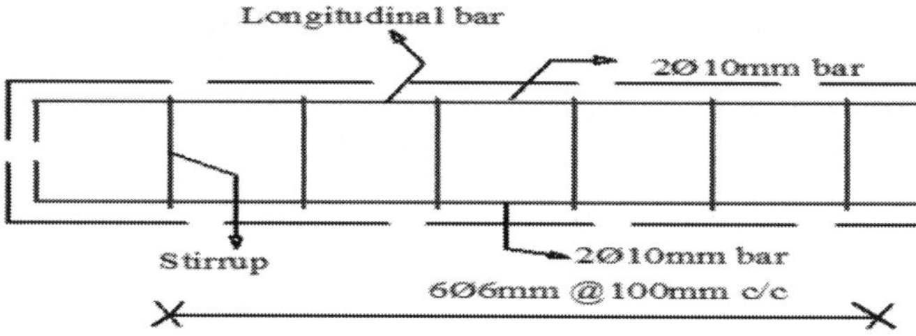

Figure - 2: Reinforcement detail of sample RC beams

4.3. Loading condition of sample beam

The reinforced concrete beam with and without fiber are loaded at the mid-span to get maximum moment and deflection shown in Figure below as per ASTM C293M. The following figure shows flexural machine, the cross section of the beam specimens with its loading condition Beam Samples are subjected for desired temperature from 1000c up to 3000c and durations starting from one hour up to two hours in the oven and air cooled to room temperature, then tested for maximum concentrated load at the mid span. The mode of the failure of the beam samples are failed in shear. If the samples were tested in hot just at a time of heat conditions without cooled to room temperature, the results will be different from results recorded of this research. The reason is reinforcing steels (rebar) will be affected by the elevated temperature when they are tested during the fire but the reinforcing steel restores when cooled.

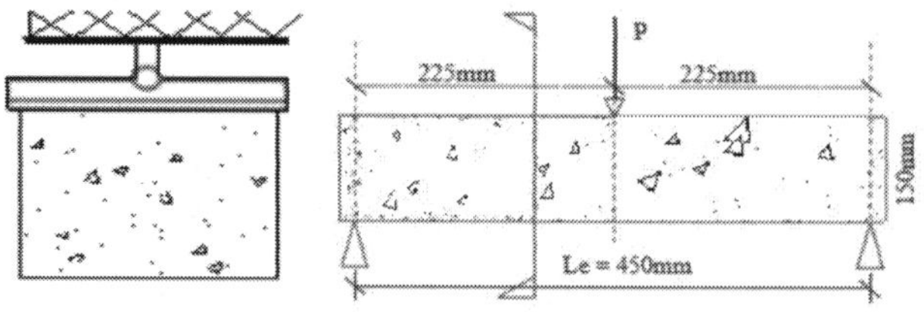

Figure – 3 : Representation of loading test of sample beam and its cross section

5. Results

5.1. Results of Concrete cube for compression

Concrete cube samples are tested for compressive strength stress with and without fiber glass in order to determine its toughness (the ability to absorb energy), relationship between concrete, fiber and fire are presented in the following table. The effect of fire on concrete structure is induces thermal expansion (tension stress) which causes spalling of concrete. Now adding fiber glass and checking fire resistance more effective to prevent this thermal stress because fiber glass used as crack arrester to reduce the expansion of concrete Martials. Finding tensile and bending stress of concrete and reinforced concrete under fire is more representative than determining the effect fiber on concrete for compression

No.	Specimen ID	No..	% of fiber	Compressive strength (MPa)
1	NCC	3	0	29.6
	NCC			24.71
	NCC			26.78
Average Value				27.03
2	FCC0.5	3	0.5,	31.5
	FCC0.5			29.12
	FCC0.5			30.3
Average Value				30.31
3	FCC 0.75	3	0.75	33.40
	FCC 0.75			30.60
	FCC 0.75			34.00
Average Value				32.67
4	FCC1.0	3	1.0	28.90
	FCC1.0			27.71
	FCC1.0			27.00
Average Value				27.87

Table -4: The Result for compressive strength of concrete with and without fiber glass.

Compressive Strength (MPa)								
% fiber weight concrete	**Room Tem.**	**Elevated Tem.(°c)**			**Room Tem.**	**Elevated Tem.(°c)**		
		1 hour				**2 hour**		
	20.5	100	200	300	25	100	200	300
0.00	27.03	23	26	22	27	22	25	19.6
0.50	30.31	27	28	25	30	27	28	23.1
0.75	32.67	28	30	27	32	29	29	25.5
1.00	27.87	25	25	23	27	26	25	21.8

Table - 5: Summary of the result for compressive strength of concrete with and without fiber glass at elevated temperature.

Compressive Strength (MPa)		
% fiber by weight of concrete	**1 hour**	**2 hours**
0.00	22.14	37.41
0.50	21	30.87
0.75	20	27.67
1.00	16.66	27.33

Table- 6: Percentage reduction of compressive strength of concrete with and without fiber exposed at 3000c for 2 hours

The relationships between compressive strength of concrete, percentage of fiber, temperature and duration for elevated temperature of cube specimens with and without fire glass are shown in figure below. Concrete with and without fiber are presented in figure.

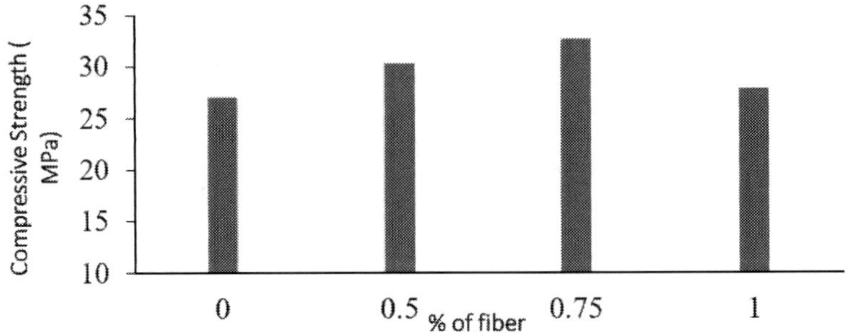

Figure 4: Variation of compressive strength of concrete with the addition of fiber glass

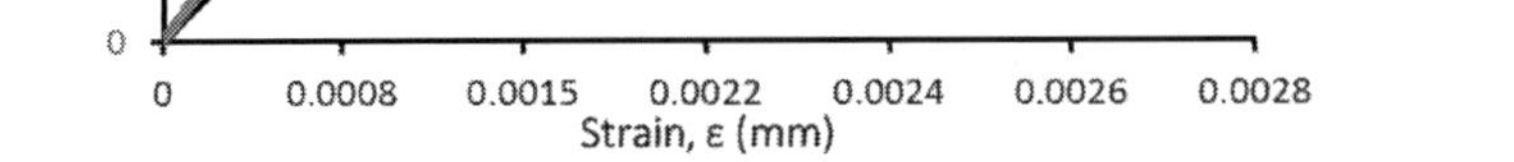

Figure 5: Stress and strain graph for concrete with and with out fiber

Figure 5 shows typical strain curves for horns with fiberglass and without fiberglass with different strengths These curves match tests on specimens which resemble the compressive pressure of the cube sample. The stress-train curves on figure 5 all increase to a maximum stress of 0.0015 to 0.0022, with a descending branch. As discussed in section, the shape of this curve results from the gradual formation of micro-cracks within the concrete structure. Fiberglass concrete has a high capacity for caring than conventional concrete. Generally, the addition of glass fiber to 0.75% by weight of concrete is stronger and more load-bearing than conventional concrete. This shows that 0.75% of the concrete is more ductile than other materials.

Figure – 6: Failure mode of normal reinforced concrete beam

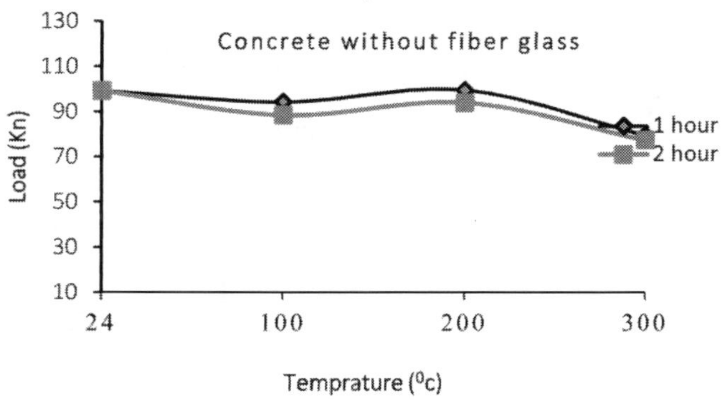

Figure-7: Variation of maximum load versus temperature for beam sample without fiber

6. CONCLUSIONS

From this research, it was established that the fiber application in concrete reduces fresh concrete workability. The results of the workingability test obtained in the standard slump test demonstrated low workability. The conclusion was that the increased fiber weight in the concrete would reduce workability. High dosage weight above 0.75 percent showed the concrete to be considerably rigid and compact. However, blood and segregation in the concrete mixture have also been reduced. With increased fiver-dose levels up to 0,75 percent, the compressive strength of concrete increases, and then decreases. The optimal percentage of fibers found in the research was therefore 0.75 percent and a continuous decrease of strength was observed when the percentage of fiber dose increases. Split concrete tensile strength increases up to 0.75%, and then it begins to decrease. The ideal percentage of fiber found in the research is therefore 0.75 per cent and a continuous reduction in strength has been observed as the percentage of the fiber dosage increases.

Fiber concrete is more resistant to the load than the fiberless cement in normal condition. The heat lasts up to two hours, the tensile spillage strength of the normal concrete decreases by

20% and the strength of the conventional reinforced beam by 24,7% with high temperatures of 3000c. Bitter with the recommended fiber (0.75%) percentage sustained for two hours at a high temperature of 3000c, decreases the initial division strength of concrete to 12.12%. Duration of exposure to a fire up to two hours lowers conventional reinforced concrete beam's load-resistant capacity to up to 28.12% when burnt at 3000c temperature. The load capacitance of the reinforced fibre-glass beam is approximately 9 percent higher than the normal reinforced beam at 3000c for 2h with 0.5 percent. • As can be seen in Figure 4.10. the strength of the tensile cylinder samples burnt for 2000c is higher than the tensile strength of the cylinder samples burnt for 1000c, this seems contrary to the results of the study because the fire temperatures increase the tensile strength of the concrete decreases. However, for beam samples, it is apparent that, due to quick drying of concrete past and porostructure, the strength of concrete at a high temperature from 100-2000 cm increases and steel increases at that temP due to residual stress. More research is needed to verify the effect of tensile strength and fire performance variation in bending times. These generally show that fibre-glass concrete takes precedence over normal concrete at room temperature and at its highest value.

REFERENCES

[1] ACI216R-89, (1994) "Guide for Determining the Fire Endurance of Concrete Elements", Reported by American Concrete Institute Committee 216.

[2] ASTM C293/C293M, (2011), "Standard Test Method for Flexural Strength of Concrete (Using Simple Beam with Center-Point Loading),agencies of the Department of Defense, Vol. 4,

[3] ASTM E119, (2000), "Standard Test Methods for Fire Tests of Building Construction and Materials", American Society for Testing and Materials.

[4] ASTM E119, (2014). "Standard Test Methods for Fire Tests of Building Construction and Materials"

[5] Awadhesh Chandramauli, Amit Bahuguna and Azher Javaid, (2018), "The Analysis of Plain Cement Concrete for future scope when mixed with Glass and Fibers", International Journal of Civil Engineering and Technology, Vol. 9, pp. 230–237, 2018. Beam for Fire", MSc Thesis, Addis Ababa University.

[6] Bažant, Z. P. (1992) Fracture mechanics of concrete structures. Abingdon, Oxon, USA, Taylor &Francis.

[7] Beitel, J. and Iwankiw, N, (2002), "Analysis of Needs and Existing Capabilities for Full-Scale Fire Resistance testing", National Institute of Standards and Technology NIST GCR 02-843.

[8] CASE, (2008) "Structural Engineer's Guide to Fire Protection, Council of American Structural Engineers,

[9] C. R. Cruz, "Elastic properties of concrete at high temperatures," Journal of the PCA Research and Development Laboratories, vol. 8, pp. 37– 45, 1966. View at Google Scholar

[10] David N., (2008). "Fire and Concrete Structures" P.E., S.E., Director, Engineered Structures and Mahmoud E. Kamara, PhD., Senior Structural Engineer.

[11] Ethiopian Building Code Standard, (1995) "Structural Use of Concrete", Ministry of Works & Urban Development, Addis Ababa, Ethiopia.K. Elissa, "Title of paper if known," unpublished.

[12] Arvindan Sivasuriyan, D.S. Vijayan, Wojciech Górski, Magdalena Daria Vaverková, Eugeniusz Koda, "Practical Implementation of Structural Health Monitoring in Multi-Story Buildings", Buildings 2021, 11, 263. https://doi.org/10.3390/buildings11060263

[13] D S Vijayan, Dr. J Revathy, Flexural Behavior of Reinforced and Pre-stressed Concrete Beam Using Finite Element Method, International Journal of Applied Engineering Research, ISSN 0973-4562 Volume 10, Number 1 (2015) pp. 717-736

[14] D.S.Vijayan, J. Revathy, Flexural Response of Fibre Reinforced Polymer Laminated Pre-stressed Concrete Beams, Indian Journal of Science and Technology, Vol 9(42), DOI: 10.17485/ijst/2016/v9i42/101824 , November 2016

STUDY ON FLEXURAL BEHAVIOUR OF HIGH-PERFORMANCE CONCRETE BEAM STRENGTHENED WITH POLYOLEFIN FIBRES

[1]B. Yuvashri, [2]A. Sattainathan Sharma, [3]K. Suganya Devi
[1]PG Student, [2,3]Assistant Professor, Department of civil engineering, SRM Valliammai Engineering College, Kattankulathur, Chennai
Corresponding Author Email id: yuvashri160498@gmail.com

ABSTRACT: This paper investigates the flexural behavior of high-performance concrete beams strengthened with polyolefin fibers. In the recent development in concrete technology, high-performance concrete is designed to give materials, exposure conditions, and applicable requirements of cost, better service life, and durability. The replacement of cement by adding the mineral admixture and chemical admixture to the concrete increases its strength and durability properties. To improve the strength and ductility of High-performance concrete, fibers are added to the concrete. This study focused on the effect of polyolefin fibers on the mechanical and flexural performance of high-performance concrete. Two types of polyolefin fibers at various percentages of 0.1%, 0.2%, 0.3% volume fraction of high-performance concrete. The flexural beams were cast and tested under two-point loading. The addition of fibers increases the first crack load, ultimate load capacity, and ductility. Polyolefin micro and macro fiber performance reinforced high-performance concrete beams compared with experimental and analytical investigation carried out using ANSYS software.

Keywords: High-Performance Concrete, Polyolefin fibers, flexural members, deflection, ANSYS

1. INTRODUCTION

Concrete plays a significant part in the development of many structures; excessive consumption of natural resources occurs through ordinary Portland cement used in the concrete binder. Its production contributes significantly to the global production of anthropogenic CO2. The concept of replacement of OPC with environment-friendly materials has gained attention. Recently, the concretes are produced, replacing cement partially with mineral and chemical admixtures to enhance the performance of concrete. The mineral admixture such as fly ash, GGBS, metakaolin, silica fume, which densifies the concrete, increases strength, safety, and durability [2]

In the production of high-performance concrete, recent concrete technology has been rapidly evolving and is primarily used in tunnels, bridges, and tall buildings. The benefits of HPC include reduced concrete members' dimensions, high frontal strength, durability, and improved elasticity coefficient. Furthermore, service life, toughness, and tensile strength have been over 100 years

[12]. High-performance concrete is achieved by adding pozzolans and low water content with a dense microstructure with high strength and low permeability. The use of high hardiness, low water-cement ratio concrete has achieved better endurance performance. Two methods reduce the capillary pores system, and the binding materials that prevent the transport of aggressive ions such as chlorides are created chemically active [4]

In contrast to average strength concrete performance, HPC is usually more fragile. Hong-Sam Kim et al[6]. It has studied the strength properties of metakaolin, which is used for the production of high-fuel concrete, as a substitute in the production of high-performance concrete for silica fume. Metakaolin has SiO2, C–S–H, and C–A–H constituting the main substances generated by the hydration. The Metakaolin binding substitute ratio has developed a higher strength than the compressive power of 60 MPa. The efficiency of strength decreases beyond a replacement rate of 10%. Metakaolin makes concrete strength due to its filler effect, which accelerates the PC hydration within 24 hours. High-performance concrete ductility can be improved by applying confining pressure on the jacket's design by incorporating new fibers. Fibers are available in various sizes, shapes from various materials such as steel, glass, polyester, aramid, polyolefin, and natural fibers. Most of the fibers are used for structural and non-structural purposes. The advantage of fibers reduces the shrinkage crack, saves cement, and increases the tensile and compressive strength of concrete. High ductility with high strength fiber reinforcement gives the best performance of flexural strength and toughness[5]

A polyolefin is a human-made synthetic fiber. Polyolefin fibers have a specific gravity less than one that is unaffected by solvents at room temperature and is swollen by aromatic and chlorinated at only high temperatures. Fibers have excellent abrasion resistance. Polyolefin fibers have many advantages, such as resistance to corrosion and high strain capacity. Due to its corrosion resistance, it is suitable for an aggressive environment to protect the structures from corrosion. Polyolefin fibers provide strong bonding in the cement matrix and cracking control over a long period. Polyolefin macro fibers have the potential to Polyolefin macro fibers have the potential to enhance the strength, durability, and post cracking behavior of the hardened concrete. Vahid Afroughsabet et al. [16] have investigated the mechanical properties and durability of high-strength concrete containing steel and polypropylene fibers. Hooked end steel fibre of 60mm length with fibre volume fraction 0.25%,0.50%,0.75%,1%. Polypropylene fibres of 12mm length with fibre content 0.15%, 0.30% and 0.45%. Fiber-reinforced concrete contained 10% of silica fume to increase the strength and durability. An increase in fiber content increases the compressive strength. A developed mix containing 0.15% polypropylene and 85% steel in tensile strength and elasticity, polypropylene fibers are much shorter and have a lower elastic modulus. Changes were made to the fiber composition of the mixed concrete, leading to a decrease in flexural strength. The incorporation of polypropylene fiber in concrete reduces its water absorption. Amin Noushini et al.[1] The result of this study focused on the impact of monofilament fibers (18mm) and fibrillated polypropylene fibers at 0.5% volume concentration on GPC. Firmness in 48 and 55mm polyfill polyolefin fibers Polyurethane fibers show the largest energy fracture due to high adhesion to low aspect ratio. Mohamed Said et al.[12] Concrete reinforced with high-performance polyvinyl alcohol had been tested PVA polymer additives increase the tensile strength and load-bearing capacity. Using PVA fibers increases durability and ductility. the flexibility of ECC beams is improved by increasing the volume of fibers

2. EXPERIMENTAL PROGRAMME AND TESTING

2.1. MATERIALS

For the concrete production, ordinary Portland cement of 53 grade, fine aggregate, coarse aggregate, fly ash, metakaolin, and superplasticizer were all used in the concrete mixes. The physical properties are conducted as per IS standards.

2.1.1. CEMENT

The cement used in this experiment is ordinary Portland cement 53 grade. The value of specific gravity of cement is 3.15, and fineness is 2. The consistency of cement is 32, and its corresponding initial and final setting time is 32 and 355 minutes. The main reason for choosing OPC 53 grade has its specific surface area and fineness for hydration and provides adequate strength.

2.1.2. AGGREGATE

The coarse aggregate passing through the 20mm sieve and retained in the 12.5 mm sieve were used. Its strength is more critical for high-performance concrete. The properties of coarse aggregate are specific gravity is 2.66, fineness modulus is 7.5 where 90% passed through 20mm sieve.
The fine aggregate used for the experiment is M sand. The primary role of fine aggregate is filling the pores between the coarse aggregate and concrete, workability, and good finishing to the concrete. The properties of fine aggregate are specific gravity is 2.58 corresponds to grading zone II.

2.1.3. MINERAL ADMIXTURE

Fly ash and metakaolin are highly active and effective pozzolanic for the partial replacement of cement in concrete. Fly ash is the by-product of coal-burning power plants having a specific gravity of 2.2. High reactivity metakaolin in a dry, dense form specific gravity is about 2.5, and its bulk density is 0.4 kg/lit.

2.1.4. CHEMICAL ADMIXTURE

Coloplast SP430 superplasticizer is used in this experiment based on sulphonated napththalene polymers, which are dispersed in the water instantly. A high range of water reducers increases the strength of the concrete.

2.1.5. POLYOLEFIN MACRO FIBRES

Polyolefin macro synthetic fiber has the potential to control post cracking and to meet shrinkage cracking. Macro fibers reinforce the concrete in its hardened state improve its strength and durability. The fiber is non-corrosive, so it has a wide range of applications, also serves as an alternative to steel fiber. The properties of the fiber are given in table 1.

2.1.6. POLYOLEFIN MICRO FIBRES

Polypropylene microfibres is a single filament that is spun individually. The filaments have different length among that 6mm and 12mm is commonly used. Microfibers are used as a secondary reinforcement as they reduce the plastic shrinkage cracking in early-stage concrete. Fibers added to the concrete can also increase fire resistance since the 1990s, polypropylene monofilament fibers used as fire resistance of concrete in tunnel linings. The properties of polyolefin fibre are listed in table 1 and shape of polyolefin micro and macro fiber is shown in fig.1.

TABLE 1. PROPERTIES OF MICRO and MACRO POLYOLEFIN FIBRES

Fibers	Length (mm)	Appearance	Specific gravity	Tensile strength (Mpa)	Young's modulus (Gpa)	Melting point (°c)	Acid and alkali Proof
MICRO	12	White	0.91	-	-	160	High
MACRO	50	White	0.91	620-758	3.5	165	High

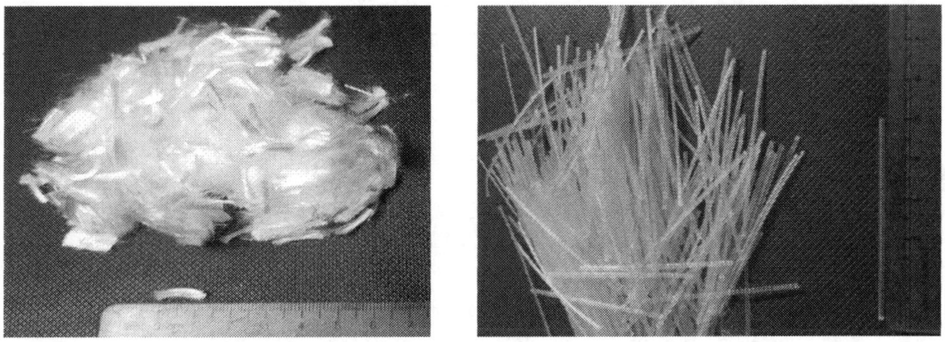

Fig 1. POLYOLEFIN MICRO AND MACRO MONOFILAMENT FIBRE

2.2. MIX PROPORTIONS

The mix proportion was carried on as per IS 10262: 2019, "Concrete Mix proportioning – Guidelines." The mix proportion for M60 is conducted. The mineral admixture such as fly ash(15%) and metakaolin(10%) is added to the concrete. Control mix is a high-performance concrete without fibers; high-performance concrete mixes with fiber dosages such as 0.1%, 0.2%, 0.3% volume fraction of concrete [7]. The details of mix proportion for M60 grade are shown in the table 2.

TABLE 2 MIX PROPORTIONS (kg/m3)

Mix	Cement	Fly Ash	Metakaolin	Fine Aggregate	Coarse aggregate	Fiber %	Water	Super Plasticizer
CB	379	76	50.5	507	1014	-	155	2.53
HPCI0.1	379	76	50.5	507	1014	0.1	155	2.53
HPCI0.2	379	76	50.5	507	1014	0.2	155	2.53
HPCI0.3	379	76	50.5	507	1014	0.3	155	2.53
HPCA0.1	379	76	50.5	507	1014	0.1	155	2.53
HPCA0.2	379	76	50.5	507	1014	0.2	155	2.53
HPCA0.3	379	76	50.5	507	1014	0.3	155	2.53

2.3. RESEARCH PROGRAM AND PROCEDURES

The experimental study is performed to detect the effect of high-performance concrete compressive strength, separated tensile, and polyolefin macro-fibers with a 0.1%, 0.2%, 0.3% of substantial volume. For the experimental investigation, various types of specimens were used [8].

2.3.1. COMPRESSIVE STRENGTH

The compressive strength test is conducted on a cube of standard size of 150x150x150mm and tested on a compression testing machine of capacity 2000kN and loading rate of 4kN/s. The test was conducted at the age of 7 and 28 days. The compressive strength of concrete without and with fibers replaced fly ash 15% and metakaolin 10%. The results of compressive strength of concrete is shown in table 3 and figure 2.

TABLE 3 COMPRESSIVE STRENGTH OF CONCRETE

MIX ID	COMPRESSIVE STRENGTH (N/mm²)	
	7days	28 DAYS
CM	32.3	63.1
HPCA0.1	32.6	63.4
HPCA0.2	35.6	65.6
HPCA0.3	38.1	68.2
HPCI0.1	32.7	63.8
HPCI0.2	36.2	66.3
HPCI0.3	40	69

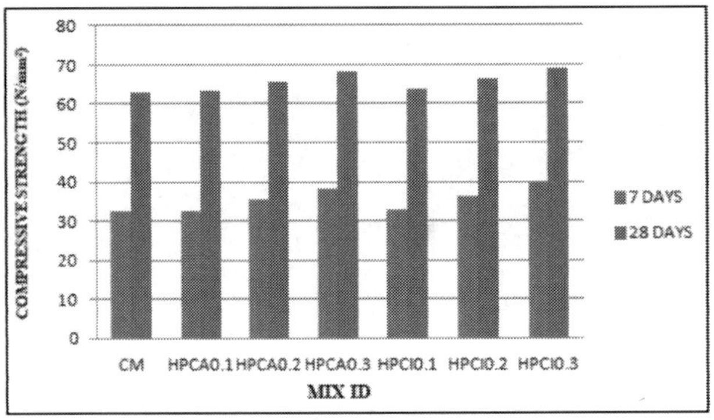

FIG 2. COMPRESSIVE STRENGTH OF CONCRETE

2.3.2. SPLIT TENSILE STRENGTH

The split resistance was achieved on standard cylinders with a diameter of 150 and a height of 300 mm. The load was continuously increased to produce 5816:1999 split tensile stress. Examples were tested at 7 and 28 days of age [9]. Split concrete tensile strength without fibers with fly ash substitution 15% and metakaolin 10%. The results of split tensile strength of concrete is shown in table 4 and figure 3.

TABLE 4 SPLIT TENSILE STRENGTH OF CONCRETE

MIX ID	SPLIT TENSILE STRENGTH	
	7 DAYS	28 DAYS
CM	3.12	4.08
HPCA0.1	3.16	4.15
HPCA0.2	3.5	4.28
HPCA0.3	3.61	4.43
HPCI0.1	3.19	4.18
HPCI0.2	3.46	4.38
HPCI0.3	3.58	4.54

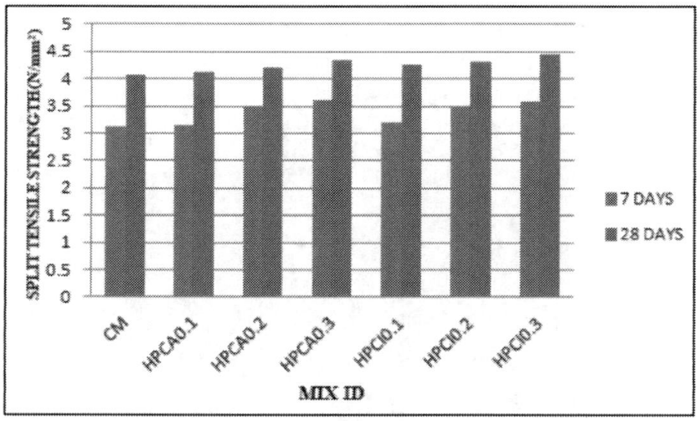

FIG 3. SPLIT TENSILE STRENGTH OF CONCRETE

3. DESIGN ASPECTS OF BEAM

The beam was designed with a cross-section of 150mm x 200mm and 1000mm. A transparent cover of 25mm was given to all beams. The grade of concrete and steel is M60 and Fe500. The control specimen with admixture is designed as B1 and takes the optimum values obtained after testing with various percentages of fibers. B2 and B3 beams with admixture and macro fibers dosages of 0.2 and 0.3% by volume fraction of concrete. B4 and B5 beams with admixture and macro fibers dosages of 0.2 and 0.3% by volume fraction of concrete [10]. Two 12mm diameter bars as hanger bars and two 16mm diameter bars in the tension zone. Two-legged stirrups of 8mm center to center distance of 150mm were provided to resist shear. Typical reinforcement and geometry of section is given in figure 4. The molds were arranged and placed on a smooth surface. The reinforcement was placed inside the mould, and the cover of 25 mm is maintained. The concrete is mixed thoroughly with various percentages of fibers until uniform mixing and pouring into the mold and compacted. The test specimen is removed from the mold after 24 hours of casting and cured for 28 days.

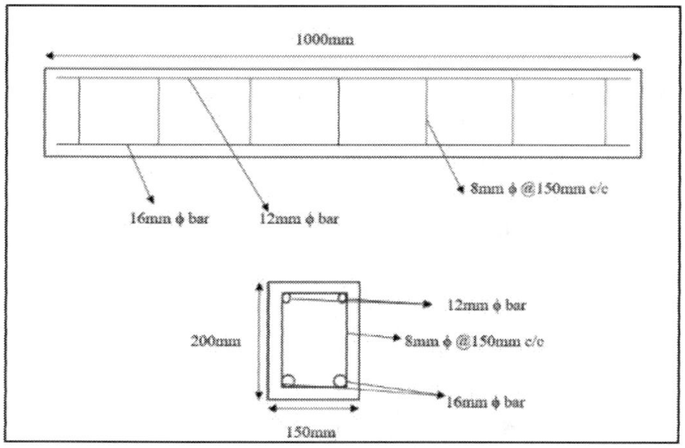

FIG 4 REINFORCEMENT DETAILS

3.1. TESTING OF SPECIMEN

All beams are tested under 2 point load setup with simply supported conditions. After the curing period, the beams were whitewashed before testing, and the support and midpoint location was marked. A loading frame of 400kN capacity and load was applied using the 50-ton capacity to increase 5kN. Deflection is measured using a dial gauge[11]. Load is applied until the ultimate load and deflection were noted. The appearance of the initial first crack and further propagation of the cracks were marked on the beam. The test setup is shown in figure 5.

Fig 5 TEST SETUP

4. EXPERIMENTAL RESULTS AND DISCUSSION

4.1. CRACK LOAD AND ULTIMATE LOAD

All beams are seen visually until the first crack appears, and the first crack load is recorded. The first crack charge increased with an increase in the fiber content due to the increase in the capacity of the concrete carrying strain in the fiber area. When fibers are added to concrete, crack spreading is stopped, leading to more energy leading to a capacity improvement [3]. All beams have a conventional beam compared to (B1). For B2 and B3 beams, the inclusion of Macro polyolefin fibers increased the crack load for B2 and B3, respectively, by 10.5% and 15%. The carrying capacity for B2 and B3 improved respectively by 7% and 17,4% at the ultimate level. The inclusion of Micro Polyolefin fibers for beam B4 and B5, respectively, increased the crack load by 21% and 26% for B4 and B5. At the final stage, the capacity for load carriage improved by 11% and 24%, respectively, for B4 and B5.

4.2.LOAD DEFLECTION CURVE

The deflection of all beams subjected to 2 point load under monotonic loading was measured. The load-deflection curve for all the specimens is shown in figure 7.

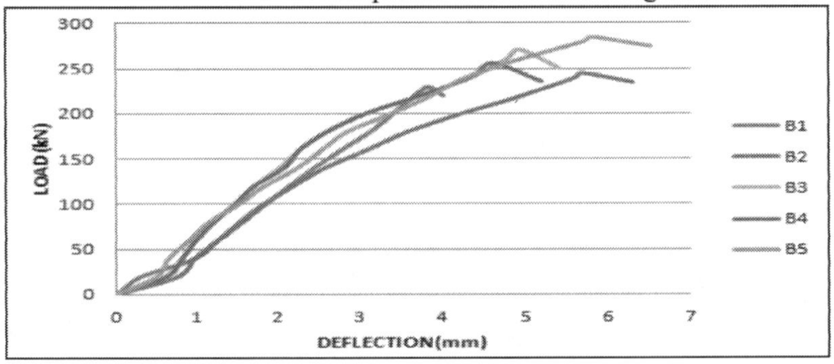

FIG.6 LOAD DEFLECTION CURVE

The load-deflection curve reacts to the initial crash. The concrete reaches its strength, and then the entire load is carried over by steel. The deflection for successive loads took place after the reinforcement was provided. Fiber beams have more significant deflection than control beams. When the beams are compared. The following measures were obtained from the load-deflection curve to assess the contribution of the micro and macro polyolefin fibers to the flexural performance of the tested beam. The experimental results of all the specimens is shown in table 6 and the values obtained from the experimental results of tested is shown in table 7.

4.2.1. INITIAL STIFFNESS

The provision of micro and macro polyolefin fibers in the high-performance concrete exhibits improvement in the initial stiffness. It is defined as the ratio of load at yield load to the corresponding displacement [12] is shown in table 7. Beams with macro polyolefin fibers and reinforced concrete enhance the stiffness for B2 and B3 by 2% and 28.8%, respectively, concerning B1. Beams with micro polyolefin fibers and reinforced concrete enhance stiffness for B4 and B5, which were 29.06 and 22.17% concerning B1.

4.2.2. ENERGY ABSORPTION

The absorption of energy of all materials from the area is obtained by its full load vs. specimen deflection curve. This is the ultimate load function shown in Table 7 as a corresponding final deflection [12] is shown in table 7.. It is noticed that energy absorption has enhanced with the addition of polyolefin fibers in high-performance concrete. Beams with micro polyolefin of 0.3% have high energy absorption capacity compared with all the beams.

4.2.3. DUCTILITY

Ductility is defined as the structure's capability to undergo inelastic deformation after the initial output without lowering its load resistance. The ductility factor of the tested beam can be defined as being presented in Table 7 as the ratio from the final level to the deflection at the yield level[12] is shown in table 7. Polyolefin fibers show improvement in the tested beam. The micro and macro fibers act as crack arrestors by bridging the crack. The cracks will not propagate in the same plane, so high energy is required to propagate cracks. Thus, the increases of ultimate load-carrying capacity.

TABLE 5. EXPERIMENTAL RESULTS

Beam ID	First Yield load(kN)	First yield deflection(mm)	Ultimate load(kN)	Ultimate deflection(mm)
B1	95	1.77	230	3.81
B2	105	1.92	245	5.69
B3	110	1.59	270	4.93
B4	115	1.66	255	4.61
B5	120	1.83	285	5.80

TABLE 6. VALUES FROM LOAD DEFLECTION CURVE

Beam ID	Initial stiffness (kN/mm)	Energy absorption capacity (kN.mm)		Ductility factor	
		Absolute	Relative	Absolute	Relative
B1	53.67	3253	1	2.15	1
B2	54.68	5962	1.83	2.96	1.37
B3	69.18	6493	1.99	3.1	1.44
B4	69.27	6095	1.87	2.77	1.28
B5	65.57	10730	3.29	3.17	1.47

4.3. Crack pattern

All the beams were tested, and its crack pattern is noted at different load levels to identify the failure mechanism and effect of polyolefin fibers in the high-performance concrete [13]. The crack pattern of all the beams were presented in the figure 7. The initial breakdown generally occurred in the middle of the bending region with all the beams and was followed by cracks from this region in the support with increasing load. As the load grew, the cracks became more profound and broader [14]. In the conventional concrete B1, its ultimate load is 230kN, the high-performance concrete beam is brittle, and the crack type was recorded as the flexure and shear crack [15]. Using macro polyolefin fiber, B2, and B3 for the tested beam delay the appearance of the first cracking concerning B1. Macro polyolefin fibers B2 and B3 reduced crack propagation, fewer visual cracks with increasing load-carrying capacity, and crack pattern is recorded as flexure and shear crack. Micro polyolefin fibers B4 and B5 show many minor cracks formed with more deflection and load-carrying capacity; the crack pattern observed is flexure and shear failure [16]. Among all the beams micro polyolefin of 0.3%, crack propagation is arrested, resulting in more energy that leads to the enhanced load-carrying capacity of the specimen.

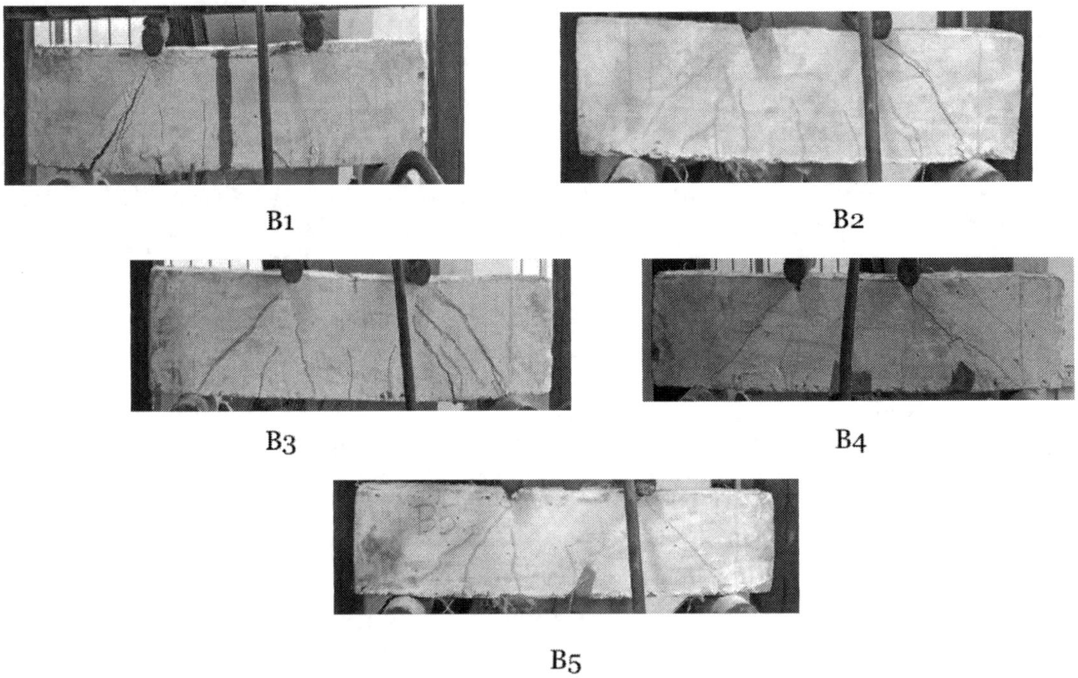

Figure 7 Crack Pattern of Specimen

5. SOFTWARE ANALYSIS

An automatic finite element analysis tool used with CAD systems and design modeler is the ANSYS workbench environment. ANSYS workbench may be a structural, thermal, and electromagnetic analysis software environment. Geometry creation and optimization, the linkage of existing geometry, fixation of a finite element model, solution, and review of results are the main aspects. The category describes how code is also used as a basic concept for simulation of finite elements and interpretation of results.

All the concrete and reinforcement material properties are based on experimental studies and developed as an ANSYS model. The geometry is created of beam cross-section is 150x200x1000mm with reinforcement and assigning of section properties. The modeling and meshing of beam is shown in figure 8. The ultimate load for the different types of beams was given, and the corresponding deflection is noted.

Journal of Physics: Conference Series **2040** (2021) 012028 doi:10.1088/1742-6596/2040/1/012028

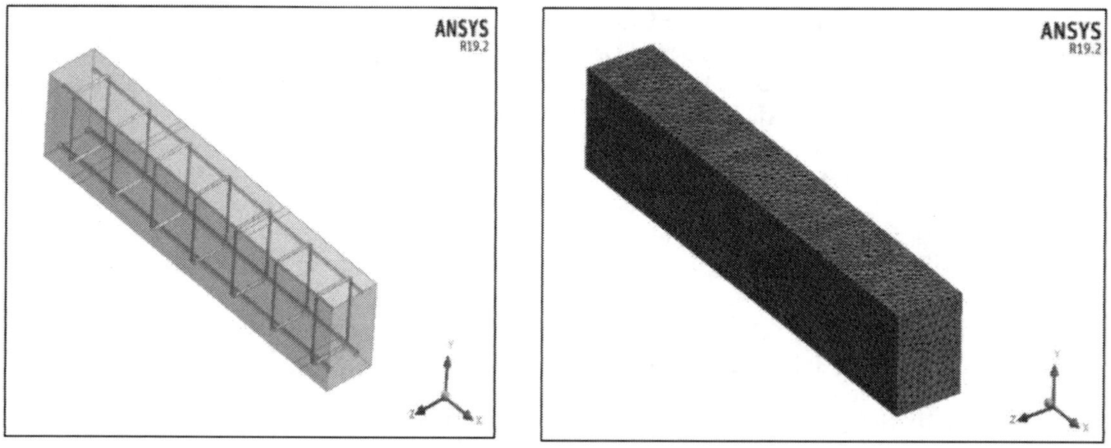

FIG. 9 MODELLING AND MESHING

5.1. RESULTS OF THE MODEL

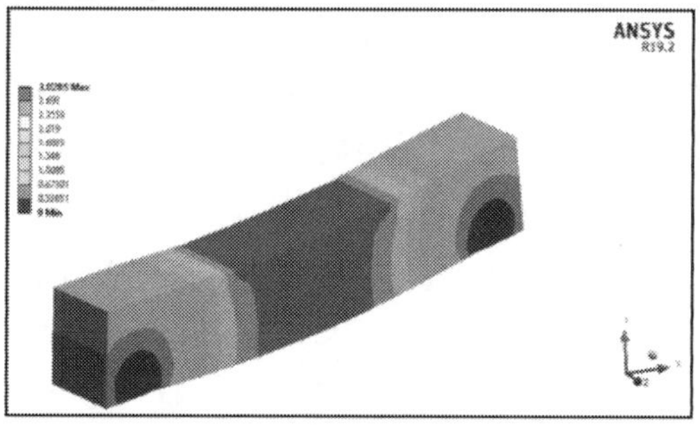

FIG.10 DEFORMATION FOR B1

FIG.11 DEFORMATION OFB3 **FIG.12** DEFORMATION OF B5

For the analytical study, the ultimate load obtained from the experimentally tested beam is given, and its corresponding deformation is obtained. The deformation of conventional concrete (B1) is shown in figure9. The deformation of high performance concrete with marco (B3)and micro

(B5) polyolefin fibres is shown in figure 10 and 11 respectively. Among all the beams, B5 is having a high load-carrying capacity and its deformation. Comparing with experimental and analytical results shows nearly similar deformation is obtained. The variation of the results in analytical results shows an 84% similarity with the experimental results.

6. CONCLUSION

Based on experiments, the following conclusion can be drawn on the flexural behavior of highly efficient concrete beams with Polyolefin fibers:

- Cracking load increased for beams with micro and macro polyolefin fibre content of 0.2 and 0.3% compared to beams without fibers. The addition of polyolefin fibers has been detected to bridge the cracking effects and delay first crack formation.
- The ultimate load carrying capacity for the addition of 0.2 and 0.3% of macro polyolefin fibers in concrete, B2, and B3 increased by 7% and 17.4%, respectively. For the addition of 0.2 and 0.3% of micro polyolefin fiber, B4 and B5 increased by 11% and 24%, respectively, with reference beams without fiber B1. Micro polyolefin fibers (0.3 %) give 6% more strength than macro polyolefin fibers (0.3 %).
- From the load-deflection curve characterization, the initial stiffness of the beams increased with micro and macro polyolefin fibers. The energy absorption capacity of polyolefin fibers beams increased for the content of 0.2 and 0.3% of micro and macro fibers. Among all the beams, 0.3% of micro polyolefin fiber exhibits higher energy absorption capacity. it is seen that ductility is improved for the beams containing polyolefin fibers concerning beam without fiber dosages
- The crack pattern observed in all types of beam with polyolefin fiber dosages shows initial crack is flexural followed by a shear crack at failure. Beams with polyolefin macro fibers control micro-cracks, less crack propagation, and less crack width. Fibers added to the concrete reduces the spalling of concrete, thus preventing the beam's sudden failure compared to the beams without fiber.
- You compare the deflection with the experimental and analytical shows 84% similar matches for all the specimens.

REFERENCE

[1] Amin Noushini et al., Mechanical and flexural performance of synthetic fiber reinforced geopolymer concrete, Construction and Building Materials 186 (2018) 454–475.

[2] Biao Li, Yin Chi et al., Experimental investigation on the flexural behavior of steel-polypropylene hybrid fiber reinforced concrete, Construction and Building Materials 191 (2018) 80–94.

[3] D. Maruthachalam et al. Influence of Polyolefin Macro-Monofilament Fibre on Mechanical Properties of High Performance Concrete, KSCE Journal of Civil Engineering (2013) 17(7):1682-1689.

[4] D. S. Vijayan, A. Mohan, J. J. Daniel, V. Gokulnath, B. Saravanan, and P. D. Kumar, "Experimental Investigation on the Ecofriendly External Wrapping of Glass Fiber Reinforced Polymer in Concrete Columns," vol. 2021, 2021

[5] Filipe R.G et al., Tensile and flexural performance of concrete members reinforced with polypropylene fibers and GFRP bars, Composite Structures 253 (2020) 112784.

[6] Hong-Sam Kim et al. , Strength properties and durability aspects of high strength concrete using Korean metakaolin, Construction and Building Materials 21 (2007) 1229–1237.

[7] Jianzhuang Xiaoa and H. Falknerb , On residual strength of high-performance concrete with and without polypropylene fibres at elevated temperatures, Elsevier- Fire Safety Journal 41(2006) 115–121.

[8] Jun-Mo Yang et al., Effect of steel and synthetic fibers on flexural behavior of high-strength concrete beams reinforced with FRP bars, Composites: Part B 43 (2012) 1077–1086.
[9] Krzysztof Ostrowski et al., Compressive behaviour of concrete-filled carbon fiber-reinforced polymer steel composite tube columns made of high performance concrete, Composite Structures 234 (2020) 111668.
[10] D. S. Vijayan and J. J. Daniel, "An investigation on the torsional effect of symmetric moment resisting frame system subjected to eccentric reinforced concrete lift wall - A finite element approach," Int. J. Eng. Trends Technol., vol. 69, no. 8, pp. 179–184, 2021, doi: 10.14445/22315381/IJETT-V69I8P222.
[11] M.G. Alberti et al., Fibre reinforced concrete with a combination of polyolefin and steel-hooked fibres, Composite Structures 171 (2017) 317–325.
[12] R. Goplalakrishnan, A. Mohan, L. Ponrajsankar, D. S. Vijayan, "Characterisation On Toughness Property of Self- Compacting Fibre Reinforced Concrete", Journal of Environmental Protection and Ecology 21, No 6, 2153–2163 (2020).
[13] P. Magudeaswaran et al., An experimental study on flexural behaviour of damaged reinforced HPC beams strengthened with CFRP wrapping, International Journal of Civil Engineering and Technology (IJCIET) (2017) Volume 8, Issue 3.
[14] Sattainathan Sharma A et al., Ductility behaviour of steel fibre reinforced concrete beam strengthened with GFRP Laminates (2019) Vol.06, No.03. PP. 6530-6539.
[15] Tomasz Drzymała et al., Effects of High Temperature on the Properties of High Performance Concrete (HPC), Procedia Engineering 172 (2017) 256 – 263.
[16] Vahid Afroughsabet and Togay Ozbakkaloglu, Mechanical and durability properties of high-strength concrete containing steel and polypropylene fibers, Construction and Building Materials 94 (2015) 73–82.
[17] D.S.Vijayan, S.Arvindan, K.Naveen Kumar, S Mohamed Javed, Seismic Performance of Flat Slab in Tall Buildings with and without a Shear wall, International Journal of Engineering and Advanced Technology (IJEAT) ISSN: 2249 – 8958, Volume-9 Issue-1, October 2019, Page no – 2672 – 2675.
[18] D.S.Vijayan, J. Revathy, Flexural Response of Fibre Reinforced Polymer Laminated Pre-stressed Concrete Beams, Indian Journal of Science and Technology, Vol 9(42), DOI: 10.17485/ijst/2016/v9i42/101824 , November 2016.
[19] Arvindan Sivasuriyan, D.S. Vijayan, Wojciech Górski, Magdalena Daria Vaverková, Eugeniusz Koda, "Practical Implementation of Structural Health Monitoring in Multi-Story Buildings", Buildings 2021, 11, 263. https://doi.org/10.3390/buildings11060263
[20] M. D. Vaverková, M. Radziemska, S. Bartoň, A. Cerdà, and E. Koda, "The use of vegetation as a natural strategy for landfill restoration," L. Degrad. Dev., vol. 29, no. 10, pp. 3674–3680, 2018, doi: 10.1002/ldr.3119.

Relationship between California Bearing Ratio and Other Geotechnical Properties of Sub grade Soils

Yimam Mohammed [1], A.Paulmakesh[2], BereketAdmasu [3], SalihaShukri[4]
[1]Lecturer, Department of Civil Engineering, WolaitaSodo University, Sodo, Ethiopia
[2]Associate professor, Department of Civil Engineering, WolaitaSodo University, Sodo, Ethiopia
[3]Lecturer, Department of Civil Engineering, WolaitaSodo University, Sodo, Ethiopia
[4]Lecturer, Department of Construction Technology and Management, WolaitaSodo University, Sodo, Ethiopia
Corresponding Author: yimam2003@gmail.com

Abstract: The California bearing ratio (CBR) is a very commonly used laboratory test for predicting the strength of subgrade layer. The aim of this study is predicting the strength of the subgrade soil from fine grained soils. The findings obtained from test results are Atterberg limit, grain size, compaction, and free swell tests. The California bearing ratios were conducted to find out the strength parameters, and correlated with liquid limit, plastic limit, plastic index, maximum dry density and optimum moisture content. The CBR values were determined in the laboratory are compared with the predicted values.

Key Words: California, Bearing, Relationship, Subgrade, Geotechnical Properties

1. INTRODUCTION

In the construction of civil engineering works, the technical properties of soil play an important role particularly in road constructions, foundations, embankments and dams. The geotechnical suitability as a building material will thus be determined. The unprecedented rate of loss of lives caused by collapsing buildings and failure on the roads in recent times calls for a solution[1].

Lateristic soils may be commonly defined as all tropical wetting items, with, or without nodules or concretions, reddish brown or dark brown, which may generally (but not exclusibly) be contained in the hardened ferruginous or hard crusts below. Many research on the engineering characteristics and conduct of laterite soils were conducted. In designing a wide variety of geotechnical and geo-environment buildings such as ground dams, retaining walls, pavements, flags and covers[5], Shear strength is an important engineering property.

The California Bearing Ratio (CBR) is a widely used laboratory test for predicting the strength of a subgrade layer in pavement design. In the 1920s, this testing procedure was first implemented in the Department of California State Highway[6].

The California bearing ratio (CBR) is a standard test in pavement design currently practised for the prediction of subgrade soil's bearing power. This method is commonly used for compact paving design worldwide due to its simplicity and relatively low cost. Nevertheless, it is difficult for highway engineers to obtain representative CBR for paving design[7].

2. STUDY AREA

These projects are found inHadiya zone, one of the administrative zones in Southern nations, nationalities and people region, Ethiopia. These soil samples were collected and used to construct these all-road projects found in HadiyaZone.

2.1 Objective of the study

The main objective of the study is predicting the strength of the subgrade soil from fine grained soils.

2.2 Specific objective of the study

- Determination of index properties of subgrade
- Determination of maximum dry density, optimum moisture content, free swell and CBR values
- Regression of California Bearing Ratio of subgrade

3. Materials and Methods

3.1 Data collection

Six soil samples were used for different road projects in this area.The visual descriptions are classified in the field. The physical properties of fine-grained soil conducted in the laboratorywere grain size analysis, Atterberg limit (liquid limit, plastic limit, plastic index), natural water content, compaction test, free swell test and California Bearing Ratio tests.

Natural humidity, specific gravity, Atterberg limited tests, grain size analysis of all seven samples were carried out in laboratory tests. In ten specified fill samples, the moisture-density ratio, CBR and CBR swelling under OMC are performed. Lastly, in the different soil groups, the impact of moisture content is assessed by (1) varying humidity in the process of preparations of CBR specimens before (dry-side) and after (wet side) OMC; (6days). Special attention has been paid to the sack of tropical residual soil products during all activities for this investigation; but no pattern exists for research into such soil types[2].

The analysis contains the following:

- Particle size distribution.
- Atterberg Limits (liquid limit: LL, plastic limit: PL, and plasticity index: PI).
- Soil classification according to the AASHTO system.
- Modified Proctor parameters (Maximum Dry Density MDD, and Optimum Moisture Content: OMC).
- Soaked CBR values

In this study simple linear, Polynomial and multiple linear regression analysis are used to correlate the CBR with other geotechnical properties of soil.

4. Results and Discussions

4.1. Natural Moisture Content

The natural moisture contents of subgrade material are determined in the laboratory. The results are presented in Table 1. The subgrade samples were collected from 20cm, 30cm and 40cm depths.

4.2. Determination of grain size

The percentages particle sizes in the six soil samples were attained by wet sieve analysis. The subgrade resistance of the soil depends on the soil type, the inner structure of the soil, and the type and mode of stress use. The shearing strength in the soil is the subgrade strength of the soil[8]. The results for grain size analysis are resented in Figure 1.

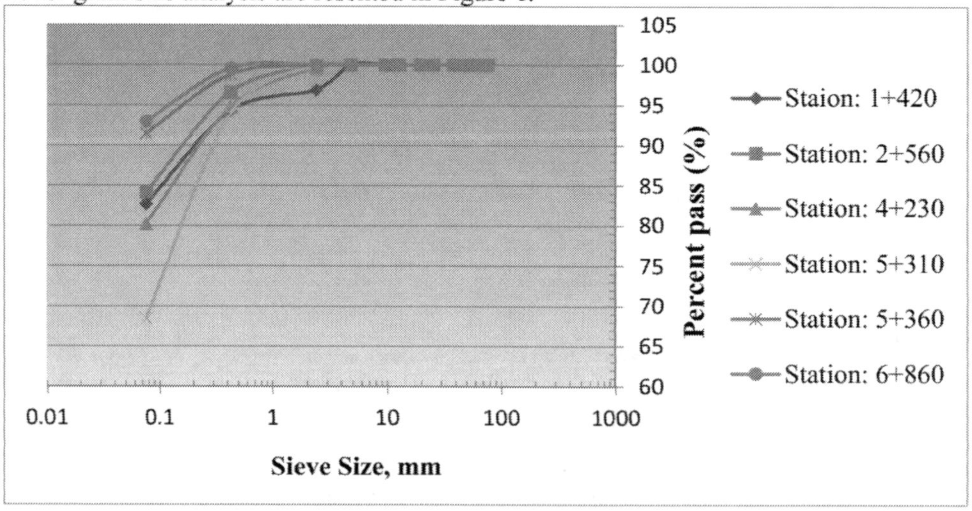

Fig.1. Grain size Analysis graph

4.3. Determination of Consistency limit

Since, the collected samples are lateritic soils, the consistency tests were conducted. Liquid limit was determined by using cone penetrometer and plastic limit was obtained by thread rolling method. Shrinkage limit was also determined. The test results are presented in Table 1 below.

4.4. Linear Shrinkage

The water content at which change in volume of subgrade soils starts to change as a result of a change in water content. The results listed below for linear shrinkage limit of the subgrade soil are lies in between 0% to 11%.

Table 1: Consistency limit natural moisture content test results

Stations (km)	Designation	Depth (cm)	Liquid Limit (%)	Plastic Limit (%)	Plastic Index (%)	Linear shirinkage	NMC (%)
1+420	S1	40		0		0	8.5
2+560	S2	20	60	38	22	11	12.8
4+230[9]	S3	40	38	27	11	6	13.1
5+310	S4	30	37	22	15	9	8
5+360	S5	30	44	33	11	6	6.9
6+860	S6	20	44	34	10	4	13.2

4.5. Compaction

The modified proctor test was applied in determination of the compaction characteristic of the lateritic soil in figure 2. The Standard Proctor compaction utilized 3 layers applying 25 blows each of a 4.5 kg rammer using $944cm^3$ mould.

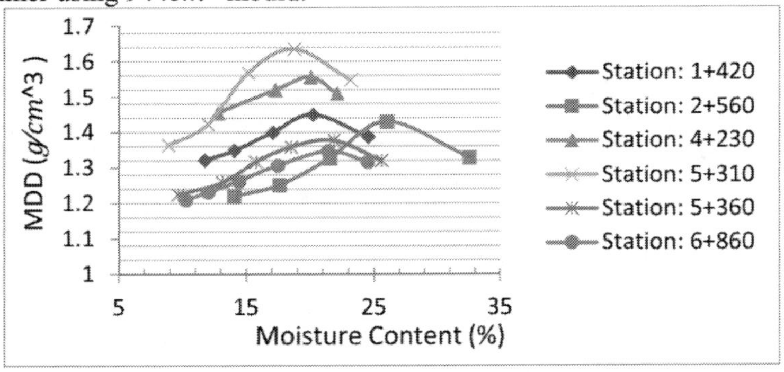

Fig.2. Maximum dry density vs Optimum moisture content graph

4.6. California Bearing Ratio

The California Bearing Ratio (CBR) test is commonly used to determine the suitability of a soil as a subgrade or sub base for highway and runway design and construction [4]. This test results aregenerally used to govern the appropriateness of a soil as a subgrade or sub base for construction of highway and other civil infrastructures.

4.7. Discussion

The engineering properties of sub-grade material used in these six highway projects are analyzed. The types of tests conducted under this investigation are Atterberg limit, compaction (maximum dry density and optimum moisture content), linear shrinkage limit, natural moisture content, grain size analysis, free swell and California bearing ratio tests.
The soil classifications were done according to AASHITO classification system. And, the results are tabulated in Table 2. below.

Table 2. AASHTO classification

Station	1+420	2+560	4+230 [9]	5+310	5+360	6+8 60
% passing 75 microns	83	84	80	68	92	93
PI	0	22	11	15	11	10

AASHITO classification	A-4 (0)	A-7-5(17)	A-6(8)	A-6(9)	A-7-5(9)	A-5(9)

4.8. Linear and Polynomial Regression Analysis of Data

The linear, Polynomial and multi linear regression analyses have been established to correlate the CBR values with consistency limit of soil. The established models are presented in the form of linear, Polynomial and multi linear regression.

4.9. Relationship between CBR and Consistency limit

The relationship graph and equations in between California Bearing Ratio and consistency limit in the forms of linear and polynomial regression were developed and presented below.

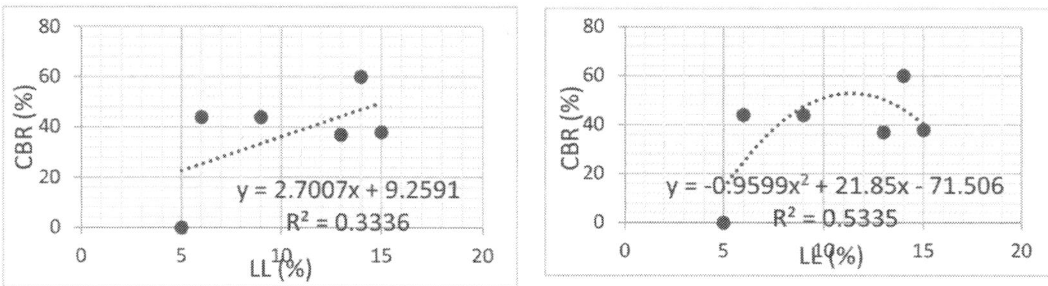

Fig. 3 Correlation graph between CBR and LL (Linear and Polynomial)

$$\text{CBR} = 2.7007\text{LL} + 9.2591 \quad (R^2{=}0.3336) \qquad (1)$$
$$\text{CBR} = -0.9599LL^2 + 21.85\text{LL} - 71.506\ (R^2{=}0.5335) \qquad (2)$$

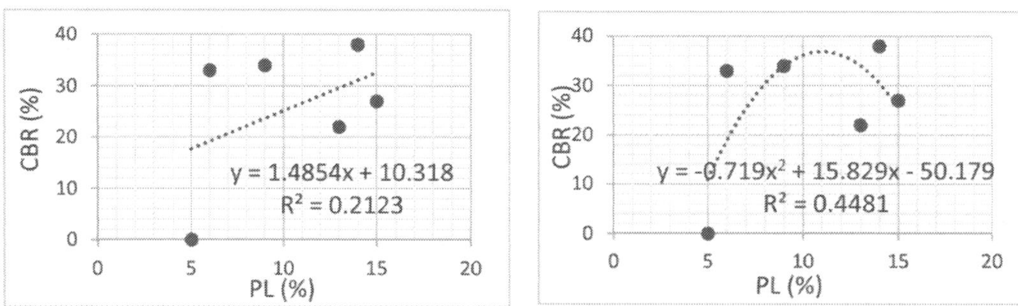

Fig. 4. Correlation graph between CBR and PL (Linear and Polynomial)

$$\text{CBR} = 1.4854\text{PL} + 10.318 \quad (R^2{=}0.2123) \qquad (3)$$
$$\text{CBR} = -0.719PL^2 + 15.829\text{PL} - 50.179\ (R^2{=}0.4481) \qquad (4)$$

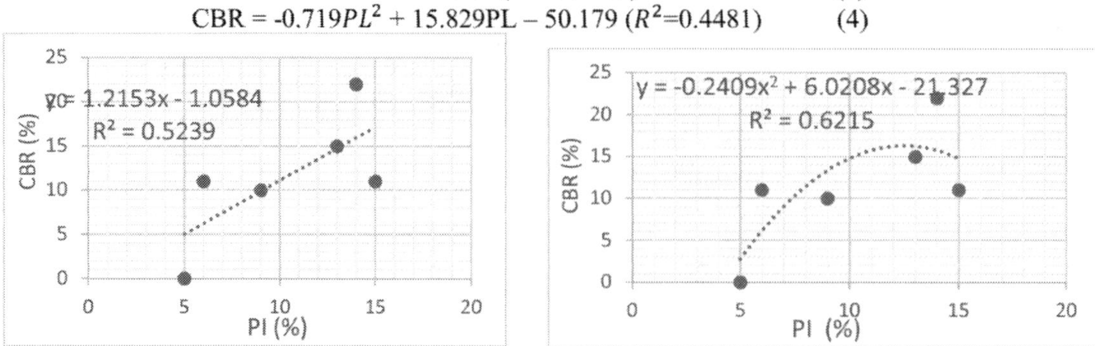

Fig. 5 Correlation graph between CBR and PI (Linear and Polynomial)

CBR = 2.7007PI + 9.2591 (R^2=0.5239) (5)

CBR = -0.2409PI^2 + 6.0208PI – 21.327 (R^2=0.6215) (6)

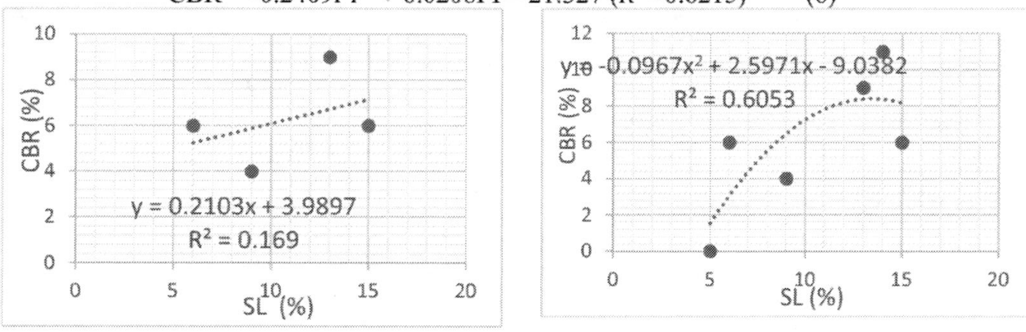

Fig. 6 Correlation graph between CBR and SL (Linear and Polynomial)

CBR = 0.2103SL + 3.9897 (R^2=0.169) (7)

CBR = -0.0967SL^2 + 2.5971SL – 9.0382 (R^2=0.6053) (8)

4.10. Compaction

The modified proctor test was applied in determination of the compaction characteristic of the lateritic soil. The Standard Proctor compaction utilized 5 layers applying 25 blows each of a 4.5 kg rammer usin-g 944cm^3 mould. The test results are presented in table 4.1 below.

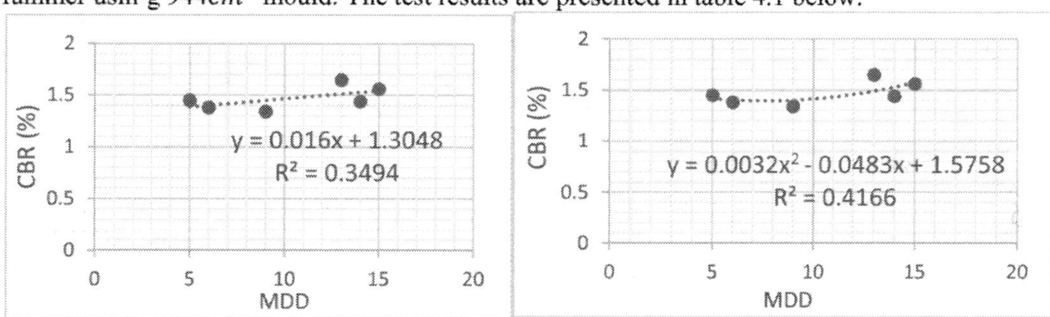

Fig. 7. Correlation graph between CBR and MDD (Linear and Polynomial)

CBR = 0.016MDD + 1.3048 (R^2=0.3494) (9)

CBR = 0.0032MDD^2– 0.0483MDD+ 1.5758 (R^2= 0.4166) (10)

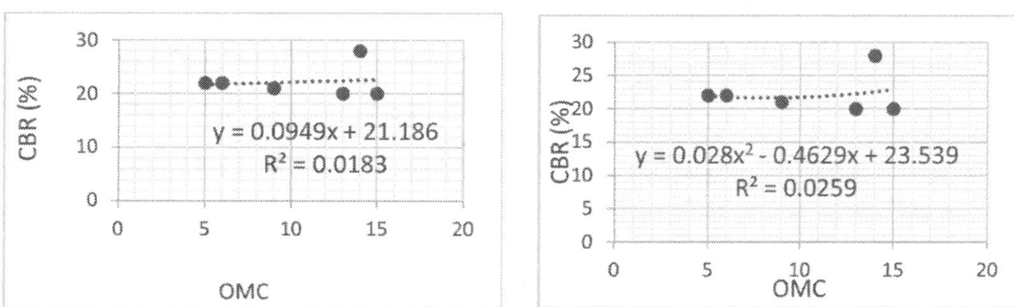

Fig. 8. Correlation graph between CBR and OMC (Linear and Polynomial)

CBR = 0.0949OMC + 21.186 (R^2=0.0183) (11)

CBR = 0.0280OMC^2– 0.4629OMC+ 23.539 (R^2= 0.0259) (12)

4.11. California Bearing Ratio

Property and CBR worth determination The compaction properties are calculated according to the AASHTO standard Proctor test. The test was conducted in a 1000 ml cylinder mould with a 2,6 kg batter with 310 mm free-fall height. Soaked soil sample CBR values have been calculated using AASHTO[3] technique. These test results are commonly used to determine the suitability of a ground as a substructure or substructure for road construction and other civil infrastructures.

During the CBR laboratory test, the appropriate force or load to cause penetration shall be reported at every 0.5mm penetration, including 2.54 mm and 5.08 mm of the load value up to 12.7 mm as a standard penetration depth. The load of penetrating resistance is then compared to the depth of permeation. The ratio of bearing to soil, which should usually be recorded at 2.54 mm, is higher than the ratio of bearing to soil 5.08 mm. Although the ratio of 5.08 mm is higher, the test is replicated fully on a fresh specimen. The design-carrying ratio would be 5.08 mm when the repeated 5.08 mm result is again greater [6]. The soaked CBR test simulates the worst possible scenario for the rainy weather in the study[10][11]. Before the experiments are conducted, it is also mandatory to soak the remoulded samples for four days. Table 1 summarizes the outcomes of the experiments

4.12. Swelling Tests

The sample soil dried up with a regular sieve of 0.5 mm was developed for free swelling ratio testing in accordance with the highway engineering test method (JTG E40-2007)[4]. The free swelling ratio is defined as the difference between soil sample volumes in water and air and air volumes in air. Table 3 lists the test results.

Table 3: Test results for CBR and Compaction

Samples	MDD(gm/cm^3)	OMC (%)	CBR	Swell (%)	Avg. Swell
Sample 1	1.45	22	5	1.97	1.87
				1.89	
				1.85	
Sample 2	1.44	28	14	0.97	0.81
				0.95	
				0.66	
Sample 3[9]	1.56	20	15	1.04	0.81
				1.01	
				0.61	
Sample 4	1.65	20	13	0.47	0.4
				0.42	
				0.37	
Sample 5	1.38	22	6	0.98	0.82
				0.9	
				0.75	
Sample 6	1.34	21	9	0.64	0.42
				0.51	
				0.34	

Table 4. List of group index of the subgrade

Sample used road	Sample	Soil type (AASHTO)	Group Index (GI)
Hambicho Ole Hawariyat Church Tutufa Shool	Sample 1 (1+420)	Dark grey clayey silty soil	0
Dabiyago Andalicho	Sample 2 (2+560)	Light brown silty clay soil	17
Fonko town Chingo[9]	Sample 3 (4+230)	Light brown silty clay soil	8
Kosha Shonkola Omoshera	Sample 4 (5+310)	Brown silty clay soil	9
Lisana town Kode Bidika	Sample 5 (5+360)	Dark grey clayey silty soil	9
HabichoSege Olewicho Boyo School	Sample 6 (6+860)	Light gray clayey silty soil	9

The results obtained from laboratory tests for liquid limit ranges from 0 - 60%, plastic limit (0-38%), plastic limit (0-22%), linear shrinkage limit (0-11), natural moisture content (6.7-13.2%), free swell (0.4-1.87), percentage passing 75 microns (68-93), maximum dry density (MDD) (1.34-1.65), optimum moisture content (20-28) and the CBR value for 95% MDD (5-15).

4.13. Multi Linear Regression Analysis (MLRA)

Development of multiple linear regression analysis model to predict soaked CBR value with consideration of MDD and OMC as the independent variables.

Table 5. Results of MLRA

Variables	Coefficients	t-stat	P-value	R-Square
Intercept	-40.9236	-1.2860	0.28869	$R^2 = 0.4778$
MDD	26.6627	1.6248	0.20266	Adjusted
OMC	0.5441	0.8591	0.4534	$R^2 = 0.1297$

Table 6. Laboratory and predicted CBR values

CBR (Laboratory)	CBR (Predicted)	Error	% error
5	9.7	4.7	94
14	12.71	1.29	9.2
15	11.55	3.45	23
13	13.95	0.95	7.3
6	7.84	1.84	30.67
9	6.23	2.77	30.77

5. Conclusion

According to the output obtained from the laboratory and prediction from regression analysis the following conclusions are drawn.

- From the laboratory liquid limit, plastic limit, plastic index, shrinkage limit, free swell, maximum dry density and optimum moisture content tests the values are ranges from (0-60), (0-38), (0-22), (0-11), (0.4-1.87), (1.34-1.65) and (20-28) respectively.
- The empirical relationships were developed for California Bearing Ratio value from consistency limit and compaction test results in the form of linear, polynomial and multiple linear regression analysis.
- The values of correlation in terms of coefficient of determination (R) are relatively weaker for linear than polynomial regression.
- In general, the overall correlation between engineering properties and California bearing ratio values of these subgrade soils are weak.

Reference

[1] Osuji O.S andAkinwamide J.T (2018), Engineering Properties of Lateritic Soils in Ado-Ekiti, South Western Nigeria
[2] DefaruKatiseDasho, Prof. R. KumerVerma (2019), Geotechnical Investigation and Effect of Moisture Content on Subgrade CBR Values; Arbaminch-Chencha Existing Road; Ethiopia.
[3] Dr. Dilip Kumar Talukdar (2014), A Study of Correlation Between California Bearing Ratio (CBR) Value with Other Properties of Soil, India.
[4] Aye AyeMyat, NyanMyintKyaw-, Htay Win (2018), Prediction Models for Estimation of California Bearing Ratio for Cohesive Soil, Myanmar
[5] OluwapelumiO.Ojuri (2013), Predictive Shear Strength Models for Tropical Lateritic Soils, Nigeria.
[6] AshenafiMulugeta (2016), Soaked CBR value determination from laboratory dcp test for Addis Ababa red clay soil, Ethiopia.
[7] Surendra Roy (2016), Assessment of Soaked California Bearing Ratio Value Using Geotechnical Properties of Soils, India.
[8] Magdi M. E. ZumrawiandHussamElnour (2016), Predicting Bearing Strength Characteristics from Soil Index Properties, Khartoum, Sudan.
[9] Lolo C., Bereket A. (2019), Geotechnical evaluation of natural subgrade soil.
[10] D.S. Vijayan, D. Parthiban," Effect of Solid waste based stabilizing material for strengthening of Expansive soil- A review," Environmental Technology & Innovation, Volume 20, 2020, 101108, ISSN 2352-1864, https://doi.org/10.1016/j.eti.2020.101108
[11] D.S. Vijayan, A. Mohan, J. Revathy, D. Parthiban , R. Varatharajan, "Evaluation of the impact of thermal performance on various building bricks and blocks: A review", Environmental Technology & Innovation 23 (2021) 101577, https://doi.org/10.1016/j.eti.2021.101577.

STUDY ON FLEXURAL BEHAVIOUR OF REINFORCED CONCRETE BEAM WITH EFFECTIVE OPENINGS

D.Sathiyapriya [1], K.Suganya Devi [2], A. Sattainathan Sharma[3]

[1]student, [2&3]Assitant Professor, Civil Engineering Department, SRM Valliammai Engineering College, Chennai, India

Corresponding Author Email: sathiyapriyad@gmail.com

ABSTRACT: Reinforced Concrete (RC) beams are essential structural elements that transfer loads from the slabs to the columns through flexure and shear. Openings in Reinforced Concrete (RC) beams are required for a variety of architectural and mechanical reasons. The purpose of this article is to investigate the flexural behaviour of an RC beam with circular openings running the length of it. The three-dimensional nonlinear finite element method was used to investigate a supported RC beam with circular web openings of varying diameters using ANSYS, a finite element software package. The study's primary characteristic is the variation in diameter and location of reinforcement around the openings. Five RC beams with simple supports were constructed, and tested were conducted under two-point loads. Beams were constructed with one conventional specimen without any openings in beams and two beams with unstrengthen circular openings located in the flexure zone with varying diameters such as 80mm and 100mm considered as a control beam. In contrast, the remaining two beams provided effective diagonal reinforcement around the openings. The load-bearing capacity and load-deflection behaviour of such openings in beams were studied and analysed in detail. To increase the ultimate strength of the beam, effective diagonal reinforcement and short stirrups at the top and bottom of openings are used to prevent the beam from collapsing prematurely.

Keywords: Reinforced concrete (RC) Beam, Circular opening, Diagonal Reinforcement, ANSYS.

1. INTRODUCTION

In today's scenario, we all have a multi-story building, which requires many pipes and pipes, to provide services such as air conditioning, telephones, electricity, fire exhaust pipes, and networks. Usually, the pipes and conduits are placed on the underside of the beam and inside the soffit. It creates an unnecessary dead space from just a few cm to half a meter. The dead space increases on each floor and increases the general height of the building with the size and depth of the pipelines. The alternative arrangements for the provision of cross-web opening in RC Beam to provide utilities shall therefore be made.

This type of web opening means that dead space is reduced and the design is compacter. This makes it more complex to include openings in RC beams by sudden changes to the beam-cross-section. The most specific beam behaviour. The provision of such openings in RC beams causes an interruption of everyday stress flux, thus reducing beam rigidity, excessive cracking, disruption, torsional strength, and beam capacity.

The web openings are randomly available in certain circumstances. If these openings are planned before building, precautions can be taken by adding further reinforcements and stirrups to the openings. However, these beams' strength and rigidity must be monitored, and necessary precautions if these web openings are boiled without any planning. The critical issue is that when these hole/openings are boiled in the supporting region, the shear capacity of the beam can be reduced considerably, and the deflection of the beam can also increase when loaded. In addition to the ductile behaviour, these opening/holes may lead to broken beams.

From the Year 1960, Research on RC beams with an opening started; most of the research results are about RC beam with Rectangular and square openings. The most adaptable openings are circular as compared with rectangular and square. Many scientists use the terms small and large without defining it or a clear-cut limit, depending on the size of the openings. Mansur and Tan [8] recommended that small openings in circular, square, or almost square shapes are, whereas the circular diameter of the opening in Somes and Corley [11] is more than 0.25 times the depth of the beam considered significant.

In the RC beam analysis with geometric discontinuities as a transversal circular opening on the Web, Mansur et al. [9] examined the model strut-and-tie. The truss model clearly explains the role of diagonal strengthening in alleviating the substantial distress at the throat area by transmitting a large amount of applied shear across the discontinuity. The effect of small circular opening on shears and the bending and ultimate strength of beams produced from standard and high strength concrete has been examined by Javad Vaseghi Amiri et al. [7]. For standard concrete, if the opening diameter exceeds 1/3 the beam depth, the ultimate capacity reduction and cracking design have changed, as has the failure mode. High concrete strength is not very strong for the ultimate strength, but the rigidity and serviceability of the beams are increased. In the course of an RC beam on his flexural behaviour, Bengi Aykac et al.[1] examined the experimental investigation into the influence of various web openings. The Vierendel action protected the web from buckling before failure, which contributed to a premature fracture of the beams around the openings. The rebars and stirrups extending down each side of the beam shaft prevented shear failures.

This study aims to investigate the amount of the effect of the presence of a circular opening in the flexure zone of the reinforced concrete beams using finite element analysis and with the consideration of the following main factors are Diameter of opening, Location of applied load, Web reinforcing bars and the type of position of reinforcing bars around the opening. Providing Proper Reinforcement Detailing around the prescribed openings can minimize Stress Concentration & Prevent Premature failures.

2. EXPERIMENTAL STUDY

Four rectangular RC beams have been designed and tested for this experimental program with openings and a single beam without openings. In the following sections, the experimental program is detailed.

2.1 MATERIALS

The properties of the materials used were experimentally obtained. The test beams were all prepared simultaneously and cured to ensure the compressive strength was virtually equal under the same conditions. At 28 days, the average testing strength was 37 MPa. Testing nine cube sizes 150 to 150 or 150 to 150 mm using a Compressive Testing Machine achieved the compressive strength (CTM). The reinforcement concrete beam, which consists of 2mm to 12 mm at bottom, 2mm to 10 mm at the top, and 8 mm stirrups, was placed at 150 mm c/c. 8mm dia reinforcement steel is used in the top and bottom of openings by providing small stirrups and around the openings, diagonal reinforcement has been used.

2.2 TEST SPECIMEN

This experimental program has included several five rectangular RC beams of the same length and the exact cross-sectional dimensions. The beams have been tested under four points bending. All of them have a length of 1200 mm, a width of 150 mm, and a height of 280 mm. The Circular opening of 80mm and 100mm opening was formed by using PVC pipes. The dimension, Reinforcement details, and location of the opening are shown in Figs 1 & 2.

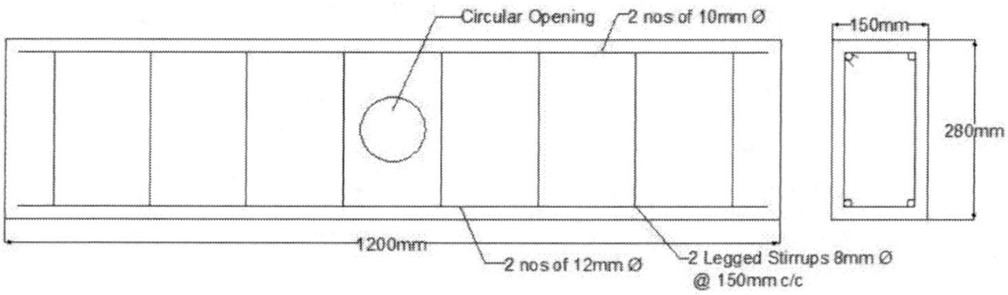

Fig 1. Longitudinal & cross-section view of RC beam with Circular Opening

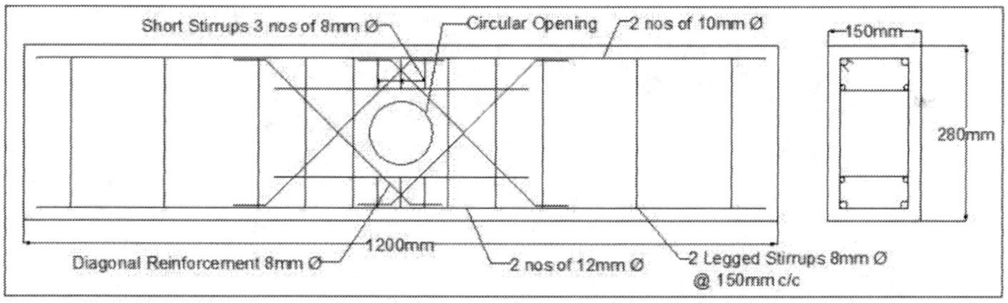

Fig 2. Longitudinal & cross-section view of RC beam with additional Reinforcement around Opening

The first beam has been used as the reference one, the conventional beam, which is a solid beam without any openings and has been referenced by (CB). 2 beams constructed with the circular opening located in flexure zone with varying diameters such as 80 mm (UB1) and 100 mm (UB2) considered as a control beam and the other remaining two beams has been referred as SB1 & SB2 by providing effective diagonal reinforcement around openings as shown in Table 1.

Table 1. DETAILS OF BEAM SPECIMEN

S.No	Beam Details	Diameter of Opening(mm)
1	Conventional Beam	-
2	Unstrengthen Beam (UB1)	80
3	Unstrengthen Beam (UB2)	100
4	Strengthened Beam (SB1)	80
5	Strengthened Beam (SB2)	100

2.3 TEST SETUP

The test of all beams was carried after 28 days. The specimen position to be adjusted on the supports with a clear span of 1000mm, and when subjected to a weight of 400kN/m bearing, the beams were found to be strong enough to bear a total load of 500kN. By hydraulics, the load was distributed until it was out of tolerance. The load setup is shown in fig.3. The deflection in the mid-span was recorded using LVDT and the crack pattern of the beams. Testing arrangement of the beam with the opening as shown in fig.4

Fig 3. SCHEMATIC ARRANGEMENT OF TEST SPECIMEN

Fig 4. TESTING OF BEAM WITH OPENING

3. TESTS RESULTS AND DISCUSSIONS

Test specimens were loaded up to their ultimate capacity. The load-deformation and failure mode of the conventional beam without openings, beams with unstrengthen circular openings, and beam with openings strengthened using diagonal reinforcement around openings were observed.

3.1 CRACK PATTERN AND MODES OF FAILURE

In Figure 5, the crack patterns and failure modes of the test specimen are demonstrated. The experimental investigation concluded that most cracks in the brittle regions continued into the pressurized regions. The original flexibility zone and increased load are correlated; both are observed in the early deformation stages. Beam UB1, which has an opening of 80mm diameter because of the cracking pattern and behavior, is similar to a solid beam. Initially, flexural cracks developed, and cracks in the opening region also lead to shear failure while increasing load. Hence it shows similar to beam type failure according to Mansur research [3].In beam UB2 circular opening with 100 mm diameter, preliminary cracks are initiated as the amount of load increased additional cracks are developed in top and bottom of the opening region, and the

beam failed suddenly in frame type failure.

Fig 5. The Crack pattern of Tested Specimen.

Beam with 100mm diameter Opening (SB2), although it initiates shear crack but its growth prolonged. Among the other beams, an 80mm diameter opening beam (SB1), which has a small stirrup on top and bottom of the opening and diagonal reinforcement, will act better than the other beams controlling the cracks attains shear failure. At loads equal to or above their carrying capacity, the beams with circular openings failed. The presence of diagonal reinforcement and short strands prevented premature failure and contributed to the bending capacity of beams.

3.2 LOAD- DEFLECTION CHARACTERISTICS

As a result of experimental studies, unreinforced beams with openings were observed in certain test specimens that conducted ductile and fragile damage. Table 2 shows the results of the test.

Table 2.TEST RESULTS

S.NO	Beam Designation	Initial Crack Load (kN)	Deflection in mm (δ_y)	Ultimate Load (kN)	Deflection in mm (δ_u)
1	CB	95	1.02	310	5.35
2	UB1	90	1.35	294	5.27
3	UB2	82	1.47	285	5.75
4	SB1	105	0.77	337	5.05
5	SB2	98	0.85	328	5.25

The experimental test results show the Conventional Beam (CB) Ultimate load is 310kN and maximum deflection is 5.35 mm. The beam with a circular opening show that it reduces the load-carrying capacity compared with other beams. While comparing 80mm (UB1) and 100mm (UB2) beams, UB1 is observed similar to that of the conventional beam. Besides, UB2 behave more brittle, and sudden failure occurs because of increased opening diameter. RC beam (SB1 & SB2) strengthened with diagonal reinforcement around opening increases the load-carrying capacity by 8.7%, and the maximum deflection is 5.25 mm. Diagonal strengthening around openings increases the beam's flexural capacity. The characteristic of load-deflection is shown in Figure 6.

The load-deflection characteristics, ductility factor, bending stiffness, and energy absorption capacity are used to evaluate the test results. the ductility ratio was defined as the ratio of the displacement value (δu) corresponding to 85 percent of the beam's maximum loading capacity to the yielding displacement value (δy). According to the literature, this ratio should be around 4–5 for a ductile beam [4].

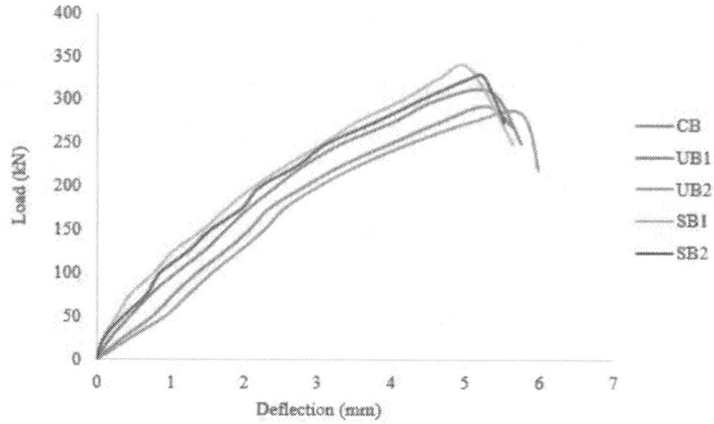

Fig 6.LOAD DEFLECTION CHARACTERISTICS

The bending stiffness of the beams was determined using their initial stiffness (rigidity). The area beneath the load-deflection curve was used to calculate the energy absorption capacity. The absolute values of ductility and stiffness and their relative values are summarised in Table 3.

Table 3.TEST RESULTS OF THE SPECIMENS

Test Specimens	Ductility factor		Energy Absorption (kN.mm)		Initial Stiffness (kN/mm)	Failure Mode
	Absolute	Relative	Absolute	Relative		
CB	5.25	1	2975	1	93.14	Shear & Flexural Failure
UB1	3.90	0.58	2718	0.91	66.67	Flexural-Opening Region
UB2	3.91	0.75	2638	0.88	55.78	Flexural-Opening Region
SB1	6.56	1.24	3170	1.06	136.36	Shear Failure
SB2	6.18	1.18	3250	1.09	115.29	Shear failure

The energy absorption decreases by including an opening in the beam, whereas using diagonal reinforcement around the opening is similar to that of a conventional beam. Besides, the stiffness also decreased when compared to the solid beam. The short strands at the top and bottom of the beam chords improved the ductility of RC beams using factor ductility.

4. FINITE ELEMENT MODELLING

ANSYS, a finite element software, is solving complex structural engineering problems faster and more efficiently. Finite element modeling is a powerful technique for calculating the displacements, stresses, and strains in a structure as a function of a given set of loads. ANSYS workbench environment is highly automated and users friendly to customize the geometric modeling according to the type of analysis or application. The central concept in finite element modeling is to generate the mesh. Meshes are used to divide a model into numerous small areas, which makes the analysis accurate.

4.1 Modelling & Deformation

The specimen models were created for conventional beam and beam with the circular opening as per the geometric required using ANSYS Software as shown in Figures 7 & 8.

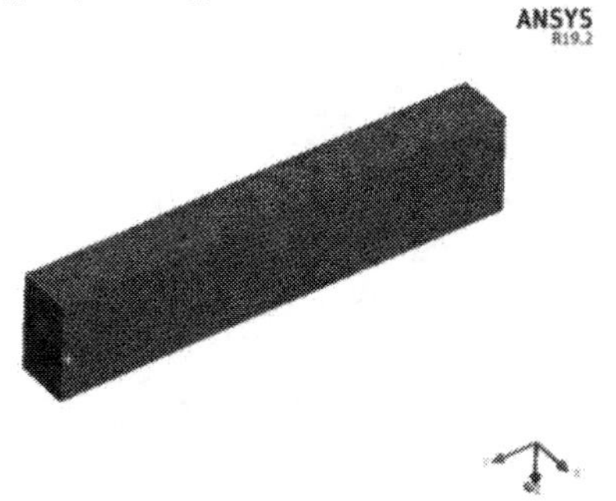

Fig 7. Conventional Beam

Fig 8. Beam with Openings

All the Specimen models are subjected to two-point loading in which load is applied at L/3 distance from both the ends and support conditions are given at 100 mm from the ends of the beam. After the load applications, the model is solved for the required results, and the failure mode of the model is obtained shown in Table 4.The deformation of the conventional beam and beam with an opening is shown in Fig 9 & 10.

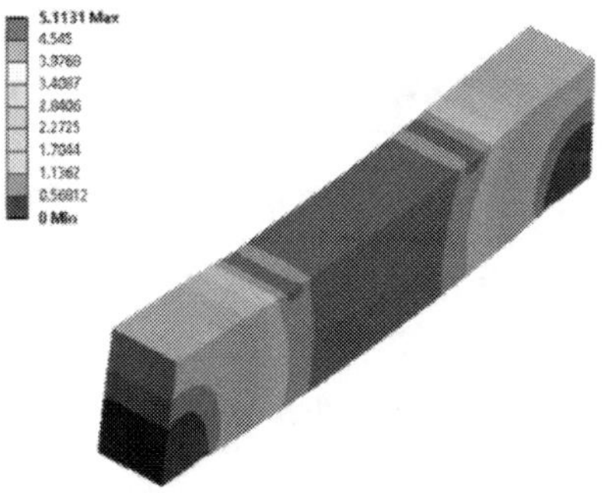

Fig 9. Deformation of Conventional Beam

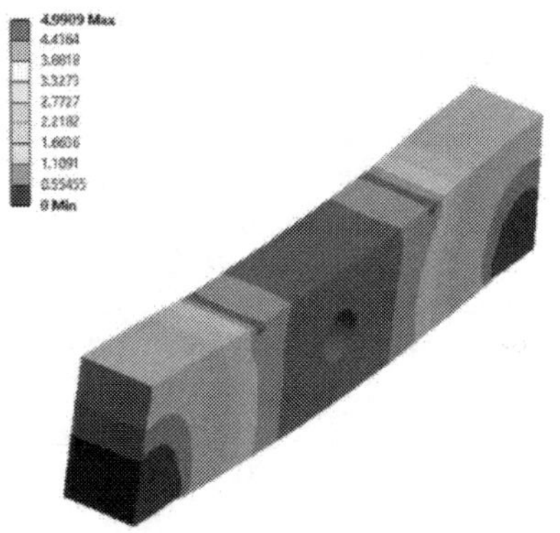

Fig 10. Deformation of Beam with Openings

From table 4, it is observed that a strengthened beam with diagonal reinforcement increases the load-carrying capacity, and correspondingly deformation is decreased compared with other beams.

Table 4. Deflection Results at Ultimate Load

S.No	Beam Designation	Ultimate Load(kN)	Deflection in mm	
			Experimental	Analytical
1	CB	310	5.35	5.11
2	UB1	294	5.27	5.02
3	UB2	285	5.75	5.19
4	SB1	337	5.05	4.99
5	SB2	328	5.25	4.83

When compared with Analytical and Experimental results, the deformation of beams varies 5-10% only. It clearly shows that the good correlation between experimental and analytical work.

5. CONCLUSION

From the above results, it is concluded that

- In this experiment, the initial flexural crack induced is independent of the presence or absence of an opening, but a large number of shear cracks will initiate sooner, corresponding to a conventional beam. As a result, when an opening is included, the beam's load capacity decreases.
- If the opening diameter is less than 0.3, the depth of the beam will behave normally; however, the location of the opening and the minimum depth of the compression chord should be considered. While the diameter of the opening exceeds the depth of the beam by more than 0.3, increasing the load causes shear cracks to develop on both the top and bottom of the opening, reducing the beam's load-carrying capacity.
- As the opening diameter in the beams increases, crack pattern and failure type change from bending to shear failure type frame or strap.
- By installing diagonal reinforced bars and the stirrups on the top and bottom of openings, the beams' ultimate strength will be increased by 8.7 percent. To increase the ultimate strength of the beam, an effective diagonal reinforcement bar and short stirrups at the top and bottom of openings are used to prevent the beam from collapsing prematurely.
- When analytical and experimental results are compared, the deflection value varies by approximately 5%-10% from the experimental deflection at their ultimate load. FE modeling with ANSYS Software can be used to predict reinforced concrete beams with openings numerically.

REFERENCES

[1] Bengi Aykac,Iller kalkan,yosuf emre,Flexural behavior of RC beams with regular square or circular web openings, Elsevier journal,Engineering Structures, Volume 56,Nov 2013,Pages 2165-2174

[2] Bashir H. Osman, Erjun Wu, Bohai Ji, and Suhaib S. Abdulhameed, "Shear Behavior of Reinforced Concrete (RC) Beams with Circular Web Openings without Additional Shear Reinforcement",KSCE Journal of Civil Engineering (2017) 21(1):296-306

[3] Concrete beams with openings-Analysis and Design a book by M A Mansur and Kiang Hwee tan(1999)

[4] Ceyhun Aksoylu a, S¸akir Yazman b, Yasin Onuralp Özkılıç c, Lokman Gemi d, Musa Hakan Arslan, Experimental analysis of reinforced concrete shear deficient beams with circular web openings strengthened by CFRP composite,Elsevier journal, Composite Structures 249 (2020) 112561

[5] Dr. Latha M S and Naveen Kumar B M, Behavior of Reinforced Concrete Beam with Opening, International Journal of Civil Engineering and Technology, 8(7), 2017, pp. 581–593.

[6] D. S. Vijayan and J. J. Daniel, "An investigation on the torsional effect of symmetric moment resisting frame system subjected to eccentric reinforced concrete lift wall - A finite element approach," Int. J. Eng. Trends Technol., vol. 69, no. 8, pp. 179–184, 2021, doi: 10.14445/22315381/IJETT-V69I8P222.

[7] Javad vaseghi amiri , Morteza , Effect of small circular opening on the shear and flexural behavior and ultimate strength of reinforced concrete beams using normal and high strength concrete, 13th World Conference on Earthquake EngineeringVancouver, B.C., Canada August 1-6, 2004Paper No.3239

[8] Arvindan Sivasuriyan, D.S. Vijayan, Wojciech Górski, Magdalena Daria Vaverková, Eugeniusz Koda, "Practical Implementation of Structural Health Monitoring in Multi-Story Buildings", Buildings 2021, 11, 263. https://doi.org/10.3390/buildings11060263

[9] M A Mansur, K Tan and W Weng, Analysis of Reinforced Concrete Beams with Circular Openings using Strut-and- Tie Model, Proceedings of The International

Conference on Structural Engineering, Mechanics and Computation, Vol 1, 2001, pp 311–318.

[10] Nasr Z. Hassan, Alaa G. Sherif, Amal H. Zamarawy, Finite element analysis of reinforced concrete beams with opening strengthened using FRP, Elsevier-Ain Shams Engineering Journal (2017) 8, 531–537

[11] Somes NF, Corley WG. Circular openings in webs of continuous beams. Shear in Reinforced Concrete. Special Publication SP-42. Detroit: American Concrete Institute; 1974; 359–98.

[12] X.F. Nie,S.S. Zhang,T. Yu Behaviour of RC beams with a fibre-reinforced polymer (FRP)-strengthened web opening, Elesevier- Composite Structures 252 (2020) 112684.

[13] R. Goplalakrishnan, A. Mohan, L. Ponrajsankar, D. S. Vijayan, "Characterisation On Toughness Property of Self- Compacting Fibre Reinforced Concrete", Journal of Environmental Protection and Ecology 21, No 6, 2153–2163 (2020).

[14] D S Vijayan, Dr. J Revathy, Flexural Behavior of Reinforced and Pre-stressed Concrete Beam Using Finite Element Method, International Journal of Applied Engineering Research, ISSN 0973-4562 Volume 10, Number 1 (2015) pp. 717-736

[15] D.S.Vijayan, J. Revathy, Flexural Response of Fibre Reinforced Polymer Laminated Pre-stressed Concrete Beams, Indian Journal of Science and Technology, Vol 9(42), DOI: 10.17485/ijst/2016/v9i42/101824, November 2016.

[16] D.S.Vijayan, S.Arvindan, K.Naveen Kumar, S Mohamed Javed, Seismic Performance of Flat Slab in Tall Buildings with and without a Shear wall, International Journal of Engineering and Advanced Technology (IJEAT) ISSN: 2249 – 8958, Volume-9 Issue-1, October 2019, Page no – 2672 – 2675.

Design of Solar Charging Case for Mobile Phones

Gladwin Antony R, Hariharan S, Hari Haran D and Clement Raj C

Department of Electronics and Communication Engineering, KCG College of Technology, Chennai 600117.

Email : antonygladwin16@gmail.com

Abstract. Every day in this modern era, a new technology is introduced in a mobile phone, and one major concern that every mobile phone user has is its battery backup. Whatever the advancements in mobile phones, the chargers we use have remained consistent since the beginning. Our project's goal is to create an integrated solar mobile charger that can be seamlessly incorporated into the protective case of the mobile phone. The proposed design traps solar energy and stores it in a rechargeable battery. This system has the ability to serve dual role, both as a protective case and act as power backup for the mobile phone.

1. Introduction

Mobile phones have evolved dramatically over the years, from simple to smart, starting from the year 1973. These mobile phones have become the primary data and correspondence centres for humans. One could argue that a cell phone is fundamental to modern life. However, the chargers used for these phones have not seen significant advancement since their inception. As mobile phones have become an integral part of people's lives, they really don't have the time to put their phones on charge and wait near a power outlet for an hour or more. The best alternative solution to this issue is to use renewable energy as a power source [1]. And the ideal renewable energy source that is compatible for this project is unquestionably solar energy. It is a highly efficient energy source that has had a significant impact on society and industry. It is possible to charge any device without using electricity with the proposed model. People who always use mobile phones for their work and people in remote areas find it hard to charge their devices due to a lack of electricity and most importantly lack of time [2]. The goal of this project is to design and implement a portable solar phone charger using PV cell and Li-ion battery.

The following objectives were created in order to achieve this goal:

- Study of various photovoltaic cells and batteries performance characteristics.
- Select ideal components for the model based on requirements.
- Designing of the circuit and prototype model.
- Construction and Testing of the prototype.

2. Methodology and Modelling

First, the photovoltaic cells absorb energy from the sun. The DC output of the solar panels is then passed to the USB charging circuit (TP4056 IC) which is generally used for charging the mobile phones in the traditional phone chargers. The output from the IC directly goes to the 5000 mAh Li – ion battery. The energy obtained from the photovoltaic cells are constantly collected and stored in the battery.

The charge stored in the battery is then sent to a DC – DC boost circuit (XL6009 IC). This circuit steps up the battery output from 3V to 5V and sends to the USB step up converter. The step-up converter provides constant 5V output in order to provide consistent charging mechanism for the mobile phone. The figure 1 depicts the simplified block diagram of the prototype.

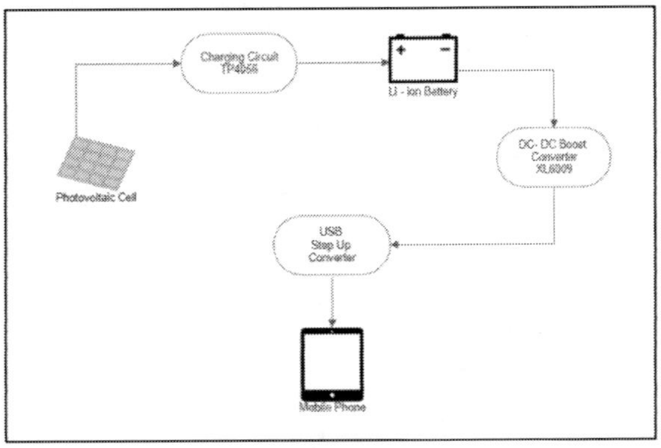

Figure 1. Block diagram of the project

The designed circuit is then arranged behind the back of the Silicon protection case of the mobile phone. The solar panels are placed above the charging and battery circuit so that whenever we turn the phone upside down, the panel captures the sunlight and stores in the battery. And whenever we want the phone to get charged, we just have to tap on a switch and the phone starts charging.

The major components used in this project are listed below:

2.1. Photovoltaic Cell / Solar Panel

The solar panel used in this project is a 3v 150mA mini–Solar Panel that has Polycrystalline solar cells which are encased and protected by a durable outer poly frame. This mini–Solar Panel is light weighted, very strong and weather resistant. These Small panels are simple to install or add to the existing product and their construction requires no frame or special modifications.

- Max output power (without load): 0.45W
- Max working voltage (without load): 3.3V
- Max charging current (without load): 150mA
- Min output power (without load): 0.45W
- Min working voltage (without load): 3V

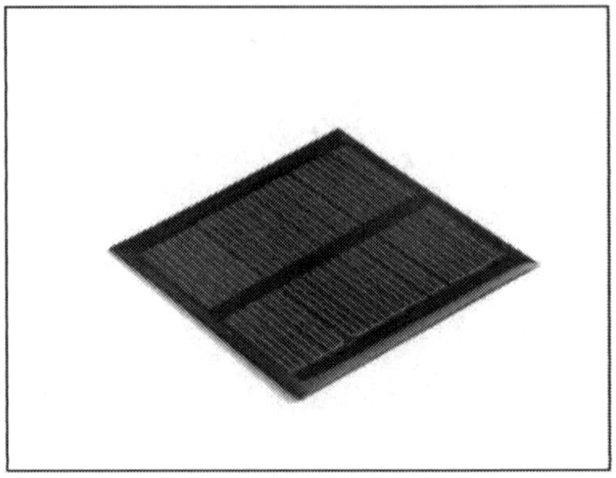

Figure 2. Mini Solar Panel / Photovoltaic Cell

2.2. USB Charging Circuit

The TP4056 is a complete constant-current/constant-voltage linear charger for single cell lithium-ion batteries. Its SOP package and low external component count make the TP4056 ideally suited for portable applications.

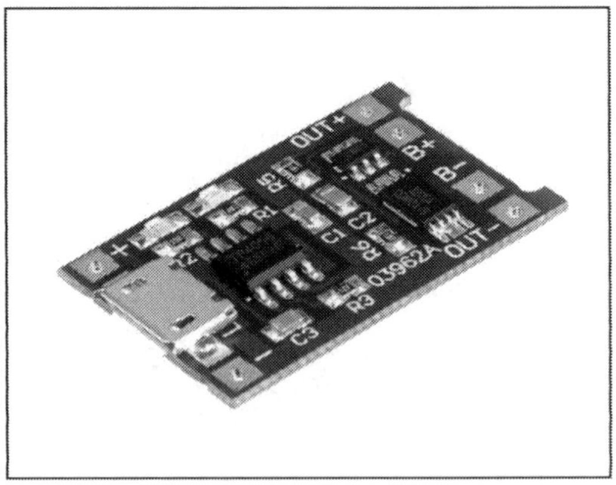

Figure 3. TP4056 USB Charging Circuit

2.3. Lithium Polymer Battery

The Battey used in this circuit is a 3.7V 5000mAH (Lithium Polymer) Lipo Rechargeable Battery also known as Lipo or Lipoly batteries. These are thin, light and powerful. This battery has a capacity of 5000mAH. These Batteries are widely used in GPS, DVD, iPod, Tablet PC, MP4 Player, Power Bank, Mobile Backup Power Supply, Bluetooth Speaker, IOT and other Industrial applications.

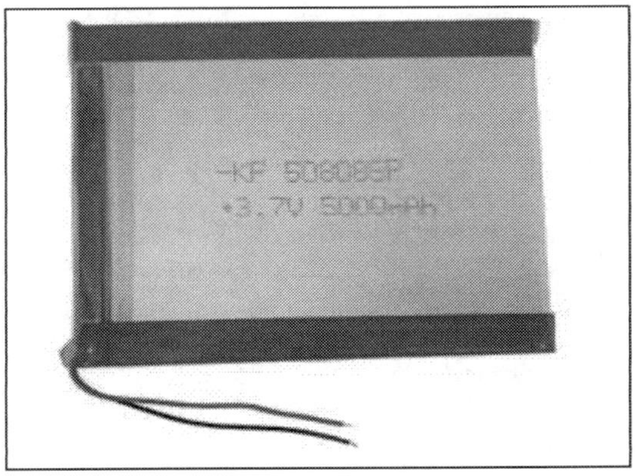

Figure 4. Lithium Polymer Battery

2.4. *DC – DC Boost Converter*

XL6009 module is a non-isolated step-up boost voltage converter featuring adjustable output voltage, high efficiency. It converts input voltage of 5-32V DC to an output voltage of 4-38V DC.

Figure 5. XL6009 DC-DC Boost Converter

3. Related Works

Users can carry mobile phone chargers with them wherever they go, but electricity may not always be available. This solution can be used to defeat this problem in many places that do not have electricity, such as rural areas and some hill stations. In this article, the author provides a solution for charging multiple devices in a public place. During the night, the stored energy is used for both lighting and charging.[3]

The author created the solar backpack to address the issue of people becoming frustrated when they are unable to contact friends due to a dead phone [4]. This backpack stores energy from solar cells on the outside that can be used to charge electronic equipment such as cellular phones and iPods. The panel is joined to the charging cycle, which contains a Voltage regulators, by means of four blocks of three solar cells in parallel which make up a panel. He also evaluated the use of two ultra-capacitors, which also save energy, to store charges and a backup battery. The result was a portable solar recharge pack. The strength of this system is the use of ultra-capacitors and backups that deliver sufficient load current. However, this system presents a challenge. Instead of a potentiometer, a special 550Ω resistor was used in the voltage controller circuit so a user can vary the voltage and charge different devices. Cell phones with a voltage need less than 5V can be charged with this backpack unit [5].

The green power charger is designed to meet the insatiable demand for power for irresistible gadgets using renewable energy resources. This design is a portable green power charger that uses solar energy to mitigate the sustainable energy challenge. They use the concept of capture, store, and charge here, with the capture coming from a green resource [6].

In this study, the input is provided by a photovoltaic cell, and the amount of voltage required is controlled by a charge controller. The received voltage is passed through an IC circuit and wirelessly transmitted to the respective device to charge it.

4. Result And Conclusion

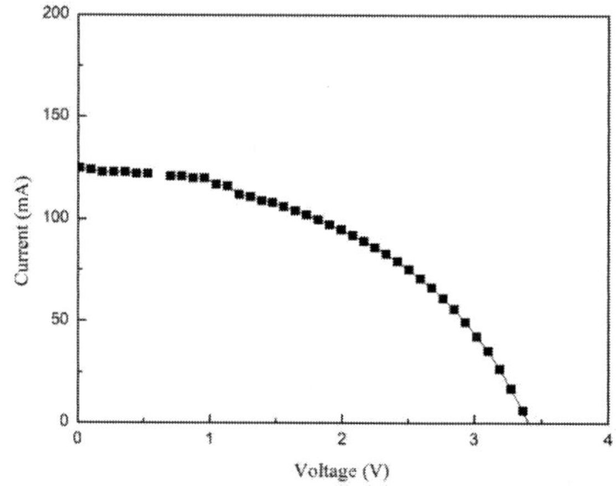

Figure 6. IV Characteristics of the solar cell used

The IV characteristics of the solar cell purchased was measured using a solar simulator with an irradiance of 1000w/m^2 at room temperature [7]. The IV characteristics of the solar cell is depicted in Figure 6. The data matches well with the specification of the solar cell purchased.

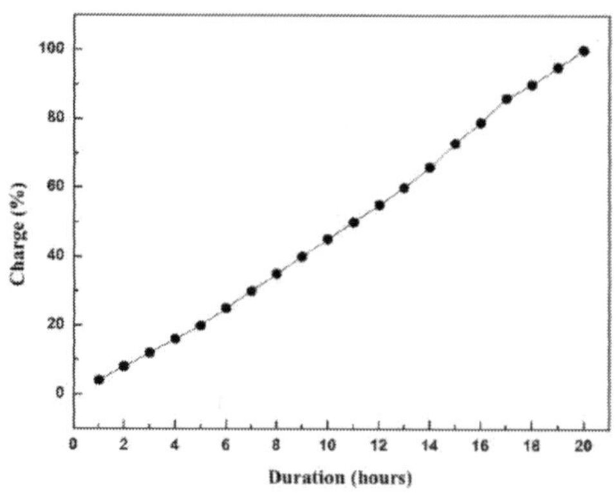

Figure 7. Charging of a 5000mAh battery using the solar powered charging case

The portable mobile Solar Charger was designed and fabricated to be compact in size and cost efficient, reliable, portable, light weight and economically viable for charging cell phones [8]. The charging of the mobile phones was tested by installing the solar based charger unit in the protective case. The charging of the mobile phone was recorded for every one hour in terms of the percentage. Since we have used two solar cells with an short circuit current of 150 mA, the complete charging of the 5000mAh battery could be completed only after 20 hours of charging.

5. Future Scope

There are still many improvements to be made to this portable solar charging case now that it has been designed.

- The first significant area for improvement to this project is the use of photovoltaic cells that can be charged not only by sunlight but also by incandescent or LED lights.
- Another major upgrade that can be done to this project is to introduce wireless charging mechanism so that the phone automatically gets charged when the phone is placed in the case which in turn reduces the wires and from the model.
- This circuit can further be modified and can be used for other electronic devices like Bluetooth headsets, speakers.

References

[1] Shufian, A., Rahman, M.M., Ahmed, K., Islam, R., Hasan, M. and Islam, T., 2019, May. Design and implementation of solar power wireless battery charger. In 2019 1st International Conference on Advances in Science, Engineering and Robotics Technology (ICASERT) (pp. 1-5). IEEE

[2] Mudi, S. Design and Construction of a Portable Solar Mobile Charger. World Academics Journal of Engineering Science, vol 7,issue 1 pp 40-44, Mar 2020.

[3] A. Barad, S. Tungar, N. Sangle, K. Bharambe and D. P. Kadam, "Solar panel based multi-mobile charger with LED illumination," 2017 International Conference on Innovations in Information, Embedded and Communication Systems (ICIIECS), Coimbatore, 2017, pp.

[4] Singh, T., P. Bhardwaj, B. Singh, and A. Kumar. "Design and development of portable power charger: a green energy initiative." International Journal of Electronics and Communication Technology 5, no. 4 (2014): 12-15.

[5] K. Liu and J. Makaran, "Design of a solar powered battery charger," 2009 IEEE Electrical Power & Energy Conference (EPEC), Montreal, QC, 2009, pp. 1-5, doi: 10.1109/EPEC.2009.5420817.

[6] Li, Chengliu, Wenyan Jia, Quan Tao, and Mingui Sun. "Solar cell phone charger performance in indoor environment." In 2011 IEEE 37th Annual Northeast Bioengineering Conference (NEBEC), pp. 1-2. IEEE, 2011.

[7] Fareq, M., M. Fitra, M. Irwanto, H. S. Syafruddin, N. Gomesh, S. Farrah, and M. Rozailan. "Solar wireless power transfer using inductive coupling for mobile phone charger." In 2014 IEEE 8th International Power Engineering and Optimization Conference (PEOCO2014), pp. 473-476. IEEE, 2014

[8] Li, Chengliu, Wenyan Jia, Quan Tao, and Mingui Sun. "Solar cell phone charger performance in indoor environment." In 2011 IEEE 37th Annual Northeast Bioengineering Conference (NEBEC), pp. 1-2. IEEE, 2011.

Intelligent model for reservoir operation through data mining-A Case study

V Ramani Bai Gopinath[1], **S Mohan**[2], **G Kangadharan**[3], **A Chun Kit**[1], **P Varadarajan**[4]

[1] Faculty of Engineering, Technology and Built Environment, UCSI University Kuala Lumpur, Malaysia
[2] *Environment and Water Resources Engineering, Indian Institute of Technology Madras, India.*
[3] Alpha Cambridge International School, Tiruchirappalli, India.
[4] *Cranes Building, Bharat Heavy Electricals Ltd., Tiruchirappalli, South India.*

Email: ramaniucsi@gmail.com

Abstract. Classical discrete mathematics could not model complex and uncertain systems effectively. These discrete mathematical equations can be replaced by a method of fuzzy logic systems wherein the operational laws are expressed in linguistic terms. Fuzzy logic means it is the form of knowledge acquisition to be suitable for notions that cannot be defined as precisely as discrete mathematics would do. But it depends upon their context, knowledge acquisition to enable the computerized devices to think like humans. The Matlab was employed to develop the operation model of the Vaigai reservoir which is in South India. Fuzzy logic control was successfully found to be a logical way for mapping between input and output variables of the operation of a reservoir. The objective of the study was to capture the experience gained over past years of operation of the Vaigai system so that the experiential knowledge was made available to the operator for assisting him inefficient management of the reservoir. The data mining helped to develop an improved database for heuristic models such as the non-crisp model for deriving an effective monthly operational policy for Vaigai reservoir.

Key words: Fuzzy models, Fuzzy logic toolbox, Vaigai reservoir, Fuzzy logic control, Takagi-Sugeno (TSK) fuzzy model.

1. Introduction

Water resources system modelling using scientific and analytical models are not the choice of engineers. This is mainly because of its ill-defined and uncertain information or the pattern existing amongst the variables. In contrast, a new intelligent approach namely, a fuzzy inference system applying "if-then-else" rules can be used to model the linguistic, human knowledge and thinking qualitative method. Moreover, this method does not require precise quantitative data, pre-process, and analyses. In fuzzy logic, the inference could be non-crisp and unclear information. For example, while the water demand is 'very high' i.e. in qualitative data form, the crisp value is essential for developing its inference system.

However, intelligent methods can be used to process this type of information as well. Most of the natural, environmental, applied, hydrogeological, sociological, biological, hydrological, and ecological processes are non-linear phenomena i.e. the relationship between the cause and effect of these processes is not clear and is of non-linear distribution. They used Stochastic Dynamic Programming (SDP) to frame the rule base in their approach. For the unsaturated zone, fuzzy control is used to represent water movement and seepage [1]. The triangle fuzzy classifier for storage intervals produces a more flexible interpretation technique than the trapezoidal classifier, which produces operating policies that mimic those defined by probabilistic stochastic dynamic programmes more closely [2]. In comparison to C5.0, NDT, and current Dam operating standards, the FIDs model has shown greater performance. This researcher utilised fuzzy logic to determine the best operating rules and allocations for the dam based on water demand [3].

The linear regression models are not sufficient to explain the uncertain and non-linear system's behaviours [4]. Artificial neural networks (ANN), Fuzzy Loci Controls (FLC), rule-based systems, genetic programming, and cellular automata are some of the most proven non-linear modelling tools and are used to a greater extent in natural process modelling than any others. For single flow control, a fuzzy rule-based equation was developed [5]. The fuzzy rule-based model gives only approximate values of the flow which can be estimated and interpolated with better accuracy than other mathematical methods. An irrigation system was modelled to estimate the system's hydraulic efficiency's optimum variables. The principal source of electricity in this system is a small-scale wind turbine, with the electricity being stored in a cell battery. A water pump is also included in the system to keep the water reservoir topped off. This pumping mechanism has the advantage of not requiring any maintenance [9].

The examination of questionnaires is a complicated procedure that must account for inherent subjectivity and ambiguity in data sets. A fuzzy signatures (FSigs)-based system was developed for analysing questionnaires with hierarchically connected (partially) unclear replies, and its usefulness a real-world problem was used to demonstrate this: a robot performing a partial analysis of current research [10]. The study aimed to use correlation, R and linear fitting regression coefficient to assess the influence between watermelon performance as well as the density of its primary pests, resulting in models for predicting crop performance based on pest density for better pest management. Watermelon is infested with a variety of pests that vary in formulation and intensity depending on the season [11].

Crop growth models of various levels of complexity were already developed to anticipate the impacts of water, soil, and nutrients and biofuel yields, as well as irrigation management, in a range of crops. For GY and BY, the Root Mean Square Error - RMSE and system performance was 0.87 to 0.96, 0.1 to 1.2, and 0.87 to 0.96, accordingly. As an outcome, the Aquacrop model is a useful tool for making agricultural development decisions for soybeans. [12]. A proportional integral was used to manage the fluid pressure in an uncertain system such as over-head tanks of non-axis symmetric objects. It does not, however, provide a precise result. Therefore, the fuzzy logic controller adds intelligence to the machine, allowing for a more accurate and effective reaction (FLC). The Kalman method, which uses fuzzy logic rules, automatically adjusts the state variables during system operation, and the controller is used to reduce error in poor circumstances.

A conical tank system is used to perform this strategy. A conical tank structure is used to perform this strategy. When using fuzzy regulators, this strategy appears to be effective, according to modelling findings [13]. It is described how a heuristic method was applied to monitor the bending strength of welds. A Mamdani-type fuzzy model was created with experimental results to predict the bending strength of welding was developed and tested in the lab. The fuzzy outputs for weld tensile strength and nugget hardness show errors of 2.13 and 1.31 per cent, respectively, when comparing to the experimental data. The developed methodology was used for real-time operation and assessment of the FSW operation within the limit of operational factors [14].

A fuzzy aided grey Taguchi technique was presented for numerous weld performance criteria. Using the fuzzy inference system, the Taguchi methodology in the early stage was employed for multi-response programming issues against its analogous issue. The input performance was altered for knowing how the outcome properties are affected. The proposed method can effectively be applied in the FSW process to maximize numerous weld quality attributes, according to a confirmation test [15]. The authors presented fuzzy operation of controllers which are self-programming and adjusting to get an optimal value [16].

An identity approach to address constantly changing operating conditions and poor ratios were modelled using other forms of fuzzy controls. The found by dynamic simulation and characteristics analysis and the grate cooler channel condition was generated through system modelling [17]. This paper uses a diary methodology to study the impacts of air pollution ratings on employees' routine physical resources and behaviour. According to multilevel data obtained over 2 metropolitan cities, affects workers badly [18]. The algorithm, and a parallel version, are detailed in the Fuzzy module of the free computer algebra software Sage Math. Solve Fuzzy System, or SFS, is an efficient technique for finding actual solutions of computation systems with symmetrical L-R fuzzy numbers as coefficients and constrained support and objective distribution functions are L and R. The real solutions of a system can be derived from the solutions of some polynomial systems with real coefficients [19].

It presented a numerical approach for generating this index as well as a way of computing them. When comparing two fuzzy sets, the comparison should be available to respond by ignoring the order. Two crisp relations with shared factors over a collection of classifiers were generated by the preferred density function. There exist a stringent order relation and the rest is a relation that reflects order indifference, based on the parameter value [20]. A distance in between the sum of upper semicontinuous fuzzy subsets of a finite-dimensional Euclidean space but also its T-convex hull has an upper limitation. As a result of this discovery, we describe an iterative approach for generating a T-convex hull of an upper semicontinuous fuzzy set. As an extension of the conclusions about t-norm fuzzy sets, we offer an example from a game-theoretic situation where t-norm fuzzy sets were utilised to handle uncertainty in a recurring two-player game [21].

2. Materials and methods

The methodology entails the fuzzy logic control (FLC) approach for tackling the problem of reservoir operation. To start with, the results of optimization were used for creating the input database of the model. Input data of Inflows (I), Initial Storage (S), Demand (D), and Release (R) were collected for the study period 1969-1997 and used in programming reservoir operation. The system has input and output engines. Hence the first task of preparing the data set was completed in this study. The logical steps involved in the development in the form of a diagram of the developed model development are shown in Figure 1. The model is to derive the operational policies for the reservoir system chosen. These systems also utilized the knowledge of experts and attempted to incorporate the imprecision and uncertainty associated with various aspects of developing rules of this integrated modelling system. The model was developed in Matlab with a Fuzzy toolbox. In this present work, explicit considerations on variability in inflow pattern, storage non-specificity, imperfection of the demand level, and fuzzy constraints on the releases were implemented in the concept of developing membership functions.

2.1. Methodology of Fuzzy Logic Model Development

The fuzzy logic system using linguistic terms is generally expressed in the form of logical implications, such as "if-then rules". The crisp constraints are fuzzified to non-crisp constraints in several water resource-related applications. But this approach is a usual refinement to conventional optimization techniques using fuzzy logic. "Tsebyschev Polynomial transformation was applied to transform the fuzzy constraints" in the study area chosen to develop a fuzzy inference system. MATLAB is a numeric computing environment hosting Fuzzy logic which has a collection of built-in functions for programming as per problem needs. The procedure for developing the fuzzy model for reservoir operation is as follows. i) Firstly, construct membership function for the variable; inflow, ii) membership function for the initial storage, iii) membership function for demand and iv) membership function for release, vi) formulation of fuzzy rules and implication. The operation of the reservoir has been represented by simulating a fuzzy model and its defuzzification of the output. The results are compared with that of the actual operation to check for suitability, adaptability, and applicability of the fuzzy logic model for reservoir operation.

The steps involved are as follows:

1. Fuzzification of Inputs

2. Fuzzy membership function derivation
3. Fuzzy Inference System (Rule Base)
4. Aggregation
5. Crisp values and output
6. Fuzzification of Inputs

For deriving operation policies in this study Type I Sugeno fuzzy inference system was used. It was attempted to show how the Sugeno type system gave freedom to incorporate a constant/linear system into a fuzzy system. The initial step of the fuzzy inferencing system is to select the input and evaluate the degree to which it will form a range of classes. Each class will then fall into different fuzzy sets for which membership functions will be developed.

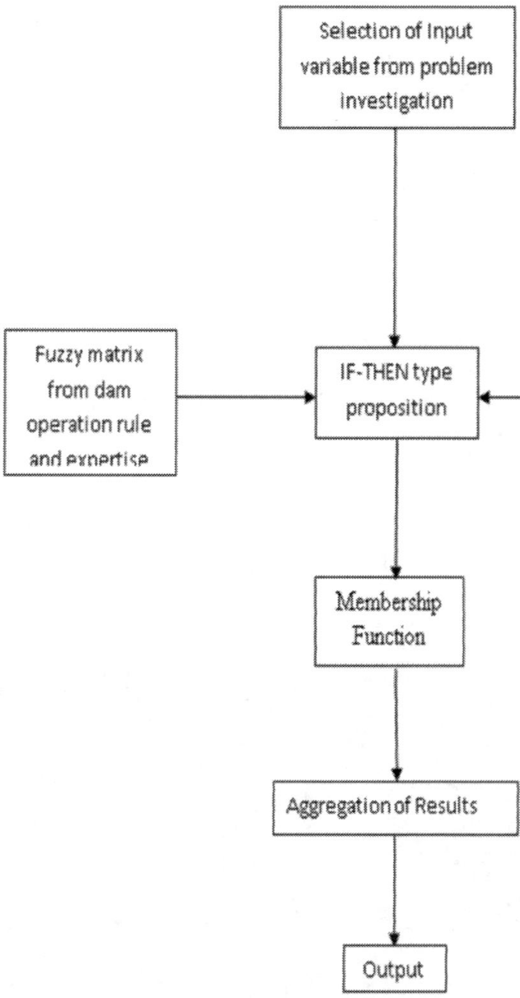

Figure 1. Flow Diagram of Reservoir Operation policy using FLC

The input parameters are grouped into many fuzzy linguistic sets of different types. The possible max and min size of the fuzzy variables (3-3-3-3, 3-5-3-3, 3-3-5-3, 3-3-3-5, 5-5-3-3, 3-5-5-3, 3-3-5-5 and 5-3-3-5) are chosen based on the testing of various topologies of fuzzy variables. When their performances of meeting the target demands were compared, 5-3-3-5 topology had given a better performance than other topologies. Hence the topology of the fuzzy model had been chosen as 5-3-3-5. The reservoir operation variables such as inflow, demand, initial storage, and release are considered in the current study. The fuzzy sets are chosen to be between the ranges of very low to very high. Some parameters are taken from low to high and some very low to very high according to their physical nature and the records. The model is developed using inputs that are a numerical value within the database and are crisps sets. For example, the June month interval was 1– 17 for inflow, 16 – 193 for storage, 33 – 43 for demand, and 7 – 52 for release in a year and vary over different months.

2.2. Fuzzy membership function derivation

It is mentioned that the fuzzy method is a more convenient way for modelling to accept the opinions of experts and is expected by the water operators [6]. A new type of Takagi-Sugeno-Kang fuzzy algorithm, known as the Takagi-Sugeno (TSK) fuzzy model, was used to create a fuzzy model for reservoir operation. The TSK fuzzy model combines linguistic and mathematical regression analysis. The TSK model's primary goal is to create fuzzy models for measured data automatically. To construct a fuzzy engine and aggregate the corresponding results, the model is used to release, inflow, demand, and storage [7]. As a result, this study relied on triangular membership functions. After fuzzification of input data, the degree of belongingness of the antecedent (membership function) may be determined [8].

2.3. Fuzzy logic modelling

Choice of membership function: The predictor's main feature was a sharp part of the subsequent rule. In general, the fuzzy logic rule base in Takagi Sugeno coding looks like this (Takagi and Sugeno 1985):

If ‘xx is “AA and ‘yy is “BB then ‘zz = “f (xx, yy) (1)

In the antecedent, A and B were fuzzy sets, while z = f(xx, yy) was a crisp function in the consequent. In most cases, f(xx, yy) was a polynomial in the xx and yy input variables. The model is a first-order polynomial as in Eq.1. This study used the fuzzy logic inferencing method for a linear degree of the Takagi-Sugeno fuzzy model with crisp inputs.

The crucial modelling fact in fuzzy logic is that this approach permits some to be partial and others to be other partial; rather than saying either everything or nothing; and thus, the degree of “fitting” to a particular fuzzy set could be quantitatively by a membership ranging from 0 - 1.0. Hence the definition for fuzzy sets for inflow, initial storage, demand, and release was obtained through historical data instead of expert’s opinion for reservoir operation and are shown in Figure 2.

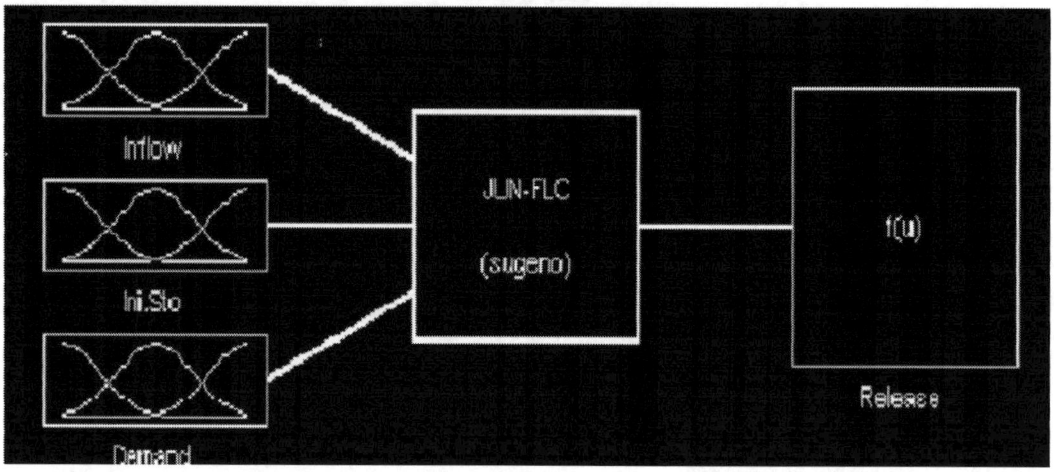

Figure 2. Fuzzy Model on reservoir operation

Fuzzy set membership functions are plotted using many different functions. The simple straight-line forming the triangular functions are shown in Figure 3- Figure 5. They are the simplest possible and are often preferred for water resources and environmental applications. In this problem, a statistical method was adopted to discretize the variables into different fuzzy sets. Standard deviations of the database were calculated. The average deviation of each set formed the pivotal value and from which further classifications of the variables were done on both sides. For example, in the discretization of the fuzzy set for initial storage, 0 was low, 112 was medium and 193 was high. "But a class size of, say 50 is in middle, i.e., it is partly belonging to two categories "low" and "medium". In fuzzy system terms, 50 had a value of 0.4 of low, 0.65 of the medium, and 0.0 of high." Similarly, other variables as inflow, demand, and release were fuzzified and are shown in Figure 3 to Figure 6.

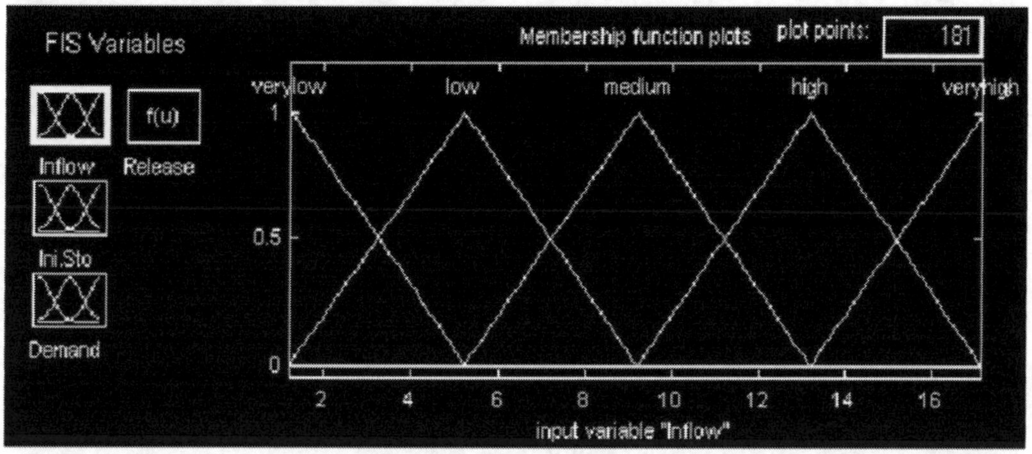

Figure 3. Membership Function for Inflow

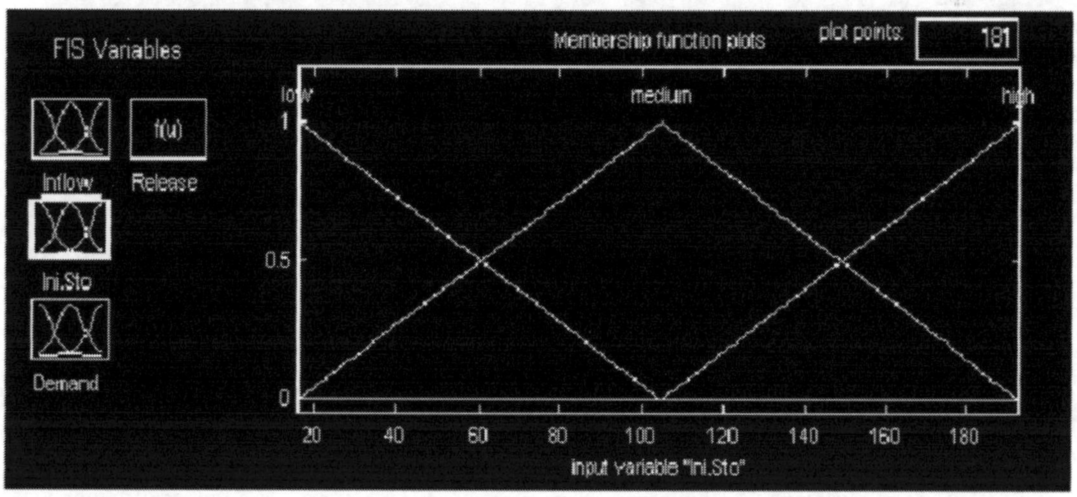

Figure 4. Membership function for Initial Storage

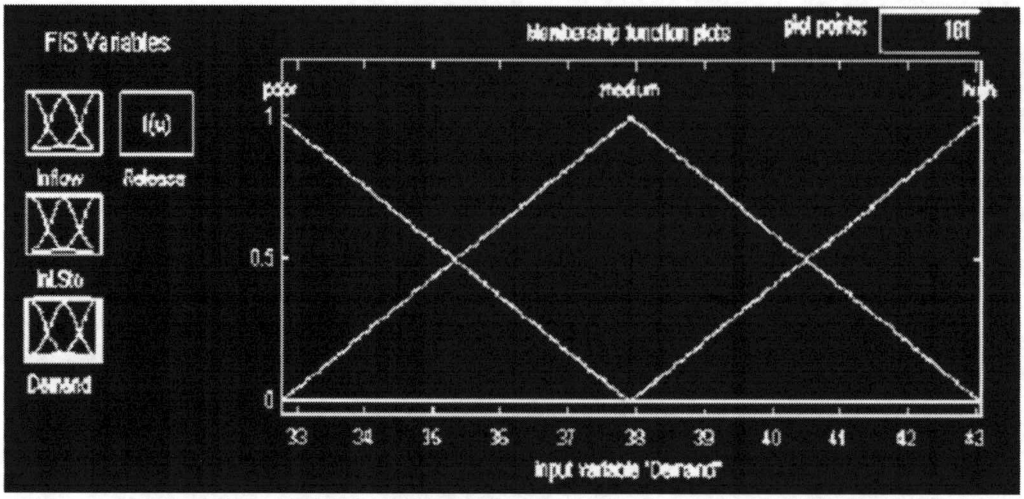

Figure 5. Membership function for Demand

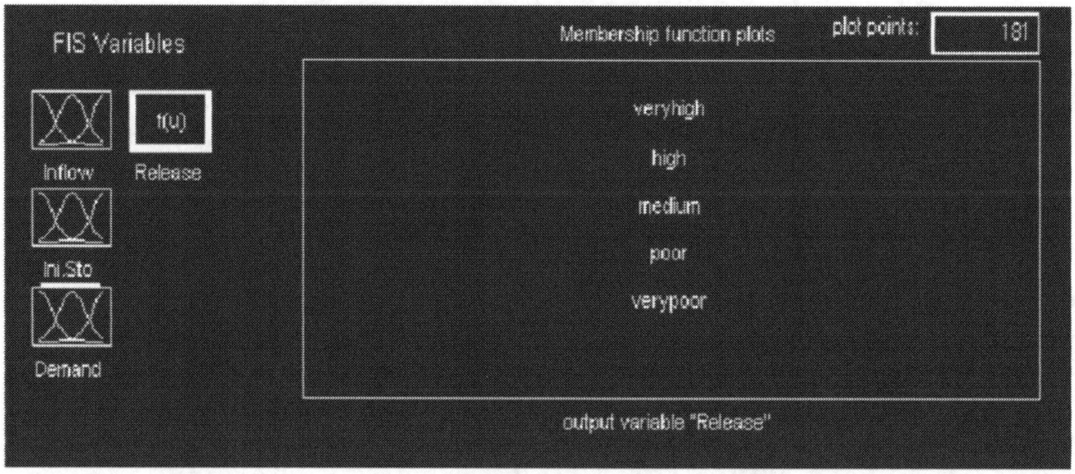

Figure 6. Membership Function for Release

2.4. The Fuzzy Inference System

Imprecise data can be modelled conveniently and diagnosed suitably in the fuzzy logic method. The operating policies developed using the Linear-programming (LP) model was used for deriving the fuzzy rule base of the system. All smaller diagrams of all fuzzy sets are used to compile for a fuzzy inference diagram. Figure 2 shows the model of the fuzzy inference system which was used in deriving the present reservoir operation policies. It could examine simultaneously all parts of the fuzzy inference process automatically. This had all the input membership functions as inflow, initial storage, and demand on one side, output membership function on the other side with fuzzy inference engine at the middle. Information flew through the fuzzy inference diagram. For example; from Figure 2, the particular inputs as inflow, initial storage and demand have helped to easily see that the minimum function was the truncation rule. The pairs of input-output sequences generated from the simulation are employed to derive the fuzzy logic rules. The rules were formed based on the pairs of input-output data sequences from the actual historic operation of the reservoir and the expert's opinion on the values of fuzzy sets. The error seemed to decrease when the membership functions for input parameter and output parameters is slowly increased. Eventually, the rule base also got changed due to this effect. The number of rule bases had increased as the number of membership functions was increased. Hence to achieve an optimal solution from the model, a trade-off was executed. The number of derived rules and their membership functions complement each other to achieve near-optimal solutions from the model.

3. Development of Fuzzy Rules

Many decisions are made by humans and hence it is necessary to use expert's opinions if a realistic model is to be constructed of a process. For capturing the knowledge and experience of senior professionals and making use of the expert's wisdom in the model a questionnaire survey was conducted for developing the rules. The methodology for building the fuzzy rules was dependent on the linear programming model and of the expert's opinion. A questionnaire survey was conducted, and the results obtained from 35 responses were aggregated suitably for developing fuzzy rules. In this study 45 rules, several different fuzzy sets were developed by resolving the inputs of each rule. For example, "inflow was very low", "inflow was low", "inflow was medium", "inflow was high", "inflow was very high", "initial storage was low", "ínitial storage was medium", "initial storage was high" and so on.

Fuzzy operation rules were constructed for all twelve months. The inputs were fuzzified based on each linguistic set before the rules could be evaluated. For example, what would be the extent of the inflow to be very low? Figure 3 shows how well the inflow into the reservoir qualified, in June (via its membership system function). In this case, the inflow rated as 9, which, had given graphical definition of low, corresponds to a degree m =0.6 for the "low" membership function. The counterpart to the operation manager will be "the inflow was low. Likewise, each set of input was studied fuzzified against the qualifying membership functions that are required by the rules. The complete sets of fuzzy system rules developed for reservoir operation are presented in the following sections.

4. Results and discussion

The fuzzy operator was chosen carefully to result in one value that represented the consequence of any rule applied. Since the antecedent of any rule has more than one output it should result in any one number. That number became an output function in the system.

4.1. Aggregation method

The output is always one true value, whereas the fuzzy operators are always few or many values from fuzzified input factors. All of the designed rules made use of the AND operation. The model, for example, has evaluated the antecedent of rule 17 and for release computation in the following month, June month. The fuzzy membership scores 0.65, 1, and 0.5 were obtained from the three different elements of the antecedent (low inflow, low beginning storage, and medium demand). The fuzzy AND operator simply chose the lowest of the three values, 0.5 (the release was very low), and therefore fuzzy logic simulation finished operation rule number 17. Similarly, all the other rules were evaluated using the AND operator. The complete sets of fuzzy system rules developed for reservoir operation are shown in Table 1.

Table 1. Fuzzy Rules Developed for Operation of Vaigai Reservoir

Rule No.	Fuzzy Rules for Vaígaí reservoír operatíon
1	Íf ÍW ís very low & S ís low & D ís low then R ís very low
2	Íf ÍW ís very low & S ís low & D ís medíum then release ís very low
3	Íf ÍW ís very low & S ís low & D ís hígh then release ís very low
4	Íf ÍW ís very low & S ís medíum D ís low then release ís low
5	Íf ÍW ís very low & S ís medíum & D ís medíum then release ís low
6	Íf ÍW ís very low and S ís medíum and D ís hígh then release ís low
7	Íf ÍW ís very low and S ís hígh and D ís low then release ís low
8	Íf ÍW ís very low and S ís hígh and D ís medíum then release ís low

9	If IW is very low and S is high and D is high then release is low
10	If IW is low and S is low and D is low then release is very low
11	If IW is low and S is low and D is medium then release is very low
12	If IW is low and S is low and D is high then release is very low
13	If IW is low and S is medium and D is low then release is low
14	If IW is low and S is medium and D is medium then release is low
15	If IW is low and S is medium and D is high then release is low
16	If IW is low and S is high and D is low then release is medium
17	If IW is low and S is high and D is medium then release is medium
18	If IW is low and S is high and D is high then release is medium
19	If IW is medium and S is low and D is low then release is low
20	If IW is medium and S is low and D is medium then release is low
21	If IW is medium and S is low and D is high then release is low
22	If IW is medium and S is medium and D is low then release is medium
23	If IW is medium and S is medium and D is medium then release is medium
24	If IW is medium and S is medium and D is high then release is medium
25	If IW is medium and S is high and D is low then release is medium
26	If IW is medium and S is high and D is medium then release is medium
27	If IW is medium and S is high and D is high then release is medium
28	If IW is high and S is low and D is low then release is medium
29	If IW is high and S is low and D is medium then release is medium
30	If IW is high and S is low and D is high then release is medium
31	If IW is high and S is medium and D is low then release is medium
32	If IW is high and S is medium and D is medium then release is medium
33	If IW is high and S is medium and D is high then release is medium
34	If IW is high and S is high and D is low then release is high
35	If IW is high and S is high and D is medium then release is high
36	If IW is high and S is high and D is high then release is high
37	If IW is very high and S is low and D is low then release is medium
38	If IW is very high and S is low and D is medium then release is medium
39	If IW is very high and S is low and D is high then release is medium
40	If IW is very high and S is medium and D is low then release is medium
41	If IW is very high and S is medium and D is medium then release is medium
42	If IW is very high and S is medium and D is high then release is medium
43	If IW is very high and S is high and D is low then release is high
44	If IW is very high and S is high and D is medium then release is high
45	If IW is very high and S is high and D is high then release is high.
Note: IW-Inflow, S- initial Storage, R- Release and D-Demand	

4.2. Crisp values and output

All three above rules of the antecedent part had been placed together to combine and aggregate the output of each into a single crisp set. FLC stands for Fuzzy logic control was found to be a convenient way to couple an input domain to an output space. The Takagi-Sugeno-type fuzzy inference system was used in this model. The fuzzy model was validated for a system of continuity equations. Its physical or boundary conditions for the reservoir operation were also incorporated in this model using fuzzy logic control. The fuzzy logic model had enhanced the generalization capability of the system by providing more reliable output when extrapolation was needed beyond the limits of the data which were used for developing them. The factors that contributed to the performance were the objective function, constraints involved, and discretisation of variables such as storage and inflows.

It was found that the fuzzy logic approach proves to be a reliable tool in analysing problems like reservoir operation, like LP, DP, and SDP alone were not essential for this approach. It was very well demonstrated that the fuzzy logic approach is a promising alternative for reservoir operation modelling among the methods discussed in previous sections.

4.3. Validation of the model

A new set of data pattern which was not used for training the Fuzzy logic reservoir operation model was subjected to the validation process. The new data set were presented to the trained FIS model, to see how well the FIS model predict the corresponding output values. The period of this validation was from 1994 -'95 to 1997-'98. The releases obtained through the FIS model were plotted against actual releases. The plot is shown in Figure 7.

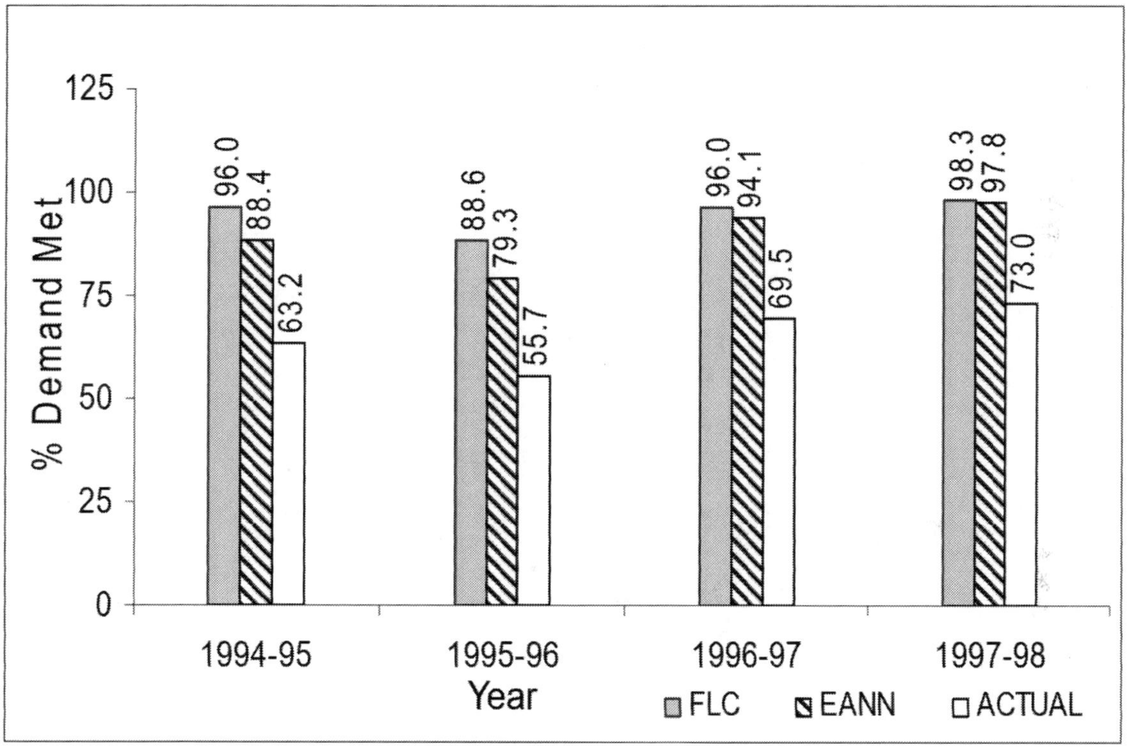

Figure 7. Performance of Fuzzy model for reservoir operation

Fuzzy releases have adequately met the demand thorough out the year whereas there were excess releases in actual practice. The modelling approach used for developing reservoir operation was much similar to many system identification techniques, but the case study model developed would support the functioning of the operation to a great extent. Further, the performance criteria such as the amount of deficit and percentage of deficit were estimated to validate the model. The results of this validation and the selection of a fuzzy model for reservoir operation are exhibited in Table 2.

Table 2. Comparison of the Performance of FLC Model with Other Models of Reservoir Operation

Method	Period	1994-'95	1995-'96	1996-'97	1997-'98	Average Deficit (%)
ACTUAL	Amount of deficit in Mm3	236.9	365.7	178.9	145.9	
	% deficit	36.8	44.3	30.5	27.0	34.6
LP	Amount of deficit in Mm3	231.4	253.5	184.2	127.7	
	% deficit	35.9	30.7	31.3	23.6	30.4
EANN	Amount of deficit in Mm3	74.7	170.7	34.4	12.0	
	% deficit	11.6	20.7	5.9	2.2	10.1
FLC	Amount of deficit in Mm3	25.1	94.0	22.7	9.3	
	% deficit	3.9	11.4	3.9	1.7	5.2

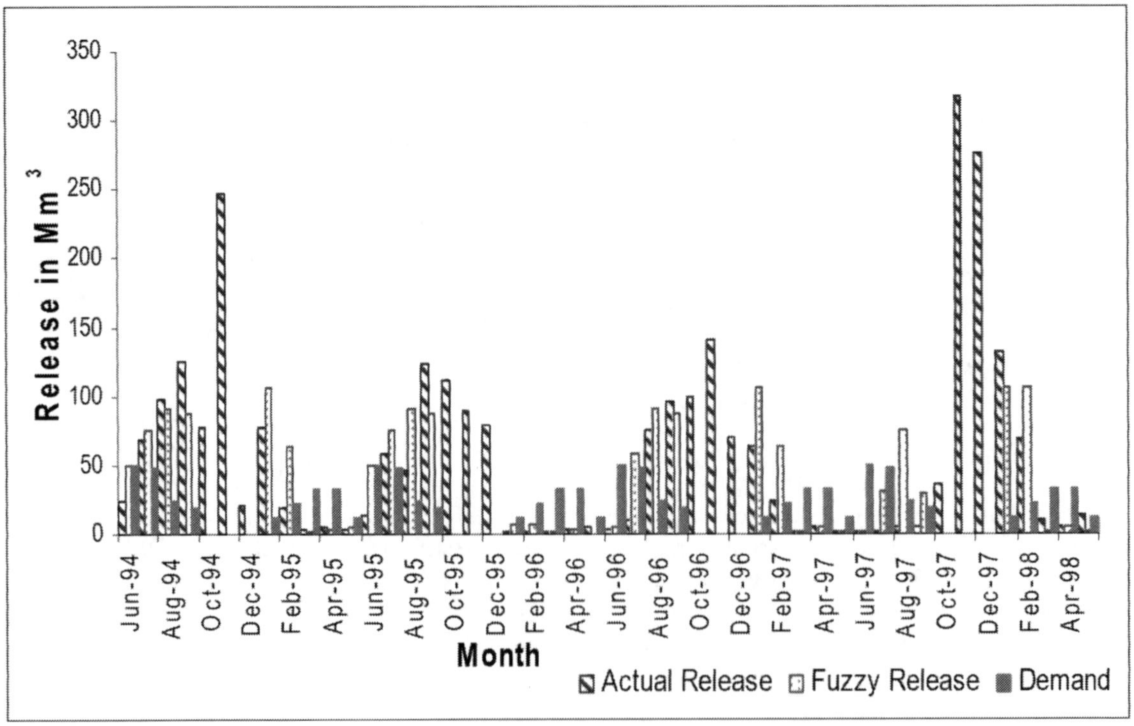

Figure 8. Comparison of performance of Fuzzy model and ANN model

The results had shown a good improvement in meeting the demand as compared with the artificial neural network model, EANN, and actual historic operation and as seen from Table 2. In addition, the percentages of demand met by these models during the validation period were compared in Figure 8. This encouraged the selection of the Fuzzy model over EANN. This has accomplished to state that this type of fuzzy model will provide guidelines for future studies. The developed model can be extended to construct a fuzzy inference system that operates between several optimal controllers. It can also be transformed into a highly nonlinear model within its operating space.

The main improvement in FIS/ FLC systems is its intelligence to imprecise data and tolerance towards insufficient information about the data. FLC can model uncertain functions of mere or arbitrary complexity as well. Hence fuzzy logic is well recommended for developing models on reservoir operation. The next approach may be the modelling of reservoir operation using Neuro-Fuzzy logic which is an integrated model of two heuristic methods namely artificial neural network and fuzzy logic system.

5. Conclusions

The developed FLC model for reservoir operation has produced better results than the system's current operational procedure. For the Vaigai reservoir in South India, FLC was developed to represent reservoir operation by carefully extracting knowledge through data mining. The consequences of a proper set of decisions or set of operating rules were evaluated using detailed data mining on developed models. The following was discovered when the results of data mining and the corresponding results were compared across various modelling methods. The fuzzy reservoir operation model produced a 5% average deficit, whereas EANN produced a maximum of a 10% average deficit. It demonstrates that fuzzy logic is more robust in processing uncertainty than artificial neural networks, with a 5 percent advantage over ANNs.

Heuristic models prescribe procedures that best achieve a pre-defined purpose, even though they describe the system in a more simplified form. They serve as starting points for a more in-depth investigation. These codes can be used in a variety of ways and with a variety of modifications. One of them will be used for the input data in the form of a variable, i.e. points generated before in the model. This will increase the operational flexibility of the proposed model. Traditional methodologies demonstrated to be a better way for extracting knowledge from data for creating optimal operational policies for the reservoir system using available data and information when reinforced by techniques such as neural networks and fuzzy logic. The model's success in capturing knowledge in the reservoir operating domain highlighted a promising study field. As a result, the inquiry will be used to build a method for dealing with different types of fuzzy and neuro-fuzzy systems, as well as combining them with other soft computing techniques such as artificial neural network systems.

Acknowledgements

The authors are very grateful for the research grant provided by the Public Works Department, Government of Tamil Nadu, South India and the Centre for Industrial Consultancy and Sponsored Research, IIT Madras, South India for pursuing the research study. The authors also thank the computer facilities for the data mining research rendered by the IT department of Indian Institute of Technology, Madras, South India.

References

[1] Bardossy, A. and L. Duckstein. Analysis of a Karstic Aquifer Management Problem by Fuzzy Composite Programming, *Wat. Res. Bullet.*, 1992; **28**, 63-73.
[2] Maqsood I, Huang G, Huang Y and Chen B. ITOM: an interval-parameter two-stage optimization model for stochastic planning of water resources systems, *Stochastic Environmental Research and Risk Assessment*, 2005; **19**, 125-133.

[3] Kindler J 1992 Rationalizing water requirements with the aid of fuzzy allocation model, *ASCE Journal of Water Resources Planning and Management*, **118,** 308-323.

[4] Mays L W and Y K Tung 1992 *Hydro systems Engineering and Management*. McGraw-Hill, New York, p. 23.

[5] Panigrahi D P and P P Mujumdar 2000 Reservoir operation modelling with fuzzy logic. *Water Resource Management*, **14**, 89-109.

[6] Russell S O and P F Campbell 1996 Reservoir operating rules with fuzzy programming, *ASCE J. of Wat. Resour. Plan. and Manage.*, **122**, 165-170.

[7] Takagi T and M Sugeno 1985 Fuzzy identification of systems and its applications to modelling and control, IEEE *Trans. On sys., Man, Cyber.*, **15**, 116-132.

[8] Wei C C and Hsu N S 2008 Derived operating rules for a reservoir operation system: Comparison of decision trees, neural decision trees, and fuzzy decision trees, *Water Resour. Res.*, **44**, 1-11.

[9] Chen Y, Chen Y R and Ma W B 2011 Land Circulating and Concession Willingness Analysis of Residents Resettlement for the Reservoir based on Logistic Model - Survey for Residents Resef settlement in Sichuan, Hunan, Hubei, *Resource Science*, **33**, 1178-1185.

[10] Koczy L T, Susniene D, Purvinis O and Zostautiene D 2020 Analyzing employee behavior-related questionnaires by combined fuzzy signature model. *Fuzzy Sets and System,* **395**, 254-272.

[11] Okrikta E, Ogunwolu E O and Odiaka, N I 2021 Modelling the Impact of Key Pests of Watermelon on its Performance Using Linear Regression Models. *Walailak Journal of Science & Technology*, **18**, 9052-11.

[12] Meysam Abedinpour 2017 Improving Crop Production by Field Management Strategies using Water Driven Crop Model. *Walailak Journal of Science & Technology*, **14**, 865-874.

[13] Tamilselvan G M and Aarthy P 2017 Online tuning of fuzzy logic controller using Kalman algorithm for conical tank system. *Journal of applied research and technology*, **15**, 492-503.

[14] Senthil Kumar S, Boopathi M, Kanish T C and Srivani A 2017 Fuzzy Logic Model for Combined Monitoring of Weld Strength and Nugget Hardness of Friction Stir Weld. *Walailak Journal of Science and Technology*, **14**, 377-388.

[15] Parida B, and Pal S 2015 Fuzzy assisted grey Taguchi approach for optimisation of multiple weld quality properties in friction stir welding process. *Science and Technology of Welding and Joining*, **20**, 35-41.

[16] Muniandy V, Samin P M, and Jamaluddin H 2015 Application of a self-tuning fuzzy PI–PD controller in an active anti-roll bar system for a passenger car. *Vehicle System Dynamics*, **53**, 1641–1666.

[17] Wang Z, and Yuan M 2012 A self-tuning fuzzy PID control method of grate cooler pressure based on Kalman filter. *In International conference on computer science and information processing (CSIP)*, IEEE, 257–260.

[18] Fehr R, K Ch Yam, W He, J T-J Chiang, W Wei 2017 Polluted Work: A Self-Control Perspective on Air Pollution Appraisals, Organizational Citizenship, and Counterproductive Work Behaviour, *Organizational Behaviour and Human Decision Processes*, **143,** 98–110.

[19] Aubry P, Marrez J, and Valibouze 2020 A. Computing real solutions of fuzzy polynomial systems. *Fuzzy Sets and Systems*, **399,** 55-76.

[20] Dombi J, and Jonas T 2020 Ranking trapezoidal fuzzy numbers using a parametric relation pair. *Fuzzy Sets and Systems*, **399**, 20-43.

[21] Bielawski J, and Tabor J 2021 Convex hull of a fuzzy set and triangular norms. *Fuzzy Sets and Systems*, **417**, 93-109.

DESIGN AND DEVELOPMENT OF AN EFFECTIVE SMART GARBAGE SYSTEM USING THE INTERNET OF THINGS

Vanitha.V1, Sangeerth Raj CV 2, Mithun Kumar KK 2, Joseph Sebastian 2

[1] Assistant Professor (GII), [2] UG Student, Department of Electronics and Communication Engineering, Aarupadai Veedu Institute of Technology, Vinayaka Mission's Research Foundation, Deemed to be University, Tamil Nadu, India.

E-mail: vanitha.ece@avit.ac.in

Abstract, Garbage management has become the most pressing challenge in today's world, particularly in countries with significant population growth rates. Excess of gas can be formed in the environment which causes production of greenhouse gas in the environment. We employed this technology to monitor the level of waste in the garbage, as well as fire and gas emissions. When there is a fire or a significant volume of hazardous gas in the waste. The location will be updated in the web and the buzzer will give an alert. The whole process used to reduce greenhouse gas in the environment. And also, it detects the level of garbage, when the level is high it shares the location to IOT.

Key words: IOT, Smart City, Garbage Monitoring

1. Introduction

The Internet of Things (IoT) has quickly become the most popular invention of our time. IoT is a technology that allows all physical devices to communicate with one another over a wireless network connection with minimal human intervention in order to maintain a clean environment. In some areas the toxic gas produced by the waste are high, to reduce gas formation and fire detection system used to reduce production of greenhouse gas in the environment. By 2030, many portions of the cosmos will have evolved and become more populous, resulting in a rise in rubbish production. As a result, the execution of this research could be beneficial in resolving this issue. This study also suggests a route tracking system that will show the truck's journey while collecting rubbish and visiting the most overflowing bins. This paper helped to reduce travel and helped in reducing the use of fuels will be saved[1]. This research presents an air pollution and monitoring model that uses a data mining method to detect pollution in the air. Sensor values from many gas sensors are detected using the sensor grid. The values from the ADC are transferred to the server using a microcontroller. The controller is connected to the client through Bluetooth, and the client communicates with the server using web services[2]. This system architecture created remote detection terminal, control master station, and mobile monitoring terminal communication protocol based on Zigbee technology, GSM technology, and supplemented by the master-slave wireless network. The site environment and gas concentration are detected via a remote monitoring terminal. Scene conditions and gas concentration state are detected using remote sense terminals. The control station is used to

Journal of Physics: Conference Series **2040** (2021) 012033 doi:10.1088/1742-6596/2040/1/012033

manage the main station's connection to the network of remote detection terminal data, as well as to provide timely alert information to your phone through GSM module monitoring terminal. In addition, data can be transferred to a computer monitor server via the serial port, allowing for data analysis and control of each remote terminal[3].

2. Design Analysis

In the current system, when a fire or dangerous gas is detected, the existing system uses a buzzer to inform the user. To transmit the data, they employ a Bluetooth module. The existing technology uses Bluetooth and can only be used for a certain distance, which is a disadvantage. In this work, we use IoT to send the data of the garbage to the corporation. The location of the garbage also sends through the IoT. Also, the buzzer will be beep when the fire is detected. It detects the level of the garbage. The benefit of this work is that when the location is discovered, immediate action will be conducted, and the gas and temperature values will be updated in real time.

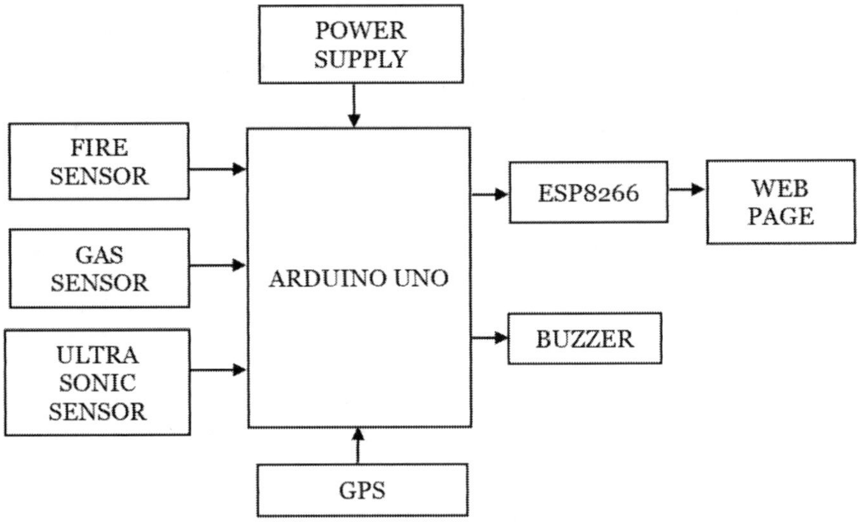

Figure 1. System Architecture

2.1 Requirements & Circuit Arrangement

Hardware used for implementation of the proposed work are Arduino Uno, Power Supply, Ultra-sonic Sensor, Fire Sensor, Gas Sensor, GPS Module, ESP8266 Module and Buzzer. Software utilized for the execution are Arduino IDE, Embedded C and PHP.

Figure 2. Circuit Arrangement

2.2 Working Plan

In this project we are developing the smart garbage system for the smart City, garbage system means you need to maintain the level of garbage and you need to check the Garbage weight of the dustbin and need to check the fire and gas level in garbage system. If u go for smart City means we need to update data of garbage like dustbin into the cooperation, so while using IOT if you provide internet to IOT means it will always store data from the dustbin level like gas fire weight of garbage level etc

For fire sensor, if the fire is detected means some door is going to open in dustbin it is why because in every smart City the dustbin is fully covered with door. Whenever we go to dustbin, we can open the door and put the dust in dustbin. When the dustbin is on fire the door will automatically open. Weight cell is there if we use ultrasonic sensor the best idea for it will call the level of dustbin and we can know the dustbin is full or empty for that we are using weight sensor.

3. Results and Discussion

After installing hardware and software, System ready for the trial working checkup, checking with gas sensor fire sensor and all other software working operations. All sensors are working smoothly, cloud storage and software running smoothly.

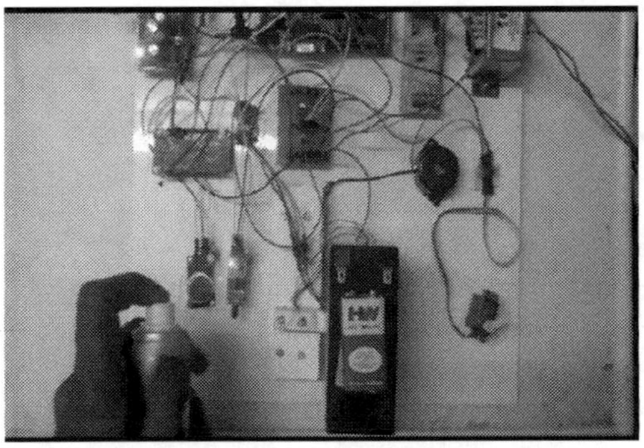

Figure 3. Gas sensing operation

Figure 4. Fire sensing operation

4. Conclusion

Finally, we completed the working prototype of the smart garbage system using IOT. All sensors and other working operation go smoothly. This small idea will help to build a smart garbage system in every developing smart city.

References

[1] Velladurai, V. S., et al. 2018 Human safety system in drainage, unused well and garbage alerting system for smart city, International Conference on I-SMAC .
[2] Raipure et al 2014 Wireless sensor network-based pollution monitoring system in metropolitan cities, Proceedings on International Conference on Communications and Signal Processing..

[3] Jian, Fang et al 2015 Harmful gases wireless network monitoring system design, Proceedings on International Symposium on Computer, Consumer and Control.
[4] Han, Fei Ling, et al.2018 An Internet of Things environmental monitoring in campus, Proceedings on International Conference on Intelligent and Advanced System.
[5] Fan, Yufeng, et al. 2018 Design and application of toxic and harmful gas monitoring system in fire fighting. Sensors, Vol.19, Iss.2.
[6] R. Al-Ali et al 2010 A Mobile GPRS-sensors array for Air Pollution Monitoring,Vol.6.
[7] Nihal Kularatna et al 2008 An Environment Air Pollution Monitoring System Based on the IEEE1451 Standard for Low-Cost Requirements, Journal of IEEE Sensors,Vol. 8.
[8] M. Abu Jayyab at al 2006 Pollumap: Air Pollution mapper for cities, Proc. IEEE Innovations in Information Technology Conference.
[9] Y. J. Jung et al 2008 Air pollution monitoring system based on geosensor network,Proc. IEEE Int. Geoscience Remote Sensing Symp,Vol.3.
[10] M. Gao et al Environmental monitoring system with wireless mesh network based on Embedded System,Proc. 5th IEEE Int. Symp. Embedded Computing.
[11] J. W. Kwon et al 2007 Design of Air Pollution Monitoring System Using ZigBee Networks for ubiquitous-city, Proceedings of In. Conf. Convergence Information Technology.
[12] H. Huang at al 2011 A Greenhouse Remote Monitoring System Based on GSM,Proc. of IEEE International Conference on information management.
[13] S.Mahendiran et al 2021 Ultra Reliable Low Latency Communication Technique for Agriculture Wireless Sensor Networks, Arabian Journal of Geosciences,Vol. 14,1246.
[14]N.Purnendu Shekhar Pandey et al 2021 A Smart Parking system based on IoT Technologies, Linguistica Antverpiensia,Vol.2.
[15]Shafiqul Abidin et al 2021 Development and Organize of Wireless Sensor Network in Home Management using IoT, International Journal of Aquatic Science, Vol.12, Issue.2.
[16]D.Vijendra Babu et al 2019 Arduino based Automatic License issuing system, International Journal of Engineering and Advanced Technology, Vol.8, Issue 6S3.
[17]V.Deepak et al 2021 An Efficient Performance Analysis using Collaborative Recommendation System on Big Data,5th International Conference on *Trends in Electronics and Informatics.*
[18] V.Chandrasekar et al 2021 Designing of Tele-Health Smart Sensor Device to assist Home care Staff, Journal of Physics: Conference Series, Vol.1964, 062106

Solar power based positive output super-lift Luo converter using fuzzy logic controller

K Pavithra[1], H Pooja[1], D Tamilselvan[2], T D Sudhakar[3]

[1]UG Student, St.Joseph's College of Engineering, Chennai, Tamil Nadu, India
[2]Research Scholar, M.I.E.T Engineering College, Trichy, Tamil Nadu, India
[3]Professor, St.Joseph's College of Engineering, Chennai, Tamil Nadu, India

E-mail: pavipoornik@gmail.com,

Abstract. Photovoltaic (PV) power generation is employed to meet the increasing demand for energy and of cleaner form. Though renewable and high energy using PV can be produced, they have certain disadvantages. PV cells have a low voltage (approx. 0.5V) rating and they have to be connected in series when higher voltage is required. These cells when connected in series should have identical electrical characteristics to avoid ripples in the output voltage and current, but it is not possible practically. To avoid this miss-match, converter circuits are employed. The major drawbacks in most power converters are reduced voltage gain and their tendency to generate harmonics in the supply system and the load circuit. To overcome these limitations, a positive output super-lift Luo converter is designed. A Positive Output Super Lift Luo converter (POSLC) is a powerful DC-DC converter where the voltage is converted from positive source voltage to positive load voltage as it produces positive voltages of comparatively higher ranges than those of conventional types. The super-lift technique also overcomes the effect of parasitic elements and thus minimizing the ripples in the output voltage and current. In addition to this, the voltage build-up can be achieved by implementing the super-lift technique where the output voltage rises in geometric progression with increased voltage transfer gain and high power density. In this paper fuzzy control is used to produce better-controlled voltage and a more refined output. The performance of POSLC with and without implementation of fuzzy control has been successfully simulated and verified using MATLAB/SIMULINK as well simulation model also developed and analyzed with Proteus package.

Keywords: DC – DC converter, High gain, POSLC, Fuzzy Logic Control

1. Introduction

In recent trends, in various industries, DC-DC conversion plays a significant role for various applications and there are various types of DC-DC conversion techniques to improve the voltage gain as well as efficiency [1]. In DC-DC conversion step-up chopper boosts the output voltage whereas step down chopper bucks the output voltage. Due to a lot of research in DC-DC conversion more topologies developed, generally, these topologies are categorized into with and without transformer. The transformerless DC-DC conversion has led to a reduction in the size, weight, cost and losses [2]-[5]. There are many voltage building techniques adopted in conversion stages, namely, voltage lift technique and super- lift technique in the case of Luo DC – DC converters. Luo converters developed from the fundamental DC – DC converter which has a simple and cheap topology as well provide high voltage gain, better efficiency and lowest ripple at the output. In the voltage lift technique, the output voltage is increased step by step in arithmetic progression. However, in the super-lift technique, the output voltage increases geometrically. Two types of super-lift techniques are Positive Output Super-lift Luo Converter (POSLC) and Negative Output Super-lift Luo Converter (NOSLC).

This paper deals with different performances of POSLC that build up the input voltage to much higher levels compared to that of other voltage building techniques [5]. Basically, in POSLC, the output DC voltage is increased step by step in geometric progression. Positive Output Super-lift Luo Converter is constructed to overcome the major drawbacks of conventional power electronic converters such as reduced DC voltage gain, harmonics, and low power factor. In POSLC, much higher voltage ranges can be obtained at a lower gain value of about 0.8. In addition to this, ripple content in the output voltage and current can be reduced and thus higher efficiency can be achieved at the output [6]. Low switching loss, large DC gain, continuous input current and higher power density are its other advantages.

In general PI, PD and PID controllers are predominantly used to obtain the precise output voltage in DC – DC conversion but in recent trends, Fuzzy Logic Controllers successfully adopted closed loop control of the conversion stage, and this method implemented in this paper to provide the output feedback to regulate the output voltage as expected.

2. Operation of POSLC

The simulation circuit diagram of the POSLC is given in Fig.1.The circuit consists of a MOSFET switch S, Inductor L_1, Diodes D_1 and D_2, capacitors C_1 and C_2, and a load resistance R. The other circuit components are Pulse generator, input DC voltage V and Scope output. POSLC produces the particular output voltage for input by introducing pulse width modulation techniques accordingly [7].

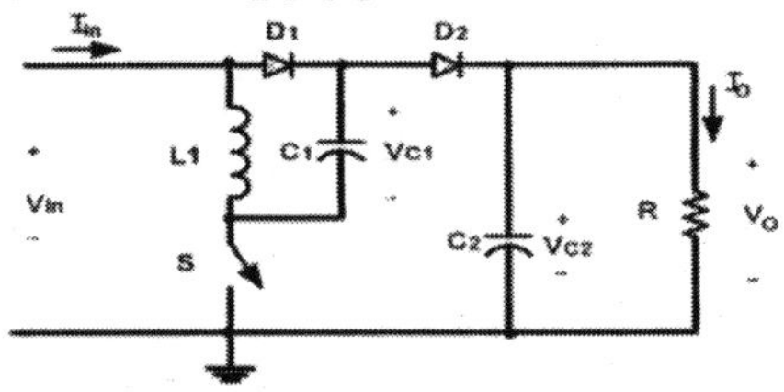

Figure 1. Circuit Diagram of POSLC

The design values for the proposed positive output super lift Luo converter [8] are derived based on the considerations as follows:
The input voltage (V_s) to the POSLC is given as 12V which is provided by a constant DC source. The output voltage (V_o) for the load resistance R of 100Ω is obtained as 36V.

The formula for transfer gain (K) is expressed as,

$$\frac{V_o}{V_{in}} = \frac{(2-K)}{(1-K)}$$

Rewritten as,

$$V_o - V_o = 2V_{in} - V_{in}K$$

$$V_{in}K - V_oK = 2V_{in} - V_o$$

Thus,

$$K = \frac{2V_{in}-V_o}{V_{in}-V_o}$$

$$K = \frac{(2\times12)-36}{12-36}$$

$$K = 0.5 \approx 50\%$$

The output current I_o is obtained from Ohm's law as,

$$V_o = I_oR$$

Rearranged as,

$$I_o = \frac{V_o}{R}$$

$$I_o = \frac{36}{100} = 0.36A$$

Then, using
Input power = Output power

$$V_{in}I_{in} = V_oI_o$$

The value of I_{in} is obtained as,

$$12I_{in} = 36 \times 0.36$$

$$I_{in} = 1.08A$$

Let us assume that 5% of ripples in the output current, then the value of

$$\Delta i_{l1} = 5\% \text{ of } I_{in}$$

$$= 0.05 \times 1.08\Delta i_{l1} = 0.054$$

The value of inductor $L_1$is calculated using the formula,

$$\Delta i_{l1} = \frac{V_o - 2V_{in}}{L_1}(1 - K)T$$

$$L_1 = \frac{V_o - 2V_{in}}{\Delta i_{l1}} \frac{(1 - K)}{f}$$

$$L_1 = \frac{36 - 2(12)}{0.054} \frac{(1 - 0.5)}{50000}$$

$$L_1 = 2.22 \text{ mH}$$

Let 5% of ripple content in the output voltage,

$$\Delta V_o = 5\% \text{ of } V_o$$

$$\Delta V_o = 0.05 \times 36$$

$$\Delta V_o = 1.8$$

The formula to find the value of capacitance C_2 is,

$$\Delta V_o = \frac{(1 - K)}{fC_2} \frac{V_o}{R}$$

Rearranged as,

$$C_2 = \frac{(1 - K)}{f\Delta V_o} \frac{V_o}{R}$$

$$C_2 = \frac{(1 - 0.5)36}{50000 \times 1.8 \times 100}$$

$$C_2 = 2 \times 10^{-6}F$$

The values of both the capacitances C_1 and C_2 are equal,

$$C_2 = C_1$$

Therefore, $C_1 = 2\times 10^{-6}F$

Simulations have been done using the parameters shown in Table 1 and the performance of the POSLC analyzed.

Table 1. Design Values for Positive Output Super-Lift Luo Converter

Parameters	POSLC
V_{in}	12V
V_o	36V
L_1	$2.22 * 10^{-3}$H
C_1	$2*10^{-6}$F
C_2	$2*10^{-6}$F
R	100Ω
K	0.5

3. Open Loop Configuration

The open loop configuration technique for the POSLC has been constructed in the Simulink model without any feedback controller by using a fixed DC source and PV as input. The open loop system with DC input is shown in Figure 2. POSLC produces the particular output voltage for input by introducing pulse width modulation techniques accordingly.

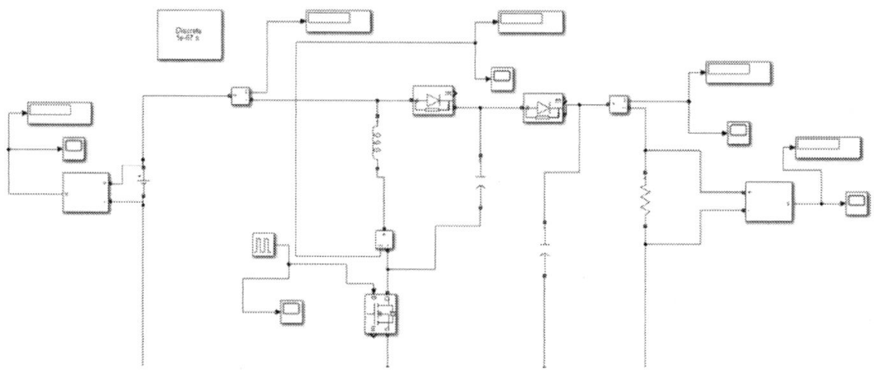

Figure 2. Open Loop System of POSLC with DC source

The POSLC is given with a constant input voltage of 12V, 50% duty cycle and switching frequency of 50kHz. The scope output for the open loop system with a fixed DC source is shown in Figure 3.

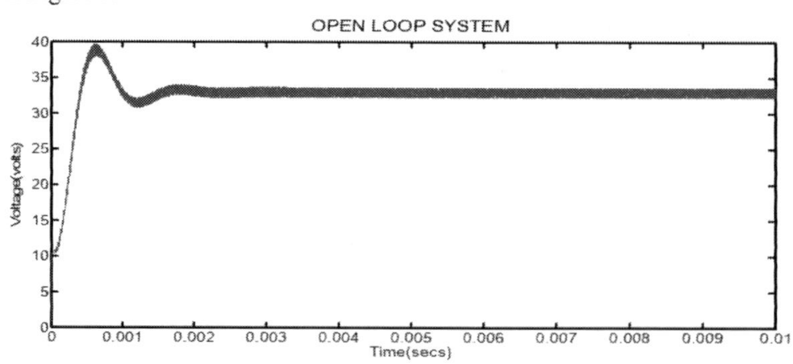

Figure 3. Output Voltage of POSLC for Open Loop with DC source

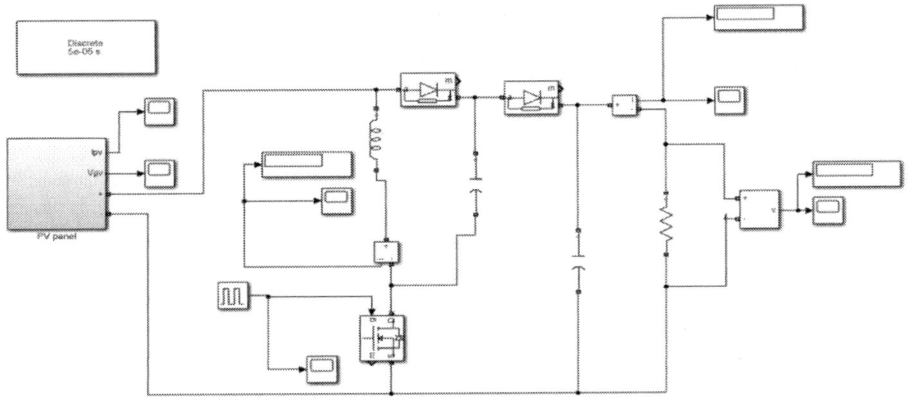

Figure 4. Open Loop System of POSLC with PV source

4. Closed Loop Configuration

The closed loop system is defined as the control action that eliminates the shortcomings of the open loop system thereby improving the overall performance of POSLC and has several benefits. This system consists of feedback that provides the switching gate pulses to MOSFET switch S. In the closed loop control of POSLC, the simulation circuit diagram is designed with the same circuit parameters along with the feedback system. The simulation of the POSLC converter with gain as the feedback controller is done. Simulation for the closed loop system is shown below in Figure 6.

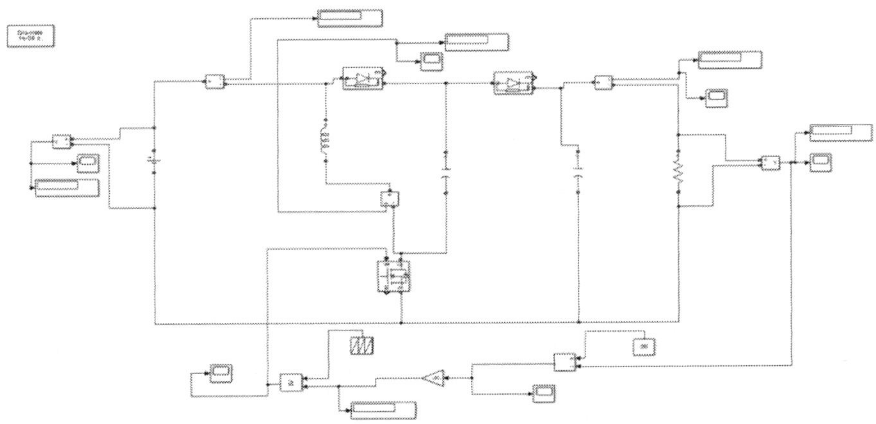

Figure 5. Closed loop control of POSLC with DC source

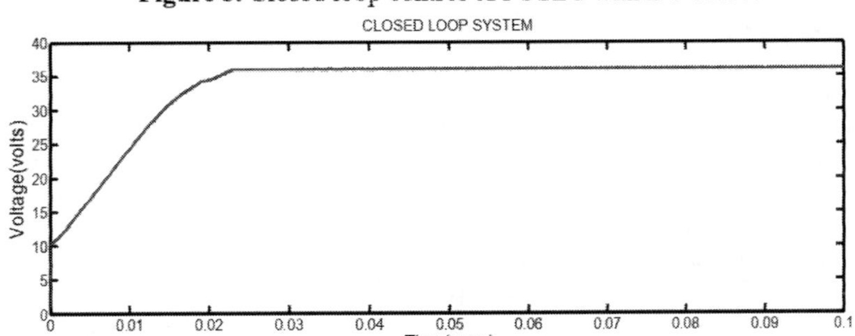

Figure 6. Output Voltage of Closed Loop POSLC with DC source

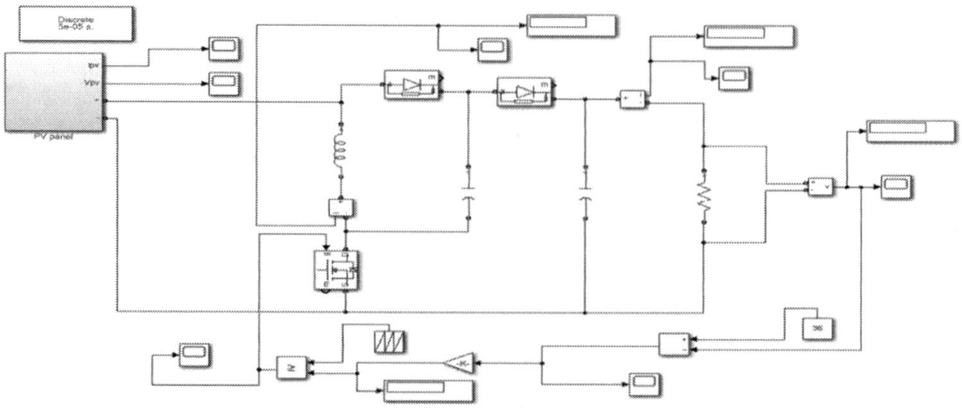

Figure 7. Closed loop control of POSLC with PV source

From Figure 7 and 9, we can infer that the errors between input and output have been reduced and the disturbances are also reduced to a greater extent compared to the open loop system.

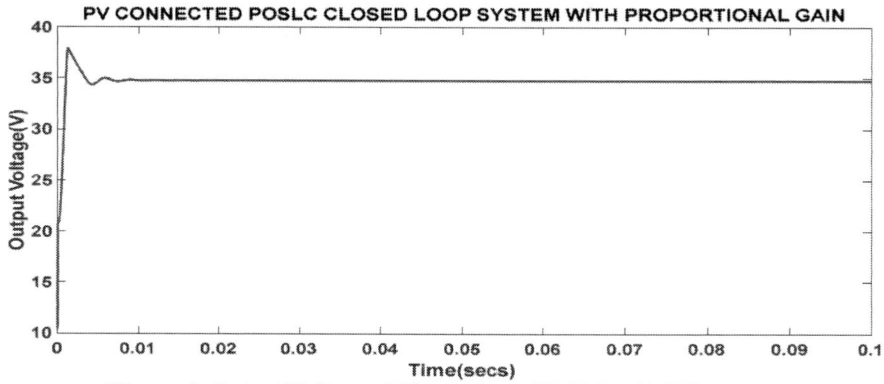

Figure 8. Output Voltage of Closed Loop POSLC with PV source

5. Fuzzy Logic Control

To obtain further precise and accurate output values, fuzzy logic (FL) control is employed. Fuzzy control systems analyze the analog input values and express them in logical variables, entirely different from digital logic. Here, the membership function is defined as a matter of degree where it is a curve drawn for input and output that depicts the point in the input space mapped to the membership value using Mamdani's fuzzy inference method [9]-[11]. The input variable given to the controller is error voltage (e) and the output variable is duty ratio (K) that defines the duty cycle. The offline implementation is employed where the lookup table is constructed based on the set of all the possible combinations of input variables and output variables. The table contains the duty ratio (K) according to the input entries (V_{ref}, V_o, Error). Using the lookup table, a set of rules is designed. There are several approaches for designing the rules, here; the IF-THEN system of rules is taken as the logical interpretation of fuzzy control.

The input variable is divided into nine groups namely: NVB (Negative Very Big), NB (Negative Big), NM (Negative Medium), NS (Negative Small), Zero (Z), PS (Positive Small), PM (Positive Medium), PB (Positive Big), PVB (Positive Very Big). In the same way, the output variables are given with the same names [12]. Having developed the set of rules and the membership function, the controlled voltage, and improved output performances have been determined using the fuzzy logic control system [13].

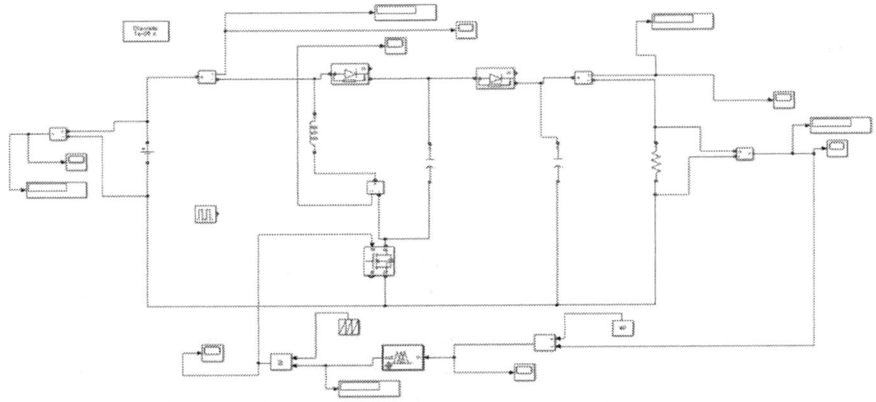

Figure 9. Fuzzy Control of POSLC with DC Source

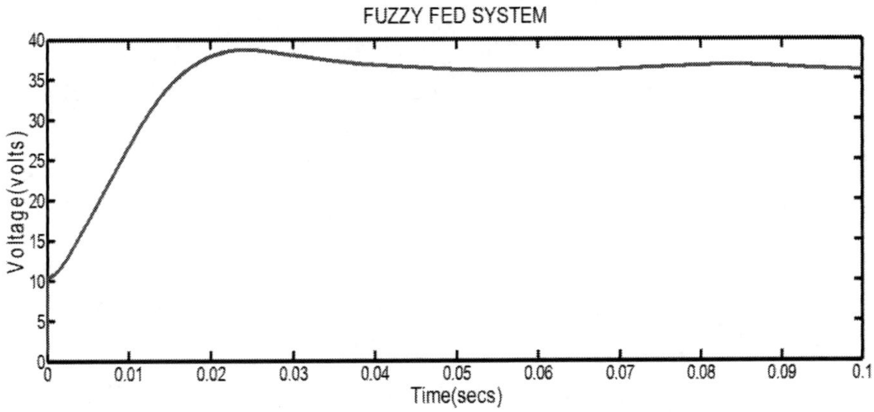

Figure 10. Output Voltage of Fuzzy Controlled POSLC with DC source

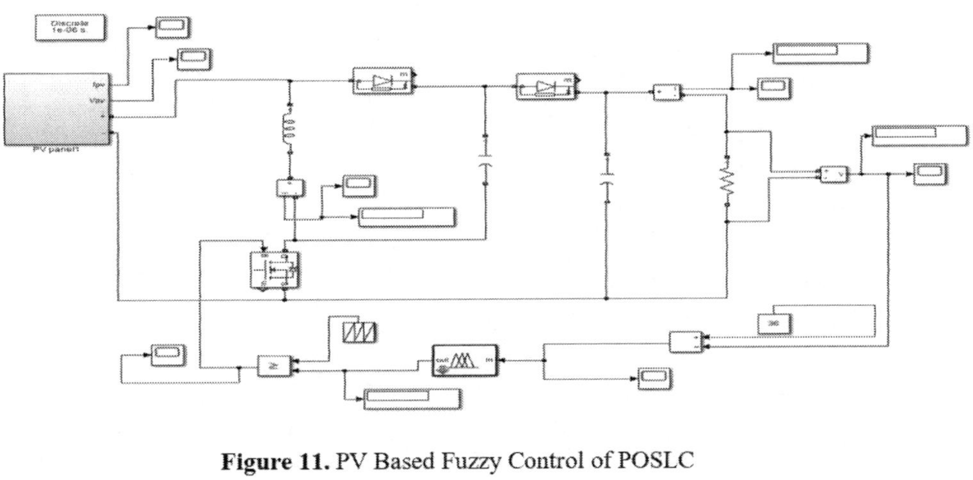

Figure 11. PV Based Fuzzy Control of POSLC

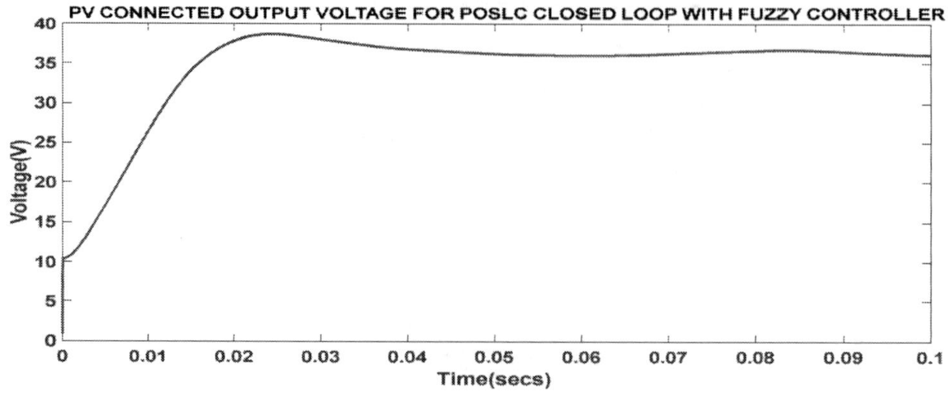

Figure 12. PV Output Voltage fuzzy controlled POSLC

Using these fuzzy rules, the output obtained from simulating is shown in Figure 11 and Figure 13. The membership function plots for both input and output variables are depicted as below:

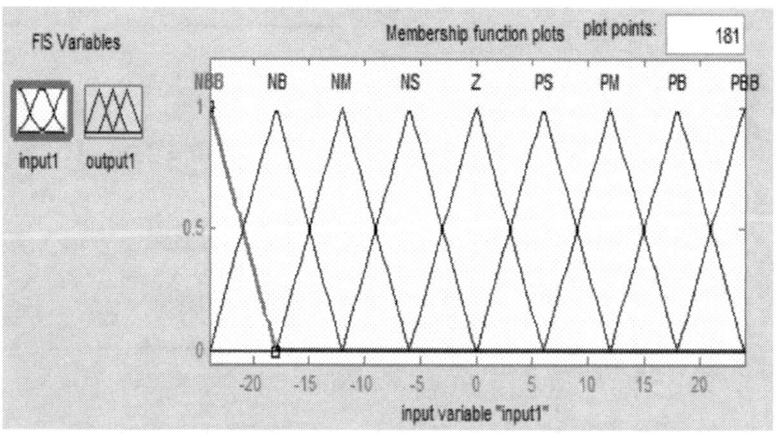

Figure 13. Ruler view: Membership function plot of error[e]

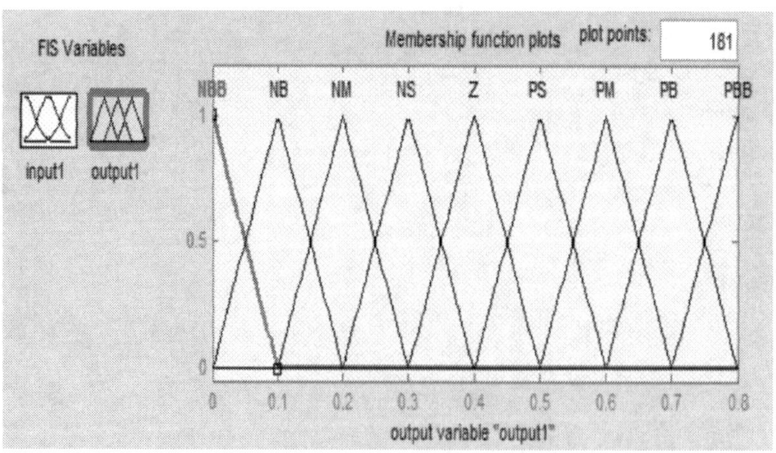

Figure 14. Ruler view: Membership function plot of gain [K]

Table 2. FIS Editor Values

V_{ref}	V_o	Error	K	FIS Editor values
36	31	5	0.3684	0.33
36	32	4	0.4	0.353
36	33	3	0.4285	0.386
36	34	2	0.4545	0.431
36	35	1	0.4782	0.467
36	36	0	0.5	0.5
36	37	-1	0.52	0.5
36	38	-2	0.5384	0.5
36	39	-3	0.5555	0.5
36	40	-4	0.5714	0.533
36	41	-5	0.5862	0.565
36	42	-6	0.6	0.6
36	43	-7	0.6129	0.602

Table 2 shows the various values of output voltage for different values of FIS along with the duty cycle.

6. IMPLEMENTATION IN PROTEUS

The Positive Output Super-lift Luo Converter for the open loop is simulated in Proteus Software to know about the real-time implementation. From the below figure 16, it is seen that the DC supply for the POSLC circuit is given by the ARDUINO UNO board which is an open-source electronics board based on easy-to-use hardware and software. The output of the circuit can be seen in Digital Oscilloscope.

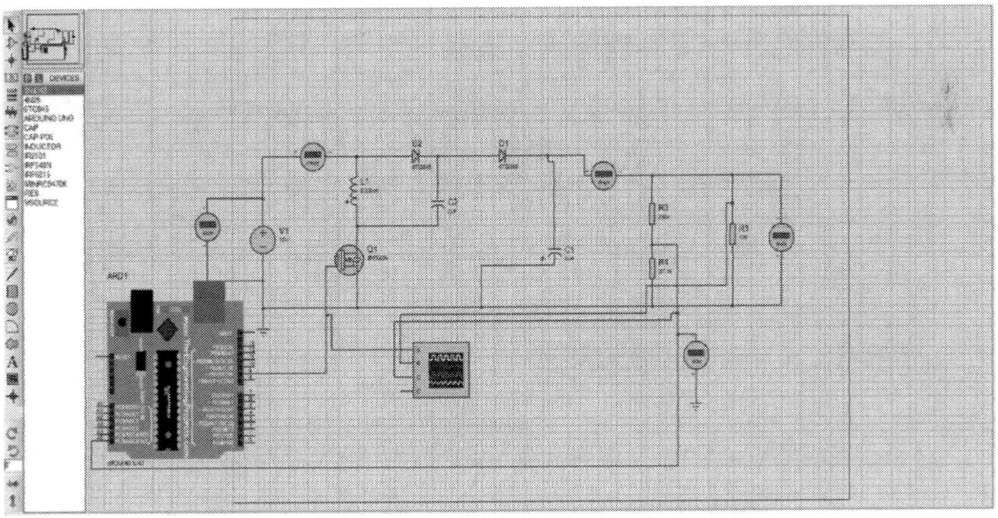

Figure 15. Open loop simulation of POSLC in Proteus

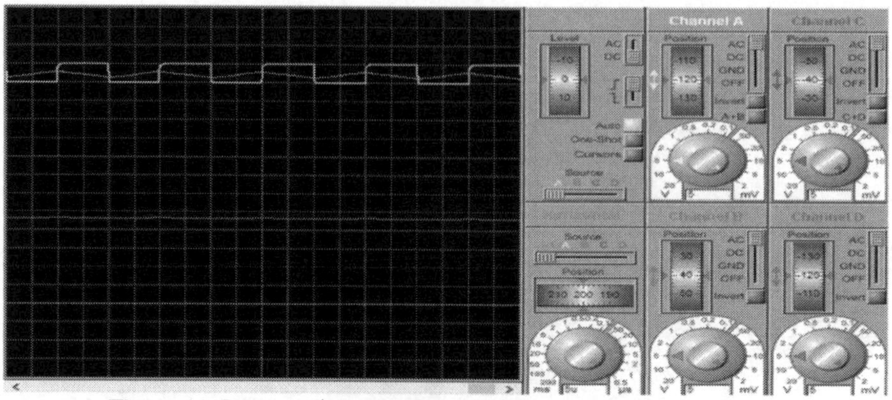

Figure 16. Proteus 1st stage output of POSLC for open loop system

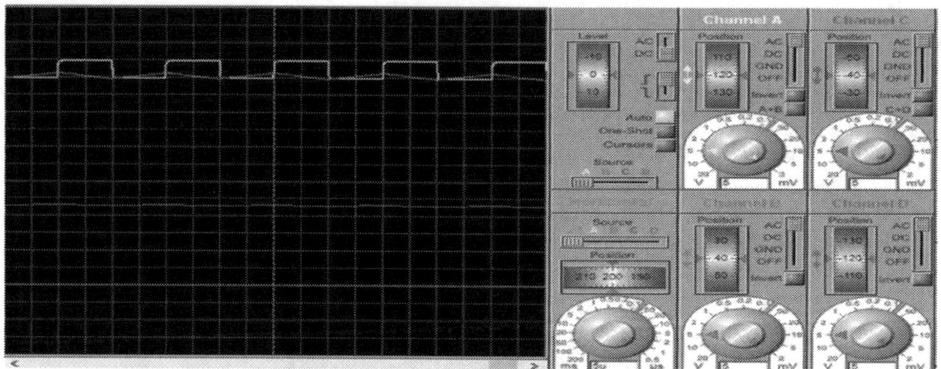

Figure 17. Proteus 2nd stage output of POSLC for open loop system

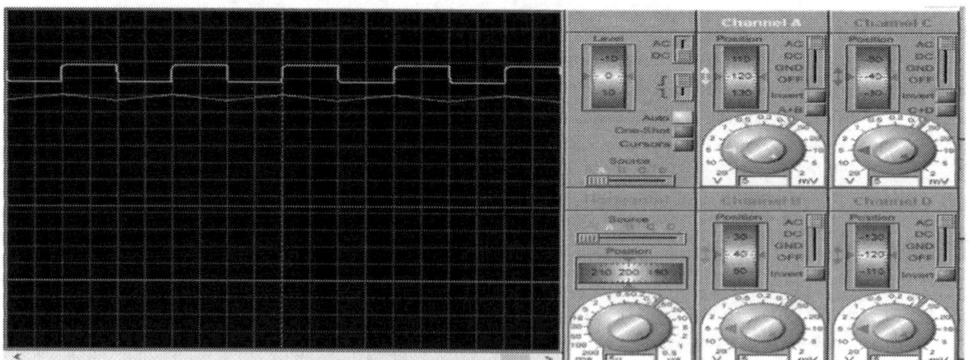

Figure 18. Proteus 3rd stage output of POSLC for open loop system

The above figures 17,18 and 19 shows the Pulse Width Modulation (PWM) to the MOSFET switch, output voltage (V_o) for the load and output current (I_o) respectively for the Positive Output Super-lift Luo Converter for open loop system.

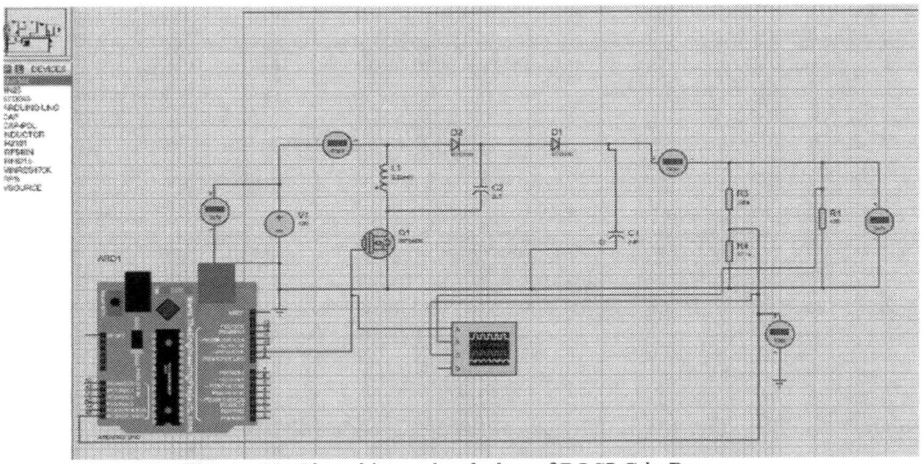

Figure 19. Closed loop simulation of POSLC in Proteus

The above figure 20 shows the Positive Output Super-lift Luo Converter for closed loop is simulated in Proteus software to obtain the required output voltage.

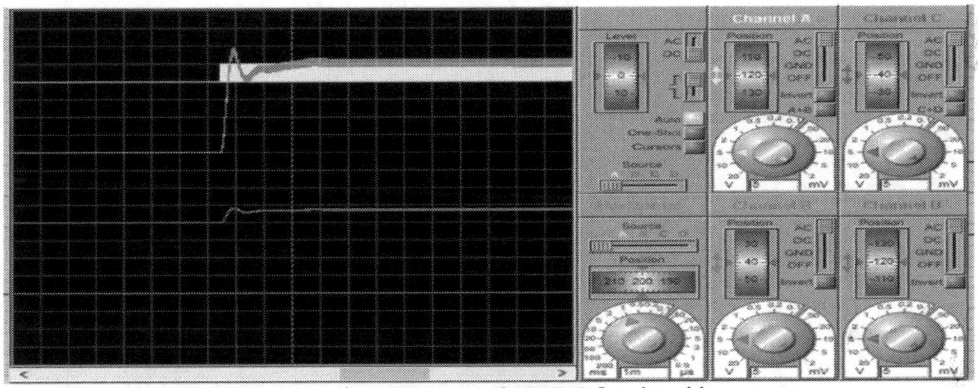

Figure 20. Proteus 1st stage output of POSLC for closed loop system

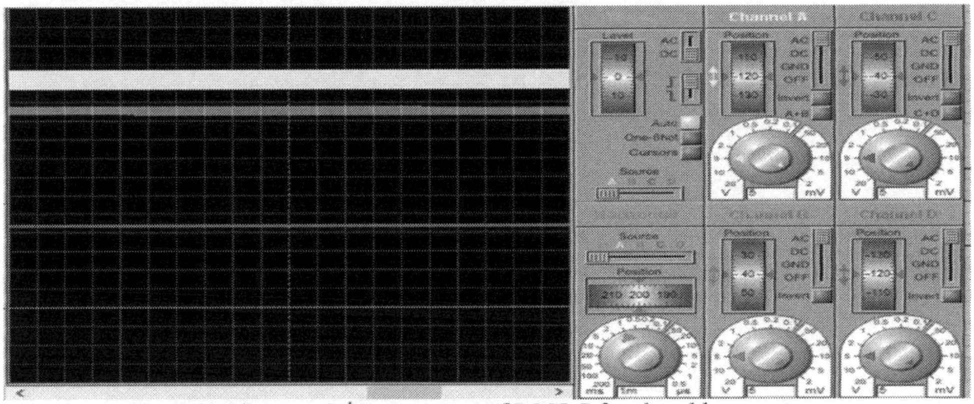

Figure 21. Proteus 2nd stage output of POSLC for closed loop system

The above figures 21 and 22 show the Pulse Width Modulation (PWM) to the MOSFET switch, output voltage (V_o) for the load, and output current (I_o) respectively for the Positive Output Super-lift Luo Converter for closed loop system.

7. Conclusion

In this paper, the simulation of open loop, closed loop and fuzzy control of the Positive Output Super-lift Converter has been successfully executed in MATLAB/SIMULINK and also the circuit model that is simulated and verified in the Proteus environment. The voltage required from the solar supply is obtained from the solar cells by connecting the cells in a series and parallel combinations. The simulation of the POSLC is done for

- Open loop
- Closed loop with Proportional gain
- Closed loop with Fuzzy Logic Controller

For various values of duty cycle output voltages measured as well the error at the output side is reduced. By using the fuzzy logic control in the closed loop provide precise and accurate voltage control, high voltage gain and therefore better output efficiency can be achieved better than the conventional converters. Fuzzy logic control-based closed loop control of positive output super lift Luo converter has been simulated and verified.

References

[1] Prasanna K, Kirubakaran D, Rahulkumar J and Rudhran J A 2015 Implementation of Positive Output Super Lift Luo Converter for Photo Voltaic System *International Research Journal of Engineering and Technology* **02**.

[2] Anushka Tekade, Rahul Juneja, Manish Kurwale and Prashant Debre 2016 Design of Super lift Luo Boost Converter for Solar Inverter *International Conference on Energy Efficient Technologies for Sustainability (ICEETS).*

[3] Varesi K, Hosseini S H, Sabahi M and Babaei E 2018 Modular non isolated multi-input high step-up dc–dc converter with reduced normalized voltage stress and component count *IET Power Electronics* **11** 1092-1100.

[4] Fang Lin Luo and Hong Ye 2003 Positive output super-lift converters *IEEE Transactions on Power Electronics* **18** 105-113.

[5] Luo F L and Ye H 2001 Chapter 17, DC/DC Conversion Techniques and Nine Series Luo-Converters of the *Power Electronics Handbook* Edited by Rashid M H, Academic Press, San Diego, USA.

[6] Shan Z-L 2012 Investigation of a super-lift Luo-converter used in solar panel system *China International Conference on Electricity Distribution* 1-4.

[7] So W C, Tse C K, and Lee Y S 1996 Development of a fuzzy logic controller for DC-DC converters: Design, computer simulation, and experimental evaluation *IEEE Transactions on Power Electronics* **11** 24-31.

[8] Chetan P. Ugale, Dhumale R B and Dixit V V 2015 DC-DC Converter Using Fuzzy Logic Controller *International Journal of Engineering and Technology* **2(4)** 593-596

[9] Raviraj V S C, and Sen P C 1997 Comparative Study of Proportional-Integral, Sliding-Mode, and Fuzzy Logic Controllers for Power Converters *IEEE Transaction on Industry Applications* **33** 518- 524.

Voice controlled shopping trolley navigation with RFID scanner and live billing using IoT

Praveen S[1] , Pugalenthiran V[2] , Jaraline Kirubavathy K[3].

Department of ECE, KCG College of Technology, Chennai, Tamil Nadu, India.

Email: - jaraline.ece@kcgcollege.com

Abstract. The current process of shopping in a mall involves figuring out and navigating to required sections and either carrying a shopping basket or pushing around a trolley. Although it might not feel so but this is a time consuming and for some people, a physically demanding task. The other main disadvantage is having to wait in queues at the cashier and have the items billed in one by one. The proposed system is a Raspberry Pi based device integrated to traditional shopping trolleys. The desired section of the mall is given to it as voice input. Necessary signals are provided to a driver IC which drives the wheels of the trolley through motors towards the required section. The very moment an item is dropped into the trolley, it is scanned by an RFID reader module. The item details and total bill amount to the customer and live update is provided at the checkout through IoT.

1. Introduction

With all the technological advancements getting updated and becoming outdated by the hour, there is a need for constant innovation in every aspect of life. Shopping is a process which has never had any kind of technological advancements in quite some time.

The entire process of traditional shopping at malls is a very time consuming and a physically engaging process [4][16]. Having to figure out the layout of a mall and locate the required sections of it can not only be time consuming [13][18][20] but can also be misleading to the customer as the customer is usually made to navigate past items that are expensive, attractive and potentially not needed to reach the items that are actually needed[6][7][23]. At times the customers forget what was required by them in the first place and end up wasting too much time and money on items that are not needed[5][24].

As time passes and the trolley keeps getting filled, naturally it becomes heavier and harder to push [19]. At the end of procuring the required items, customers have to wait in usually long queues to get their items billed [3][12]. Since the cashier has to scan every single item of every single customer [1][14], it can prove to be a very demanding and an error prone task [2]. Technology is required in many instances [8] during the entire shopping process to benefit both the customer and the cashiers [25].

2. Literature survey

In shopping malls, RFID technology is also used for billing mostly during purchasing decision, and IOT is used for bill management through the ESP module.The payment details are getting to be sent to the server by which the central billing unit will affect the customer's payment. The ESP module goes to be operating as a short-distance Wi-Fi chip for wireless communication. But there is a drawback which includes constraints like distance and interference. The server goes to be busy if customers are high and internet connectivity should be stable for finishing the tactic. [7]

Sainath (2014) developed an automated supermarket trolley for a supermarket billing system, that uses barcode technology to scan goods. The bill is going to be forwarded to the central billing system where the customer can pay them by showing a singular id. The limitation of barcode scanning needs a line of sight for scanning and it ought to be fastened inside its boundary. [9]

Budic (2014) developed a cash register line optimization system using RFID technology. The system was developed for smart shopping using RFID. The RFID is employed for scanning products and also the info is kept within the database which will be paid online or during a central bill. It also uses an internet application to take care of entire shopping details. It requires the upkeep of an internet application server. No necessary steps are taken for the merchandise that's accidentally dropped into the trolley by the client. [10]

In Smart shopping trolley using RFID paper. They implemented a sensible way of a shopping trolley with RFID and Zigbee by which bill is generated by a scan of products within the reader and bill transmitted to central billing department by that bill could also be paid at the counter that would be a serious issue for the client. [11]

Prateek Aryan (2014) presented a smart shop cart with automatic billing and Bluetooth, which is a method for billing to be done while in a trolley and then uploaded to the user's Android device through Bluetooth.. Every customer can't be expected to possess a Smartphone and Bluetooth can have connectivity issues and the range is a smaller amount. [15]

In Smart RFID-based Interactive Kiosk Cart using wireless sensor nodepaper , they used RFID technology for the smart automated shopping. They used a fanatical website for billing maintenance and for user interaction. Every user with a singular id accesses the webserver for the bill payment and invoice information. Internet service is mandatory during this sort of service. therefore the method could fail thanks to internet instability and server error issues may additionally occur due to high load. [17]

Vinutha (2014) built an automatic billing with server end employing RFID technology for shopping and automatic billing.This scans product by frequency identification then the bill is generated at the server finish that's then communicated to the client. This needs server maintenance and internet connectivity each for the client and shopkeeper. [21]

Smart handcart with customer-oriented service by Hsin-Han Chiang (2016), established a thought of an automatic billing system and programmed shopping trolley where they used face recognition for client authentication. It's not an easy straightforward method as face recognition of shoppers throughout shopping hours won't be easy and proper as malls are often crowded. Many errors are potential whereas using recognition for authentication. [22]

3. Methodology

The proposed system primarily aims to automate the existing shopping process as much as possible. A smart device is fitted to a traditional trolley. The device takes the customer's required section of the store or mall as voice input. As per the input, the trolley automatically moves towards the desired section, eliminating the need to push the trolley while searching for the needed section. Whenever an item is dropped onto the trolley, it is scanned and updated in the trolley's audio output and also at the designated billing system at the cashier. When an item is removed from the trolley, it is removed from the live bill both in the trolley and the billing computer.

The many advantages of this system like automation of trolley movement, voice input, automatic billing and updation make this a very practical and efficient way to shop.

4. Implementation

The various technologies, software and hardware used in the implementation of the prototype include Raspberry Pi, Wi-Fi module, Bluetooth module, RFID Reader and tags, driver IC, Motors with battery power supply, IOT, headset for audio output, Python to code and program the trolley.

Journal of Physics: Conference Series **2040** (2021) 012035 doi:10.1088/1742-6596/2040/1/012035

The very first step in this approach to shopping is to give the trolley the desired section of the store as a voice input. This is done through an app on the phone as it is much more convenient than bending down to the trolley and speaking out the command. This voice command is sent to the device's brain, Raspberry Pi through Bluetooth.

Once the command is received by the Raspberry Pi, it gives out an audio acknowledgement and then sends the instructions to the driver IC to power the motors attached to the trolley's wheels and navigate to the desired section. The directions are pre-programmed using python to the Raspberry Pi using the layout of the store. Once the trolley reaches the destination, a prompt is given to the user to point them to the precise shelf where the required item is kept.

The entire process is based on the fact that every item in the store is RFID tagged. When the customer drops the item to the trolley, the RFID reader scans the item's RFID tag and gives out a voice prompt informing the customer of the details of the item dropped in the trolley and the total bill amount. In case of removal of any item, the action is recorded and the item gets removed from the live bill which also provides a voice prompt for the same action. Simultaneously the same details are being updated at the billing section at designated billing lane's billing computer online, thereby implementing IOT to make shopping better.

On completion of shopping, the customer can check out at the designated counter for the trolley after making the payment for the bill which was created in real time during the entire duration of the purchase.

There are numerous advantages to the proposed system like,

- Ease of shopping
- Convenient way to find the required items
- Automation of trolley thereby minimizing physical involvement only to drop items inside the trolley
- Live billing
- Time and energy saving approach
- Practicality
- Ease of installation and maintenance

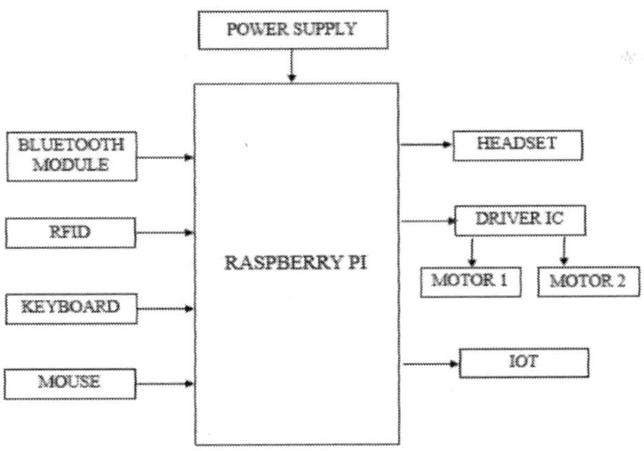

Figure 1. Block diagram

Figure 2. Prototype

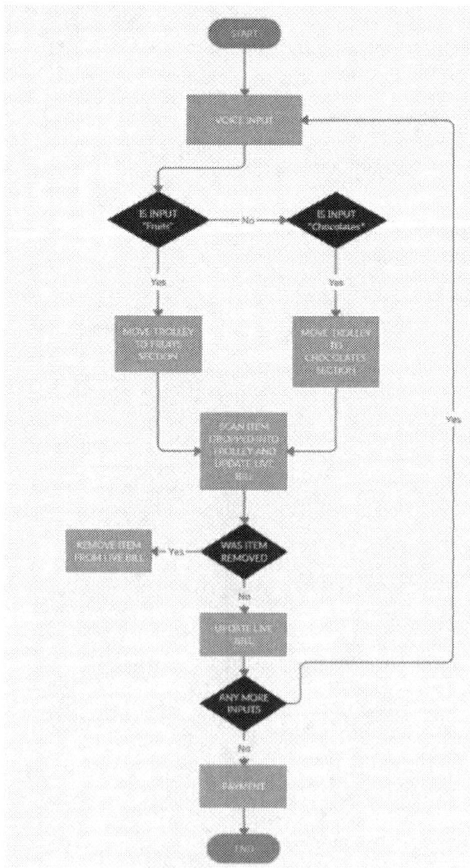

Figure 3. flow chart

```
if (str(x,('utf 8'))=="chocolates" or str(x,('utf 8'))=="c"):
        print("working")
        ser.write(b'w')
        os.system("espeak ' " + "TROLLEY MOVE TO CHOCLATES SECTION" + " ' ")
        time.sleep(1)
        os.system("espeak ' " + "TROLLEY MOVE FORWARD" + " ' ")
        time.sleep(1)
        GPIO.output(motorA,1)
        GPIO.output(motorB,0)
        GPIO.output(motorC,1)
        GPIO.output(motorD,0)
        time.sleep(3)
        GPIO.output(motorA,0)
        GPIO.output(motorB,0)
        GPIO.output(motorC,0)
        GPIO.output(motorD,0)
        time.sleep(1)
        os.system("espeak ' " + "TROLLEY REACHED FRUITS SECTION" + " ' ")
        time.sleep(1)
        os.system("espeak ' " + "IN THE RIGHT SIDE TRACK FIVE STAR, DAIRY MILK AND MILKYBAR CHOCOLATES" +
" ' ")
        time.sleep(1)
        os.system("espeak ' " + "IN THE LEFT SIDE TRACK PERK, ECLAIRS, AND ALPENLIBLE CHOCLATES" + " ' ")
        time.sleep(1)
        os.system("espeak ' " + "START PURCHASE" + " ' ")
        time.sleep(1)
        while(1):
                x=serl.readline()
                print(x)
```

Figure 4. Main Program

Python is installed in the Raspberry PI. This Main program is coded in python and this will of thecontrol all modules which are connected to the Raspberry PI.

Figure 5. Raspberry PI

Raspberry PI is a series of small single board computer.it is a Quad-core 64-bit ARM Cortex A53 clocked at 1.2 GHz.it is used to get the input from RFID reader and gives the output to driver IC, Node MCU, headset and IOT.

Figure 6. Bluetooth

The Bluetooth module has a GFSK modulation and the sensitivity is ≤-84dBm at 0.1% BER.It is used to get input(which product customer wants) from the mobile phone.

Figure 7. RFID Reader

EM-18 RFID Reader is used to select the products and remove the products.

Figure 8. L293D Driver

L293D Driver is used to control the motors. It will get the input from the Raspberry PI and it gives the output to the motor.

Figure 9. DC Motor

DC Motor is used to move the trolley from one location to another. It get the output from the L293D Driver.

Figure 10. Node MCU

Node MCU is a SDIO 2.0, SPI, UART and it has a 32-pin QFN package. Integrated MAC/baseband processors. It is used for billing the products. It get the input from the Raspberry PI(which products the customer selected) and it gives the output to the billing counter.

5. Output

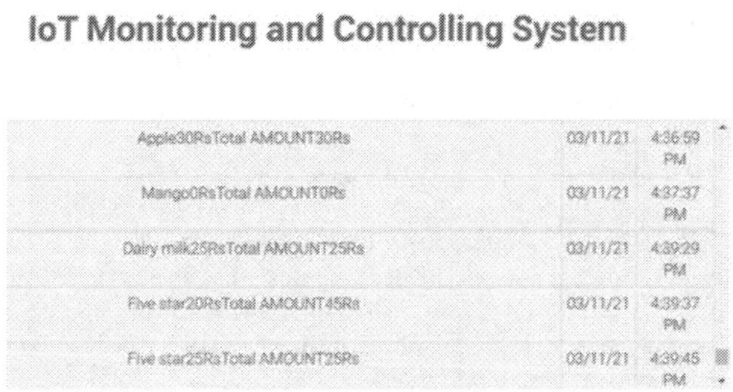

IoT Monitoring and Controlling System

Apple30RsTotal AMOUNT30Rs	03/11/21	4:36:59 PM
Mango0RsTotal AMOUNT0Rs	03/11/21	4:37:37 PM
Dairy milk25RsTotal AMOUNT25Rs	03/11/21	4:39:29 PM
Five star20RsTotal AMOUNT45Rs	03/11/21	4:39:37 PM
Five star25RsTotal AMOUNT25Rs	03/11/21	4:39:45 PM

Figure 11. Live Bill

We can bill the products automatically with this output and it is a time saving process and also the customers no need to wait in the bill counter.

```
b''
b''
b''
b''
b''
b'260065FSEF59'
bread:Rs.50
total amount: 50
b''
b''
b''
b''
b'2600660186C7'
 sugar 1kg:Rs.100
total amount : 150
b''
b''
b''
b''
b'0800AEACFAFO'
malk: o
egg: o rice: 0
bread: 1
salt: o
```

Figure 12. Simulation output

6. Conclusion

The proposed system doesn't just focus on current difficulties in shopping like the physical requirement of pushing a trolley or long queues at the billing section, but also aims to make innovative advances using technology to make the entire shopping experience a short but efficient one. The system is also very practical and even helpful for physically challenged people equally as it is useful for any other customer. In the growing age of technology, there isn't a place in the world that lacks the need to be updated and technologically improved. By using the various technologies like IOT, Cloud and RFID, the project aims to technologically advance the process of shopping as far as possible. The result of the project is not just to help the customer save time and shop efficiently but also cover all aspects of shopping like eliminating long queues at the billing section, reducing the cashier's work, increasing the number of customers at the mall etc. Furthermore, this alternative to traditional shopping does not prove to be harmful for the environment and does not contribute in any type of pollution. In fact, the energy source used is electric batteries, which can be recharged and reused several times. Considering the very negligible cons of the project (Need to recharge the batteries), the many advantages make this a perfect solution to the problem that is the method of traditional shopping.

References

[1] M. R. Mane, N. G. Amane, S. D. Patil, and A. L. Lakesar, ''Electronic shopping using barcode scanner,'' Int. Res. J. Eng. Technol., vol. 3, no. 4, pp. 1–5, 2016
[2] D. N. Sanjay and S. Pushpalatha, ''All-in-one intelligent shopping trolley with automatic billing and payment system,'' Int. Res. J. Eng. Technol., vol. 4, pp. 59–62, Jul. 2017.
[3] M. A. Lambay, A. Shinde, A. Tiwari, and V. Sharma, ''Automated billing cart,'' Int. J. Comput. Sci. Trends Technol., vol. 5, pp. 148–151, Mar./Apr. 2017.
[4] S. Jadhav, S. Kamath, S. Yadav, A. Rajput, and P. K. S. Sakure, ''Smart shopping application using NFC,'' Int. Res. J. Eng. Technol., vol. 5, no. 3, pp. 1521–1524,

2018.
[5] D. Choi, C. Y. Chung, and J. Young, ''Sustainable online shopping logistics for customer satisfaction and repeat purchasing behavior: Evidence from China,'' Sustainability, vol. 11, no. 20, p. 5626, 2019.
[6] Dr. P. MuthuKannan, Anupriya Asthana, "Automatic Retail System Using RFID" , Volume 1, Issue 5, October 2013, International Journal of Advance Research in Computer Science and Management Studies.
[7] D. Budić, Ž. Martinović, D. Simunic, "Cash register lines optimization system using RFID technology"
[8] Satish Kamble, SachinMeshram, Rahul Thokal, Roshan Gakre, "Developing a Multitasking Shopping Trolley Based on RFID Technology" , International Journal of Soft Computing and Engineering (IJSCE) ISSN: 2231-2307, Volume-3, Issue-6, January 2014
[9] Sonali J. Mane "An Automated Shopping cart using RFID" International Journal of Engineering Science and Computing, Vol. 8 Issue No.4 , April 2018
[10] J. Thangakumar "Automated Shopping Trolley for Super Market Billing System" International Journal of Computer Applications (0975 – 8887) in 2014.
[11] Dhavale Shraddha D, DhokaneTrupti J, Shinde Priyanka S "IOT Based Intelligent Trolley for Shopping Mall" International Journal of Engineering Development and Research, Volume 4, Issue 2, 2016
[12] anhaviIyer, Harshad Dhabu, Sudeep K. Mohanty, "Smart Trolley System for Automated Billing using RFID and the ZIGBEE" International Journal of Emerging Technology and Advanced Engineering Volume 5, Issue 10, October 2015
[13] Ria Singh, SatyamVarma ,Ms.krithi,"RFID and IR based Smart Shopping Mart Management System" , 2018 International Conference on Advances in Computing, Communication Control and Networking (ICACCCN),01 July 2019
[14] K P Vidya, K M Swathi, D Chaitra, S H Jayalakshmi, M V Manoj Kumar, H R Sneha, Likewin Thomas, B H Puneetha "Virtual Cart: Novel Approach for Revamping Smart Shopping Experience" , 2018 IEEE Distributed Computing, VLSI, Electrical Circuits and Robotics (DISCOVER), 28 March 2019
[15] Vinutha M L, Harshitha P Bale , Sushma R , Suchitra M "SHOPPING AND AUTOMATIC BILLING USING RFID TECHNOLOGY", International Journal Of Electronics And Communication Engineering & Technology (IJECET), Volume 5, Issue 8, pp. 132-138, 2014
[16] Mr.P.Chandrashekar,Ms.T.Sangeetha "Smart shopping cart with automatic billing system through RFID and ZIGBEE" ISBN No: 978-1-4799-3834-6/14, 2014 IEEE
[17] Hsin-Han Chiang, Wan-Ting You, Shu-Hsuan Lin"Development of smart shopping carts with customer-oriented service", IEEE International Conference on System Science and Engineering (ICSSE), 2016
[18] Megha R. Mane1, Nilam G. Amane2,SunitaD. Patil3, Archana L.Lakesar4, "Electronic Shopping Using Barcode Scanner", International Research Journal of Engineering and Technology (IRJET), Volume: 03 Issue: 04 | Apr-2016.
[19] B. N. Arathi and M. Shona, "An Elegant Shopping using Smart Trolley" India Journal of Science and Technology, Vol. 10, Issue 3, Jan 2017.
[20] Vrinda Gupta, NiharikaGarg, Analytical Model for Automating Purchases using RFID-enabled Shelf and Cart,2014
[21] J.C. Narayana Swamy, D. Seshachalam, Saleem Ulla Shariff "Smart RFID based Interactive Kiosk cart using wireless sensor node", IEEE International Conference on Computation System and Information Technology for Sustainable Solutions (CSITSS), 2016
[22] Komal Ambekar, Vinayak Dhole, supriya sharma,Tushar Wadekar "Smart Shopping Trolley Using RFID"International Journal of Advanced Research in Computer Engineering & Technology (IJARCET) Volume 4 Issue 10, 2015
[23] V. Lobo, R.Prabhu, Polu Bal Kumar Reddy, International Journal of Innovative

Technology and Exploring Engineering (IJITEE) ISSN: 2278-3075, Volume-8, Issue-11S2, September 2019

[24] Ramya P , Aravind C , Mouriya N , Pavithra S "Smart shopping for vishually impaird using RFID" International Journal of Information And Computing, Volume 6, Issue 3, March 2019 ISSN NO: 0972-1347, Volume 6, Issue 3, March 2019.

[25] Lakshmi Narayanan, Dhanya Sudhakaran, Subhika Grandhe, Namra Iqbal, Jimcymol James " A Deep Learning Enabled Smart ShoppingCart", Biosc.Biotech.Res.Comm. Special Issue Vol 13 No 13 (2020) Pp-247-251.

Study on Engineering behavior of conventional cement concrete by partially replacing sea shell as Fine aggregate

D.Parthiban[1], D.S.Vijayan[2], Sachu andrews[3], Sangsangrach sangma[3], Arif mohammed[3]
[1]Assistant Professor, Department of Civil Engineering, Aarupadai Veedu Institute of technology, Chennai, India
[2]Associate Professor, Department of Civil Engineering, Aarupadai Veedu Institute of technology, Chennai, India. Email: vijayan.has.siva@gmail.com
[3]UG Final year Students, Department of Civil Engineering, Aarupadai Veedu Institute of Technology, Vinayaka Missions Research Foundation, Paiyanoor, Chennai – 603104.
Corresponding author: Parthi92bhde@gmail.com

ABSTRACT. In this study, it was investigate about the engineering behavior of conventional concrete of grade M25, by replacing the fine aggregate by partially with crushed sea shell. On because of progression in the field of infrastructure, the dearth of construction material becomes upsurge day by day. All the investigator was observe to find the substitute construction materials to meet the demand of construction in world wide. The reason behind is, while continuous utilizing of the natural raw material for construction decline the resource of those materials and further, it leads to affect the routine function of the earth and made drastic change in the life cycle of living beings. Therefore, it is obvious to find out the possible alternative material to suit for all type of construction. In this study, the sea shell was used as a fine aggregate in the method of crushed powder in an apparent proportion such as 10%, 20% and 30% replacing in the place of fine aggregate sand. The compression and split tensile behavior of sea shell aggregate mixed concrete comparatively with plain sand concrete was studied under different curing periods such as 7 days, 14 days and 28 days. It was shows that at 20% replacement of crushed sea shell aggregate provide the compressive strength of 35 N/mm2 at 28 days of curing nearly with the design strength of M25 grade conventional concrete. It was also noticed that, the percentage increase of crushed sea shell decline the strength of concrete as radically. Hence it was acclaimed that, the percentage of crushed sea shell as fixed equal or below 10 % to get a better strength of concrete.

Keywords: sea shell, Concrete, natural material, Distractive test

1. INTRODUCTION

Environmental pollution was the major problem faced almost by all the Developing countries. In order to control the pollution, all the countries have developed their own strategies to protect the environment by implementing strict regulations. In our India, to preserve the availability of natural resources, it was banned to extract the gravel and sand from the major rivers of the nation. But, it was estimated that annually more than 15 million tons of aggregate resources like processed rock, gravel and lime stone are required to meet the demand in construction industry[1]. Hence, it is necessary to find the alteration of new material in the replacement of aggregates. Generally the aggregates which are applied with the concrete are categorized in to coarse aggregate and fine aggregate. Usually the gravels and sand were used as aggregates from the ancient time. Out of which, the sand is consider as a major element in production of concrete [2]. Due to infrastructure growth, the necessity of sand becomes increases, but while concerning environment and scarcity of sand, it is recommended to use the alteration material [3] . Before selecting the suitable alteration material, it is important ot discuss the Already, so many industrial waste such as glass, slag and demolition wastes are utilized instead of sand by partially or fully. In that sequence, we are investigating the influence of replacing crushed sea shells as a fine aggregate in the conventional concrete.

Globally, it was estimated that the about 23% of mollusk a kind of sea shell organism was utilized for food, oil and medicine [4]. Our India was considered as the 9th largest exporter of marine food products in Asia and 16th largest exporter in the world. As per the latest information of fishery department of India, a billion tons of shellfish was harvested every year in the field of aquaculture industry, mostly they are clam shell or cockle shell based [5]. Usually, the shell are split in to two halves are joined with the hinge portion. This type of shell fishes are avail abundantly on west and east coastal regions of our India. After extracting the flesh, the shells are remaining as waste [6]. Due to poor disposal practices, the shells are become as a environment polluted materials and may cause several serious issues in future. Hence the usage of those sea shells in the form of crushed aggregate with concrete can reduce the problem of disposal and being a ecofriendly or sustainable material in the field of construction.

2. MATERIAL AND METHODS

The following are the materials which are utilized in this study

2.1 Cement

Cement is an essential element in the concrete; the major role of cement is to make the concrete as heterogeneous one by combined all the ingredients together and provides the strength to the concrete [7]. For this study, the OPC grade- 45 cement was used and its specific gravity was measured as 1.818.

2.2 Aggregate

Aggregates are consider as the packing material of the concrete, For this study two kind of aggregates are used such as sand and Crushed sea shell [8,10]. The sand, which is collected from the estuaries of Pallar river, near Kalpakkam, Tamilnadu and the Sea shells were collected from the coastal region of Mahabalipuram, Tamil nadu as shown in Figure 1. The specific gravity of collected sand and aggregate were analyzed under the standard specification and found the values such as 2.74 for sand and 2.148 for crushed sea shell respectively.

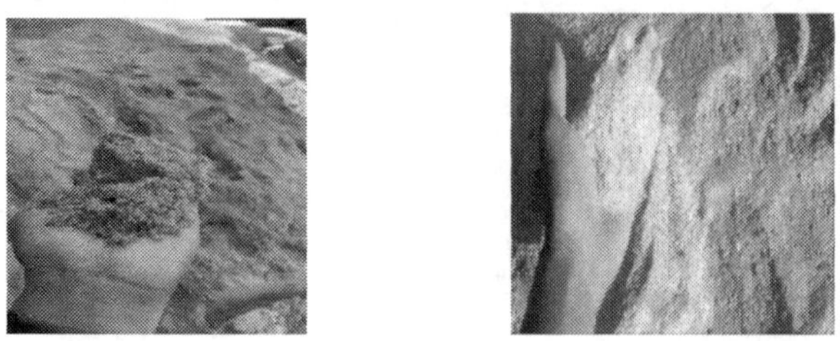

Pallar River Sand **Crushed Sea Shell**

Figure 1. Sample of River sand and Crushed sea shell

3. MIX DESIGN

As per IS 10262:2009 guidelines, by using the specific Gravity value of materials the mix design of M25 Grade concrete was arrived and find the proportion was arrived in the form of ratio as 1:3.23:3.85 with the standard water cement ratio of 0.4 [9]. The target mean compressive strength was calculated as 31.6 N/mm2 achieved at 28 days of curing. The crushed sea shell was introduced at 0%, 10%, 20% and 30% with the concrete by replacing the sand. The weight of crushed sea shell at each proportion was defined in the table 1 below.

TABLE 1. Proportion of each materials under different Proportion

Sno	Materials [kg/m^3] 1 : 3.23: 3.85	0%	10%	20%	30%
1	Cement	269	269	269	269
2	Sand	834	750.6	667.2	583.8
3	Crushed shell	-	83.4	166.8	250.2
4	Coarse aggregate	891	891	891	891

4. RESULT AND DISCUSSION

After the preliminary investigation of all raw materials and the proportion of design of mix, the concrete has to be tested at hardened form to find out its compression and tensile behavior. In this context, it is help to understand the behavior of crushed shell admixed concrete comparatively with conventional sand concrete based on the following test results.

4.1. Compressive strength of Cement Mortar Mixes

Before Preparing any type of concrete, it is advisable to check its compression behavior by using cement mortar cube of standard size 70.6mm x 70.6mm, compatible to IS : 10080-1982. For this study the cement mortar cubes were prepared under the ratio of 1:5 and Soaking the hardened demolded specimens in water for the period of 28 days. Later, it was tested under the series of compression load and the following results are shown in the Figure 2 below.

From the following results, it was understand that, the addition of crushed sea shell with cement mortar at different proportion can moderately reduce the strength of mortar mix when compared to conventional cement mortar mix. However, it was suggested that, to check the compressive behavior of each ratio of material when react with concrete as individually.

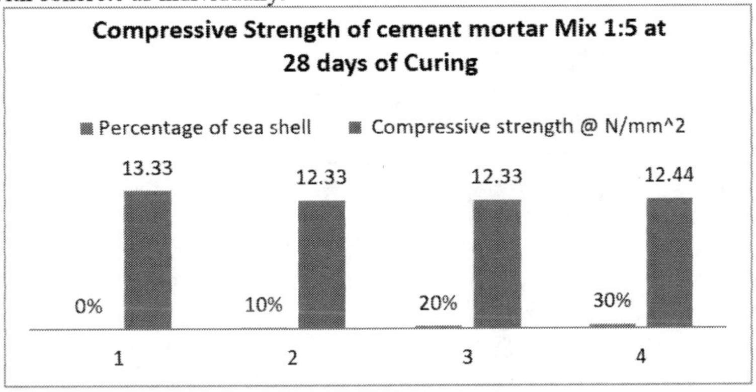

Figure 2. Compression behavior of Crushed sea shell in Cement mortar mix

4.2. Comparison study of compressive strength of cement concrete at different curing periods

The compression behavior was considered as the very essential parameter of any type of concrete. The stability of any structure was decided also by means of using compressive strength. Basically, the compression behavior of concrete was tested using the standard cubes of size 150mm x 150mm x 150mm as per the specification of IS456:2000. Here, the cube are tested under the controlled condition with the universal testing machine of capacity of 2000kN was shown in Figure 3 below.

Figure 3. Compression of Cube Specimens

4.3. Compressive strength of cement concrete at 7 days curing periods

The Figure 4 below, which depicts the comparative average compression strength of conventional concrete and the crushed sea shell replaced concrete at the curing period of 7 days. From the results, it was noticed that replacing 10% of sea shell in the conventional concrete could provide almost equivalent result comparatively with the conventional concrete. But, while increasing the percentage of replacement of sea shell, have decline the strength of concrete drastically when compared to pure conventional concrete. It was suggested that, to monitor the variation of strength in further curing periods.

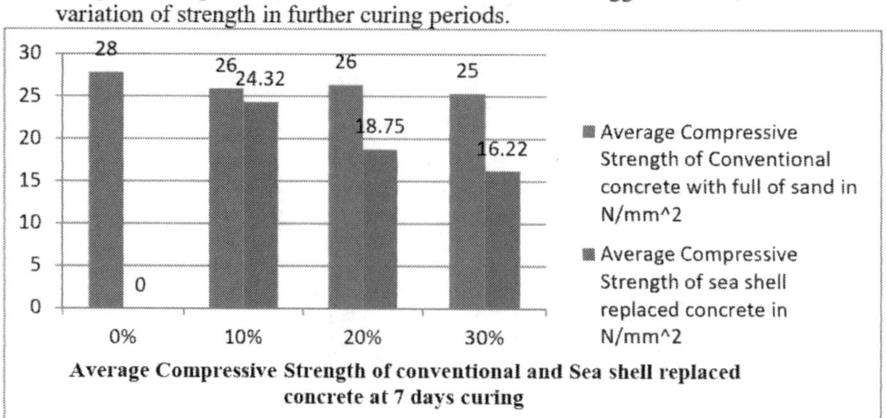

Figure 4. Average Compression strength of both concrete at 7 days curing period

4.4. Compressive strength of cement concrete at 14 days curing periods

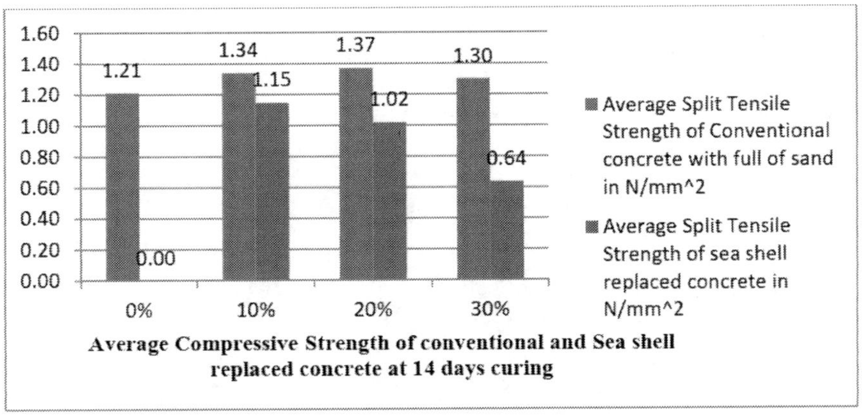

Figure 5. Average Compression strength of both concrete at 14 days curing period

The Figure 5 above, which represents the comparative average compression strength of conventional concrete and the crushed sea shell replaced concrete at the curing period of 14 days. From the results, it was noticed that replacing 10% and 20% of sea shell in the l concrete could provide noticeable result comparatively with the Design strength of conventional concrete. But, while increasing the percentage of replacement of sea shell, have decline the strength of concrete drastically when compared to pure conventional concrete. It was suggested that, to monitor the variation of strength in further curing periods.

4.5. Compressive strength of cement concrete at 28 days curing periods

The Figure 6 below, which represents the comparative average compression strength of conventional concrete and the crushed sea shell replaced concrete at the curing period of 28 days. Through the results, it was observed that almost all the replacing 10%, 20% of sea shell in the concrete was provide some promisable result comparatively with the Design strength of conventional concrete But not more than conventional concrete. Hence, it was suggested that, the addition of sea shell at the 20% was provide better results than other type of replaced concrete.

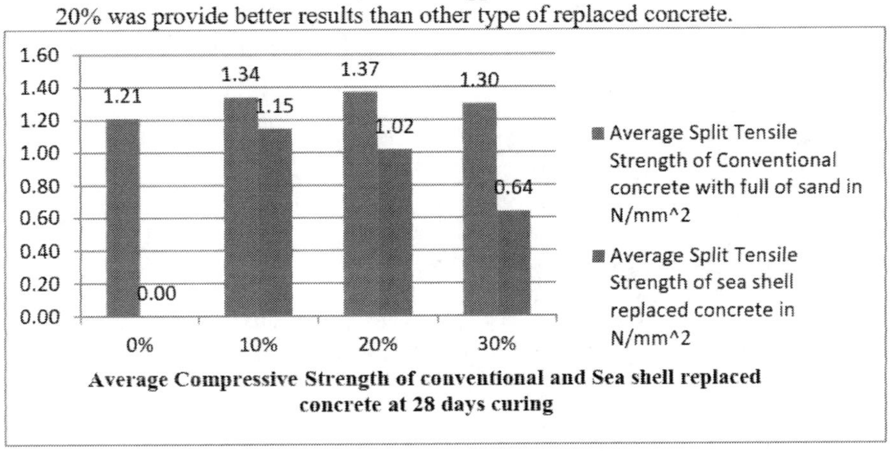

Figure 6. Average Compression strength of both concrete at 28 days curing period

4.6. Split Tensile behavior of Concrete

Even though the concrete is good in compression, it also need to resist the Tensile force which acting over it. In the RC structures, the major portion of tensile force are carried out by the steel, However it is mandatory to carry the minimum percentage of tensile force by the concrete to avoid the failure of the structure. In this study, the Split tensile behaviour of concrete was examined using standard cylindrical specimen of length 300mm and diameter is 150mm by the ratio of 1:2. Figure 7 which shows the Split tensile testing of cylindrical specimen at 28 days of curing.

Figure 7. Split Tensile Test of Cylindrical Specimen

4.7. Split Tensile strength of cement concrete at 28 days curing periods

The Figure 8 below, which explains the comparative average split strength of conventional concrete and the crushed sea shell replaced concrete at the curing period of 28 days. While Concluded the results, it was observed that almost all the replacing 10%, 20% and 30%of sea shell in the concrete was not provide some promotable result comparatively with the Design strength of conventional concrete But during 10% addition of Crushed sea shell may provide some comparative result with conventional concrete. Hence, it was suggested that, the addition of sea shell at the 10% was provide better results than other type of replaced concrete.

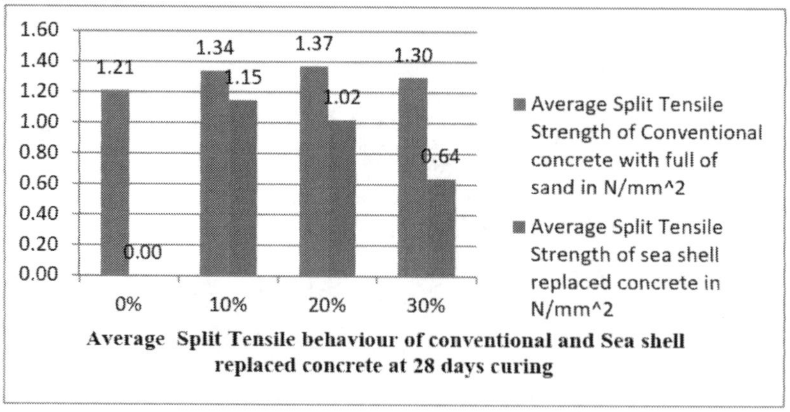

Figure 8. Split Tensile Test of Cylindrical Specimen at 28 days of curing

5. CONCLUSION

The following points are recommended as conclusion based on the justification of results so far discussed in this study.

- The utilization of crushed sea shells provide promotable result when compared to conventional sand based concrete
- The replacement of 20% of crushed sea shells were provide appropriate design strength of concrete, but not same as conventional concrete strength
- The replacement of 10% of crushed sea shells were improve the tensile behaviour of concrete, but not meet the expectation of conventional concrete
- Based on result, it was suggested that the replacement of crushed sea shell at the minimum percentage can provide expected result and control the rate of exploitation of construction raw materials.
- By utilizing the crushed sea shell, may reduce the disposal problem of those wastes and control the overall cost of construction project as well as growth the economy of aquaculture industry.

Reference

[1] Vijayan D.S "Natural aggregates used for Light weight concrete - A Review" IOP Conference Series: Materials Science and Engineering, 2020, 993(1), 012042

[2] Leema Rose "Experimental studies on strength and durability of sustainable concrete using bottom ash by replacement of fine aggregate" Journal of Green Engineering, 2020, 10(9), pp. 6938–6949

[3] R Sanjay Kumar, "Effect of silica fume on strength of glass fiber incorporated concrete" AIP Conference Proceedings, 2271, 030020 (2020); https://doi.org/10.1063/5.0024775

[4] R.Abirami "Mechanical behaviour on Replacement of Aggregate by E-waste in Concrete" International Journal of Innovative Technology and Exploring Engineering (IJITEE), 2019/12, 303-305.

[5] Parthiban.D "Performance evaluation of Fly ash based GPC with partial replacement of RHA as a cementitious material" Materials Today: Proceedings, 2020, 33, pp. 550–558

[6] Parthiban.D "Study on Stress-Strain effect of reinforced Metakaolin based GPC under compression" Materials Today: Proceedings, 2020, 22, pp. 822–828

[7] Nivetha.C "Green composite form of eco-friendly concrete by adding pva fiber" Journal of Green Engineering, 2020, 10(6), pp. 3084–3101

[8] Aravindan.S "Characteristic study of concrete by replacing glass cullet and ceramic tiles over conventional aggregates" International Journal of Scientific and Technology Research, 2019, 8(10), pp. 1802–1805

[9] Saravanan.B "A research on replacement of red-mud in M30 grade of concrete" International Journal of Innovative Technology and Exploring Engineering, 2019, 9(1), pp. 4177–4180

[10] P Subathra " Experimental investigation on concrete with marble dust and steel fiber" AIP Conference Proceedings, Volume 2271, Issue 1, 10.1063/5.0024779

DISCERNMENT AND DIMINUTION OF BLACK HOLE ATTACK IN MOBILE AD-HOC NETWORK USING ARTIFICIAL INTELLIGENCE

[1] K Srinivas, [2]V Harsha Shastri, [3] Vinay Kumar Nassa, [4]Gudapati Syam Prasad, [5]Prathipati Ratna Kumar,

[1]Professor, Department of Computer Science and Engineering, Koneru Lakshmaiah Education Foundation, Hyderabad - 500075.Email: srirecw9@klh.edu.in

[2]Assistant Professor, Department of Computer Science and Engineering, Loyola Academy, Secunderabad-500010, Telangana. Email: harshashastri28@gmail.com

[3]Principal/Professor, Department of Computer Science & Engineering, South Point Group of Institutions, Sonepat-131001(Hr.) Email: vinay_nassa@rediffmail.com

[4]Professor & HOD, Department of IT, Narasaraopeta Engineering College, Narasaraopeta, Guntur District. Email: syamprasad.gudapati@gmail.com

[5]Assistant Professor, Department of Computer Science and Engineering, Koneru Lakshmaiah Education Foundation, Hyderabad – 500075.
Email: rk30111972@klh.edu.in

Corresponding Author Email ID : srirecw9@klh.edu.in

Abstract: The MANET is an interesting research topic that has movable nodes. The mobile ad hoc system is an important research area. The nodes are connected by wireless connections. Thus it is vital to have a stable network because information and interaction are crucial in diverse fields such as security and catastrophe emergency response. Because of the decentralized system's vibrant existence, these channels are vulnerable to various threats, including BHA, GHA, SHA. One of the well-known security issues in MANET is the black hole assault. Each node has a routing table containing information from the target node. The number of assaults in MANET is usually protocol networking assaults. The study focused on black hole attacks for discernment and decrease. A black hole attack node wrongly picks up all packet and collects them without passing them to an endpoint by taking a fresh path to the destination nodes. This paper presents BHA defence from assaults with the principle of Artificial Neural Network and calculates its efficiency according to the set of elements, such as efficiency, PDR and delay and energy utilization. This paper aims to devise an artificial neural network method for black hole discernment and reduction in MANET.

Keywords: MANET, Black hole attack, ANN (Artificial Neural network).

1. Introduction

MANET is a self-configuring system without a specific router. Because the node is not centralized and each node acts like a router, it is possible to do this. Each node has a limited range for system data transfer, and communication takes place from one node to another[1]. The MANET routing means that the whole nodes in the network are regular, unless no changes are made to the information and path of the router. MANET finds use in a variety of areas such as the handling of disasters, automotive computing, and more, and uses constructive and hybrid routing mechanics for transmitting data. The connectivity details of points is saved in the table in a dynamic network and updated when the path is shifted. As the node says, the route is generated in a reactive routing protocol if the base station transmits information to the destination network, it operates on request [2]. The final protocol consists of a combination of the benefits with the above MANET Routing Protocol.

Two methods of MANET processing, path exploration, and data transmission are involved. These phases can be attacked in numbers. At first, the rival could stop discovery of the path by copied manipulated control traffic[3]. The assailants can then obstruct the dissemination of legitimed controlling traffic routes and incorrectly exploit basic knowledge of neutral nodes. Two types of threats usually occur in the MANET, one is Passive and the other is Active.

The hacker listens to the contact channel passively during a passive attack without changing or deleting packets of data. But the attacker may alter or delete the actual information in an Active Attack. MANETs are suited to emergencies, such as the nature of disaster response, clinics, battlefields, conventions and practical surveillance, because of limited setup and rapid deployment[3]. Therefore, data transfers must be secure among two nodes. However, active attacks such as a BHA, a RA and a WHA have a major effect on the cable network results.

There is no need for MANET to have fixed networks like access points, so it is an appealing network choice for fast and casual connection of mobile devices. Dynamic topology network, highly variable throughput, multi-hop forwarding, self-organization, self-adaptation, and configuration make it a great choice for a wide network environment, especially for military, personal, meeting, and disaster-area systems. MANETs have unique attributes: self-contained interface, fast deploying, complex topology, fluctuating bandwidth, and limitations on capacity, lack of fixed infrastructure and services, self-assessment, distributed service and personal safety [4]. Five core security priorities establish an ad-hoc network atmosphere that is safe and stable.

Confidential information, honesty, accessibility, security and non-repudiation are all. Attack can be categorized as two major types of MANETs: passive and aggressive threats. Passive assault forms are different: snooping, location data and monitoring of flows. Sleep starvation, wormhole assault, BHA, SHA, gray hole assault , rash assault, Sybil assault and DDoS assault [11], are active threats. The goal of this research is to present an artificial neural network (ANN) mechanism for detecting a black hole attack in MANET [5]. ANNs are an artificial intelligence framework which can serve as a powerful instrument to identify malicious activity in MANETs[15].

2. Review of Literature

Ashish et al. [7] Presented a Fuzzy approach based on an enhanced binary connection to decrease the effects of the AODV-routing method Black Hole Attack. Fuzzy classification model logic has been used to distinguish the real node from the assailant frame. Fuzzy Logic operates on a principle collection, which is the biggest downside. Therefore, if we have a seems at of several nodes, there are many laws that cannot be developed by the creator and become complicated.

Rao et al. [8] Concentrated on reducing false responses from database algorithms from malignant entities. Any such structure relies on only detecting the module of assault of the Black hole.

Poonam Yadav et al [4] The framework suggested on the basis of a fuzzy node theory is contaminated or not by a strike of a black hole. The analysis provided solutions to transmission damage in the future of a black hole network intrusion. The first step is the detection of the black hole node using the fluid law. The fuzzy law is applied on node contact reaction time. It can only manage the transfer aimed only at this point, rather than just transmitting information to that node.

Pravin Kshirsagar et al[6] has defined various neural networks to accomplish their aims, such as the activation functions of one layer, layered perceptions, Rbf, Ffnn, neural probabilism networks (Pnn), Grnn, etc. This standard neural network has been considered useful for a limited quantity of data and structures that need modest runtime. But the volume of data to be processed and the system design implementation period are decreased with digital technological advancements. This prompted many researchers to amend the traditional neural networks, which can accommodate a lot of data and have less time to converge. Neural networks may describe or generate the effects of complicated and unreliable data.

Wahane et al. [9] Suggested the amendment to the Protocol on Vector Routing Ad Hoc on Variables Required. The suggested study proposes two new principles, Data packet header table preservation and network cross checking. The proposal for a security procedures can be used to recognize many black hole nodes on a MANET and thus to define a safe forwarding path from the source node to the given destination, thus eliminating black hole nodes.

Mitra et al [10] Suggest a Black Hole automatic node discovery tactic based on Artificial Neural Network (ANN). Dynamic in design is the emerging ANN-based structure. Enacted Black Hole node interconnectivity processes allow to further model different the route table as they operate from both sides, CRC and TTR. The inter process communication scheme is installed.

3. Security Measures in MANET

MANETs have high challenges to protection because there are no predetermined limits, no system opponents, no unified controls and restricted natural resources. Multiple hazards and assaults impact MANETs. Different, but not confined, black hole, buggy, gray hole, snooping, man-in-the-center assaults, grey hole, etc. seriously disrupt the mechanisms of routing and weaken ad hoc network execution[12]. This assailants either are aggressive or passive. One of the deadly assailants identified in this investigation was the Black Hole Attack.

Essentially, two approaches exist, constructive and reactive, to defend from MANET assaults. The constructive solution seeks to deter hackers from initiating crypto assaults in the early stages, while the responsive procedure applies the analytical procedure and thus reacts to protection threats[13]. A full strategy for safety combines all interventions and consists of three areas, e.g. identification, avoidance and response. Protected networking could be assured by different methods for prevention and elimination of safety including accessibility, secrecy, authorization, verification, honesty, non-reproofness and confidentiality. The security risks in MANET are seen in Figure 1.

3.1 Routing in MANET

In MANET, nodes will exchange information if they fall within the contact range of the radar of each other. Otherwise they will interact as intermediary hops to other networks and so data transmission cooperation between nodes is needed[14]. The nodes serve as adapters to communicate and transmit information between an intermediate nodes and a target node. Multi - hop networking system's key goals are to minimize the high cost, battery power and PDR consumption. The protocols are, in a general sense, split into 3 following groups

3.1.1 Re-active or on demand routing protocol:

In this classification, paths are only accessed if a message is to be sent to a particular location and there are no cached paths. At first, there are no roads in the system and only then is a road exploration protocol begun, which would find an optimum intermediate node when a node has to transfer data to another point. The trail is recorded for future use for some time.

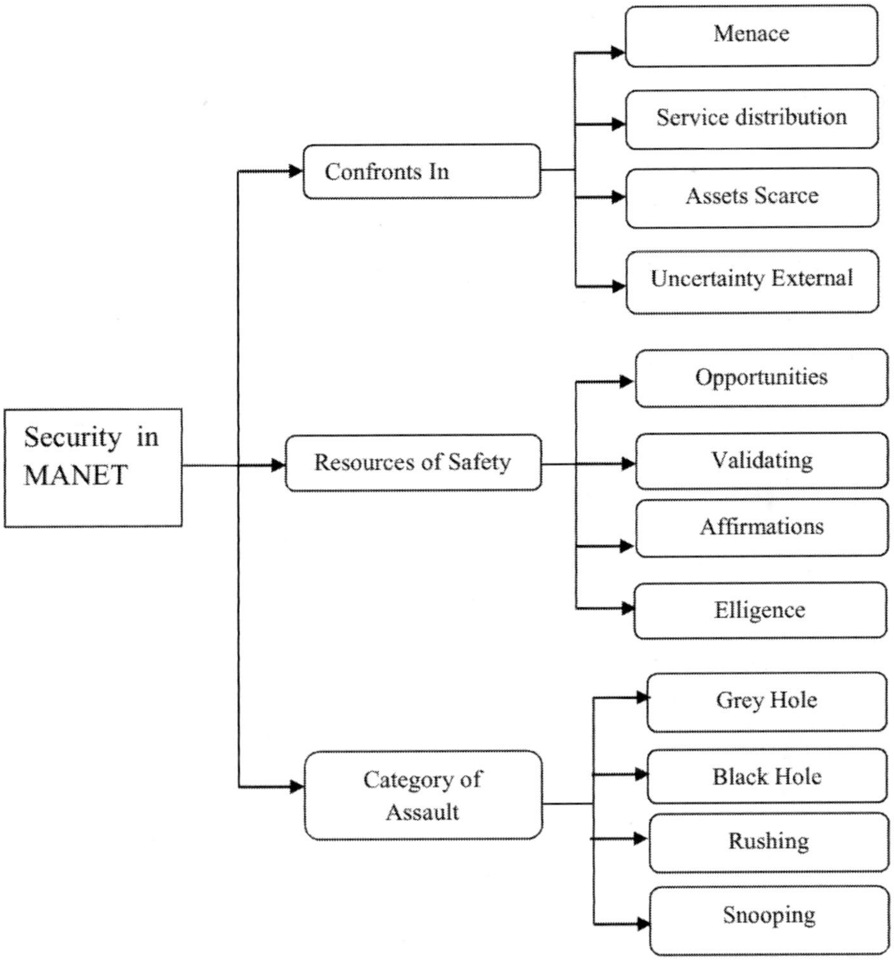

Figure 1: Security Measures in MANET

3.1.2 Pro-active or table driven routing protocol:

All paths are identified in this configuration group, and all routing are accessible and managed in the channel by all the nodes. Each node maintains a table that maintains this always updated with nearby nodes when exchanging the routing table. If a node has to send data to a different node, it scans its route cache and determines next hop along the way. The Optimized State Connection Routing Protocol (OLSR) rank among these categories as Destination Sequence Distance Vector (DSDV) routing protocol.

3.1.3. Hybrid routing protocol:

This method group is a combination of both responsive and constructive methods. Proactive techniques are being used for nearby neighborhoods and path vector mechanisms will be used for areas further out. ART protocol (ZRP) is a type of protocol for hybrid routing. The use of either of these Protocols depends entirely on the device implementer's requirements and needs as well as on the cable network expectations of goals, actions and efficiency.

4 Black hole attack

Black hole attack is a networking level assault where information from another device is rotated. Packets are generally sent on different nodes and packets are dropped on a forwarding node. The assault targets the network architecture. Black hole attacks with a significant impact on the network's virtual mesh[16]. The active black hole assault on the DOS. It is harder to identify Black hole attack; it's often seen in transient networking, such as wireless servers.

The black hole strategy is going to have a strong effect on the system results. When the source node attacks with a black hole, it receives an answer response from the defect nodes. Failure sender forwards answer response to the client and transfers it to confusion in two responses. This makes the fault node the sender node and the whole data it receives. In this, the transmitted signal lowered packets of data completely.

In Figure 2, give node 1 to any neighbouring node with an overwhelming size of RREQ message. If the RREQ request is sent from an error device, the RREP message will be sent to a non-real node that also indicates the shortest route to the device. The sender would then recognize the data packet from the non-real single node the flaw nodes and forward the packets. It is regarded as an invasion by the black hole.

A fault node acknowledged by source nodes is not taken into account in black hole attack and all the connections are lost. This nap derivative attack is also known. This intrusion is split into two categories, namely black hole assault internally and externally. The following are the reasons for these assaults:

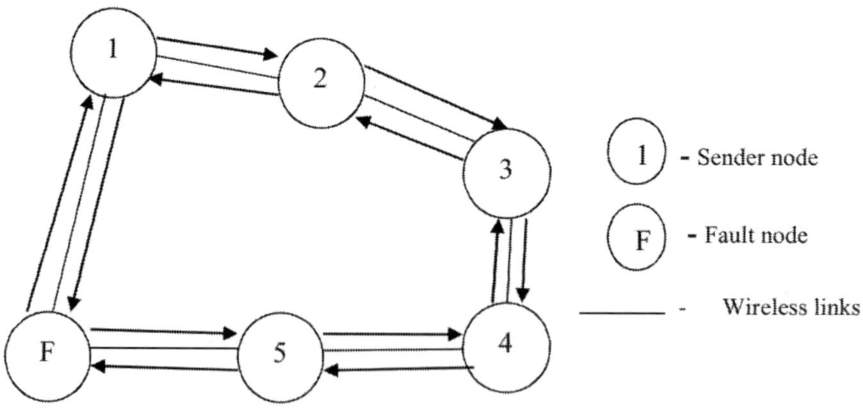

Figure 2: Configuration for Black Hole Attack

4.1Internal black hole attack:

It happens simultaneously in the system. This ensures that the internal node becomes the defect node and Move from source node to destination node.

In Figure 3 send RREQ per each point, the transmitter node 1 returns replies out of each device, and approved networks make up the entire network [16]. Unexpectedly, a defect node is authorized and a black hole attack happens internally.

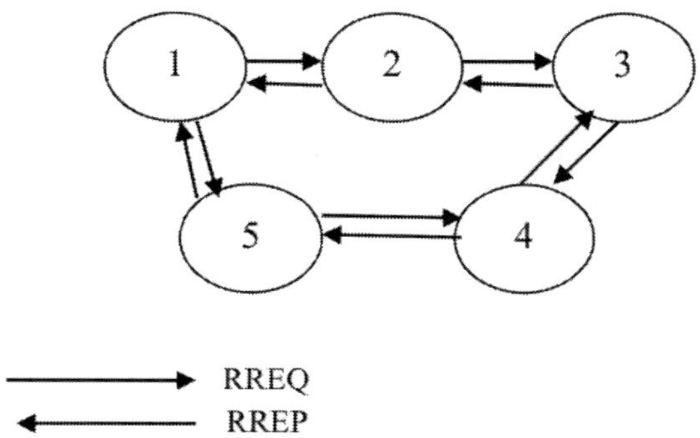

Figure 3: Internal black hole attack

4.2 External black hole attack:

This assault takes place from outside device. The DOS (reject to service) assault is often referred to. The system utilizes data traffic in this intrusion and is breaking down the whole network. External defect node is used and the internal node is worked as well. A few of the moves below will be followed:

- The failure node is enabled and transforms into the node of recipient.
- Transfer Rreq packet to the Defect Node, which indicates the least path becoming a member of the channel. REP text
- It accepts all encrypted messages from the system source nodes.
- The RREQ and RREP messages sequence take place and information exchange is completed and a black hole attack takes place.
- The information receives the default node and the connection will drop 100 percent of the messages.

4.3 Discernment of Black hole Technique:

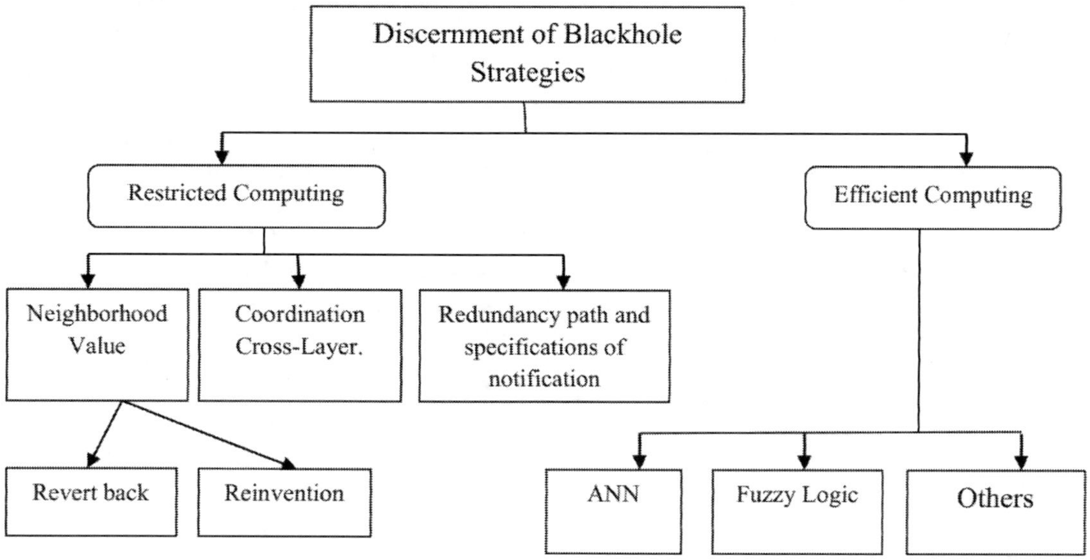

Figure 4: Discernment of Blackhole Technique

Testing procedures computationally constrained and computational cost complex depending on the type of calculation their do. It is seen in Figure 4.

Advanced computing technologies focus on soft computing methods such as Fuzzy Logic, ANN. In comparison, computerized methods focus on basic infrastructure modulation specifications for the identification procedure. The structure of identification is yet another aspect that influences the difficulty and operational needs of the prediction model. Centrally controlled, decentralized and mixed detector technologies can be categorized.

The clustered systems typically need a strong central location that can analyze data gathered and stable traffic to the base station. In the clustered approach, several nodes or all nodes are used to detect. The centralized technique helps to overcome other pitfalls induced by a centralized solution, such as the centralized time of frustration or undue energy usage at one device in the network or need of strong networks, in addition to proper identification and elimination of additional delays in transmitting parameters among endpoints. Some methods use a mixed technique to obtain all the benefits of dispersed and centralized techniques by initiating central nodes through other devices in the system for the completion of a detecting process[17].

4.4 Diminution of Black Hole Attack Technique:

When there is no direction in the route cache for the endpoint, the usual node will transmit the route discovery process. The intermediary legitimate node tests whether the path concepts defined is received. It verifies the whether path to the target is valid[18]. If a correct route exists, the actual response packet will be sent out instead, the path query message will be released, but malignant node will leverage the vulnerability of the network applications that are usually programmed to mutually cooperate between the entities and will provide the wrong control messages to unleash the BHA. The black hole attack may deny service attacks and falls in under the connection falling attack category that disrupts contact between both the base station and the endpoint node after the route has been taken by sometimes making false statements to the base station in the route repair message and then starting to drop packets.

In case of total packet error assault, the intruder drops all network packets and does not truly engage in path and sends erroneous data to the base station in the server responds so as to draw congestion to itself, while the attacker user reduces a number of packets if a partial constant bit event occurs. It may honestly be involved in the development of road exploration and falls specific database shipments referred to as smart grey hole attack. The intruder may even send incorrect information, and then pick the datagram's to drop, while having a correct path to the destination, named a grey hole control messages assault as seen in Fig. 5

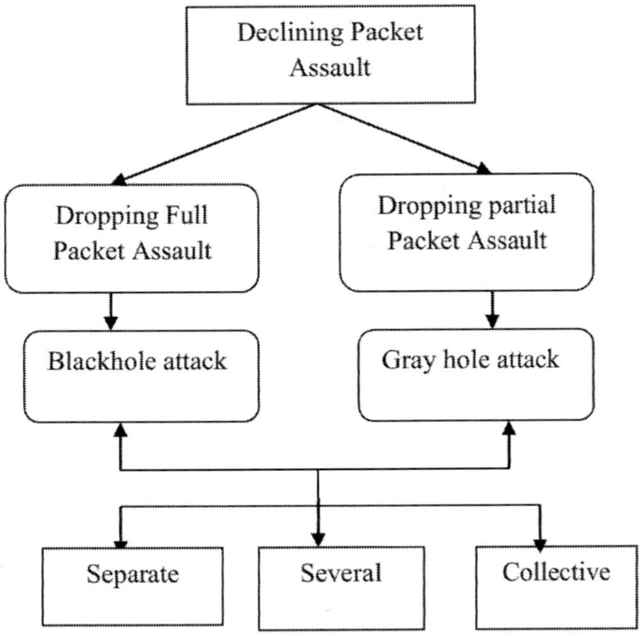

Figure 5: Diminution of Black Hole Attack Technique

5. Research Methodology

5.1 Artificial Neural Network:

An input data, one or two hidden layers and output layers is used in the structure of the ANN. In each network layer a series of interconnected modules consists of nodes. These neurons associate with each other on a weight-based basis. In the next layer, each neuron is connected to all neurons. The information is seen in the neural network input layer[21]. For the required input data, the neural network output shall be shown by the output layer. This connected devices can be calculated with hidden layers between input and output. Design of Artificial Neural Network is shown in Figure 6.

The number of hidden must be selected in the severity of the problem. Normally, a hidden layer is required to discuss much of the problem. Based on the size of the problems, the quantities of neurons in the secret layer is determined from the minimum and then raised by experiments. The neural network process includes a set of the input and destination output values. For output effect on the aquatic, input parameters were selected. The procedure for the back replication of genetic disorders has been commonly used [19].

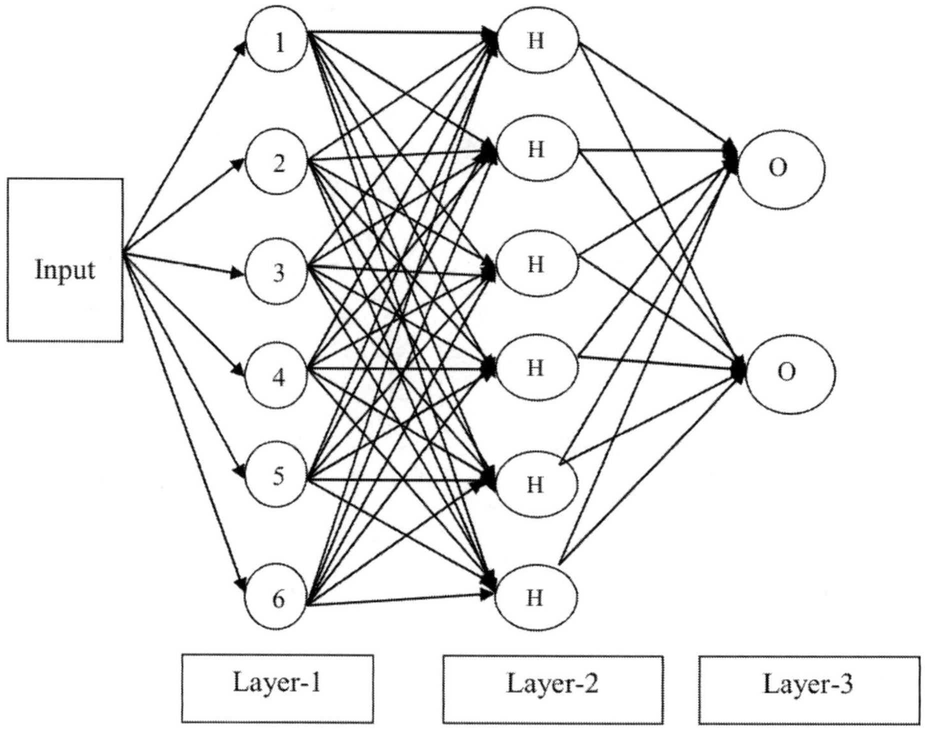

Figure 6: Architecture of Artificial Neural Network

The reverse propagation is essentially a conditional reduction mechanism to lower the system error model:

$$E = \sum_{j=1}^{K} \sum_{i=1}^{n} (e_i\,(j) - t_i(j))$$

where ei(j) and ti(j) are simultaneously calculated and focused. The performance samples quantity is 'n,' and the training samples number is 'k.'

Until the ANN instruction begins, weights are randomized automatically. Values are adjusted on the basis of mistake reproduction

$$\Delta W_{ij}(n) = \propto \frac{dE}{dW_{ij}} + \eta \Delta W_{ij}(n-1)$$

Wij(n) and Wij(n − 1) Where The training rate and stimulus factor improve in the nearest increments between the I and J nodes, and α and β. Adequate analysis and modification of the training data is necessary for the efficient training of the recurring neural network.

ANN is a method in artificial intelligence that is used by the human brain definition to overcome the dynamic sequence issue. ANN consists of several simultaneous components. ANN uses weight mainly. At first, ANN produces a wild assumption for the current problem. The network then checks the target outcome and makes corresponding weight adjustments[20].

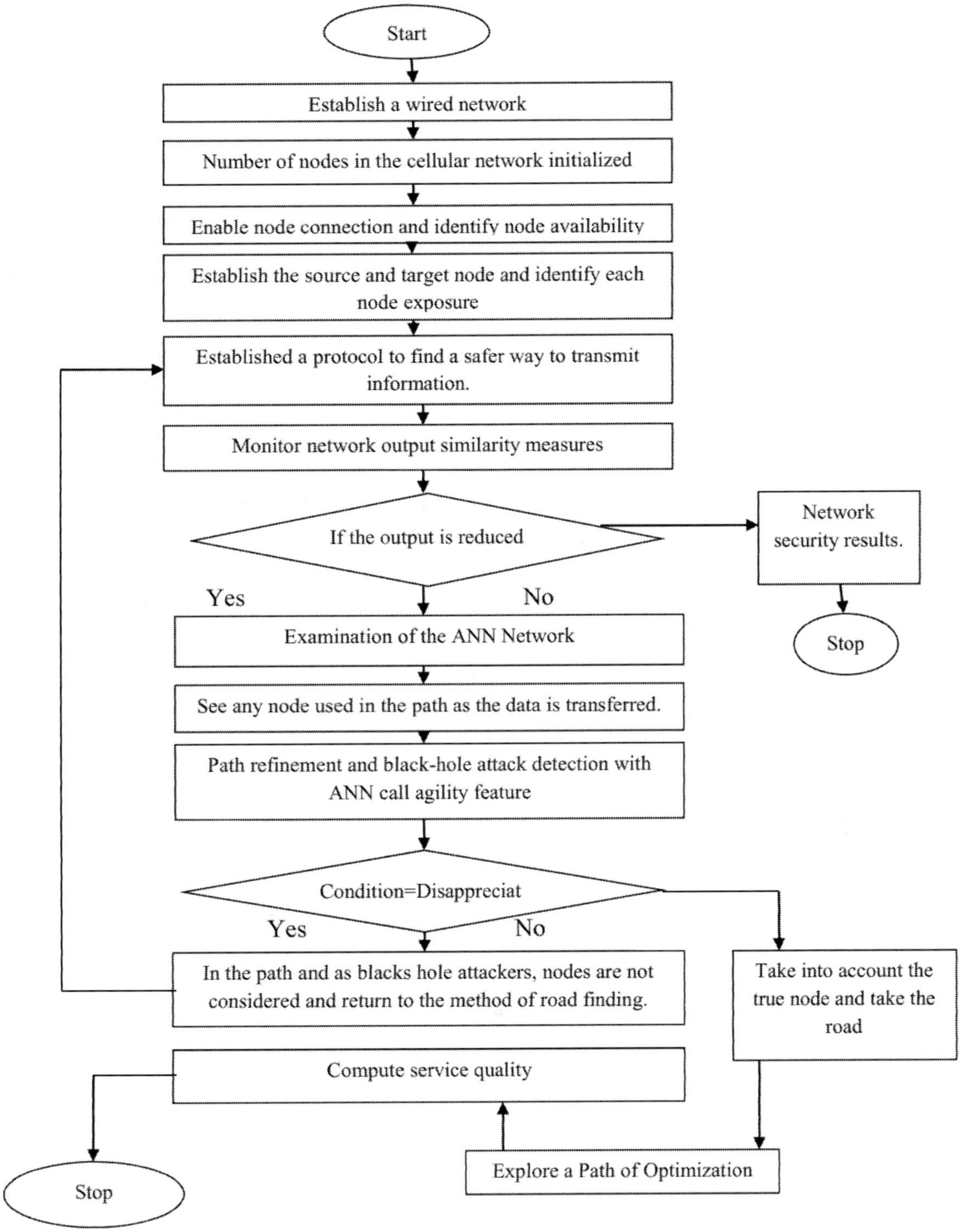

Figure 7: Flowchart of proposed work

In this research, the nodes' optimum characteristics are given as the input to the ANN input layer, which is learned from ancient times by the neural network to distinguish Black hole nodes from legitimate nodes[24].

ANN is used in the same analysis to investigate and protect the Black hole node of the legitime node. Adjust to life used to identify Black hole attack as seen in Figure 3[23]. Figure 3. The ANN arrangement used in this report consists of three major layers in the neural: the input, the mystic and the output layers. The number of nodes in the secret layer is chosen so that neural network neither overlap nor underlay exists. In addition, per the hidden layer, the neurons number in the output layer is automatically created. In the unseen layer, the neurons are adjusted with a bias tension. This prejudicial approach is applied so as to achieve the optimal output[25]. Figure 7 displays the conceptual flow chart.

6. Performance Metrics

The criteria used to calculate the work's output are:

6.1 Throughput:

The total data packets transmitted during the simulation time is listed. That is the cumulative amount of data transferred in an acceptable time period from input to output. The output can be measured as percentages in terms of Mbps, Kbps and Gbps.

Throughput can be described as

$$Throughput = \frac{\sum Packets\ sent}{Total\ data\ packets} \quad \text{--------------(1)}$$

6.2 PDR calculation:

The ratio is determined with equation 3 between both the entire number of data packets achieved and the sum of packets received.

$$PDR = \frac{Number\ of\ received\ packets}{Number\ of\ sent\ packets} \times 100 \quad \text{--------------(2)}$$

6.3 End-To-End delay calculation:

The default rate is measured whether the information consume a while to reach its target base station of the transmitter and the fixed time period using equation 4.

$$End\ to\ end\ delay = \frac{Received\ packets\ time}{Sent\ packets\ time} \quad \text{---------------(3)}$$

6.4 Energy Consumption:

It is measured as the total power a network uses when transmitting packet information from the source node to the target node. It is possible in Joules to quantify it. It can be described mathematically by:

$$Energy\ consumption = E_{TX} + E_{RX} + E_{AMP} + E_{Amp} + E_{Agg} + E_{prop} \text{----------(4)}$$

7. Result and Discussion

7.1 Throughput calculation:

Nodes	Throughput (Kbps)	
	Fuzzy logic	Artificial Neural Network
12	53	56
22	51	54
32	48	52
42	45	48
52	43	45
62	38	44
72	32	42
82	31	38

Table1: Evaluation of throughput

The performance was calculated when the attack happened and then when the system was blocked from attacking Black hole. The improvement in percentage estimated following the Black hole network prevention using the suggested Artificial Neural Network algorithm is11.12 percent. The results of the proposed ANN model could be seen from figure 8 and table 1, where the Black hole nodes are 12 - 52 percentiles. By the number of Black hole nodes, output was dramatically increased and 70% of Black hole nodes became the greatest difference.

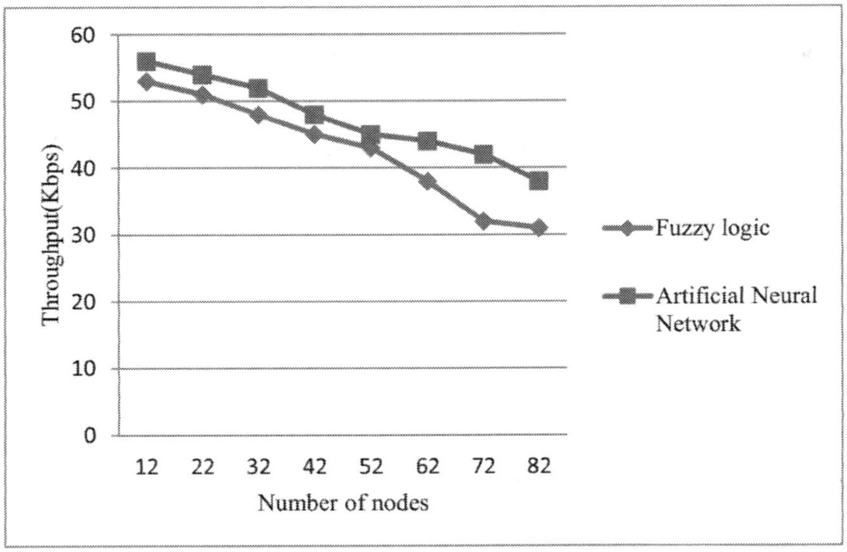

Figure 8: Evaluation of throughput

7.2 PDR Calculation:

Nodes	**PDR (%)**	
	Fuzzy logic	**Artificial Neural Network**
12	91	93
22	90	92
32	79	82
42	70	73
52	52	59
62	50	54
72	47	52
82	30	35

Table 2: Evaluation of PDR

The chart among PDR and Black hole nodes in table 2 and figure 9 shows a decline of the packet sent to the network as the percentages of Black hole nodes are increased. The overall PDR value of 63.88 percent and 68 percent for both the ANN and current job measures. So, the ANN algorithm is used with an increase of 4.68%. Figure 9 shows that up to 22 percent of the Black hole node stays stable in PDR and that the platform's networks then decline dramatically in excess of 22 percent. It's because the Black hole nodes are strictly losing the data packet.

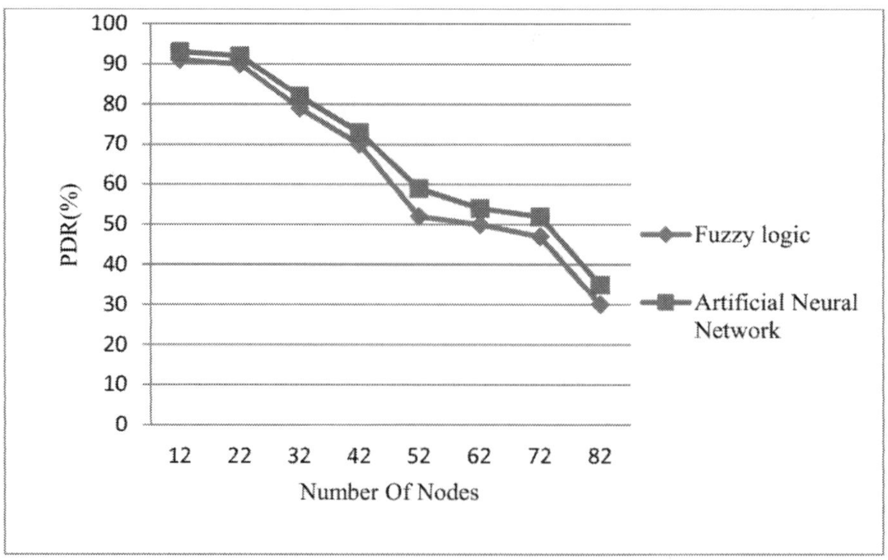

Figure 9: Evaluation of PDR

7.3 End To End Delay Calculations:

Nodes	End - End Delay (s)	
	Fuzzy logic	Artificial Neural Network
12	9	7
22	13	11
32	17	15
42	30	28
52	33	31
62	40	38
72	55	53
82	65	62

Table 3: Assessment of delay from end to end

The schematic diagram of the time limit calculated for ANN models and current work was seen in Table 3 and Fig. 10. The graph shows that the delay in the arrival of the packet data grows as a proportion of Black hole nodes too. In the trial, both the AB NN model and the existing work are 31.26 and 32.88% for average percentages of end-to-end time. The wait was therefore shortened by 4.93% of the research currently under way. The analysis of the Black hole node delay (12% to 32%) is smaller, and 32% to 702% of Black hole node.

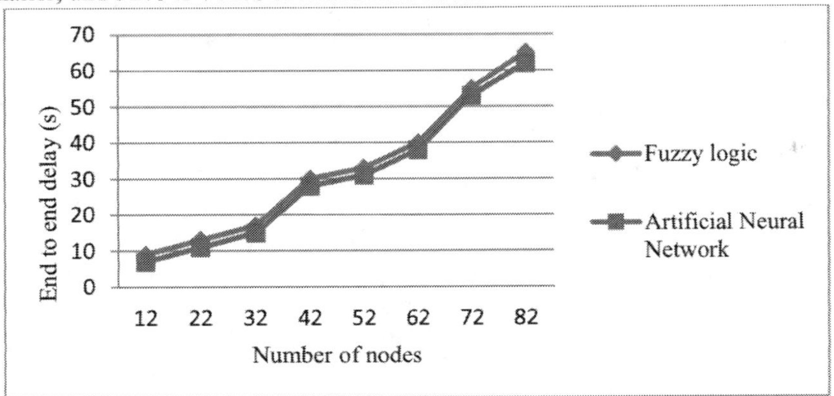

Figure 10: Assessment of delay from end to end

7.4 Energy Consumption:

No. of Iterations	Energy consumption (J)	
	Fuzzy logic	Artificial Neural Network
1	72	62
2	52	42
3	32	28
4	62	58
5	48	38

Table 4: Energy Consumption

The ANN model is estimated in five rounds as the external energy usage parameter. Figure11 and Table 4 display the power consumption values that are calculated with or without avoidance algorithms. The total energy value absorbed by networks with no algorithms for prevention and avoidance algorithms is respectively 53.38 J and 44.92 J, an increase of 15.85% when the Black hole System is mitigating the attack. This means that Black hole nodes waste a lot of resources in the network, and thus keep the network from attacking Black hole and saving energy.

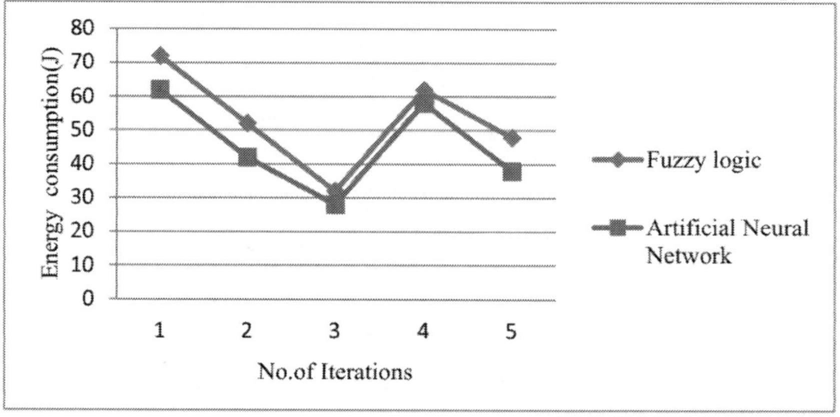

Figure 11: Energy Consumption

8. Conclusion

For any connectivity network, secure communication is a crucial prerequisite. Further security issues may be created through contact via air media (wireless communication). MANETs that configure themselves with portable devices enable improved cyber threats. The safety of the MANETS is an imperative step with a massive rise in cyber attacks. The assault by Black hole is one of MANET's best-known attacks. A protection framework is introduced using the principle of artificial intelligence to protect the network from Black hole attacks (ANN). The current research shows that almost all researchers use fuzzy logic as a classification technique, leading to a strong measurement, as it takes a lot of parameter settings and a huge storage space. ANN is used to resolve these issues and also provide a stable Black hole assault network. In this article, ANNs modeling was examined for the detection of black hole attacks. The input data for neural network training where using throughput, packet distribution ratio and end-to-end delay. They should adjust the parameters of the training and validation parameters. It has been shown that the developed framework can efficiently be used to locate black hole nodes. Therefore, the network performance was significantly increased.

References

[1] T. Singh, J. Singh, and S. Sharma, May 2017 ``Energy efficient secured routing protocol for MANETs," *Wireless Netw.*, vol. 23, no. 4, pp. 1001-1009.

[2] E. O. Ochola, L. F. Mejaele, M. M. Eloff, and J. A. van der Poll, 2017``MANET reactive routing protocols node mobility variation effect in analysing the impact of black hole attack," *SAIEE Afr. Res. J.*, vol. 108, no. 2, pp. 80-92.

[3] Shahabi, S., Ghazvini, M., & Bakhtiarian, M. (2016). A Modified algorithm to improve security and performance of AODV protocol against black hole attack. Wireless Networks, 22(5), 1505-1511.

[4] Yadav Poonam, Kumar Naveen, Gill R.K., "A Fuzzy Based Approach to Detect Black Hole Attack" in International Journal of Soft Computing And Engineering (IJSCE), ISSN: 2231-2306, Volume-2, Issue-3, July 2012.

[5] Sujatha K.S. , Dharmar V. ,2012, Bhuvaneswaran R.S., "Design of genetic algorithm based IDS for MANET", IEEE, pp.28-33.

[6] P. Kshirsagar, S. Akojwar, Nidhi D. Bajaj , "A hybridised neural network and optimisation algorithms for prediction and classification of neurological disorders" International Journal of Biomedical Engineering and Technology ,vol. 28,Issue 4,2018.

[7] Ashish., Tokekar V. and Shrivastava S., 2018 ,"Security Enhancement in MANETs Using Fuzzy-Based Trust Computation Against Blackhole Attacks", in *Information and Communication Technology*, pp. 39-47, Springer, Singapore.

[8] R. L. Rao, B. Satyanarayana, and Kondaiah, 2018 ,"Performance of CBIDS on AODV Routing Protocol against Blackhole attacks in MANET", *International Journal of Scientific Research in Computer Science and Information Technology*, vol. 3, no. 3, pp. 1637-1644.

[9] G. Wahane, A. Kanthe, 2014,"Techniques for detection of cooperative Black hole Attack in MANET" in *IOSR-JCE.*

[10] A. Mitra, R. Ghosh, A. Chakraborty, D. Srivastva, 2013 ,"An Alternative Approach to Detect Presence of Black HoleNodes in Mobile Ad-Hoc Network Using Artificial Neural Network" in IJARCSSE.

[11] Pravin Kshirsagar And Sudhir Akojwar- July 2016 **"**Hybrid Heuristic Optimization For Benchmark Datasets**"** *International Journal Of Computer Applications* 146(7):11-16.

12] T. A. S. Srinivas and S. M. Manivannan, "Preventing collaborative black hole attack in IoT construction using a CBHA– AODV routing protocol," International Journal of Grid and High Performance Computing, vol. 12, no. 2, pp. 25–46, 2020.

[13] S. Shrestha, R. Baidya, B. Giri, and A. Thapa, "Securing blackhole attacks in MANETs using modified sequence number in AODV routing protocol," in in 2020 8th International Electrical Engineering Congress (iEECON), pp. 1–4, IEEE, 2020.

[14] Shashi Gurung, Siddhartha Chauhan, A review of black-hole attack mitigation techniques and its drawbacks in mobile ad-hoc network, in: International Conference on Wireless Communications, Signal Processing and Networking (WiSPNET), IEEE, 2017, pp. 2379–2385.

[15] P Kshirsagar and V Deshmukh, -2013, **"** National Conference On Innovative Paradigms In Engineering & Technology". Proceedings Published By *International Journal Of Computer Applications®*(Ijca),

[16] Dr.Parminder Singh,Damanpreet Kaur," Approach to Improve the Performance of WSN during Wormhole Attack using Promiscuous Mode", volume 73,international journal of computer application, july 2013,pp 26-29.

[17] Lakshmi, S., & Radha, S. (2013). Selfish Scheduler For Packet Scheduling Based On Packet. Online, 9(1), 37-42.

[18] Dorri, A. (2016). An EDRI-based approach for detecting and eliminating cooperative black hole nodes in MANET. Wireless Networks.

[19] R. M. Aghav, S. Kumar, and S. N. Mukherjee, "Artificial neural network modeling in competitive adsorption of phenol and resorcinol from water environment using some

carbonaceous adsorbents," *Journal ofHazardous Materials*, vol. 188,no. 1-3, pp. 67–77, 2011.

[20] Poongodi, T., & Karthikeyan, M, "Localized secure routing architecture against cooperative black hole attack in mobile Ad hoc networks," Wireless Personal Communications, vol. 90, issue 2, pp. 1039- 1050,2016.

[21] Pravin Kshirsagar and Sudhir Akojwar, "Prediction of neurological disorders using optimized neural network" International Conference on Signal Processing, Communication, Power and Embedded System (SCOPES),2016.

[22] M. V. Anand & S. Hariharan, "Bee-mimetic DSR Based Secure Routing in MANET using Artificial Bee Colony Optimization", *International Journal of Pure and Applied Mathematics*, vol. 119, no. 17, pp.1337-1350, 2018.

[23] P. Gupta, P. Goel, P. Varshney, and N. Tyagi, "Reliability Factor Based AODV Protocol: Prevention of Blackhole Attack in MANET", in *Smart Innovations in Communication and Computational Sciences*, pp. 271-279). Springer, Singapore, 2019.

[24] Pravin R. Kshirsagar, Pranav P. Chippalkatti, Swagat M. Karve ,2018," Performance Optimization of Neural Network using GA Incorporated PSO" *Journal of Advanced Research in Dynamical and Control Systems.*

[25] N. Panda, and B. K. Pattanayak, "Energy aware detection and prevention of blackhole attack in MANET", in *International Journal of Engineering and Technology* (UAE), vol. 7, no. 26, pp. 135-140, 2018

[26] A Seenu, M R Narasingarao, U S S Padmajyothi, International Journal of Advance Research in Computer Science, 2014, "A System for Personal Information Management Using an Efficient Multidimesional Fuzzy Search".

[27] Gaikwad Kiran P, Dr. C M Sheela Rani, International Journal of Scientific & Technology Research, 2020, "Regression Model with Modified Linear Discriminant Analysis Features for Bimodel Emotion Recognition".

[28] Mr. Prashant MininathMane, Dr. C M Sheela Rani, IJPAM, 2018, "Triple Data Encryption Algorithm based Multiple Authority Access Control in Cloud System Using Optional Threshold".

[29] D V Manasa, M R Narasinga Rao, A S Lalitha, International Journal of Applied Engineering and Research, 2014, "Analyzing the Services and Privacy Conflict Resolutions of Shared Data in OSNs".

[30] Swathi Karkarlapudi, M R Narasinga Rao, International Journal of Science and Advanced Technology, 2011, "A Novel Facial Recognition Model Using an Artificial Neural Networks"

[31] Sunanda Nalajala, Kanekar Akhil, Venkat Sai, Divyendu Shekhar, Praveen Tumuluru, 2019, "Light Weight Secure Data Sharing Scheme for Mobile Cloud Computing".

Role of Machine Learning in Software Project Management

R. Mamatha[1], Dr K. G. Suma [2]
[1]Research Scholar, Department of CSE, KoneruLakshmaiah Education Foundation, Aziz Nagar, Hyderabad, Telangana State, India and Assistant Professor, Department of CSE, G.Narayanamma Institute of Technology and Science, Shaikpet, Hyderabad, Telangana State, India. mamatha.racharla@klh.edu.in
[2]Associate Professor, Department of CSE, KLEF, Hyderabad.kgsuma@klh.edu.in

ABSTRACT: Monitoring and control project work is both critical and challenging. The greatest asset of any organization is its human resource that ensures the success to achieve the goals and objectives of the company. Managers have to manage and organize the team to enhance productivity. The effective human resource enables their employees to contribute effectively and achieve desired growth of an organization. The assignment of role to a process depends on the skills and competence of an individual. If these details of an individual are known to the manager, then it will an added point to him while assigning a new task. It is observed from the previous study that managers spend a lot of time in allocation which is performed according to the experts experience rather than following some specific criteria which lead to inefficient utilization of the resource available, so if a model is developed and trained using machine learning algorithms to handle this routine task so that the manager can spend time in other developmental activities rather than spending time on routine tasks.
Keywords: Resource allocation, Productivity, Machine Learning.

I.INTRODUCTION

According to the Project Management Body of Knowledge (PMBOK), three factors include time, cost, and scope that are used to analyze the quality of work in a project [5][6][7]. However, only these three factors are not always sufficient in a real environment to decide the success of a project as other factors contribute to successful projects like team member's collaboration, their satisfaction with their position in projects [8]. The failure of projects is always due to improper human resource allocation. The main objective of human resource allocation is to decide what work needs to be carried by whom. For organizing and completing a project it is needed to schedule the project carefully. Effective project scheduling provides success. To keep the project on track it is necessary to allocate resources appropriately and manage the quality to reduce errors. This will typically increase the project's productivity and decreases the cost. Ultimately, a machine learning system will save time by efficiently using historical project data for allocation. It provides a service that increases the overall productivity by predicting the right task to the right team member. Early project management software's focused on managing a team by supporting the team in a specific area rather than dealing with complexities involved in managing the project. To provide even more value to a project, project management with machine learning needs to evolve.

Despite the work done on task allocation. Little research has been done to understand the factors that influence human resource allocation, this can be seen in the section 3 of the paper. The main aim of this paper is to identify the factors to be considered while allocating the task which ensures that important points are not missed when considering the human resource allocation in a team.

Journal of Physics: Conference Series **2040** (2021) 012038 doi:10.1088/1742-6596/2040/1/012038

2. RELATED WORK

Human resource allocation is first considered in a social network of individuals and has been known as highly computationally intractable by T. Lappas et.al [9], who explained the team formation problem in the presence of a network of individuals. Team formation is a practice, as it the combination of people working together, for example, development team formed with needed programmers and software engineers for building the project, marketing team with needed communication skills are formed for marketing the product, so it is required that the team is built with all required skilled person for the success of the project [6][11].

Idri, et al. [1] used K-nearest Neighbour Algorithm (KNN) this paper conducted studies on management of missing values (MV) in data systems of software engineering (SE) The study has observed increase use of Machine Learning (ML) techniques in dealing with the missing values in software engineering data.

Alsalemi and Yeoh [2] has given the summary of Machine learning techniques used to predict requirement volatility. It is observed that researchers used machine learning techniques like Fuzzification, Bayesian network, Binary classification, Artificial neural network. Most of the software development use iterative models, every type of project encounter requirement volatility where the requirements keep changing. The research show that change in requirements causes project failure Machine learning methods like ANN and logistic regression uses independent variable which are extracted from requirement document and apply regression analysis to find the estimation of correlation of independent variables.

Sharma and Singh [3] It is observed that Machine learning models are proven to provide high accuracy due to their learning nature. In this paper machine learning algorithms like Random Forest (RF), Multi-Layer Perceptron (MLP), Support Vector Machine (SVM) are used to predict the effort put in the projects.

Shivhare andRath [4] According to this paper Effort estimation is done based on on non-quantitative data Machine learning techniques. The estimation is carried out by applying Naive Bayes classifier. Performance of these techniques is evaluated and compared based on three parameters such as Mean Magnitude of Relative Error (MMRE), Mean Absolute Error (MAE), Root Mean Squared Error (RMSE) and it is observed that among all, Naïve Bayes classifier achieves better results when compared with other Neural Network techniques

3. RESEARCH QUESTIONS

In recent studies, numerous papers identified problems and solutions to solve this problem related to project management. Many studies have focused on the development phase of the project development life cycle but inadequately covered the project management phase which plays an important role in project success. Thus, the main objectives of this paper are to identify and classify the methods and predictors for resource allocation in project management planning assessment. In particular, focused on the recent studies which discovered the current attributes that were used for prediction of resource allocation in project management to improve the way of predicting task to assign a team member. Thus, the objective of this paper is to understand how traditional project management planning copes with the problem of human resource allocation methods they utilize. Identification of factors that influence task allocation will help the researchers to know what factors have more impact while assigning the task to a team. Few research questions are formulated to understand the current situation of project development issues.

The questions are framed to understand the impact of resource allocation problems in different types of organizations so that decisions can be taken to solve the problem in a better way.

RQ 1: Where does your team operate from?

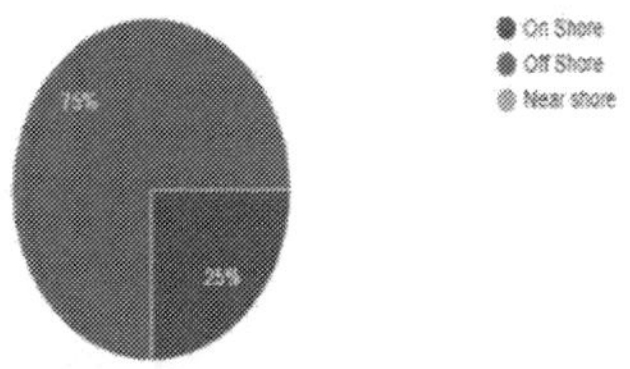

RQ 2 Which type of software development process do you use?

RQ 3: What is the immediate team size you are managing?

RQ 4: How much time do you spend in a day on each of the project managementactivities.

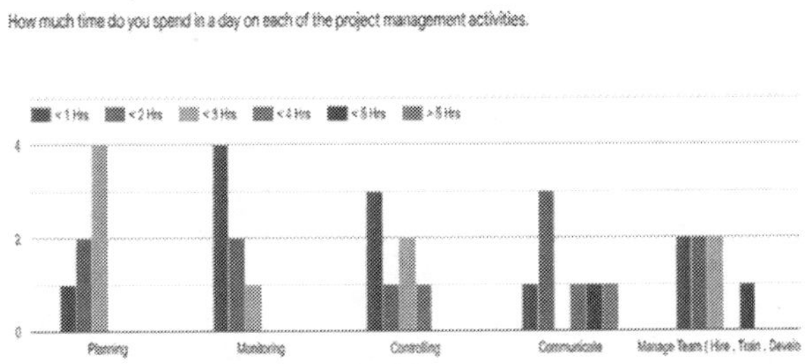

RQ 5: Which project management areas do you think can be automated?

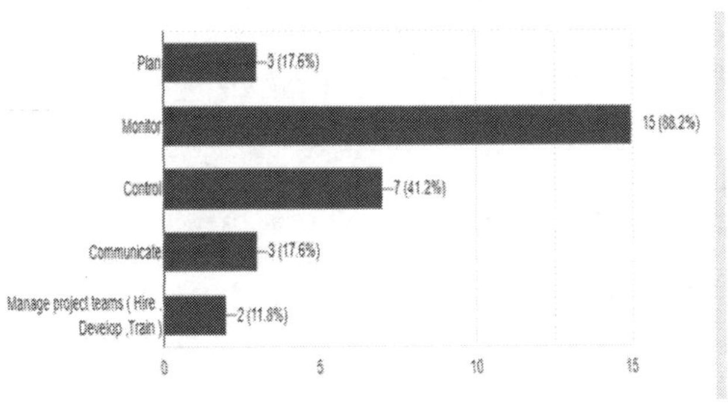

RQ 6: What tool do you use to assign the task to the employee?

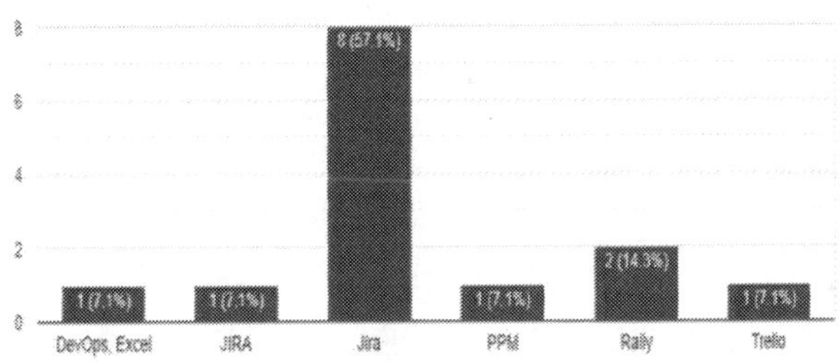

RQ7: How is task allocation done in your organization?

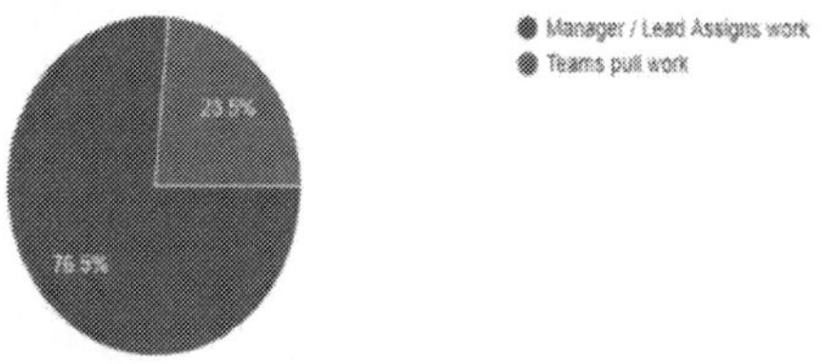

RQ8: If the Manager / Lead assigns work, what is the most important factor you consider in assigning the task to an individual.

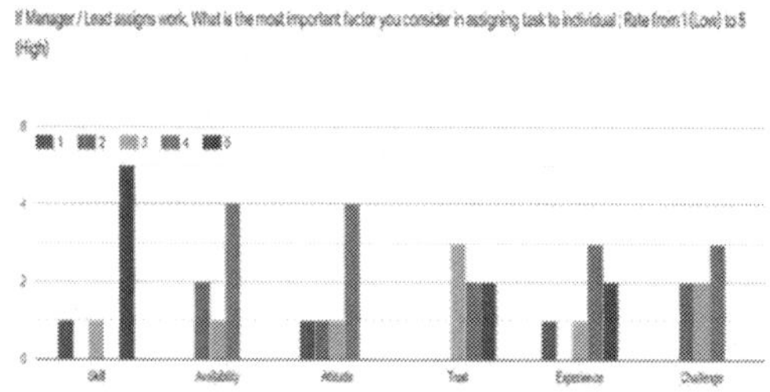

The results were considered from both product-based companies and service-based companies. As the human resource allocation criteria have a different impact on these two types of organizations. In Product based companies the resources need to be utilized more efficiently when compared with service-based where not much impact as just to provide service to the product.

Many of the responses from the survey resulted that Machine learning in project management will result in great progress in the project. It is evidenced that till today the resource allocation in a project is done using online resource management software like Instagantt, Team Gantt, Asana, Click up, Wrike, etc... This helps to track, create user stories and issues, plan sprints, distribute tasks across the teams, and provide complete visibility of teams' work eventually improve the team's performance. There are also paid software available in the market like JIRA, Monday.com, LeanKit, etc. which provides bigger space, can extend to a greater number of comprehensive-time storage, comprehensive-time and task management also provides security and support for the project information. It is observed it will manager seat use to software manager to automate routine tasks like task allocation will help them to invests their time in innovation and to focus on increasing teams' productivity.

4. CONCLUSION

The research present in this paper has emphasized machine learning in project management, which will have a great impact on the overall increase in achieving the organization's goals and objectives. A depth study of the project attributes is required for implementing a machine learning model. It is a big challenge to get the right dataset related to project resource allocation to implement a machine learning model. ISBSG is one of the non-profitable benchmarking standard group which was founded in Australia in the year 1997 by a group of national software metric association which promotes the use of IT industry data in research and to improve the software process and products. It depends on the researcher to identify relevant data and to identify attributes in the dataset which will have more impact on the resource allocation process and come up with the best model to automate this process. If done in right way it has a great impact on overall increase in productivity of the software project.

REFERENCES

[1] A. Idri, I. Abnane, and A. Abran, "Systematic mapping study of missing values techniques in software engineering data," in 2015 IEEE/ACIS 16th International Conference on Software Engineering, Artificial Intelligence, Networking and Parallel/Distributed Computing (SNPD), 2015, pp. 1-8.

[2] A. M. Alsalemi and E.-T. Yeoh, "A Systematic Literature Review of Requirements Volatility Prediction," in 2017 International Conference on Current Trends in Computer, Electrical, Electronics and Communication (CTCEEC), 2017, pp. 55-64.

[3] P. Sharma and J. Singh, "Machine Learning Based Effort Estimation Using Standardization," in 2018 International Conference on Computing, Power and Communication Technologies (GUCON), 2018, pp. 716-720.

[4] J. Shivhare and S. K. Rath, "Software effort estimation using machine learning techniques," in Proceedings of the 7th India Software Engineering Conference, 2014, pp. 1-6.

[5] D. Haughey. (Dec. 12, 2011). Understanding the Project Management Triple Constraint. [Online]. Available: https://www.projectsmart.co.uk/ understanding-the-project-management-triple-constraint.php

[6] M. Cross, ''Multi-skilling brings cost and productivity benefits,'' Process Eng., vol. 67, p. 67, 1986.

[7] N. Thorne. (Apr. 18, 2016). Project Management Triangle, Nick's Digital Solutions. [Online]. Available: http://www.nicksdigitalsolutions. com/project-management-triangle/

[8] G. Jin, S. Séverine, and G. Philippe, ''Management of the design process: Human resource allocation in factories of the future,'' Insight, vol. 20, no. 4, pp. 19–22, 2017.

[9]T. Lappas, K. Liu, and E. Terzi, ''Finding a team of experts in social networks,'' in Proc. 15th ACM SIGKDD Int. Conf. Knowl. Discovery Data Mining (KDD), vol. 9, 2009, pp. 467–475.

[10]A. PMI, "guide to the project management body of knowledge (PMBOK guide)," in Project Management Institute, 2013.

[11]L. J. Kirsch, "Software project management: An integrated perspective for an emerging paradigm," Framing the Domains of IT Management: Projecting the Future... Through the Past, pp. 285-304, 2000.

[12]A. M. Aladwani, "IT project uncertainty, planning and success," Information Technology & People, 2002.

[13]G. R. Cates and M. Mollaghasemi, "The project assessment by simulation technique," Engineering Management Journal, vol. 19, pp. 3-10, 2007.

[14] A. Idri, I. Abnane, and A. Abran, "Systematic mapping study of missing values techniques in software engineering data," in 2015 IEEE/ACIS 16th International Conference on Software Engineering, Artificial Intelligence, Networking and Parallel/Distributed Computing (SNPD), 2015, pp. 1-8.

Study of Self Depuration Capability of River Pamba

Aravind Raj P S[1]*, Divahar R[1], Arun R Nair[2], Aswin S[2], Bipeesh B[2]
[1]Associate Professor, Department of Civil Engineering, Aarupadai Veedu Institute of Technology, Vinayaka Mission's Research Foundation, Paiyanoor, Chennai.
[2]U.G.Student, Department of Civil Engineering, Aarupadai Veedu Institute of Technology, Vinayaka Mission's Research Foundation, Paiyanoor, Chennai.
(*Corresponding Author, Email: aravindraj.civil@avit.ac.in)

Abstract. The natural waterbody such as rivers and lakes are able to do the purification process themselves, which has the primary dependant of absorption capacity and dissolution capacity of atmospheric oxygen from the surface of the water body. The various pollutants such as biological pollutants and chemical pollutants are broken-down into insignificant strength over a period of time by the growth of certain bacteria. This bacterial growth in the water is possible only with the absorbed and dissolved oxygen content in the water. This absorbed and dissolved oxygen content in the water bodies are based on the velocity of the stream, depth, discharge rate and temperature of the water, thus the self-depuration capacity of the water bodies are depended on the natural profiles and environmental factors of the waterbody and its location. This explains that the turbulent water will get purified by itself in much higher rate than the stagnant water, which tends to become septic because of the oxygen scarcity. The degree of self-depuration of a natural water body is based on the physiochemical and biological activities occurring in the system. Various parameters like basic physical properties, chemical properties such as pH, hardness, dissolved solids, mineral content, BOD, dissolved oxygen for the samples were taken along the Pamba River. The samples are taken from three zones of the Rivers viz. 1) Forest zone, 2) Residential zone and 3) Industrial zone. The water quality factors from the samples are validated against the desirable values as per IS 10500: 2012 and justifies the current status of the River and recommendations for its wellbeing.

Keywords: Self-depuration, Dissolved Oxygen, Contamination, Environmental degradation.

1. Introduction

The competence of a River to self-depurate is based on its current pollution stage and the available environmental factors which supports to enhance the dissolved oxygen (DO) level, where the DO is the only source to develop the bacterial content in the water that breaks down the chemical, biological pollutants and toxics in the waterbody. The DO is naturally incorporated into the waterbody through the turbulence. Even after sufficient turbulence effect, if the water quality contamination rate is uncontrollably higher, the self-depuration duration will become singularity. The desired DO content is about 2.0 mg/L to maintain the life forms in the water, but, generally when the DO drops below the 4mg/L the life forms begin to reduce below the undesired level.

Metha (2010) studied the different samples of ground water in Vadgam taluk of Gujarat state for the properties such as turbidity, electrical conductance, pH, etc. and compared with the standard drinking water parameters as per ICMR and EU (1998). He concluded with possible measures to be followed to control the ground water quality as per standards. Gabriel (2010) performed an analysis on the quality of groundwater used for drinking and irrigation in the Yola district of NE Nigeria by correlating the TDS and electrical conductivity (EC) values. Also, the salt content and heavy metal content of the water samples were studied and groundwater quality was predicted using linear regression equations. Reddy & Lakshmi (2019) developed a statistical predictive model for the quality profile of groundwater in the Prakasam district, Andra Pradesh. Biological, physical and chemical parameters as per standards were studied for the collected and compared by the correlation coefficient. The results are compared with the water quality standards as per WHO (World Health Organization). The impact of heavy metals in waste water, methods to determine and identify the various heavy metals in the waste water, such as optical plasma spectrometry, mass spectrometry, etc. and, the process and procedures to remove those heavy metals from the waste water were described (Baysal et al. (2013), Reddy et al., (2012)). The physiochemical properties of the water samples were studied at various test points and resulted with the compatibility of the water for drinking and irrigation usages as per the World Health Organization recommendations. (Yadav & Rajesh (2011) , Heydari et al. (2013)). Tian et al. (2011) performed research on the self-purification capacity of the Juma River in China by sampling the water and aquatic plants for analysing the pollution level. The authors conclude by recommending an aquatic plant with greater organic pollutant absorbing capacity to control the river's pollution. Ostroumov (2005) studied the basic theories of water self-purification in fresh and marine eco-system. Studies on the impact of toxic pollution in river water due to domestic, industrial and agricultural pesticide runoff, were evaluated and compared with various factors of polluted water such as BOD, total suspended solids, toxicity level, etc. with respect to the WHO standards (Raghav & Shrivastava (2016), Schulz & Peall (2001)). Jain & Shrivastava (2014) compared the localized physical and chemical properties of Katraj Lake in Pune with standard values as per authorities and sensitized the awareness among the people to save the lake from reaching the state of the septic.

Nath et al. (2018) studied the Brahmi River in Odisha by multivariate statistical method such as factor analysis, principal component of analysis for the samples taken at various places along the stream. Comparing the complex relationship among the pollutant factors and arrived the ways to improve the quality of the water of the river. Kumaraswamy & Sivanandham (2009) estimated the current level of pathogenic pollution indicators in the River Cauvery of Tamilnadu by collecting samples from 16 sites along the river during various climates throughout the year. It was concluded with various reasons that causes pollution and intensity of the pollution in the river. Divahar et al. (2019) studied the impact of textile effluent pollution in the Kalingarayan canal of Erode district arrived with the water samples and compared the water quality recommendations as per BIS with the values. The author concludes that the excess discharge of the industrial effluent is the reason for the deviation of the water parameters form standards. Divahar et al. (2020) assessed the ground water quality of the areas around the Kalingarayan canal in Erode, by collecting nine samples from the downstream side of the canal in the period around the year of 2014-2016. It is suggested that the ground water has enough hardness and toxicity to make it not recommendable for edible purposes (Divahar et al. (2020)).

The scope of this study is to find out the self-depuration competence of the River Pamba. The project will estimate the ability of the river for self-purification with respect to the current polluted state. The Pamba River is the third longest river in the South Indian state of Kerala. The Pamba Originated originates at a Pulachimalai hill in the Peerumedu plateau in the Western Ghats at an altitude of 1,650 metres (5410 ft) and flows through Chitttar, Ranni, Keezhukara, Arattupuzha, Edanad, Puthencavu, Mannar, Kadapra, Melpadom, Thevery, Pallathuruthy and one of its ends is at Vembanad Lake and the other end is at Thottappally Spillway via Karuvatta district. The major length of River Pamba is flowing through the forest region, the profile of the river bed consists of lot of pebbles and stones as per the observations. Urbanization, industrialization and population increase around the basin of Pamba are considered as the major cause for the excessive pollution. Different physicochemical and biological test on water samples collected from various zones and the test results are evaluated with respect to the Bureau of Indian Standards (IS 10500:2012).

2. Experimental Programme

2.1 Collection of Water Sample

In order to study the real characteristic of the pollution stage of the River under the scope of the study, the samples were collected along the river at various zones such as forest zone, residential zone and industrial zone. The location of the sample collection spots in the respective zones are given in Table 1.

Table 1. Zone categorization for sample collection

Sl.No.	Zone	Station 1	Station 2
1	Forest	Ranni	Chittar
2	Residential	Aranmula	Kozhencherry
3	Industrial	Chengannur	Edanad

2.2. Tests on samples

Water samples collected were tested for various physiochemical parameters. The characteristic values of the samples collected were compared with the respective limitations for the parameter prescribed by Indian Standard (IS) 10500. The physical properties studied were colour, odour, temperature, turbidity, taste form a batch of samples and chemical properties such as pH, total hardness, dissolved solids, dissolved oxygen (DO) and biological oxygen demand (BOD) are studied. The mineral contents such as calcium, fluorine, magnesium, copper, sulphate, nitrate and aluminium were also analysed from the water samples sorted. The recommended limits of the physical and chemical properties as per IS 10500: 2012 are shown in Table 2.

Table 2. Recommended Physical and Chemical Properties as per IS 10500: 2012

Properties	**Desirable limit**	**Properties**	**Desirable limit**
Colour	<5 u	Fluoride	<1 mg/l
Turbidity	1- 5 NTU	Chloride	250 mg/l
pH	6. 5- 8. 5	Hardness	200 mg/l
Conductivity	5- 50	Residual Chlorine	0. 1- 0. 2 mg/l
Acidity	200 mg/l	Dissolved oxygen	>13 mg/l
Alkalinity	200 mg/l		

2.3. Results

The physical properties such as colour, odour, taste and turbidity of the water samples collected from different zones of the river are shown in Table 3.

Table 3. Physical Properties of Samples

Sl. No.	Parameters	Zone 1		Zone 2		Zone 3	
		Station 1 Ranni	Station 2 Chittar	Station 1 Aranmula	Station 2 Kozhencherry	Station 1 Chengannur	Station 2 Edanad
1	Colour	1 HU unit max	3 HU unit max	5 HU unit max	5 HU unit max	6 HU unit max	7 HU unit max
2	Odour	unobjectionable	unobjectionable	objectionable	objectionable	objectionable	objectionable
3	Taste	unobjectionable	unobjectionable	objectionable	objectionable	objectionable	objectionable
4	Turbidity	1 NTU max	2 NTU max	6 NTU max	4 NTU max	7 NTU max	9NTU max

The chemical and physiochemical properties such as pH, total hardness, iron (Fe) content, Chloride (Cl) content, dissolved solids, calcium (Ca) content, Magnesium (Mg) content, copper (Cu) content, Manganese (Mn), Sulphate (as SO_4), Nitrate (as NO_3), Alkalinity (as $CaCO_3$), Boron (B), Arsenic (As) content, Cyanide (CN) content, Phenolic compounds, Zinc (Z) content, Chromium (as Cr^{+6}), Fluoride (F) content and Aluminum (A) content of the water samples collected from different zones of the river are shown in Table 4.

Table 4. Physical Properties of Samples

Sl. No.	Parameters	Zone 1		Zone 2		Zone 3	
		Station 1 Ranni	Station 2 Chittar	Station1 Aranmula	Station 2 Kozhencherry	Station 1 Chengannur	Station 2 Edanad
1	pH	6.5	6	7.5	7.9	7.8	8.2
2	Total Hardness	16 mg/L max	21 mg/L max	213 mg/L max	220 mg/L max	316 mg/L max	329 mg/L max
3	Fe	Not detected	Not detected	0.3 mg/L max	0.2 mg/L max	0.5 mg/L max	0.53 mg/L max
4	Cl	Not detected	Not detected	261 mg/L max	256 mg/L max	308 mg/L max	316 mg/L max
5	Dissolved Solids	16 mg/L max	29 mg/L max	200 mg/L max	216 mg/L max	425 mg/L max	453 mg/L max
6	Ca	Not detected	Not detected	100 mg/L max	108 mg/L max	120 mg/L max	110 mg/L max
7	Mg	Not detected	Not detected	36 mg/L max	39 mg/L max	40 mg/L max	46 mg/L max
8	Cu	Not detected	Not detected	0.03 mg/L max	0.02 mg/L max	0.05 mg/L max	0.04 mg/L max

Sl. No.	Parameters	Zone 1		Zone 2		Zone 3	
		Station 1 Ranni	Station 2 Chittar	Station1 Aranmula	Station 2 Kozhencherry	Station 1 Chengannur	Station 2 Edanad
9	Mn	Not detected	Not detected	0.08 mg/L max	0.06 mg/L max	0.2 mg/L max	0.1 mg/L max
10	SO_4	Not detected	Not detected	190 mg/L max	198 mg/L max	218 mg/L max	232 mg/L max
11	NO_3	Not detected	Not detected	38 mg/L max	44 mg/L max	48 mg/L max	60 mg/L max
12	Alkalinity as $CaCO_3$	15 mg/L max	22 mg/L max	190 mg/L max	196 mg/L max	234 mg/L max	256 mg/L max
13	B	Not detected	Not detected	Not detected	Not detected	0.8 mg/L max	0.7 mg/L max
14	As	Not detected	Not detected	Not detected	Not detected	0.04 mg/L max	0.04 mg/L max
15	CN	Not detected	Not detected	Not detected	Not detected	Not detected	Not detected
16	Phenolic Compounds	Not detected	Not detected	0.02 mg/L max	0.01 mg/L max	0.2 mg/L max	0.4 mg/L max
17	Zn	Not detected	Not detected	Not detected	Not detected	4 mg/L max	Not detected
18	Cr^{6+}	Not detected	Not detected	Not detected	Not detected	Not detected	Not detected
19	F	0.3mg/L max	0.4 mg/L max	1.8 mg/L max	2 mg/L max	1.1 mg/L max	1.8 mg/L max
20	Al	Not detected	Not detected	0.01 mg/L max	0.01 mg/L max	0.5 mg/L max	0.1 mg/L max

The Biological oxygen demand and dissolved oxygen values of the samples collected from various zones and stations as classified are given in Table 5. The variation of the average BOD and DO across the zones are represented in Figure 1 and Figure 2 respectively.

Table 5. Test Results of DO and BOD

Parameters	Zone 1		Zone 2		Zone 3	
	Station 1: Ranni	Station 2 : Chittar	Station 1: Aranmula	Station 2: Kozhencherry	Station 1: Chengannur	Station 2: Edanad
DO	5.0 mg/L	5.4 mg/L	2.2 mg/L	3.2 mg/L	2.9 mg/L	3.5 mg/L
BOD	124 mg/L	102 mg/L	193 mg/L	147 mg/L	242 mg/L	158mg/L

 IOP Publishing
2040 (2021) 012039 doi:10.1088/1742-6596/2040/1/012039

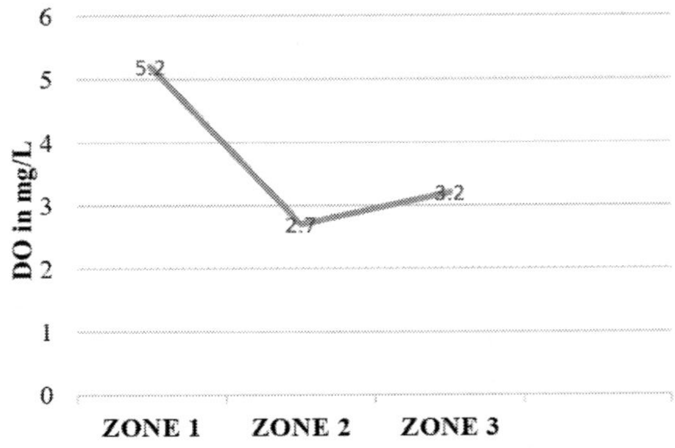

Figure 1. Variation of DO across zones

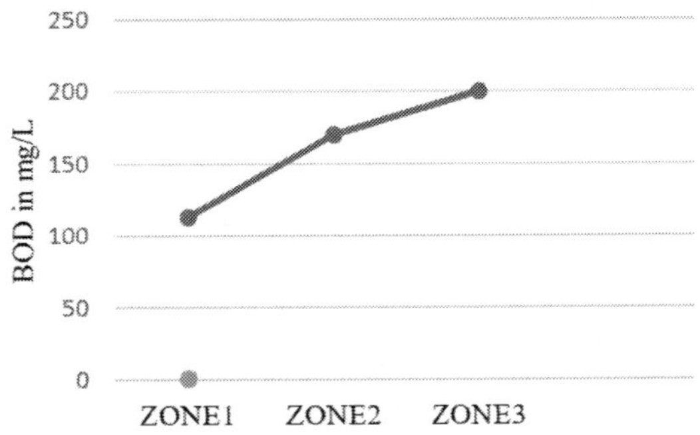

Figure 2. Variation of BOD across zones

3. Discussions

Based on the physical, chemical and biological characteristic analysis of the water samples of River Pamba from various zones, following discussions are arrived.

3.1. Forest zone

Compared to the residential and industrial zones, forest zone is least polluted. The water samples exhibited no bad odour and had least turbidity. The average pH value of water sample denotes that the water is slightly acidic in nature. As compared to the other two zones the DO level was higher and the BOD level was lower, this indicates that this zone contains least amount of organic matter. In this zone the river bed was characterized by a combination of course sand, fine sand and pebbles, where the impurities are held between spaces of media and also the course the of river helps in aeration of the water. Abundant fine sand content was observed in the river bed of the forest zone and the water flow was found to be vigorous with clear water.

3.2. Residential zone

This zone lies between the forest zone and industrial zone in terms of pollutant levels. The water samples gathered at this zone exhibited bad odour caused due to the dense population around the basin. The water was observed to be alkaline and consisted of chlorine, calcium and magnesium due to the anthropogenic activities and discharge of sewage from nearby sewers.

Both river stream in this zone had floating non-biodegradable wastes and debris such as plastic materials. The DO level was least in this zone because of the accumulation of various particles in the river bed. This zone is also subjected to heavy discharge from municipal sewers. The riverbed in this zone possess very small amount of broken stones and some amount of fine sand which is not sufficient for straining the impurities. The fluctuating flow rate of the river stream in this zone slows down the self-purification process.

3.3. Industrial zone

This zone is the most polluted among all the zones as per the characteristics results. The water from the zone had a foul smell of ammonia, light brown in colour and was more turbid than the water samples from other two zones. The average pH of the water in the zone was found to be 8 which indicates that the water is alkaline in nature. Despite having waste water treatment systems in the industries like brick and tiles manufacturers, the discharge consisted of chlorine, calcium, magnesium, sulphur and certain other elements. The BOD level in this zone is higher than other two zones, this indicates the presents of organic matter present in the water. The flow velocity of this region is average in nature due to the deviation of the river, also the river bed in this region possess very least amount of coarse sand and fine sand compared to the other two zones. Aquatic plants such as algae were also present in the water.

4. Conclusions

Self-depuration competence of a natural stream has its ability to purify itself by removing the impurities present in the water by natural factors. After conducting the test and evaluating the results it is evident that there is self-purification process happening at a longer distance than in the expected sampling point. The flow velocity of river Pamba is varying in each zone, this velocity affects the self-purification efficiency of each zone. The forest zone has higher self-purification efficiency than residential and industrial zones, this is because in forest zone there is less inclusion of waste water and also this part of the river flows through the forest region. It is also found that quality of water is in a deteriorating condition at residential and industrial zones. Compared to industrial zone, residential zone has a high rate for self-purification because the river bed in residential zone possess broken stones, fine and course sand these act as a filter media for straining the impurities. In residential and industrial zones proper inspections should be done so as to ensure that the discharge sewage. Artificial methods can be adopted in industrial zone to enhance its purification process. Hence the self-purification process would be achieved if the river possess a moderate flow velocity, which reflects the outcome at some distance away from the station points.

5. REFERENCES

[1] Metha K V 2010. Physicochemical characteristics and statistical study of groundwater of some places of Vadgam taluka in Banaskantha district of Gujarat state (India). *Journal of Chemical and Pharmaceutical Research* **2**(4): 663-670

[2] Obiefuna G I and Orazulike D M 2010. Physicochemical Characteristics of groundwater quality from Yola Area, Northeastern Nigeria. *Journal of Applied Sciences and Environmental Management* **14**(1): 5–11

[3] Reddy M A and Lakshmi K S. Statistical predictive models of ground water quality in and around textile cluster of Chirala, Prakasam (Dist), Andhra Pradesh, India. Statistical predictive models of ground water quality in and around textile cluster of Chirala, Prakasam (Dist), Andhra Pradesh, India. *International Journal of Scientific Research and Reviews* **8**(2): 3259-3270

[4] Baysal A, Ozbek N and Akman S 2013 Determination of Trace Metals in Waste Water and Their Removal Processes, pp 145-71 . In: Asli Baysal, Nil Ozbek and Suleyman Akman (eds). *Waste Water - Treatment Technologies and Recent Analytical Developments* IntechOpen Publisher. DOI: 10.5772/52025

[5] Reddy V H, Prasad P M N, Reddy A V R and Reddy Y V R 2012. Determination of heavy metals in surface and groundwater in and around Tirupati, Chittoor (Di), Andhra Pradesh, India. *Der Pharma Chemica* **4**(6): 2442-2448

[6] Yadav S S and Rajesh K 2011. Monitoring Water quality of Kosi River in Rampur District, Uttar Pradesh, India. *Advances in Applied Science Research* **2**(2): 197-201

[7] Heydari M M, Abasi A, Rohani S M and Hosseini S M A 2013. Correlation Study and Regression Analysis of Drinking Water Quality in Kashan City, Iran. *Middle-East Journal of Scientific Research* **13**(9): 1238-1244

[8] Tian S, Wang Z and Shang H 2011. Study on the Self-Purification of Juma River. *Procedia Environmental Sciences* **11**(C): 1328-1333

[9] Ostroumov S A 2005. On Some Issues of Maintaining Water Quality and Self-Purification. *Water Quality and Protection: Environmental Aspects* **32**(3): 337-346.

[10] Raghav N and Shrivastava J N 2016. Toxic Pollution in River Water and Bacterial Remediation: An Overview. *International Journal of Current Microbiology and Applied Sciences* **5**(4): 244-266

[11] Jain N and Shrinivas R K 2014. Comparative Review of Physicochemical Assessment of Katraj Lake, Pune. *International Journal of Science and Research* **3**(8):40 402-405.

[12] Nath T K, Tripathy B and Das A. 2018. A Study of Water Quality of River Brahmani, Odisha (India) to Assess its Potability. *International Journal of Engineering Research & Technology* **7**(7): 301-311

[13] Kumaraswamy P and Sivanandham V 2009. Enumeration and Identification of Pathogenic Pollution Indicators in Cauvery River, South India. Research Journal of Microbiology **4**(12): 512-558

[14] Schulz R and Peall S K C 2001. Effectiveness of a Constructed Wetland for Retention of Nonpoint-Source Pesticide Pollution in the Lourens River Catchment, South Africa. *Environmental Science & Technology* **35**: 422-426

[15] Divahar R, Raj P S A, Sangeetha S P and Kavitha M T 2019. Impact of Industrial Wastewater Disposal on Surface Water Bodies in Kalingarayan Canal, Erode district. Indian Journal of Ecology. 46 (4): 823-827

[16] Divahar R, Raj P S A, Sangeetha S P, Kavitha M T and Meenambal T 2020. Dataset on the assessment of water quality of ground water in Kalingarayan Canal, Erode District, Tamil Nadu, India. *Data in Brief* **32**: 106112 1-10

Design and Analysis of Zigzag Classifier in Food Industry Applications

Yella Harshavardhan Reddy*, Purushothaman K, Sreenivasan M, Sivakumar E R

Department of Mechanical Engineering, PACE Institute of Technology and Sciences, Ongole – 523272, India.

Email : yharshureddy@gmail.com

Abstract. In a zigzag classifier the material is being entertained utilizing varied channel velocities and mass load of the air. Specialty of zigzag classifier is it is being used in separation of particles having about the same size, but differing specific gravity, Separation of bulk materials which cause challenges and expensive costs if sieved. In this Zigzag classifier, sugar, salt, flour and grain, starch, milk products; spices; tea, gelatin, and vegetables are included in a typical application for the food sector. Zigzag classifiers are usually maintenance free, wear-resistant, overload-resistant and not sensitive to changes in the composition of the feed material. Due to these properties, the classification system can be operated without any oversight for lengthy durations. Even under overload situations, the goal output is clean particles or lightweight material.

Keywords: Zigzag classifier, Velocity, Food Industry Applications, Materials

1. Introduction

In recent years, serious food safety accidents have been often experienced, people are highly concerned about excess metal and additive content, residues of pesticides, and food safety has become a worldwide topic [1]. The classification of particles is significant and developed as the key control parameter for most industrial solid processing activities, based on the terminal deposition velocity. This adjustment velocity depends on typical particle features such as size, density and form. The other major property that controls the separation efficiency is turbulent particle diffusion [2].

The theory of gravitational classifiers is founded on the idea that, under the influence of various forces, particles suspended in fluid streaming commonly air or water migrate to various places so that they can be distinguished [3]. A straight, rectangular channel with several steep turns consists of the divider. Air is blasted from the ground to the top, while portion is fed to the channel through a feed aperture around halfway upwards. The movement of particles is caused by the interaction between gravity, fluid forces and collisions with other particles and the walls of the device [4]. These particles are not controlled by quality; rather, they have been subject to circumstances. Accordingly, the performance from an operational point of view is generally stated as alternating with the air velocity for the vertical air classification of solid waste particles [5].

In the design and operation of a waste to power generation facility, the essential characteristics of zigzag air classification designs are examined. The construction of an operating scope determined performance assessment approach is studied. In the past, the air classification in mineral extraction, calcareous sizing and seed and grain cleaning has been used in the industries and in agriculture. However, due to the complex and changing character of the waste, air classification is adapted to resource recovery and residue in power generation installations. A number of setups are tested which offer a continuous array of components for the zigzag classification. To assess its separation and operating air speed sensitivity, each configuration shall be studied. Results show that the Zigzag classifier setup does not affect its maximum separation efficiency. However, the results show clear restrictions on operating parameters that result in the highest efficiency of the various combinations. These numbers show the air classifier's sensitivity to air fluctuation variations. One notable observation is that smaller particles are affected by changes of configuration and larger particles don't seem to influence the distribution of the particles found in the classifying material: A novel method is created and deployed to classify performance assessment [6].

Kaiser's first attempt to define the classification of particles in a zigzag canal in 1963 was based on the concept of particles making a random move from stage to stage. Subsequently no modelling was completed until Senden and Tel's work was published in 1978 and 1979. They examined the behaviour of individual particles at very low particle concentrations in zigzag air classifiers and calculated the transition probabilities of these particles (i.e. the probability of moving to the next level). These probabilities were found to rely on the "history" of the particles. There could be two different sorts of probabilities of particle transfer the possibility for particles entering the stage in a dropping stream that will increase the likelihood for particles entering the stage in an increasing stream the following higher level. In addition to their experiments, fine particulate density and coarse particle density were independent at these low particle input rates from the classifier phase [7-9].

The release of building debris through the commenting generates a primarily mineral mixture with a density ·s distribution of 1.8 - 1.7 g/cm3. The requirements for sharpness of the method used to separate the partially released aggregate and concrete brick scrap are high because of the relatively small density range. A test rig consisting of the zigzag canal, fan, air cyclone, filter and particle feeding system was constructed for the gravity separation. The specific mass flow levels 3 to 16 t/(m2.h) for the transverse area of the device and the accompanying mass energy consumption were achieved between 0.2 and 8 kWh/t. The separation function for the efficiency is determined and is compared to a multi-stage turbulent cross-flow separation theoretical model. Due to the well known separation sharpness and geometric flexibility for a zigzag device, this separation principle may be demonstrated to be well suited for the separation gravity of mineral materials [10].

In the single stage behaviour of particles, and the interplay of stages, the separation performance of zigzag air classifiers with an angle of section 90°, 120°, and 150° were tested with low particle levels. On the basis of an examination of air flow pattern and trajectories, the influence of stage geometry on the single-stage behaviour has been explained. A stochastic one-step-memories model that ties multi-stage performance to single-stage characteristics has been described as the particle interchange between stages. The separation of particles is significant and is designed as the main operating parameter for most industrial solid processing applications based on the terminal settling speed. This settling velocity depends on the qualities of particles such as size, density and shape. Multi-stage sand and gravel separation studies were done in this study employing various channel velocities and air mass loads. In the separation functions and characteristic parameters of separation sharpness and product quality, the performance has been studied and commented [11].

Classification is essentially a process of separation of particle size. When air is commonly utilized as a fluid medium, classification can be described as separating solid particles into two or more products according to their velocities. Classification takes place every time the zigzag tube changes in direction, and allows extremely high, step-less adjustable cutting accuracy [12]. This classification is a classifier that is upstream. The zigzag design of the classification tube guarantees that the raw material gets washed again and again by the air. This leads to a multi step grading and the grader obtains a very high cutting precision. The cutting point can be set between 0,1 and about 10 mm. The classifier is used mainly for the sorting and separation of various materials. For sharp separations in range of approx. 0.1 – 10 mm, the Zigzag classifier is excellent.[13-15].

Classifiers are usually maintenance-free and wear-resistant, overload-resistant and insensitive to changes in the composition of feeding stuffs. As a result of these features, the classification system can be operated without any supervision over long time periods. Even under overload situations, the goal output is clean particles or lightweight material. In this paper, design and experimental analysis of zigzag classifier for food industry applications has been discussed. Based on the experimental results, fine and coarse particles of materials with various size ranges plotted the graphs. It is suitable for commercial applications like food industries.

2. Materials and methods

In industry, hoppers must be constructed such that they can be loaded easily in order to protect and store powdered products. More importantly, hoppers need to be constructed to be unloaded easily. The design of the trunk impacts the flow rate of the powder out of the trunk when it is at all flowing. Also, the way the hopper is structured influences the amount of material stored that may be released and if solid sizes or dead space are mixed to limit the effective hopper holding capacity. These issues are vital to take into account in the design of Feed Hoppers (Figure 1).

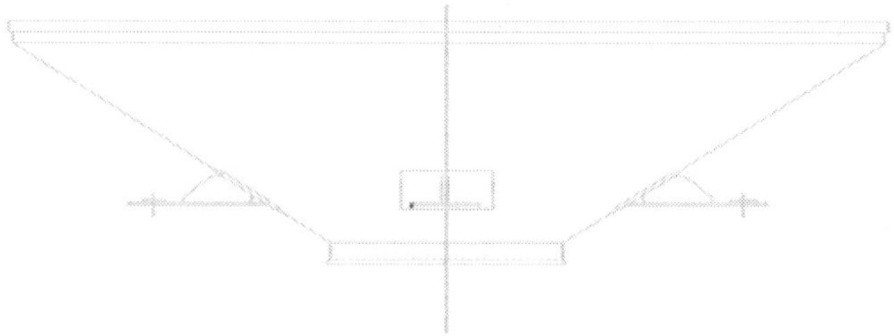

Figure 1. Design layout of Hopper

This system is designed to continuously feed bulk solids, e.g. powders, grains, chips, flakes and fibres, gravimetric- and gravimetric-size. The motivating source or power, at the appropriate angle, is attached to the feeding tray. Because of the physical features of the goods, this angle is different. The entire feeder is moved forward or upward, either hanging or insulated. The complete feeder moves the material forward and up. Then the tray returns to its proper place. However, because of the slower gravity action, the material does not go backwards. This gives a slowly advanced position to the substance before the procedure is repeated, bringing the material forward in a number of quick jumps undetectable to the sight (Figure 2).

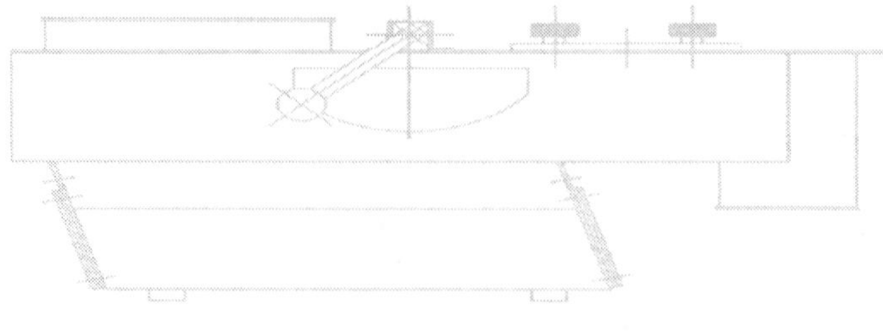

Figure .2 Design layout of Vibrating Feeder

The major duty is to control the flow of material from one room to another while ensuring that the airlock is good. The material or product is frequently flowing with dry powder, dust or grains. A Rotary Air-lock Valve's major role is to control the flow of material from one room to the next, while maintaining a healthy air condition. Usually, the material or product is dry, free powder, dust or granules. The engine and the drive chain turn the rotor shaft in operation and rotate the rotor inside the box and the headboards. As the blades rotate a defined volume of material goes from the material intake into the areas between neighbouring blades (called rotor pockets) and is sent to the material exit in the pockets.

Rotary Air-lock Valves are used to load materials at a controlled pace or act as an airlock on the bottom of the containers, cyclones, powder collectors or feed hopper. They are also utilized in positive or negative transport systems for the introduction of material. A rotary feeder/Airlock valve comprises of a rotor that revolves close to its housing at a certain RPM to keep the material flow rate consistent and provides a screening. The Rotary Air-locks offer high pressure, high temperature and other serious service requirements reliable services. Valves can be operated in several industries like food, plastics, chemistry, asphalt, mining, baking, cement and painting. The material is supplied from hopper or bin into rotor bags (a gap between two vanes). The shaft vanes revolve as well. The material in the rotor pockets shall be moved from the airlock valve inlet to its outlet due to the spinning of the valves (Figure 3). The pneumatic conveyor system will be driven from the output.

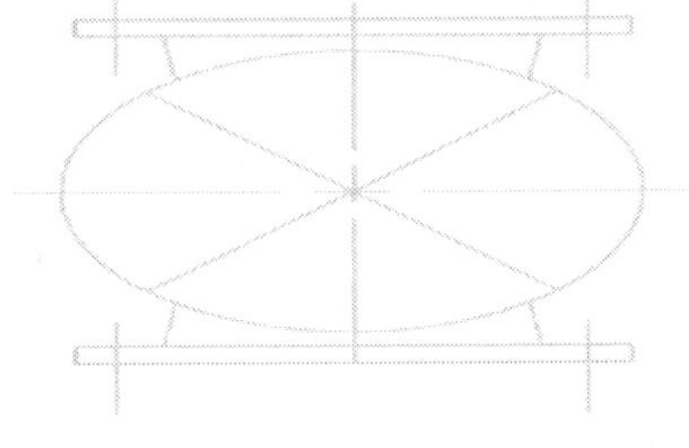

Figure .3 Design layout of Rotary Valve

The blowers are formed by a dynamically balanced impeller that revolves in a cum-filtering chamber with very fine clearances between them, and the pulls are directly mounted on the motor shaft and the pressure opening is situated at the external periphery of the stator, thereby reducing the impedance of air flow (Figure 4).

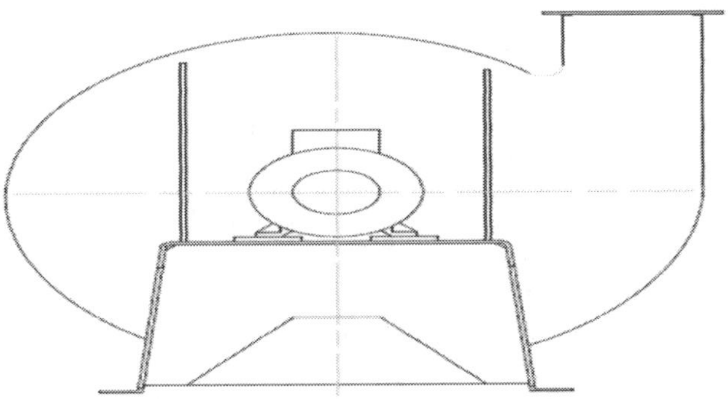

Figure 4 Design layout of Blower

The blower is a positive displacement machine which does not develop a constant volume pressure within the blower; the system releases pressure accordingly vary in order to meet the pressure and the flow volume of the systems is nearly constant, with fluctuating pressure except a slight increases in the slip as the pressing pressure increases by horsepower.

Cyclones provide for relatively low costs or removal from exhaust gas streams of particulate particles. Cyclone dividers are a bit more complex than just gravity control systems and so their removal efficiency is significantly better than the dividing chambers (Figure 5).

However, cyclones are commonly deployed as pre-cleaners in front of those efficient devices and not as efficient as electrostatic precipitators.

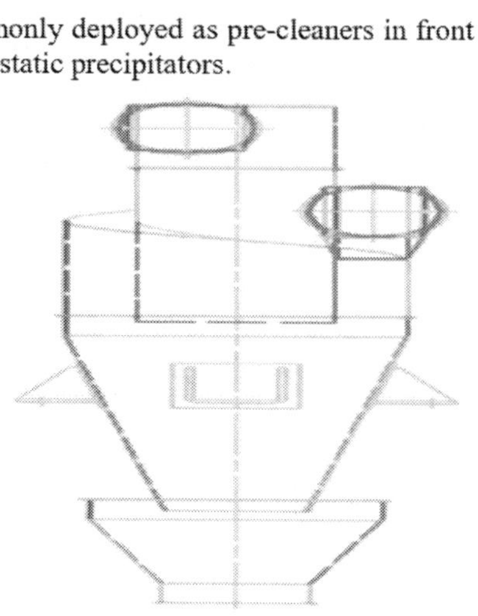

Figure 5 Design layout of Cyclone Separator

The fundamental principle of isolation stays the same. Particles streaming gas into the gadget the gas stream is forced to turn, however, larger particles are more dynamic and cannot swing with the gas these larger particles affect the cyclone wall and tumble down and are gathered in hopper. Indeed, gas stream swirls in a helical pattern a number of times, much like a tornado funnel, the repeated turns allow particles to flow through streamlines, which hit the slamming cyclone wall. The range of particle size gathered in a cyclone depends on the relative size of the device throughout all diameters. Cyclones can be stored in series or in parallel to improve overall efficiency of the collectors.

The entry of the cyclone, the dust discharge and the discharge all affect the total cyclone efficiency. The inlet which is instrumental in the development of the vortex is directing gas into cyclone. The particle matter is driven to the wall in the cyclone body. The gas continues down to the cone of the cyclone body, giving the gas sufficient rotating speed to retain particles at the wall. Gas shifts the direction from downhill to upward at the bottom of the cone. The ascent turbo enters a rope extension which is called the turbo finder in summary times and leaves the cyclone.

The collected particles in the meantime fall into a hopper where they are removed periodically or permanently. The cyclone removal effectiveness is largely reliant on the dimensions of the cyclone for a given size particle. Diameter most affects the efficiency of a given volumetric flow rate. The total length of the vortex dictates the number of turns. The inlet width and longitude or also, as the smaller the inlet, the more the inlet speed is extended the more efficient the inlet speed is, but the more pressure loss is increased

A butterfly valve is a valve which can be used for isolating or regulating flow. The closing mechanism has the form of a disk. The operation of a ball valve is similar, which enables rapid shut-off. Butterfly valves are often favored since they have lower costs and also less weight, which means that less support is necessary for other valve types. The disk is located in the center of the pipe, a rod connected to an actuator outside the valve is passed through the disk. The actuator rotates parallel to or perpendicular to the flow of the disk. In contrast, the disc is always present in the flow; therefore, no matter the position of the valve, a pressure drop in the flow is always induced.

3. Working principle

Once the product is recognized, the hopper is initially filled with a small amount needed for a dosing test. The dosing test can provide us the feeding rate in which the product is charged via vibratory supplies to the zigzag classification. This will also give us an idea of the zigzag classifier's fine cutting performance.

Again, the vibration feeder is used for the specific dosage test performance for our tests. Then add product to the hopper and keep it ready. The rotary vacuum valve 1 and the rotary vent 2 start one by one from the Blower. Check the air velocity in the zigzag tube and adjust to the speed at which the product is separated by the exhaust butterfly valve and air inlet butterflies valve.

When all parameters are ready and the vibration feed starts, the product is entered from above via a rotating airlock valve into the zigzag tube. The air stream from the bottom is rinsed or cleaned. The lighter particles will be transmitted by the air to a cyclone separator at the setting speed that we have maintained in the system. The tumultuous particles that can withstand the upstream air draws the force out of the zigzag classifier and falls down. The lighter particles are separated from the air in the cyclone separator and collected on the bottom via a relative airlock valve. And the air coming out of the cyclone separator is re-cycled to the zigzag classifier again and the process of classification continued.

4. Design of Zigzag Structure

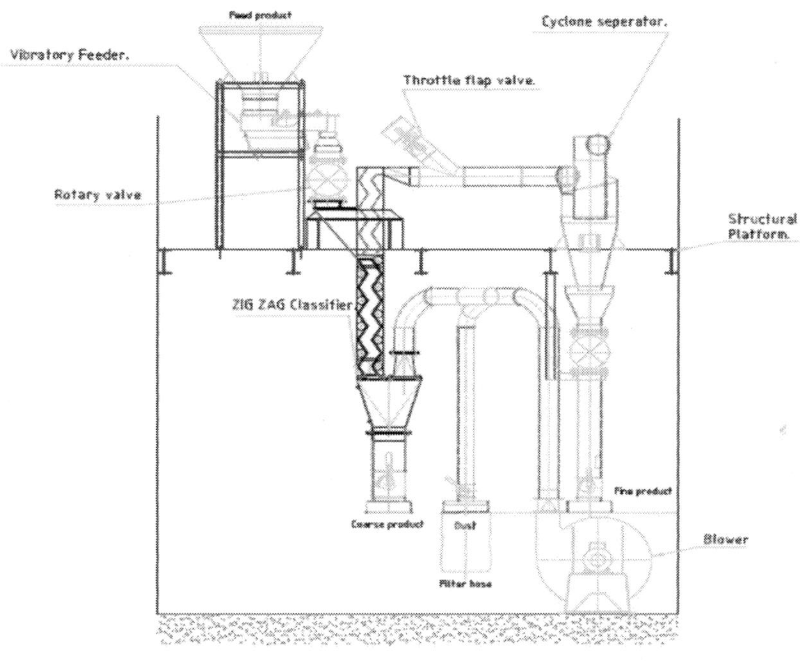

Figure 6. Schematic layout of Zigzag Classifier

Zigzag classifier contains a simple, vertical, indoor tube with a rectangular cross-section, tilted to the right and to the left at the same angle to the vertical (Figure 6). A flow channel divided into 15 uniform height segments h = 250 mm, width b= 1980 mm, and depth t = 250 mm comprises of a zigzag classifier utilized. These segments are connected under a bending angle α = 120 °. The air flows upwards and generates eddies in each corner of the channel. This leads to an additional particle dispersion and enhances separation. Each stage acts as a single cross-flow unit (Figure 7).

Figure 7. Coarse fraction collection and Zigzag classifer.

5. Results and Discussion

The studies were undertaken to remove tiny contaminants which are present along with the plastic granules sample. Each time in the zigzag classification the experiment was done three times with variable air velocity. After each experiment, the material was recovered from its proper place in both the coarse and fine fractions. The laboratory was supplied with a modest amount of coarse and fines, and the distribution of the particles were assessed. Air jet sieve analyzer has been used to detect particle size. It is essentially a screwdriver with the vacuum press for sewing.

5.1 Experiment 1

All tests have maintained a constant feed rate in order to observe how the air velocity in the zigzag tube solely varies. The cutting point was considered to be d97=2mm and the air velocity calculated as described in the previous chapters. Settling speed Vs = 4 m/s is the result. With both butterfly valves in exhaust as well as in the air inlet in the zigzag classifier system, the airflow was controlled. At a place from which the air enters the zigzag classifier, a pilot tube has been used to measuring air velocity. The results are estimated at least d97<2mm if the cut point is set at 2 mm, i.e. 97 % of the finished goods are smaller than 2 mm and light-weighted contaminants. Vice versa, the products in the coarse fraction are larger than 2mm in size.

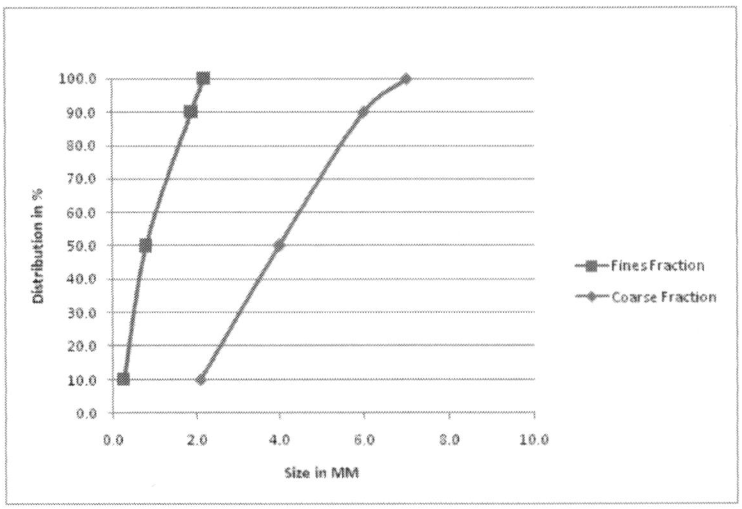

Figure 8. Performance chart of Experiment 1

The figure 8 is relatively limited that the overlap between the curve of the fines and the curve of the coarse fraction. And the zigzag classification is therefore particularly efficient for product segregation at the settling velocity.

5.2 Experiment 2

All tests have maintained a constant feed rate in order to observe how the air velocity in the zigzag tube solely varies. Settling speed of Vs = 6 m/s is obtained. With the assistance of both butterfly valves, the air flow in the exhaust and the inlet was controlled in the zigzag classifier system. The air speed was monitored using a pilot tube where the air enters the zigzag classification. The results are expected to at least d97<4 mm if the cut point is set at 4 mm, that is, 97% of the products contained in the fines are smaller than 4 mm and the impurities are slightly weighted. The coarse proportion of the products will be larger than 4mm vice versa.

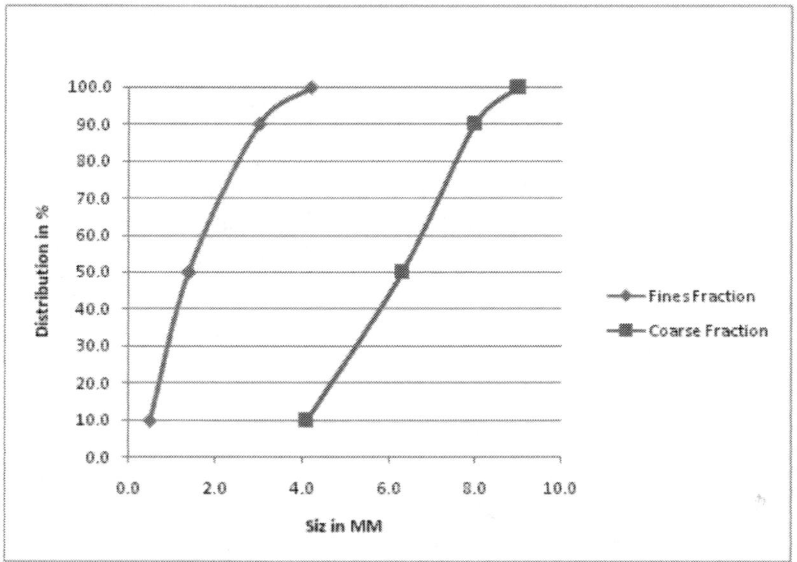

Figure 9. Performance chart of Experiment 2

Figure 9 has extremely marginal overlapping between the curve of the fine fraction and the coarse fraction curve. The zigzag classifier is hence particularly efficient for distinguishing items at the settling velocity.

5.3 Experiment 3

The feed rate was kept as constant for the entire experiment to investigate the effect that air speed only varies within the zigzag tube. The settling velocity (Vs) results = 8 m/s. With both butterfly valves on exhaust and air inlet, the airflow was controlled in the zigzag classifier system. The air velocity was measured using a pilot tube from where the air flowing into the zigzag classification system. If the cutting point is 6 mm, the results should be d97 < 6 mm at least, i.e. 97% of the products in the finished fraction should be smaller than 6 mm and impurities should be light weighted. The products in the coarse portion are greater than 6mm in size vice versa.

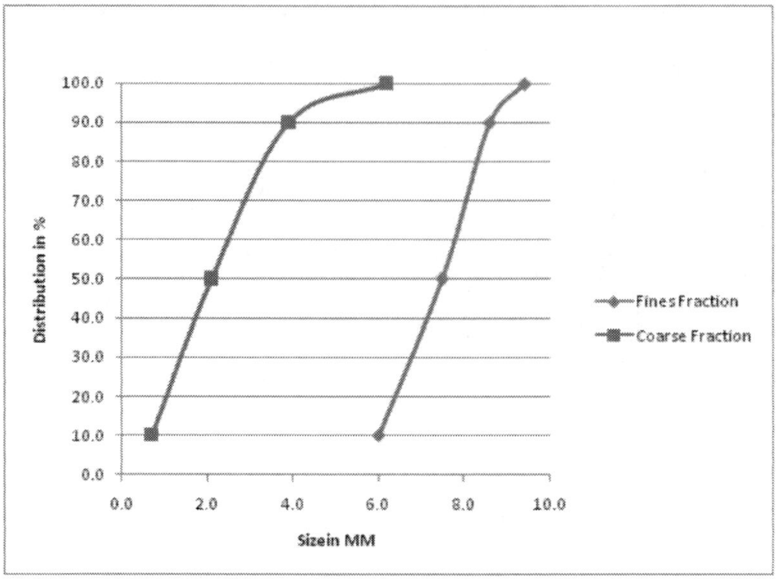

Figure10. Performance chart of Experiment 3

The overlap between the curve of the thin fraction and the crude fraction is quite marginal from figure10. The zigzag classifier is hence particularly efficient for distinguishing items at the settling velocity.

6 Conclusions

The 15 vertical tube multi-stage zigzag classifier has been successfully designed and produced with an angle $\alpha = 120$ degrees. Effective product breakdown tests were also carried out on the same basis. The findings of trials demonstrate that finer and coarser of the product have only a minimal overlap. The tests can study various drag forces operating on the product. We may also learn the capacity of the machine to handle by means of the dosing test. The zigzag classifier can be employed as a high capacity separator in accordance with the dimension of the zigzag tube. The entire experiment was effectively carried out and readings with various speed were conducted for different divisions. The only parameter that we affected was throughout the experiments velocity. All other factors such as feed rate remain unaltered. The manufacturing and assembly process is correctly conducted at all levels with all the safety measures needed.

References

[1] Gao, H., Yan, C., Wu, W., & Li, J. (2020). Application of Microfluidic Chip Technology in Food Safety Sensing. *Sensors*, *20*(6), 1792.

[2] Lukas, E., Roloff, C., Mann, H., Kerst, K., Hagemeier, T., van Wachem, B., & Tomas, J. (2020). Experimental Study and Modelling of Particle Behaviour in a Multi-stage Zigzag Air Classifier. In *Dynamic Flowsheet Simulation of Solids Processes* (pp. 391-410). Springer, Cham.

[3] Dong, J., Zhang, P., Wang, W., Li, J., & Chen, G. (2021). An Experimental Study on Performance and Structural Improvements of a Novel Elutriator. *Processes*, *9*(3), 478.

[4] Lukas, E., Roloff, C., van Wachem, B., & Thévenin, D. (2020). Experimental investigation of the grade efficiency of a zigzag separator. *Powder Technology*, *369*, 38-52.

[5] Yoshida, T., Hirai, Y., & Sakai, S. I. (2021). Meta-model of vertical air classification: A unified understanding of different separation curve models. *Powder Technology*, *383*, 522-535.

[6] Cederholm, C., (April 1978),"Paper Fiber Recovery System," Flakt Engineering, Vol. 2, PP 2-8.

[7] Colon, F. J. and Kruijdenberg, H., March (6-8, 1978), "The Mechanical Separation of Municipal Refuse into Useful Components and Their Application as Raw Materials in Industry," Proc. First World Recycling Congress, , Basel, PP. 3.15.i-3. 15. ix.

[8] Fastov, B. N., et al., (May 1975), "Testing of the Zigzag Classifier for Granulated Materials," Khimicheskoe iNeftyanoe Mashinostroenie, No.5, PP. 44-45, Plenum Publ. Co. 1976, UDC 621.928.6.001 .5.

[9] Hoberg, H.and Schulz, E., (1977), "Das Aachener Ver fahren zur Hausmtillaufbereitung," Aufbereitungstechnik,No. 1, PP. 1-5.

[10] Joergen Tomas., (1975), " Gravity separation and test rig and classification of building rubble " PP 25-30.

[11] Kaiser. E., (1963), " Classification of particle in a Zigzag channel " PP. 75,85.

[12] Pierce, J. and Wittenberg, N. (1984), "In their study of Zigzag configurations and Air classifier performance" PP. 100-105

[13] Senden, M.M. G., (March 1979), "Stochastic Models for Individual Particle Behavior in Straight and Zigzag Air Classifiers," Ph. D. Thesis, Eindhoven University of Technology, Eindhoven, Holland. PP 3.5, 5.8.

[14] Senden, M. M.G. and Tels, M., (1978), "A Stochastic Model of Multistage Separation in Zigzag Air Classifiers," Journal of Powder and Bulk Solids Technology, Vol. 2, No. 1, PP. 16-35.

[15] Worrell, W.A., (1978), "Testing and Evaluation of Three Air Classifier Throat Designs," report from Department of Civil Engineering, Duke University, Durham, North Carolina. PP 8.6, 9.

Wearable Technology for Visually Challenged People Commutation using Ultrasonic Mapping

D.Vijendra Babu[1], R.Karthikeyan[2], Sai M Bharadwaj[3], Belly Ajay Sai[3]

[1]Professor & Vice Principal, [2]Assistant Professor (Grade-II), [3] UG Student
Department of Electronics & Communication Engineering, Aarupadai Veedu Institute of Technology, Vinayaka Mission's Research Foundation, Deemed to be University, Tamil Nadu, India.

E-mail: drdvijendrababu@gmail.com

Abstract. 253 Million People live visual impairment Globally according to World Health Organization (WHO).Assistive technologies plays a major role for the improvement in the quality of life for Visually challenged perosns.Visually impairment Blind people can lead their life either by using manual or technological support. Vision is the dominant sense that assist in execution of day to day activities effectively. The proposed prototype takes care of transition of manual support to technological support with the help of a Wearable device. The Wearable device is being attached with the belt to the top wear of the person and Ultrasonic mapping is used for the commutation of Visually challenged/Blind people without the help of other Human. The Ultrasonic wave is used to detect the obstacle in front and back. It indicates while the obstacle is close enough. If the obstacle is within a reach of 100cm radius, the vibrator connected to it will vibrate and alert the User.

Key words: Visually challenged, Blind, Wearable device, Ultrasonic Sensor.

1. Introduction

Across Globe, the people affected with near/distance vision impairment is 2.2 Billion out of it 50%(1 Billion) are affected by various types of symptoms such as Cataract, Diabetic Retinopathy, Glaucoma and others.There is an gigantic Global financial burden due to Vision impairment poses a productivity losses appraise to US$ 250 Billion[1].Assistive Technology, subspace of Technology aids in improve the quality well-being of the needed. In the absence of assistive technology, people with disabilities will be isolated and leads to poverty and burden on morbidity. Assistive products include hearing aids, spectacles, wheelchairs, walking frames, prosthetic legs and pill organisers and also with the use of ICT.In 2015, United Nations (UN) Member States endorses Sustainable Development Goals (SDG), where Goal 3 address on Ensure Health Lives and promoted well-being for all Ages [2].

2. Literature Survey

P. E. Lanigan [3] enumerates on the assistive technology for the Blind in Grocery shopping which a cost-effective prototype is named Trinetra.A. Johnson [4] discusses about a Wearable device converting visual information into tactile signal for the Blind /visually impaired people as it improves the Quality of Life by restoring their ability to self-navigate. D. Yang [5] elaborates on a framework in assisting Blind people to see the World by auditory pathway based on Embedded block coding with optimized truncation (EBCOT) which operates in real-time. EBCOT algorithm [15, 19] applies two-tier coding and optimal Wavelet base with its ability to code with minimum rate-distortion. It compares the various Wavelets with SPIHT and EZW. H.Jabnoun [6] discuss about Object recognition for Blind people based on Feature extraction. It provides an overview of various visual substitution systems.

G. A. Farulla [7] enumerates on the designing an effective and low cost Navigation system for Blind and visually impaired persons named as Orientoma.F. Puente-Mansilla [8] discuss upon a Wearable UV Sensor for blind people for offering an appropriate skin care recommendations based on Ultraviolet Index (UVI), exposure time and skin color. To obtain Vitamin D for the Human body is achieved from Ultraviolet radiation from the Sun. C. S. Silva [9] discuss about a multi-sensor fusion approach for an electronic navigation aid for the blind and visually impaired persons. F. Ahmad [10] address on a Wearable bionic Kinect device by Haptic & Voice feedback for Visually challenged persons. The prototype is light weight & portable.
Ali Jasim Ramadhan [11] describes on a Wearable Smart System to help Visually Impaired Persons (VIPs) walk by themselves. J. Bai [12] enumerates on a prototype to help the blind people walk to the destination efficiently and safely in indoor environment with a novel Wearable navigation device. G. Sareeka [13] discuss about designing a Wearable device which identifies the texts in the surroundings of the wearer and convert it to voice message which will be given as feedback. S. Khan [14] enumerates in developing a smart and intelligent guiding mechanism at indoor and outdoor locations for blind and visually impaired people (BVIPs).

3. Proposed System

Main objective is to develop a small, comfortable and user friendly device to detect obstacles ahead of a Blind/visually challenged person. The idea is to develop a Jacket with the belt to help to make the Blind/visually challenged people commutation easier and safe. The Jacket is comes with the Ultrasonic sensor in its Waist for detecting the obstacle while movement and the vibrator connected to the sensor picks up the signal and indicate to the blind/ visually challenged person moving through vibration signal. It process the data received from Sensor [16, 20] and measures it sends it to vibrator for indication. It also detects different vibration levels for the obstacle gets closer in regular intervals and alerts the blind person, helps to improve the mobility constraint in real-time.

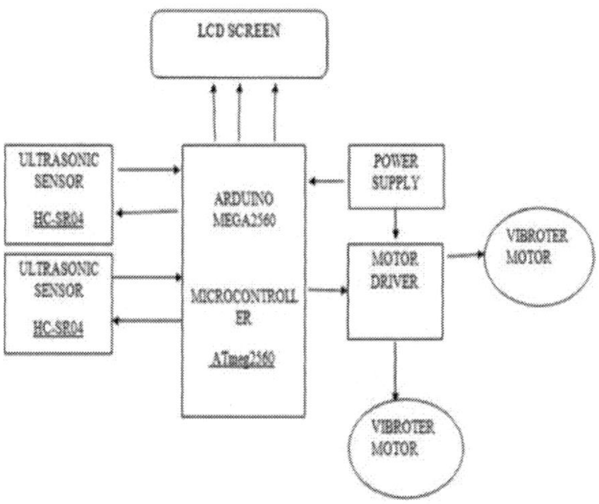

Figure 1. Block diagram

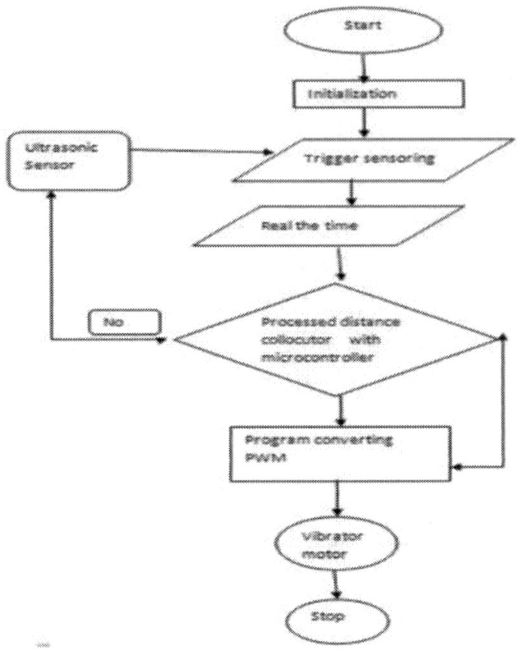

Figure 2. Process Flow Chart

The proposed model consists of four modules integrated together to perform the operation namely, Arduino Power Source, Microcontroller Unit [17, 18], Display Unit & Software Unit.ATMega2560 Microcontroller is used for Signal transmission and also used to monitor the messages or signals send from the sensor and it gives the desired message to the vibrator motor to vibrate. Figure 1 displays the block diagram of the proposed model. The main components consists of Arduino Mega 2560, Ultrasonic Sensor, LCD Screen & Motor driver. LCD is used to display sensor values for improving the functionality of the prototype and used to represent the distance for the analysis purpose and to calibrate sensor and to measure the performances. In this unit compile the program for execution of operation in proper manner.Codec is used in will corresponds with the Arduino programming to drive the board and its peripherals in a synchronized way. Figure 2 displays the Process flow chart. Figure 3 displays the block diagram of the proposed model.

4. Hardware Module

Figure 3 displays the Circuit diagram for the proposed model. Figure 4 displays the block diagram of the proposed model. Ultrasonic Sensor implemented in the prototype senses the distance between Obstacle and user. When the Obstacle is very close to the Visually challenged/ blind, then the sensor detects the Obstacle and sends this message to vibrator motor.

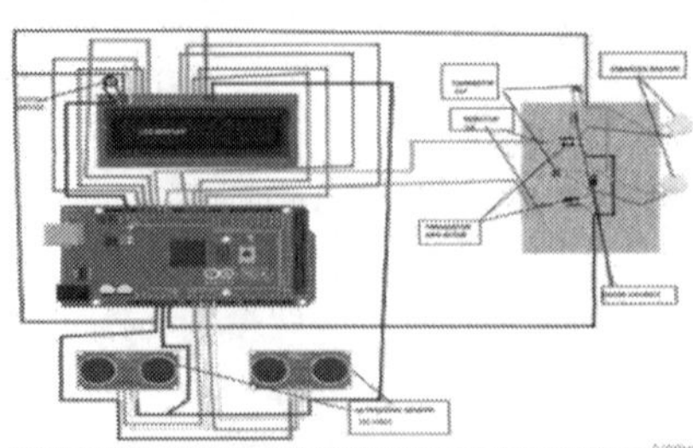

Figure 3. Circuit diagram

Figure 4. Snapshot of Protype

5. Results and Discussions

The various responses from Front Sensor A are displayed in Fig.5 based on the distance of the Visually challenged/Blind people to the obstacle. Front Sensor for 100 cm and above, the vibrator motor does not work for safety. Front Sensor A displays "ALERT" if any obstacles crosses between 50 cm-100 cm and Pulse mode (150) analog signal vibrations will turn "ON". Front Sensor A displays "CLOSE WARNING" if any obstacles crosses between 2 cm-50cm and Pulse mode (200) analog signal vibrations will turn "ON".

Figure 5. Snapshot of response from Front Sensor A

The various responses from Back Sensor B are displayed in Fig.6 based on the distance of the Visually challenged/Blind people to the obstacle. Back Sensor for 100 cm and above, the vibrator motor does not work for safety. Back Sensor B displays "ALERT" if any obstacles crosses between 50 cm-100 cm and Pulse mode (150) analog signal vibrations will turn "ON". Back Sensor B displays "CLOSE WARNING" if any obstacles crosses between 2 cm-50cm and Pulse mode (200) analog signal vibrations will turn "ON".

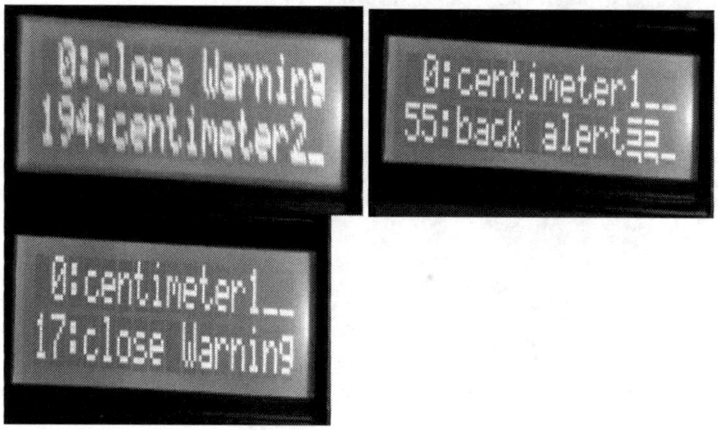

Figure 6. Snapshot of response from Front Sensor A

6. Conclusion

The designed prototype of Wearable Technology assist for Visually challenged/Blind people Commutation using Ultrasonic mapping based on Ultrasonic Sensor. The proposed model has the benefits over the conventional methods in terms of design time, production Cost & power consumption. The prototype can be further enhanced with inclusion of sound alert system instead of vibration alert and also inclusion of GPS and GSM to detect the current location of the Visually challenged/Blind people.

References

[1] https://www.who.int/news-room/fact-sheets/detail/blindness-and-visual-impairment

[2] World Health Organization, Geneva EBM,Improving access to assistive technology-The Need for Assistive Technology. 2016.https://apps.who.int/gb/ebwha/pdf files/WHA71/A71 21-en.pdf

[3] P. E. Lanigan et al 2006 Trinetra: Assistive Technologies for Grocery Shopping for the Blind, *10th IEEE International Symposium on Wearable Computers.*

[4] A. Johnson et al 2006 A Navigation Aid for the Blind Using Tactile-Visual Sensory Substitution, *International Conference of the IEEE Engineering in Medicine and Biology Society.*

[5] D. Yang et al 2007 From Image to Sound: the Study of Assistive System for the Blind Based on EBCOT, *IEEE/ICME International Conference on Complex Medical Engineering.*

[6] H. Jabnoun et al 2014 Object recognition for blind people based on features extraction, *International Image Processing, Applications and Systems Conference.*

[7] G. A. Farulla et al 2015 ORIENTOMA: A novel platform for autonomous and safe navigation for blind and visually impaired, *10th International Conference on Design & Technology of Integrated Systems in Nanoscale Era.*

[8] F. Puente-Mansilla et al 2016 A wearable UV sensor and accessible smartphone application for blind people, *IEEE International Symposium on Consumer Electronics.*

[9] C. S. Silva et al 2016 Sensor fusion for visually impaired navigation in constrained spaces, *IEEE International Conference on Information and Automation for Sustainability*

[10] F. Ahmad et al 2016 Bionic Kinect device to assist visually impaired people by haptic and voice feedback, *International Conference on Bio-engineering for Smart Technologies.*

[11] Ali Jasim Ramadhan 2018 Wearable Smart System for Visually Impaired People, *Sensors*, **Vol.18, Issue 3**

[12] J. Bai et al 2018 Virtual-Blind-Road Following-Based Wearable Navigation Device for Blind People,*IEEE Transactions on Consumer Electronics*,**Vol. 64,No. 1.**

[13] G. Sareeka et al 2018 pseudo Eye — Mobility assistance for visually impaired using image recognition, *2nd International Conference on Inventive Systems and Control.*

[14] S. Khan et al 2021 Analysis of Navigation Assistants for Blind and Visually Impaired People: A Systematic Review, *IEEE Access*, **Vol. 9.**

[15] D.Vijendra Babu et al 2008 EBCOT using Energy Efficient Wavelet Transform, *International Conference on Computing, Communication & Networking*

[16] S.Mahendiran et al 2021 Ultra Reliable Low Latency Communication Technique for Agriculture Wireless Sensor Networks, *Arabian Journal of Geosciences*, **Vol.14.**

[17] D.Vijendra Babu et al 2019 Home automation through Eyeball motion for Disabled persons, *International Journal of Engineering and Advanced Technology*, **Vol. 8, Issue 6S3.**

[18] D.Vijendra Babu et al 2016 Biometric facial image encryption for secure Image storage, *Journal of Chemical & Pharmaceutical Sciences*, **Special Issue 9.**

[19] D.Vijendra Babu et al 2014 A Novel Morpho Codec Medical Video Compression based on Lifting Wavelet Transform, *Asian Journal of Scientific Research*, **Vol.7, No.1.**

[20] V.Chandrasekar et al 2021 Designing of Tele-Health Smart Sensor Device to assist Home care Staff, *Journal of Physics: Conference Series,* **Vol.1964, 062106**

Optimization of Dissimilar aluminum alloy by Friction stir welding Process Control variables with Multiple Objectives

S.D.Kumar1 R.Pugazhenthi2 S. Ajith Arul Danial3, G. Swaminathan4

1Research Scholar, 2Professor , 3Assistant Professor, 4Assistant Professor

1, 2, 3 Department of Mechanical Engineering, Department of Mechanical Engineering, Vels

Institute of Science and Technology and Advanced Studies, Chennai-117.

4Department of mechanical Engineering, SRM Institute of Science and Technology, Ramapuram

Campus, Chennai-89

E-mail: sdkumar1973@gmail.com

Abstract. The progress of FSW has offered a different approach for fabricating superior class weld. This paper deals with optimization of process control variables influencing weld features in customized friction stir butt welding having multiple objectives of 6 mm thick dissimilar plates of AA7075 and AA6101 using Taguchi grey relational approach. The L27 orthogonal array has been employed to Obtain the experimental trails and the joints have been fabricated in a laboratory stage friction stir welding equipment by changing tool rotation speed, worktable translational speed, tool plunge force and tool pin shape. After welding, tensile and impact strength of the weld have been evaluated and found that optimum ultimate tensile strength of 129 Mpa., maximum yield strength of 119.21Mpa, maximum % of elongation 24.64% and impact strength 23Mpa simultaneously in FSW process variables of Tool rotation speed is 1200 rpm, transvers speed of 30 mm/min and axial force is at 5 kN. Based on the experimental results optimum levels of process control variables were obtained using grey relation rating and compared with confirmation test.

Key words: Friction stir welding, mechanical properties, multi objective, dissimilar aluminum alloy, Grey Relation Grade.

1. Introduction

In technology growth, the scientists and technologists are handling extremely challenging problems in the area of metal joining technology. The problem with the conventional joining methods mainly be attributed to novel metals with low welding strength. Scientists in the field of material science are novel frontier metals with a high strength, hardness, toughness and other diverse behaviors. The welding of metals with these properties is more difficult and affect their basic properties by conventional welding methods. In last two decades, an innovative solid-state joining method generally called as FSW was invented and patented by The Welding Institute (TWI) in United Kingdom in the year 1991[1] and its need was addressed in the first paragraph as FSW was environmentally friendly and accessible to materials with a high strength to weight ratio as well as with no welding defects such as hot cracks and porosity [2, 3]. Genetic programming (GP) is a comparatively new approach to advanced computation, with the key benefit of this method being the estimation of efficient predictive mathematical models or equations without any assumption as to the potential type of functional relationship [4]. Owing to low precipitate dispersal, and/or rather than grain size in the weld, the hardness decreased with increased tool traverse speed. The R2 values for the projected model of all the properties were obtained nearly 90%, it showed a good committment between the independent variables and the response data [5]. Many combinations of process control variables were formed using L18 orthogonal array. By using Principal Component Analysis (PCA) it was noticed that weightage of 45.36 %, 44.51% and 10.11% hardness tensile strength and power consumptions respectively. Optimal process parameters were obtained using Multi-Objective Ratio Analysis (MOORA) optimization, which results in the increasing of tensile strength and hardness lower the power consumption [6]. The difference between the higher and lower value of the gray relation grade of the variables of the FSW operation is as follows: 0.2756 for tool rotational speed, 0.14171 for welding speed and 0.08436 for axial force. By comparing these values, the most active variable influencing quality attributes is ascertained. The correlation will provide the degree of importance over the various quality characteristics of the input process control variables. Here, the maximum value of 0.2756 indicates that the tool rotational speed has the biggest impact among the other process control variables on the quality characteristics [7]. The plot graphically assesses the impact of each input welding parameter on the efficiency of the weld. On the Basis of main impact plot, input parameters are found to be important as they reach their center point, while their lower and higher levels have not affected the consistency of the FSW specimens significantly. But the tilt angle of the FSW tool is considered important when it is between the middle and the higher level [8]. The interactions between the rotational speed of the tool and the shoulder base angle closely, it becomes clear that the impact of the shoulder base angle on the UTS depends on the tool rotational speed. If it rotates at 1250 rpm, the strength is minimized at 5°, however, at 1000 rpm the shoulder base angle yields the maximum strength at 5°. The shoulder base angle and rotational speed have a more significant impact on the surface hardness compared to other interactions [9]. Among the numerous common evolutionary algorithms, the Non-Dominated Sorting Genetic Algorithm-II (NSGA-II) is widely adopted as an effective method for improving product quality in all manufacturing activities such as machining, shaping, and welding. NSGA – II uses random genetic operations to scan a whole design space for global optimum operation through different design points [10]. The mean absolute percentage error (MAPE) is applied to conduct error analysis of the implemented decision-making processes used in this report. To do this, 30 separate computer code runs were performed, and a final solution was obtained for each trial using the Shannon entropy and decision-making methods of TOPSIS. The value of each target (ultimate tensile power, elongation, and minimum hardness in the HAZ zone) was then equated with the better solutions obtained by each method after 30 runs [11]. The optimal setting of process parameter represents the relationship between the reference sequence and objective sequence, therefore greater fuzzy grey relational grade reveals the objective sequence has a stronger relationship than the reference sequence. Accordingly, the optimal setting of process parameters is larger fuzzy – GRG is desirable for obtaining larger

UTS and TE of fabricated FSW specimen. The FSW joints made from square pin had the strongest Fuzzy GRG. Since flat-faced Pin profiles are correlated with eccentric material flow. This eccentricity of material flow allows for the movement of an incompressible material flow across the pin profile. The interaction between the static volume and the dynamic volume determines the direction from the leading edge to the trailing edge of the revolving tool for the distribution of fleshy material [12]. It is observed that increase in tool rotational speed increases frictional heat generation and lower heat input condition prevails at lower tool rotational speed like 700 rpm which are also associated with lack of stirring. The net result is poor due to consolidation of materials and it lead to poor wear resistance at lower tool rotational speeds. Higher tool rotational speed like 1200 rpm led to higher heat generation than required and release excessive stirred materials. Micro voids appear at higher tool rotational speed led to poor wear resistance [13]. It is evident from the graphical data that mean values of S/N ratio increase with increase in 'Tool rotational speed'. It means that the weld joint efficiency increases with increase in 'Tool rotational speed [14]. Alloys of Aluminum and magnesium are majorly finding their applications in industries like automobile, aerospace and sports applications. The joining of these dissimilar alloys is very difficult by using traditional fusion welding technique inherently because of the "brittle Inter Metallic Compounds (IMC)" like Mg2Al3 and Mg17Al12. In order to enhance the joint strength, the formation of inter metallic compounds due to friction stir welding has to be characterized first. In the current paper some important welding parameters and their effects on weld quality are discussed along with the mechanical properties of the weld joint" and to identify the optimized process variables by multi-objective optimization techniques.

2. Methodology

The materials utilized for this examination are aluminum amalgams AA6101 and AA7075. With the help of a power hacksaw machine the rolled plates of 6mm in thickness were sliced into the essential size (100mm x 50 mm x 6 mm) and squaring the butting faces with the help of the milling process. Before FS welding process, butting edges of the weld specimens were cleaned by using a wire brush. Edges to be welded were arranged with the goals that are completely parallel to one another. This ensure that there is no uneven hole between the plates that may not give good properties to welded joints. In addition, surface arrangement was also performed in such a way that the surfaces of both plates are of same size and balance. The compound structure in terms of weight rate tabulated in Table 1 and Mechanical properties of the base metals employed in this investigation at atmospheric condition is recorded in Table.2.

Table I Chemical Compounds (wt %) of parent metals

Element	Al	Si	Fe	Cu	Mg	Mn	Zn	Ti	Cr
AA6101	95	0.8	0.7	0.4	1.2	0.15	0.25	0.15	0.35
AA7075	87.5	0.4	0.5	2.0	2.9	0.3	6.1	0.2	0.28

Table 2 Mechanical behaviors of parent metals

Element	UTS (MPa)	YS (MPa)	% of Elongation	Hardness
AA6101	135	118	19	70
AA7075	622	573	10	195

3. Design of Experiments

Design of Experiments (DOE) is a formal, coordinated method for determining the relationship between factors that influence a process and performance of the process. Three operation control variables of FS welding are considered factors of control are tool TRS, WS, and AF. Each variable has 3 different pitches like high, medium and low represented by 3, 2 and 1 respectively. Due to the wide range of influential factors, it was decided to use three factors, three levels, and a central composite design matrix to prescribe the (33=27) 27 number of experiments. If three operation control variables and 3 pitches for each L27 orthogonal array parameters should be used for conducting tests, based on the Taguchi method. FS welding operating variables and their levels take in to account for conducting tests is presented in Table 3.

Table 3 Process control variables and Levels

Process Control Variables	Levels		
	1	2	3
TRS (rpm) – A	1000	1100	1200
WS (mm/min) – B	30	45	60
AF (kN) – C	4	5	6

4. Experimentation

The Friction Stir Welding (FSW) Machine setup is shown in Figure 1. In this work, Butt welding of AA7075 and AA6101 dissimilar alloy materials is carried out at different process parameters. For welding of 9 samples of AA7075 and AA6101, all specimens were prepared to a size of 100*50*6 mm. All specimens were positioned and firmly clamped with help of backing plates to avert separation of the attached butting edges. The forces are fairly large during the tool's initial plunge and additional alerts were needed to verify the plates were not separated in a butt arrangement. The tool mounted in tool holder with tilt angle of 1.5° and the tool pin was throwing to a predestined deepness at the edges of the butting surface of the plates to be joined. The tool was transversed forward after residing time at the end of which the joint was formed by a single pass. After the weld is finished, the tool is released from work piece and allowed to get cooled.

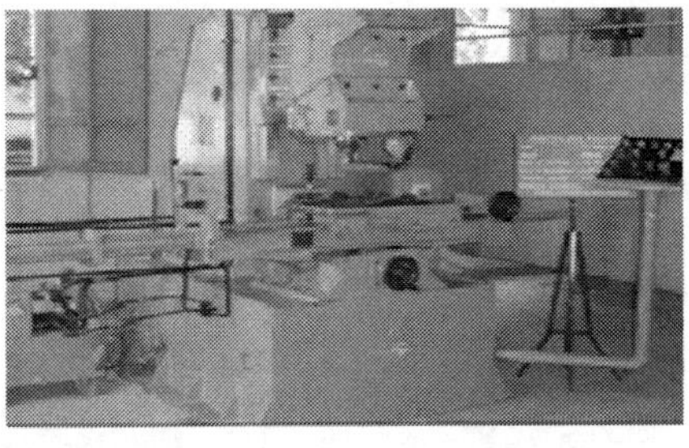

Figure 1. Friction stir welding machine

4.1 Process Response Measurement

The tensile behavior of FS welded joints were determined using the UTM (Make: FIE & Model: UTN 40). The trail samples were sliced from the fabricated joints and according to ASTM E8 dimension machined as in Figure 2. Three identical specimens were tested to acquire the average tensile strength. The camera image of the FS welded specimens after tensile fracture is shown in Figure 3. The impact toughness was measured using pendulum type impact test machine (Make: FIE & Model: IT 30 ASTM). The three specimens were extracted across the weld line from friction stir welded joints and machined as per ASTM E23 standard size displayed in Figure 4. FS welded joint specimens fractured realistic images after charpy test are exposed in Figure 5. Table 4 shows the L27 orthogonal array along with the experimental results of ultimate tensile strength, yield strength, percentage of elongation and impact strength of welded joint.

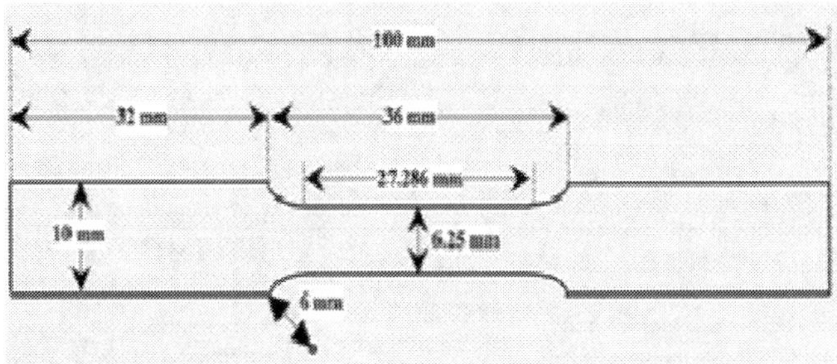

Figure 2. ASTM E8 Standard Tensile test specimen

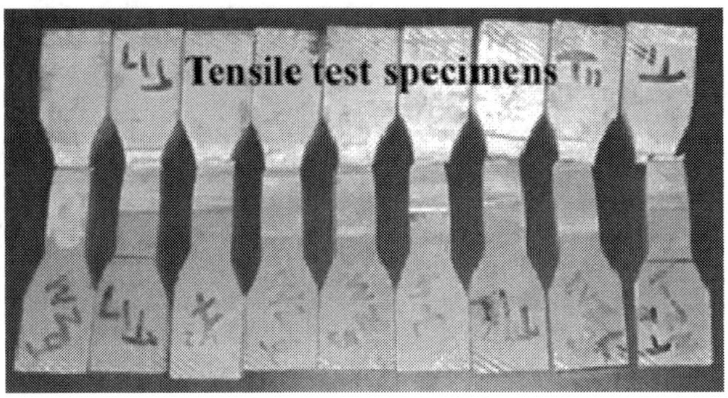

Figure 3. Tensile test specimen after fracture

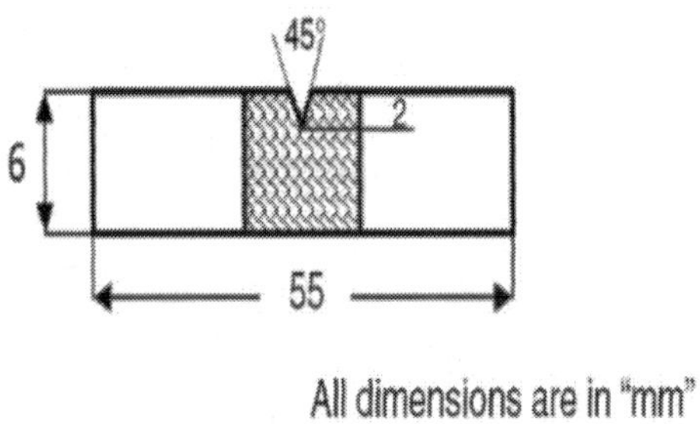

Figure 4. ASTM E23 Impact test specimen

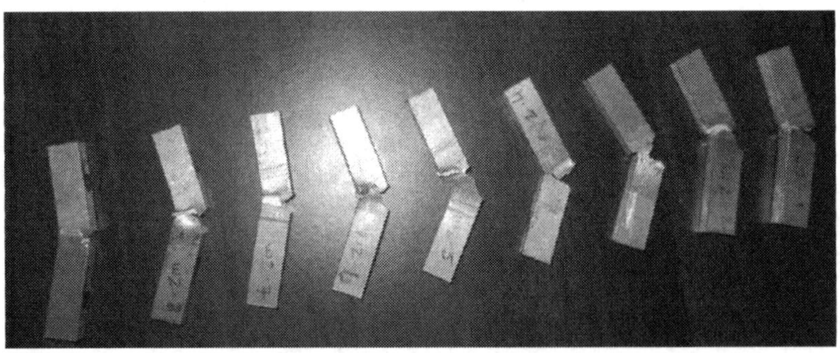

Figure 5. Impact test specimen after fracture

Table 4 Taguchi's L27 Orthogonal Array with Experimental Results

Trial	Tool Rotational Speed (rpm)	Axial Force (KN)	Welding Speed (mm/min)	Tensile Strength (Mpa)	Yield Strength (Mpa)	% of Elongation	Impact Strength (joules)
1	1000	4	30	120.24	112.76	18.46	17
2	1000	4	45	119.45	111.68	17.45	18
3	1000	4	60	116.64	110.64	14.24	16
4	1000	5	30	121.1	113.47	19.14	16
5	1000	5	45	120.18	112.21	18.21	16
6	1000	5	60	119.28	110.49	17.42	17
7	1000	6	30	116.73	107.5	14.21	17
8	1000	6	45	116.52	106.51	14.71	16
9	1000	6	60	114.72	104.23	14	16
10	1100	4	30	128.76	122.41	24.18	23
11	1100	4	45	126.37	121.31	23.46	21
12	1100	4	60	125.23	119.02	21.42	21
13	1100	5	30	128.42	121.98	24.21	23
14	1100	5	45	127.65	120.46	22.61	20
15	1100	5	60	124.02	119.37	20.41	19
16	1100	6	30	126.42	119.54	18.62	20
17	1100	6	45	126.21	121.03	23.33	19
18	1100	6	60	123.12	117.65	21.46	19
19	1200	4	30	124.68	118.2	21.42	18
20	1200	4	45	124.2	117.56	20.33	18
21	1200	4	60	122.16	116.7	20.14	18
22	1200	5	30	125.46	119.45	23.42	19
23	1200	5	45	123.76	116.98	21.46	18
24	1200	5	60	121.78	116.42	19.33	18
25	1200	6	30	124.24	118.61	22.16	17
26	1200	6	45	122.78	116.23	20.15	18
27	1200	6	60	123.04	116.78	21.68	16

5 Results and Discussions

5.1 Grey-Taguchi Technique

In this work, Grey-Taguchi technique was employed for finding better combination of process control variables to join AA7075 and AA6101 alloy materials by friction stir welding method and it is one of the best practices for multi objective optimization problems. Generally, Taguchi method is supportive for planning of experiments and finding of optimal setting individually for each output response, but in the present research there are different output responses for tensile properties and impact strength. There is necessity to find out the supreme combination of process control variables for all the output responses simultaneously. The step-by-step procedure in Grey-Taguchi technique is shown in Figure 6.

5.2 Normalization of Experimental Results

In the Grey-Taguchi test the initial step is to normalize the experimental results of tensile properties and impact strength. out once the signal-to-noise ratio of the required quality criteria is obtained as shown in Table 10. This process conveys the original sequence to a comparable sequence and the experimental results are normalized in the range between zero to one due to different measurement unit [15]. For normalizing tensile properties (ultimate tensile strength, yield strength, % of elongation) and impact strength 'Higher-the-better' (Equ. 1) criterion is used. The normalized data for tensile properties and impact strength is given in Table 5.

$$X_j(k) = \frac{y_j(k) - \min y_j(k)}{\max y_j(k) - \min y_j(k)} \quad (1)$$

Where,
Xj (k) = value after normalizing data/Grey relational generation value,
min yj(k) = smallest value of yj(k) for kth response,
max yj(k) = largest value of yj(k) for kth response.

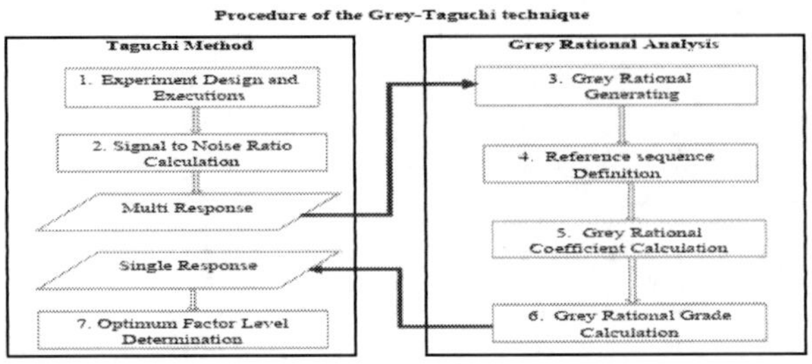

Figure 6. Procedure of the Grey-Taguchi Technique

Table 5. Normalized data for UTS, YS, % of elongation and impact strength

Trials	Normalization Values			
	Tensile Strength (Mpa)	Yield Strength (Mpa)	% of Elongation	Impact Strength (joules)
1	0.39316	0.4692	0.43683	0.21429
2	0.33689	0.40979	0.3379	0.35714
3	0.13675	0.35259	0.02351	0.07143
4	0.45442	0.50825	0.50343	0.07143
5	0.38889	0.43894	0.41234	0.07143
6	0.32479	0.34433	0.33497	0.21429
7	0.14316	0.17987	0.02057	0.14286
8	0.12821	0.12541	0.06954	0
9	0	0	0	0.07143
10	1	1	0.99706	1
11	0.82977	0.93949	0.92654	0.78571
12	0.74858	0.81353	0.72674	0.71429
13	0.97578	0.97635	1	1
14	0.92094	0.89274	0.84329	0.57143
15	0.66239	0.83278	0.62782	0.5
16	0.83333	0.84213	0.4525	0.64286
17	0.81838	0.92409	0.91381	0.5
18	0.59829	0.73817	0.73066	0.42857
19	0.7094	0.76843	0.72674	0.35714
20	0.67521	0.73322	0.61998	0.35714
21	0.52991	0.68592	0.60137	0.28571
22	0.76496	0.83718	0.92262	0.42857
23	0.64387	0.70132	0.73066	0.28571
24	0.50285	0.67052	0.52204	0.35714
25	0.67806	0.79098	0.79922	0.21429
26	0.57407	0.66007	0.60235	0.35714
27	0.59259	0.69032	0.7522	0.07143

5.3. Grey Relational Coefficient

Once calculate normalized tensile properties like ultimate tensile strength, yield strength, % of elongation and impact strength, the next step is estimation of grey relational coefficient values for tensile properties and impact strength. The grey relational coefficient εj (k) can be estimated by using Equ. (2). the grey relational coefficient values for tensile properties and impact strength are given in Table 6.

$$\varepsilon_j(k) = \frac{\Delta_{min} + \varphi\Delta_{max}}{\Delta_{oj}(k) + \varphi\Delta_{max}} \quad (2)$$

Where,

$\Delta oj(v) = Xoj(v) - Xj(v)$,
Δmax = larger value of Δoj,
Δmin = smaller value of Δoj,
Xj (v) = value after normalizing data/Grey relational generation value,
Xoj(v) = Ideal value = 1 and in general assumed φ=0.5.

Table 6. Grey Rational Coefficient for UTS, YS, % of elongation and impact strength

Trials	Grey Rational Coefficient Values			
	Tensile Strength (Mpa)	Yield Strength (Mpa)	% of Elongation	Impact Strength (joules)
1	0.45174	0.48506	0.47029	0.38889
2	0.42988	0.45863	0.43026	0.4375
3	0.36677	0.43576	0.33864	0.35
4	0.4782	0.50416	0.50172	0.35
5	0.45	0.47123	0.4597	0.35
6	0.42545	0.43265	0.42917	0.38889
7	0.3685	0.37875	0.33797	0.36842
8	0.36449	0.36375	0.34954	0.33333
9	0.33333	0.33333	0.33333	0.35
10	1	1	0.99416	1
11	0.74601	0.89205	0.8719	0.7
12	0.6654	0.72837	0.64661	0.63636
13	0.9538	0.95483	1	1
14	0.86347	0.82337	0.76137	0.53846
15	0.59694	0.74938	0.57327	0.5
16	0.75	0.76003	0.47733	0.58333
17	0.73354	0.86819	0.85297	0.5
18	0.5545	0.65632	0.6499	0.46667
19	0.63243	0.68346	0.64661	0.4375
20	0.60622	0.65208	0.56817	0.4375
21	0.51542	0.61419	0.5564	0.41176
22	0.68023	0.75436	0.86599	0.46667
23	0.58403	0.62603	0.6499	0.41176
24	0.50143	0.60279	0.51127	0.4375
25	0.60832	0.7052	0.71349	0.38889
26	0.54	0.59528	0.55701	0.4375
27	0.55102	0.61753	0.66863	0.35

Journal of Physics: Conference Series **2040** (2021) 012042 doi:10.1088/1742-6596/2040/1/012042

5.4. Grey Relational Grade (GRG) and Order

Grey relational grade (GRG) for each investigational run is the normal of grey relational coefficient value for a particular investigational run. GRG can be calculated by using Equ. (3). Larger value of grey relational grade specifies the top value, so highest grade value provides the higher order. The GRG and their position are given in Table 7.

$$\gamma_j = \frac{1}{n}\sum_{k=1}^{n} \varepsilon j\ (k) \tag{3}$$

Where,

n = No of process responses,
εj (k) = Grey relational coefficient

Table 7. GRG and order

Trials	GRG	RANK
1	0.44899	20
2	0.43907	21
3	0.37279	24
4	0.45852	19
5	0.43273	22
6	0.41904	23
7	0.36341	25
8	0.35278	26
9	0.33750	27
10	**0.99854**	**1**
11	0.80249	3
12	0.66919	7
13	0.97716	2
14	0.74667	4
15	0.60490	9
16	0.64267	8
17	0.73868	5
18	0.58185	12
19	0.60000	11
20	0.56599	14
21	0.52444	17
22	0.69181	6
23	0.56793	13
24	0.51325	18
25	0.60397	10
26	0.53245	16
27	0.54679	15

Journal of Physics: Conference Series **2040** (2021) 012042 doi:10.1088/1742-6596/2040/1/012042

5.5. Multi Objective Optimization

In order to investigate the significant control variables on tensile properties and impact strength. ANOVA was performed. ANOVA table for GRG are shown in Table 8., it is shows that tool revolving speed has utmost dominating process control variables which is about 70.88% influence on grey relational grade and succeeding with transverse speed and axial force has effect on grey relational grade with contribution of 10.63% and 4.93%. The interaction effects of TRS x AF, TRS x WS, AF x WS on grey relational grade with contribution of 2.80, 4.70, 2.76 % respectively. Table 9 & Table 10 shows the response table for grey relational grade which gives the average of each process responses (Means and S/N ratio) for each level at each response. The ranks and delta values show that rotational speed have high effect on grey relational grade as well as it is followed by transverse speed and axial force in that order.

Table 8. Analysis of Variance for GRG

Source	DF	Adj SS	Adj MS	F-Value	P-Value	% Contribution
TRS	2	0.54698	0.273489	85.94	0.000	70.88
AF	2	0.03805	0.019025	5.98	0.026	4.93
WS	2	0.08206	0.041029	12.89	0.003	10.63
TRS*AF	4	0.02160	0.005400	1.70	0.243	2.80
TRS*WS	4	0.03623	0.009058	2.85	0.097	4.70
AF*WS	4	0.02129	0.005323	1.67	0.248	2.76
Error	8	0.02546	0.003182			3.30
Total	26	0.77167				

Table 9. Response table for GRG (means)

Level	TRS	AF	WS
1	0.4028	0.6013	**0.6428**
2	**0.7513**	**0.6024**	0.5754
3	0.5718	0.5222	0.5078
Delta	0.3486	0.0802	0.1350
Rank	1	3	2

Table 10. Response table for GRG (S/N ratio)

Level	TRS	AF	WS
1	-7.951	-4.786	-4.288
2	-2.632	-4.735	-5.099
3	-4.888	-5.950	-6.085
Delta	5.318	1.215	1.797
Rank	1	3	2

As grey relational grade 'higher the better' type response, it can be seen from Figure 7, that the third level of rotational speed, third level of force and first level of transvers speed offers extreme value of grey relational grade. Hence, Tool rotational speed is 1100 rpm, force is at 4 KN and transverse speed of 30 mm/min is the finest combination of process control variables for obtaining utmost tensile properties and maximum impact strength simultaneously in FSW process. Figure 8 also suggest the same combination of process control variables.

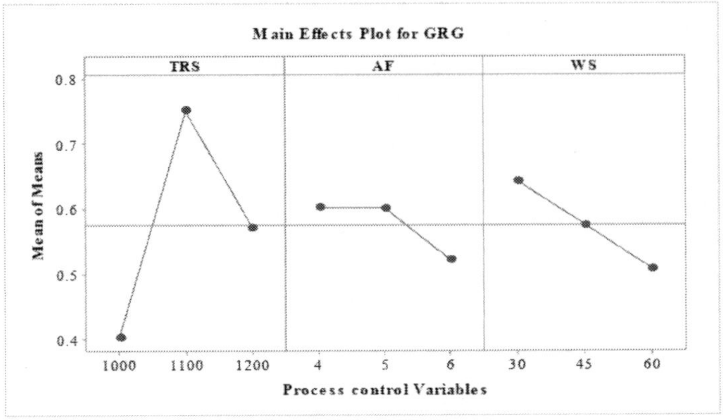

Figure 7. Main effect plot for GRG

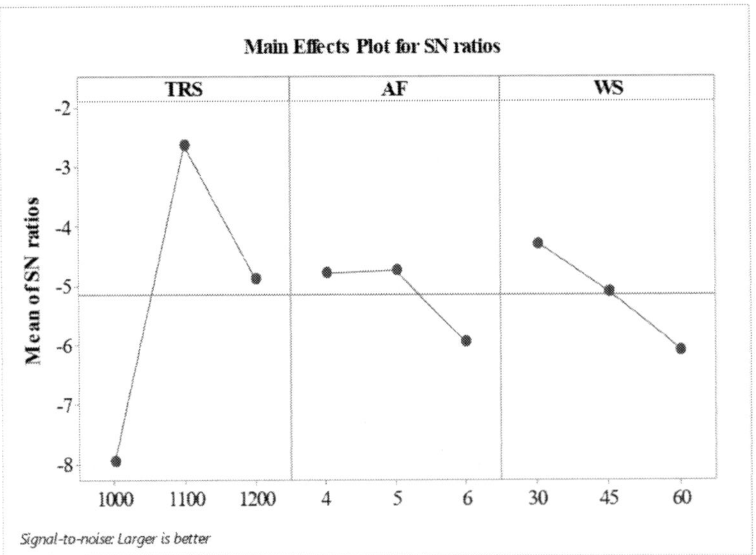

Figure 8. S/N Ratio for GRG

5.6. Anticipated optimum condition

Based on trials, the optimum level setting was found at tool rotational speed of 1100 rpm table transverse speed of 30 mm/min and axial force of 4 kN. So, the anticipated grey relation grade can be determined as:

$$\hat{\gamma} = \gamma_m + \sum_{k=1}^{m}(\hat{\gamma}_k - \gamma_m) \quad (4)$$

Where

γ_m is the total mean GRG, $\hat{\gamma}_k$ is the mean GRG at the optimum level, and n is the number of process control variables that affect the quality characteristics.

So, anticipated grey relation grade = 0.8067.

The validation experiment is not necessary here, as the optimized experiment at factor level now available within the planned trialing. The actual grade of gray relation at optimal condition is 0.9735; while the grade of gray relation expected is 0.8067. So the gap is just 0.16 (approx.) and it arises due to ignoring the nonlinear effects in three factor three level Taguchi L27 orthogonal array.

6. Conclusions

In this investigation, to find the optimum combination of friction stir welding process control variables to join AA7075 & AA6101 alloy materials, Taguchi based grey analysis was applied. The important conclusions from the present research work are summarized as follows:

- The finest combination of process control variables for obtaining optimum ultimate tensile strength of 129 Mpa., maximum yield strength of 119.21Mpa, maximum % of elongation 24.64% and impact strength 23Mpa simultaneously in FSW process is found at Tool revolving speed is 1200 rpm, transverse speed of 30 mm/min and axial force is at 5 kN.

- ANOVA results shows that process control variables too revolving speed shows major effect on the output responses in FSW process and axial force shows less influence on the output responses.

- The current Gray Relationship Grade at the optimum condition is 0.9735, while the predicted Gray Relationship Grade is 0.8067. Thus, the difference is only 0.16 (approx.). This variation occurs due to the neglect of the nonlinear effects in the orthogonal 3-level Taguchi L27 array.

References

[1] W. M. Thomas et al 2020 Improvements to Friction Welding', GB Patent Application No. 9125978.8.
[2] R. S. Mishra et al 2005 Friction stir welding and processing, Material Science and Engineering: R: Reports, Vol.50, pp 1-78.
[3] K. E. Knipstrom et al 1997 Friction stir welding process goes commercial, Welding Journal, Vol. 76, pp 55 – 57.
[4] Mohammed Yunus et al 2018 Mathematical Modelling of a Friction Stir Welding Process to Predict the Joint Strength of Two Dissimilar Aluminium Alloys Using Experimental

Data and Genetic Programming, Modelling and Simulation in Engineering, Vol 2018, pp 1-19.

[5] Srinivasa Rao Mallipudi et al 2018 Optimization of Process Parameters for FSW of Al-Mg-Mn-Sc-Zr Alloy Using CCD And RSM, Journal of Mechanical Engineering, Vol.68, pp 195-224.

[6] V. K. Parikh et al 2017 Multi-Objective Optimization of FSW Process Parameter using MOORA method, International Journal of advances in Production and Mechanical Engineering, Vol 3, pp 19-24.

[7] Saurabh Kumar Gupta et al 2014 Multi-Objective Optimization of Friction Stir Welding of Aluminium Alloy Using Grey Relation Analysis with Entropy Measurement Method, Nirma Univeristy Journal of Engineering and Technology, Vol. 3, pp 29-34.

[8] Shanavas Shamsudeen, John Edwin Raja Dhas 2018 Optimization of Multiple Performance Characteristics of Friction Stir Welded Joint with Grey Relational Analysis, Materials Research, Vol.6, pp.1-13.

[9] Sajjad Khaki et al 2019 Optimizing Friction Stir Welding Process for Enhancing Strength and Hardness using Taguchi Multi-Objective Function Method, International Journal of Advanced Design and Manufacturing Technology, Vol.12, pp.25-33.

[10] D.Vijayan et al 2018 Multi Objective Process Parameters Optimization of Friction Stir Welding using NSGA – II', IOP Conference Series: Materials science and Engineering, Vol 390, pp 1-7

[11] Mehran Tamjidy, et al 2017 Multi-Objective Optimization of Friction Stir Welding Process Parameters of AA6061-T6 and AA7075-T6 Using a Biogeography Based Optimization Algorithm, (MDPI) Materials, Vol.10, pp 1-19.

[12] D.Vijayan, V. Seshagiri Rao, et al 2017 Optimization of friction stir welding process parameters using RSM based Grey–Fuzzy approach, Saudi Journal of Engineering and Technology, Vol.2, pp 2-25.

[13] R.Palanivel et al 2012 Prediction and optimization of wear resistance of friction stir welded dissimilar aluminum alloy, Procedia Engineering, Vol.38, pp. 578-584.

[14] Wali Muhammad et al, 202 optimization of Friction Stir Welding Parameters of AA2014-T6 Alloy using Taguchi Statistical Approach", Journal of Welding and Joining, Vol.38, pp 493-501.

[15] Assefa Asmare et 2020 Experimental Investigation of Friction Stir Welding on 6061-T6 Aluminum Alloy using Taguchi-Based GRA", Metals, Vol. 10, pp 1-21.

Design and fabrication of drainage cleaning equipment

S.Kalyanakumar[1], R.Praveen[2], Arun Shaji [3], Diwakar [3]
1,2 Assistant Professor Gr II, Department of Mechanical Engineering, Aarupadai Veedu Institute of Technology, Vinayaka Mission's Research Foundation, Deemed to be University, Tamilnadu, India.
3 UG Student, Department of Mechanical Engineering, Aarupadai Veedu Institute of Technology, Vinayaka Mission's Research Foundation, Deemed to be University, Tamilnadu, India.
E-Mail : kalyanakumar@avit.ac.in

Abstract: Water could also be variety one necessity of humans and each one residing beings. There is a number of water on the earth but it sincerely is not suitable for human use. Clean water is extra important and is used for a couple of purposes. The impurities observed in water can purpose dangerous diseases. Impurities in drainage water are often like empty bottles, polythene bags, papers etc. House drains empty themselves into the foremost drains which run beneath the foremost streets and beneath many lanes. Solid topics which is perhaps created thru human or animal activities, and which is perhaps disposed of due to the reality they'll be dangerous or useless are mentioned as sturdy waste. Most of the sturdy wastes, like paper, plastic containers, bottles, cans, or even used motors and virtual gadgets aren't biodegradable, this suggests that they're not getting lessened through inorganic or herbal processes. Thus, after they collect they purpose a hazard to people. Decaying wastes moreover enchantment to have circle of relatives pests and convey about metropolis areas becoming unhealthy, dirty and unsightly places to be dwelling in. Moreover, it moreover motives damage to terrestrial organisms on the equal time as moreover reducing the uses of the land for specific extra useful purposes. Therefore, this problem goals immediately remedial measures. These impurities observed in drainage water can purpose blockage or the system.

Keywords: Hydraulic motor, water, empty bottles, polythene bags, papers

1. Introduction

The device might be cleaned sometimes manually or this sort of device might be designed with the intention to robotically throw out wastages and should maintain the water smooth. This task enables to scrub water within the device whenever any wastage seems which is an green and straightforward manner of cleansing the device and stopping the blockage. It additionally reduces human energy and improves the standard of water it truly is cleaned. The drainage structures are cleaned whilst there could also be no water in them i.e. whilst it's now not raining, however whilst it's raining the drainage structures cannot be cleaned due to the difficult situations of the rain which nobody might volunteer to verify that rubbish wouldn't input into the drainage structures. Our device makes use of a gutter/drain cleansing device that we could fluids accompany the drift thru it however catches massive stable waste like bottles, plastic and accumulates it. So gutter cleaners need to simply smooth those gutter cleansing structures established at factors as against cleansing complete gutter floors. We are travelling to enforce a tool that creates use of a sequence force device pushed through a motor which includes battery. Drainage is that the herbal or synthetic elimination of floor and subsurface water from a neighbourhood. As lengthy because the draining device is taken into consideration the characteristic of the principle drainage device is to gather, shipping and obtain obviate the water thru an outfall or outlet. Impurities in drainage water could also be best like empty bottles, polythene bags, paper etc. These impurities found in drainage water can reason blockage or the drainage device. The drainage device could also be cleaned sometimes manually or this sort of device could also be designed so as to robotically throw out wastages and will maintain the water smooth.

Figure 2.1 Drainage cleaning equipment assembly

A hydraulic motor has two holes, one is inlet and another outlet. Pipeline is joined to the inlet of the motor and outlet is given to a cylinder which has holes at the bottom Which we pass water to inlet of the motor, it will rotate the hydraulic motor shaft which is connected to cutting blade and then comes out through the outlet and get into the cylinder with holes at the bottom and comes out of the holes. While coming out it comes out with a force which will give a thrust for the equipment to move forward. By this method we can get both drilling and the thrust to move forward in single force. The water is pumped into this equipment with force. The force of pumping the water will directly reflect in motor rotation and the moving thrust. This is the drainage pipe cleaning equipment which is about to develop.

Journal of Physics: Conference Series **2040** (2021) 012043 doi:10.1088/1742-6596/2040/1/012043

Figure 2.2 Drainage cleaning equipment full setup

This consists of a propeller blade at the front which will be in the form of a spring. The product consists of a water jetting system and a rotating shaft which is surrounded by screw-like fins. The figure shown below is the drainage cleaner inserted into the cylinder head. The rotating shaft and water nozzles will enter completely into the cylinder head. As shown in above figure now for cleaning the pipe the entire equipment is inserted into the pipe and at first the water jetting process is done. After that the block inside the pipe will be removed. For collecting the remaining waste materials the rotating shaft starts to rotate and collect the waste. By this process the entire solid waste inside the pipe will be removed. The objective of our project is to remove this scattered debris from the pipe after water jetting which will give more time for the next block inside the pipe. Just adding a small modification to the water jetting equipment can improve the cleaning work and reduce the work effort of workers

2. Work Methodology

Hydro Jetting This inexperienced drain cleansing answer makes use of in most cases high-stress water and a unique nozzle to unblock the clog. A bendy hose connects thru an adjustable nozzle that emits water to fall apart sludge, soap, grease, calcified deposits, and everyday family objects. Even tree roots which have invaded the joints of your pipes gets flushed away to the municipal sewer system. The water waft is extraordinarily stepped forward and the pipelines are smooth and geared up for fast utilization once more following the cleansing session. Unlike industrial drain cleaners, hydro jetting will now no longer reason capability damage including environmental hazards, contamination, or superior corrosion as it is based on natural water to get the activity done. Most of all, hydro jetting wontt require widespread panorama digging and excavation. When utilized by our professional technicians, it wont reason harm in your pipes. Itâ€™s relevant to each residential and industrial sewer and drain lines. The system is essentially transportable and may be delivered withinside the van for solving drains in homes, restaurants, and offices. The complete technique simplest takes some hours to be finished and also you wont need to spend days with out water predominant get entry to. Traditional Methods For touchy piping materials, our technicians can nonetheless convey out conventional drain cleansing equipment including snakes and augers. When used nicely and withinside the proper hands, those equipment can remedy clogs to keep away from premature emergencies. Drain snakes are regularly used to restore clogged bathrooms and sinks. It can remedy small blockages including a clump of hair, tissue, etc. and to interrupt up larger particles. The lengthy steel stick can healthy thru the drain and poke the clog, permitting water to return back thru. Expert technicians regularly use snakes alongside augers for speedy clearing up the blockage. The snake pokes the clog even as the auger pulls it the usage of a hooked end, bringing into floor tough particles including jewelry, toys, and different items. Here at SPT, we think about the scope and nature of the drain hassle to provide you with the first-rate answer. The technique is constantly the same, whether or not the problem happens withinside the kitchen drain, the rest room drain or the basement drain. Our sewer line inspections require truly no digging. We can see what is in poor health your pipes the usage of one get entry to point. Call us and get the first-rate drain cleansing offerings in Maryland today. Sludge Withdrawal Half every year or every year desludging of septic tanks is desirable. Normally, the tanks want to be wiped clean whilst the sum of the intensity of the scum and the sludge is found to exceed 1/2 of the intensity of the tank. A part of sludge now no longer much less than 25 mm intensive have to be left at the back of withinside the tank backside which acts because the seeding cloth for the clean deposits. Manual coping with of sludge have to be avoided. Spreading of sludge at the floor withinside the region have to now no longer be allowed.

3. Conclusion

In the latest beyond there was a great deal studies completed for waste water management. Our challenge additionally stands one in all them with ideology & new tech. Many unique empirical research were completed and classes together with drainage cleansing machine and its automation were studied to a excellent depth. We consciousness greater on making the machine cellular with inside the drainage.

- The purifier functioned circulate successfully at some stage within the heavier rains which had greater quantity of walking water with rubbish and excessive velocity
- The machine can circulate with inside the drain to collect the floating waste in order that it'll lessen human labour.
- Within the remedy machine of drainage Waste water manipulate with the help of using the motor, curler chain and sprocket, lifter and therefore the accumulating bin to accumulate semiautomatic manipulate of sewage waste water remedy

Reference

[1] Ganesh U L,et.al. "Automatic Drain For Sewage Water Treatment Of Floating Materials", International Journal of Research in Engineering and Technology, Vol No-05, Jul-2016.

[2] S D Rahul Bharadwaj, Shraddha R Jogdhankar, "Automatic sullage water treatment process to reduce global warming" International Journal of Environmental Science: Development and Monitoring, Vol No- 2 (2013)

[3] Dr .K.Kumaresan et.al., "Automatic waste Cleaning Equipment", International Conference on Explorations and Innovations in Engineering and Technology , 2016.

[4] Nitin Sall, et.al., "Drain Waste Water Cleaner",Global Journal of Researches in Engineering: J General Engineering Vol No- 16, 2016.

[5] R.Sathiyakala et.al., "Smart waste Cleaning System" International Journal of Innovative Research in Computer and Communication Engineering, VolNo4, February 2016

[6] Nikhil S. Pisal, et.al., "Structural Analysis Of Motorcycle Chain By Using C.A.E. Software", International Journal Of Advance Research In Science and Engineering, Vol No-4, Special Issue (02), February 2015.

[7] James C. Conwell, G.E. Jhonson, "Design, Construction and Instrumentation of a Machine to Measured Tension and Impact Forces in Roller Chain Drives", December 1989

[8] N. Prbhushankar et.al.," Design and Fabrication of Drainage Cleaning Machine with Spring Return Pneumatic System", International Journal For Scientific Research and Development, Vol No.4, 2016.

[9] Gregor Burger, et.al, "Designing and Implementing a Multi-Core Capable Integrated Urban Drainage Modelling Toolkit: Lessons from CityDrain3", Advances in Engineering Software 100, 2016.

A Survey on Automatic Text Summarization Techniques

S Hima Bindu Sri[1], Sushma Rani Dutta[2]

[1]Assistant Professor, [2]Assistant Professor
Department of Computer Science and Engineering, Koneru Lakshmaih University,
Department of Computer Science and Engineering, Koneru Lakshmaih University.

E-mail: sushamadutta@klh.edu.in

Abstract. In recent years, there is a tremendous explosion in the amount of text data on the internet and in the archives of news articles, scientific papers, legal documents and even in online product reviews. Text summarization is playing an important role in automatic content creation, minutes of meeting generation, helping disabled people and also for quick online document reading. To achieve these, several automation techniques have been proposed in various researches. In this regard, performing an exclusive survey on different methods, approaches of automatic text summarization which are published in different articles in most recent three years.

Key words: Text summarization, Supervised, Unsupervised, Optimization, Natural Language Processing.

1. Introduction

The aim of the text summarization is to extract the summary of large documents of text. The great advantage of using summary is to reduce the reading time, reduces effort and cost. The process involved in this text summarization is to identify the important meaningful information available in many documents. This is to be done in 2 ways: Extractive and Abstractive. Extractive summarization can be done by selecting the important sentences from the original document. An abstractive summarization can be done by having awareness of the main concepts of the document and interpret them in a new way. This can be achieved through several supervised and unsupervised techniques such as CNN, RNN, LRC and several deep learning techniques were used which will be mentioned in different articles. Machine learning is also used to review the reviews of any e-commerce website like Amazon, it helps to know about the trends and behaviors of users to serve to the proper items and also provides reminders which are relevant [1] [2]. Then it will use those outcomes to display advertisements which are relevant to them [3][4]. The main advantage to use ML is, we need not to superintend every step of our project, instead it provides the machine to have the ability to make predictions [5] [6] [7].

The organization of the paper is as follows. Section II describes about Literature review and section III will discuss regarding machine learning techniques used for text summarization, after basics, we underlined all available automatic text summarization techniques and their performances in Section IV. Section V presents conclusion and future work and finally, Section VI presents references.

2. Literature Review

2.1 Machine learning techniques for text summarization

Supervised and unsupervised learning [8] are two types of machine learning techniques. Supervised learning means machine is learning under a supervisor and is trained with data which is already verified and labelled correct. In this learning technique, result is already known and confirmed. Once machine learns from true training data, then new unknown data is given to predict outcomes. It helps in optimization of various results according to criteria. They are mainly used to solve real world complex computational problems. Supervised learning is further classified into two techniques i.e., classification technique and regression technique. In regression technique, a single continuous output value is produced using training data. In classification technique, as the name says, it classifies or categorize input into some output category. Classification can be named as binary or multiclass classification depending on output categories. Author Cheng-Zen Yang in his paper proposed a supervised leaning technique to produce bug report summary.

Unsupervised learning technique, unlabelled data is given as input and algorithms are applied in order to get output. It is used in order to find out all possible patterns or classes in data. Model is not supervised in this technique. It is further classified into two different techniques, namely, clustering and association. Clustering technique will take input data, and after processing it, it will classify the data into various clusters based on common features. Association technique helps in connecting data objects or associating them in order to discover some relationship in a very large data set or database. Author Senthil Kumar proposed a method to produce bug report summary using unsupervised learning technique in his work "AUSUM: Approach for unsupervised bug report summarization".

2.2 Evaluation parameters of generated summary:

- Precision - Precision is the ratio between the positive predictions which are predicted correctly and all the total predicted positive observations.

  Precision = TP/TP+FP

- Recall (Sensitivity) - Recall is the ratio between the positive predictions which are predicted correctly and all the observations in its class.

  Recall = TP/TP+FN

- F1 score - F1 score is the weighted average of precision and recall. Therefore, this score considers false positives as well as false negatives in to consideration. F1 score is more useful if we have an uneven class distribution.

  F1 Score = 2*(Recall * Precision) / (Recall + Precision)

- ROUGE- The abbreviation of ROUGE is Recall-Oriented Understudy for Gisting Evaluation. A set of measures are used for evaluation of machine translation and automatic summarization of texts. It evaluates by comparing the summaries which are produced automatically or the translations against a group of summaries references.

3. Automatic Text Summarization Techniques

Author AKukkar [9] introduced one effective approach to produce a flexible and productive bug report summary as well as to minimize load and work of developer. Author used particle swarm optimization (PSO) approach for searching the effective semantic text. Author tried to address four central points that are, extractive bug report summarization, to increase the ROUGE score by selecting effective semantic text, sparsity of data and reduction of information. Proposed methodology used collection of comments and some feature extraction methods to generate or produce the bug report summary. Multiple summary subsets are produced and optimal summary subset evaluated by PSO optimization technique. Author compared the proposed approach with existing Email Classifier (EC) and Bug Report Classifier (BRC). ROUGE score was selected as one of the evaluation criteria and was calculated for all approaches. At the same time, the ROUGE score was compared with three human-generated summaries of 10 bug reports of Rastkar dataset. As a result, PSO approach summary subset was less redundant, and included all important points needed to be present in a bug report.

Author Beibei Huai and team [10] gave new Intention-based bug report summarization approach, alias IBRS which is based on intention taxonomy. This work considered sentences intentions in order to generate summary report. Sentence intentions were classified according to their taxonomy levels into seven categories: bug Description, fix solution, opinion expressed, information seeking, information giving, meta/code and emotion expressed. Now, sentences are categorized in specific intention with the help of pattern matching and machine learning model. Finally bug report summary is produced. This summary was compared with BRC (Bug Report Classifier) and found better in terms of precision (5% improved), recall (3% improved), F-score (3% improved) and pyramid precision (5% improved).

Creating summary is selecting important topics of sentences as well as recognizing relevant relationships among those concepts which are mentioned in that text. The key problem is generalization which is identified by ATS task. Stating an example: summarization financial or medical reports are conceptually different from summarizing news articles. To solve the above issue, to achieve more relevant summary, this paper, author Hernández-Castañeda, Ángel [11] proposes (EATS).EATS is based on clustering technique which holds by GA i.e. Genetic Algorithm to get relevant topics in the proposed document. To identify key sentences in the clusters, this method includes Topic Modeling Algorithm (LDA) which is based on keywords those are generated automatically. This clustering technique needs LDA and Doc2Vec to map text to numeric vectors along with tf-idf and n-grams. This method is tested on DUC02 dataset to achieve the goal of producing summaries as close to as human generated summaries. TUCII dataset is used to test with multi domain and multi-language frameworks. This system achieves more effectiveness than earlier methods based on the evaluation results such as skip grams(ROUGE SU) and also (ROUGE-1), bi-gram (ROUG-2). As a result, the obtained summaries showed matches of unique words as well as appends context through matching the adjoining words.

ShubhraGoyal Jindal et al.[12] suggested rapid automatic keyword extraction and term frequency-inverse document frequency method which helps to get proper keywords and key-phrases with an applicable score. For extraction of sentences, fuzzy C-means clustering can be utilized to bring out sentences. For sentence selection rule engine is used. Hierarchical clustering is used for redundancy removal, re-rank generated summary and to enrich the summary. The suggested approach is assessed on Apache Project Bug Report Corpus

(APBRC) and actual Bug Report Corpus (BRC) achieves an average value of 78.22%, 82.18%, 80.10% and for precision 81.66%, f-score recall, and pyramid precision consequently.

Author Ashima Kukkar's [13] worked on the bug report summarizing as well as analyzed whether the bug report summary helps in finding the appropriate severity of bug. Unsupervised bug report summarization system was followed and the approach used in this work is swarm intelligence. In order to generate bug report summary, step by step process is done. Initially, pre-processing of text was done which further includes tokenization, stop-word removal and stemming. After preprocessing, all unwanted data is removed and features are extracted. Then sentence score was calculated using n-gram technique. In the next step, sentences were combined to generate summary subsets and for every subset, a relative subset score was calculated. Particle Swarm Optimization (PSO) approach is used to select optimal summary subset. Again, features scoring was done of selected optimal summary subset and relative score was calculated using Ant Colony Optimization (ACO). Model training is also done using these optimized features. This data result was compared with three benchmark datasets and proved better results in terms of Precision, Recall and F-measure.

To provide the developers with the important information, authors CHENG-ZEN YANG et.al and YU-HAN CHUNG [14] have proposed a two-layer semantic model (TSM). TSM uses two layers to identify the informative sentences for summary. TSM model is trained based on the textual features which are generated from two-layers. The first layer uses an ENR model to maintain the semantic information. It gives informative sentences by using a rule-based-Algorithm called algorithm1. The second layer uses BRC textual model to get the textual features by using the informative sentences which are generated from first layer. Based on the textual features which are generated form second layer a supervised logistic regression model is used for training the TSM. The datasets which are used by this TSM summarizer is the BRC dataset which has 36 bug reports in total. By comparing the TSM model with the BRC and AUSUM, it is proven that the TSM is outperformed and has highest value of F1 measure that is 0.530.And the F1 values of others are 0.40 and 0.50 respectively. The relative improvement achieved by TSM when compared with BRC and AUSUM are 34.3% and 7.3% respectively.

Bug report summarization has some characteristics which effect the performance of the summarization technique. To resolve this problem the author Xiaochen Li and his team [15] have proposed an unsupervised approach called DeepSum. To get the effective summary from bug reports DeepSum summarizer works based on deep learning. The main part of the DeepSum is a stepped auto-encoder network, which concludes the bug report summary established from the hidden layers of the network. The data sets which are used for evaluating the summarizers are summary data set and authorship data set. DeepSum achieved good results in terms of metrics like F-Score and R-1 when compared with comparative algorithms. DeepSum outrun the other algorithms by up to 0.119 and 0.092 in terms of F-score and R-1 respectively when tested on SDS data set. And achieves equal results on ADS dataset when compare with comparative algorithms. And the F-score and R-1 values of DeepSum are 0.462 and 0.563.

In this paper the authors Akalanka Galapathi and John Anvik [16] extends the Rastkarel'smodel. They have find out an extractive way to generate summaries for bug reports without using the complex features which are used in the Rastkar's model. They have find out how precision and recall varying when using and not using complex featues:These features are Length and Lexical. For sentence classification they have used a logistic regression model. For example they have used sklearn package for classification. To observe the performance of both individual and combined features, they have used different types of logistic regression models. By comparing author's with Rastkar'smodel,They have oserved the following. Author's model Precision,recall and F-Score values:44% , 24% and 29%. Rastkar's model Precision,recall and F-Score values:57%,35% and 40%. There is decrease in the performance due to usage of fewer features.

The authors of this paper are BeomjunKim ,Sungwon Kang and Seonah Lee [17].Generally bug reports are the lengthy conversations. While reading this, developers face many problems. To solve this problem, the authors have used weighted page-rank algorithm to maintain multiple relationships. They have used Relation based Bug Report Summarizer (RBRS) with duplicates, blocks and depends on relations for automatically generating the bug report. For weighting they have used page rank algorithm. RBRS page rank score to each sentence in the bug report .They have used the bug report corpus as data source. Duplicate relations are used for improve the quality of the bug report. By comparing proposed model(RBRS) with the existing model Page Rank based Summarization Technique (PRST) .They have observed the following for precision ,recall and F1-score for bug reports without duplicates relationships are 50.91% ,50.93% , 50.92% and 44.50% ,44.79% and 44.64% respectively. By Comparing proposed model (RBRS) with the Existing model (PRST), the following is observed precision, recall and F1-score for) bug reports with duplicates relationships the values are 51.69% ,51.82%,51.76% and 49.71% , 49.77%,49.74% respectively.

Generally, in software industry, software maintenance for fixing the bug reports, 45% of software development will be spent. A familiar Bug report called Bugzilla repository which had around 12,36,000 bug reports. The tasks like bug fixing, duplicate bug identification and bug-traing are tough tasks which will be time consuming. To perform above said tasks effectively He Jiang Et Al[18] suggested an efficient approach that is constructing attributes which are hardly influence the presentation of supervised algorithms for summarization of bug report. In this paper, to reveal the existing methods for constructing attributes, they first conduct a survey on the authors of 40 papers. Secondly, they suggested a new method called Crowd-Attribute (CA) to fully conclude attributes for bug report summarization in crowd sourcing. Thirdly, to achieve more accurate summaries of bug reports, they construct 11 new attributes successfully by trying the new method CA and suggested a recent approach named Logistic Regression with Crowd sourced Attributes (LRCA). In LRCA, 11 attributes are assessed to feed the statistical model called as logistic regression, over the training set. To predict its summary for a new bug report, the attributes of each sentence are calculated and fed into the trained model. The LRCA is tested to evaluate its effectiveness on a large-scale datasets with 1,05,177 bug reports and even experimented on the dataset SDS with 36 bug reports which are annotated. LRCA improved the result by 10.11%, 1.33%, 8.94%, and 5.89% with regard to recall, precision, F-score and Pyramid of supervised approach (BRC), respectively.

Surbhi Bhatia et al. [19] is providing the comparative knowledge about abstract techniques and extractive techniques. Abstractive type of summarization uses graph based technique for duplicate sentences. It follows a sequence of operations like constructing graphs, ensuring sentence correctness by using some of the constraints and finally scoring the sentences individually by combining sentiments using SentiWordNet. Extractive type of summarization uses the technique of (PCA) Principle Component Analysis which reduces the number of dimensions of data and finds the summary based on the ranking of most relevent sentences. The Opinosis dataset (Garmin Nuvi 255W GPS). The present abstractive technique achieved 13% of performance improvement. The results of both the techniques are evaluated through ROUGE score. The future challenge for the researchers is to employ semi abstractive, extractive techniques to be carry out the benefits of both approaches. Also, we can study excessively on threshold values, how they are affecting performance.

JIANLI DING et al. [20] addresses the problems of insufficient summary precision, insufficient use of semantic information and sematic loss in the generated summary. They proposed seq-to-seq architecture through dual-encoder is used to provide ample of semantic information. Here position embedding and word embedding technology. The information retrieval techniques such as Tf-Idf, parts of speech, key score are added to word's feature. In this research, to solve the problem of semantic loss Gain-Benefits gate structure is designed in the encoder. The results are tested against LCSTS and SOGOU datasets (News headlines). The future enhancement could be, the performance of this model needs to be still improved when it is dealing with special names and also familiar nouns. So, further research can be done in area.

HAOJIE ZHUANG et al. [21] proposed a model which consists of three components. They are – encoding long input text to shorter form using a generator. Secondly used a discriminator to train the generator to produce summaries which are human-readable. This research focused on single-document summarization. The data set they have used is CNN/DailyMail. In the present years of research, abstractive methods which are based on deep learning have shown promising results, specifically the methods that are based on supervised learning CNN to encode the input, another feed forward neural network is used to generate the summary. RNN to balance between extractive and abstractive approaches hybrid pointer-generator network is used. Pointer network for copying words directly from input text and generator network generate new words and phrases. The proposed model GAN (Generative Adversial Network) involves 2 opposing networks: First generator- which generates data samples. Secondly- discriminator- which distinguish network real data from input text and the data samples which are generated from generator. The generator should be able to generate samples that are "most similar" to real data when they well-trained. The future work is to combine the proposed model GAN with LSTM-CNN based deep learning and also the generator to be replaced with a pre-trained BERT model and processing the input phrase-by-phrase my help further to improve the accuracy.

4. Different text summarization techniques

Table 1 Different text summarization techniques

Author/Ref	Processing Steps	Features	Approach	Dictionary	Evaluation measure	Result	Future scope
A Kukkar [9]	Collection of comment, feature extraction methods, summary generation optimal summary subset selection	Collection of comments	Particle swarm optimization (PSO)	Rastkar dataset	Rouge-1 Rouge-2 Rouge-3	79.14 75.83 75.17	To achieve more accurate summary, other optimization techniques will be applied
BeibeiHuai [10]	Sentence intentions to generate summary	Intentions of Sentences	Intention-based Bug Report Summarization approach	BRC	precision recall Fscore Pyramid precision	5% 3% 3% 5%	Need to improve intention classifier characteristics
Ángel Hernández-Castañed [11]	Feature extraction	Word unigrams, Bigrams, features	Tf-Idf,D2V,LDE ,OHE	TAC11	F-score	0.249	To improve performance apply more classification Techniques
Shubhra Goyal Jindal[12]	Preprocessing Keyword and sentence identification Clustering Rule designing Clustering	Keyword and Sentence identification	Tf-idf RAKE, Fuzzy-C means Heirachical clustering	BRC, APBRC	F-score Precision. Recall	80.1% 78.22% 82.12%	Analyzed the impact of different clustering algorithms. ROUGE evaluation metric is used for documents which includes text.
Ashima Kukkar [13]	Feature Extraction and sentence score calculation	Feature Score Calculation	PSO and Ant Colony optimization	Benchmark datasets	precision recall Fscore	79.74 78.79 79.76	More study needed to be done on relation between bug summarization and severity classification.
Cheng Gen Yang[14]	Extracting the important information using Anthropogenic and Procedural classes	Textual features And informative sentences	TSM uses ENR and BRC textual model	BRC	F1	0.530	Some informative sentences are classified into others that needs to be improved. A dictionary is needed for word processing

Xiaochen Li[15]	Vector generation Tokenization	Sentence selection Word weighting Sentence weighing	Deep sum a stepped auto-encoder network	Summary Data Set and Authorship Data Set	F-score R1	0.462 0.563	In the future there is more needed to be studied for automated model for examine the length of bug report
Akalanka Galapathi[16]	Selecting highest probability sentences	Complex features Length and lexical features	A logistic regression model	BRC	Precision Recall F-Score	44% 24% 29%	Some complex features with high variability are explored
Beomjun Kim [17]	Analyzing duplicate, blocks or depends-on graph generation sentence score	Selecting top scored sentences	Relation based Bug Report Summarizer(RBRS)	Bug report corpus	Precision Recall F-Score	51.69% 51.82% 51.76%	In future focus must be on larger corpus. And also needed more study on preprocessing techniques to improve the quality of bug report summaries
He Jiang[18]	Constructing newAttributes	Constructed 11 attributes by CA approach	Logistic Regression with Crowd sourced Attributes (LRCA)	SDS-36 bug reports	Recall, Precision, F-score and Pyramid of supervised approach (BRC)	10.11%, 1.33%, 8.94%, and 5.89%	-
Surbhi Bhatia [19]	Ranking the sentences based on most relevancy.	Constructing graphs, ensuring sentence correction based on some constrints	SentiWordNet , Principle Component Analysis	Opinosis dataset	Precision, Recall, F-Measure	0.49, 0.27,0.35	Semi abstractctive, extractive techniques will be working together to get the benefits of the 2 approaches, Study excessively on threshold values, how they are effecting performance the summary.
JIANLI DING[20]	Tf-Idf, Parts -of - speech	keyscore,	Dual-Encoder	LCSTS & SOGOU dataset	Rouge-1, Rouge-2, Rouge-L	0.41 0.22 0.37	The performance is to be improved further when it is dealing with special names &nouns which are familiar.
HAOJIE ZHUANG [21]	Encoding long input to shorter form by using a generator	Train the generator to generate summaries which are human readable	GAN model in which we have Pointer network & generator network	CNN/DailyMail	Rouge-1, Rouge-2, Rouge-L	40.11 16.23 37.18	Teaching the generator to compose / construct abstractive sentences properly is still a challenging problem to continue this research. If the generator to be replaced with a Pre-trained BERT model and processing the input phrase-by-phrase may

							help further to improve the Accuracy.
Yang Gao , Yue Xu[22]	Representing the sentences as vectors	Topic, sentence predictions	CCTSenEmb	DUC2005, DUC2006	Rouge-1, Rouge-2	38.65 7.73	The future work involves in upgrading the topic learning process to broaden CCTSenEmb usability.

5. Conclusion and future work

Researchers depends on effective summaries from large documents of text. The purpose of summaries is to produce a quick review of any topic and to work in detail on more relevant and appropriate paper. Developers work on different platforms but they need to deal with bug reports. While reading the generated and assigned bug report, a developer should understand what a bug is, why is it raised and how to deal with it to get the expected result. So, to resolve the bug issue, an appropriate summary of bug report should be available.

This paper includes survey of different bug report summarization techniques which help developers to deal with their domains. Also, a performance analysis is done for every technique and different metrics like precision, recall, F-score and ROUGE score were calculated. This work gives a better clarity to understand which summarization technique should be used to get more effective summary. In future, we will work upon various combinations of techniques in order to obtain better results.

References

[1] Dutta, Sushama Rani, et al. "IoT in Autism Detection in Its Early Stages." Internet of Things: Enabling Technologies, Security and Social Implications. Springer, Singapore, 2021. 47-58.

[2] Dutta, Sushama Rani, Sujoy Datta, and Monideepa Roy. "Using cogency and machine learning for autism detection from a preliminary symptom." 2019 9th International Conference on Cloud Computing, Data Science & Engineering (Confluence). IEEE, 2019.

[3] Dutta, Sushama Rani, et al. "A Machine Learning-Based Method for Autism Diagnosis Assistance in Children." 2017 International Conference on Information Technology (ICIT). IEEE, 2017.

[4] Giri, Soumyajit, Sushama Rani, et al. "A real time epidemic alert generation system for rural areas using WBANs and kiosks." 2017 International Conference on Information Technology (ICIT). IEEE, 2017.

[5] Dutta, Sushama Rani, and Monideepa Roy. "EDOT: Context-aware tracking of similar data patterns of patients for faster diagnoses." 2017 Second International Conference on Electrical, Computer and Communication Technologies (ICECCT). IEEE, 2017.

[6] Dutta, Sushama Rani, and Monideepa Roy. "Providing context-aware healthcare services using circular geofencing technique." *2016 3rd International Conference on Computing for Sustainable Global Development (INDIACom)*. IEEE, 2016.

[7] Dutta, Sushama Rani, et al." Handling Scarcity of Data in Autism Identification using Binary Imputation Method" International Journal of Recent Technology and Engineering (IJRTE) ISSN: 2277-3878, Volume-8, Issue-1, May 2019

[8] Donalek, Ciro. "Supervised and unsupervised learning." Astronomy Colloquia. USA. 2011.

[9] Kukkar, Ashima, and Rajni Mohana. "An optimization technique for unsupervised automatic extractive bug report summarization." International Conference on Innovative Computing and Communications. Springer, Singapore, 2019.

[10] Huai, Beibei, et al. "Mining Intentions to Improve Bug Report Summarization." *SEKE*. Vol. 2018. 2018.

[11] Angel Hern´andez-Casta˜neda et al. "Extractive Automatic Text Summarization Based on Lexical- Semantic Keywords". In: IEEE Access 8 (2020), pp. 49896–49907.

[12] Shubhra Goyal Jindal and Arvinder Kaur. "Automatic Keyword and Sentence-Based Text Summarization for Software Bug Reports". In: IEEE Access 8 (2020), pp. 65352–65370.

[13] Kukkar, Ashima, Rajni Mohana, and Yugal Kumar. "Does bug report summarization help in enhancing the accuracy of bug severity classification?." Procedia Computer Science 167 (2020): 1345-1353.

[14] Yang, Cheng-Zen, Cheng-Min Ao, and Yu-Han Chung. "Towards an Improvement of Bug Report Summarization Using Two-Layer Semantic Information." IEICE TRANSACTIONS on Information and Systems 101.7 (2018): 1743-1750.

[15] H. Jiang, X. Li, Z. Ren, J. Xuan and Z. Jin, "Toward Better Summarizing Bug Reports With Crowdsourcing Elicited Attributes," in IEEE Transactions on Reliability, vol. 68, no. 1, pp. 2-22, March 2019, doi: 10.1109/TR.2018.2873427.

[16] Galappaththi, Akalanka. Automatic sentence annotation for more useful bug report summarization. Diss. Lethbridge, Alta.: University of Lethbridge, Department of Mathematics and Computer Science, 2020.

[17] Kim, Beomjun, Sungwon Kang, and Seonah Lee. "A Weighted PageRank-Based Bug Report Summarization Method Using Bug Report Relationships." Applied Sciences 9.24 (2019): 5427.

[18] Jiang, He, et al. "Toward better summarizing bug reports with crowdsourcing elicited attributes." *IEEE Transactions on Reliability* 68.1 (2018): 2-22.

[19] Bhatia, Surbhi. "A Comparative Study of Opinion Summarization Techniques." IEEE Transactions on Computational Social Systems (2020).

[20] Ding, Jianli, et al. "Generative Text Summary Based on Enhanced Semantic Attention and Gain-Benefit Gate." IEEE Access 8 (2020): 92659-92668.

[21] Zhuang, Haojie, and Weibin Zhang. "Generating semantically similar and human-readable summaries with generative adversarial networks." IEEE Access 7 (2019): 169426-169433.

[22] Gao, Yang, et al. "Jointly learning topics in sentence embedding for document summarization." *IEEE Transactions on Knowledge and Data Engineering* 32.4 (2019): 688-699.

Fuzzy Logic Algorithm and GSM IoT Based Fire Fighting Robot

P Vasanthkumar[1], P V Arunraj[1], N Mohammed Bilal Khan [2], A V Akash[2], R. Mukunthan[2], R. Harish Babu[2]

[1]Assistant Professor, [2] UG Student

Department of Mechanical Engineering, SRM Institute of Science and Technology, Ramapuram, Chennai, Tamil Nadu, India.

E-mail: pvasanthme@gmail.com

Abstract. Artificial Intelligence (AI) covers many areas related to mechanical, electrical, psychology and philosophy. GSM IoT is the new method and idea in a simple application of fire detection. In our project, we are using IR sensor to detect a fire and it sends the information to a microcontroller, which is programed based on a fuzzy logic algorithm that helps to detect the fire and moves the nozzle head of the Fire Fighting Robot. The stepper motor is used for a 360 degree clockwise and anticlockwise rotation. The D.C motor controls the valve opens of the nozzle. The fuzzy logic algorithm is helps to detect the fire and to turn the nozzle in the direction of fire.

Key words: Fire Fighting Robot (FFR), Fuzzy Logic Algorithm, Internet of Things (IoT), Global System for Mobile Communication (GSM)

1. Introduction

In our day-to-day life, we hear news about many fire disasters in flats, textile shops and some chemical industries. To overcome these disasters, we invented a fire fighting robot using GSM and IOT based. We introduced GSM and IOT in our project to make our project more efficient. Whenever, the fire occurs in the flat or industries in the absence of the property owner our robot helps to detect the fire and extinguish the fire. Simultaneously, the GSM sends a message to the concern owner of the property and to the fire-station. So that, the fire can be extinguished before it gets complicated. Many textile shop and manufacturing companies doesn't have proper fire extinguisher to extinguish the fire in the absence of people.

To overcome this, we replace our fire fighting robot in the place of the fire extinguisher. By replacing this, we can avoid the 95% of the fire accident occurring across the city. We use fuzzy logic algorithm to detect the fire and instruct the robot's nozzle to turn in the correct direction. Our project also reduces the human risk. The presence of a IR alerts the robot when the temperature exceeds the threshold temperature and the robot will sprays the water available in its tank.

A fire fighting robot is a preprogramed device that performs tasks normally performed by humans or machines that are programmed to perform repetitive or repetitive tasks. After so many studies we analyse that the fire fighting robot can be useful in medical purpose [1], reawakening [2-6], rescue-function [7, 8] and sector [9]. In various industries robot have been introduced many years ago. Commercial robots are versatile controlled designed for specific objects, partitions, devices with several system gesticulations to perform numerous functions [10]. Mechanical study has also aroused attentiveness in robotics, but only a fraction of the latest developments in robots can be connected with machine intellects. A recent project for the growth of robots has incorporated machine intellects algorithms [11-15] to improve intellect in robots.

Due to this, it reduces electricity waste over time and cost, at the same time the productivity also increases in the industries. Research into the use of humanoid-like robots is actively conducted to reduce the wound and demise of fire-fighters and to improve productivity, security, efficiency and quality of a given job [16].The design includes a range of sensors and a two-node MCU. The fire fighting robot is wirelessly connected to the node MCU. If the node fires, it notifies the central node to the MCU. The Central Node MCU sends information to fire safety officers and launches a robot to carry out fire fighting operations and launches pumps to extinguish fires. MHZ-14 CO2 sensor values are analysed using graphs. If the CO2 value is in the lower limit on the graph, the robot will start pumping CO2. Otherwise, it uses a water pump. At the receiving end, the interface for the two-motor L293D motor driver. For the movement of robots in all direction these two motors are used. Batteries are used to supply power. The storage system is used to store the values of the graphs analysed by the sensors. The node MCU receives all values from the various sensors and commands the robot to start the pump.

1.1 Fuzzy Logic Algorithm

Fuzzy from is a new technology that enhances traditional system models of engineering expert and using fuzzy logic we can reduce the need of rigorous mathematical modelling. Fuzzy logic is a real explanation of traditional logic, and fuzzy logic controller is a real explanation to the resembling control model. The point to successful use of fuzzy logic is intelligent combination of traditional methods. In addition, the fuzzy system is time-unchanging and decisive. Therefore any validation and fuzzy logic can also perform stability analysis method.

The rapid increase of numbers have seen in last few years and applications of fuzzy logic. There are some major uses that was increasingly visible from the field of user products, control techniques and industry systems. Uses for information processing, accusatory, human machine interfaces, cross check and decision support systems are less visible, but of increasing importance.

1.2 Development Tools

A software development tool suitable for fuzzy logic must meet the following criteria:

- Support all development stages (design, emulation, optimization, validation, and implementation)
- Industry standard interfaces (using MS-Windows, standard fuzzy logic programming language support, DLL / DDE / OLE and interfaces to control simulated software tools).
- Using fuzzy logic is a good approach. These recompiles basically translate the text of a fuzzy logic system into a standard programming language such as ANSICC.

1.3. Hardware Platform for Fuzzy Logic Systems

Fuzzy logic is a universal technique; Its applications cover the entire spectrum of electronic control. Therefore the target hardware platforms are different. The Todd tool allows you to run fuzzy logic systems on almost any hardware, from 8-bit AT89C52 to process control systems that use distributed 64-bit RISC processors.

1.4 Third Generation Integrated Fuzzy Processors

Integrated dimmer processors overcome second-generation errors by integrating dimmer processor functionality into standardAT89C52. The functionality that processes fuzzy logic systems can be implemented in a variety of ways. One way is to integrate complete fuzzy logic coprocessor units on a chip with a AT89C52 so that the two work together. Its advantage is that the communication overhead is much smaller than the two over chip solutions, because the dimmer logic coprocessor can use the same registers that handle data exchange.

Expanding the teaching set of a standard MCU is a different solution. The easiest way is to expand the micro code and use the same ALU architecture. It speeds up the calculation of fuzzy logic three times and compresses the code twice based on MCU architecture and fuzzy logic system size. Common sense, greed for human thought and judgment fuzzy logic, ultimately brings fuzzy logic to the party. This is a question that has been under discussion for the past few years and will be disputed quite soon.

2. Methodology

The design includes a range of sensors and a two-node MCU. The fire fighting robot is wirelessly connected to the node MCU. If the node fires, it notifies the central node to the MCU. The Central Node MCU sends information to fire safety officers and launches a robot to carry out fire fighting operations and launches pumps to extinguish fires. MHZ-14 CO2 sensor values are analysed using graphs. If the CO2 value is in the lower limit on the graph, the robot will start pumping CO2. Otherwise, it uses a water pump.

2.1 Materials Required

The design includes a range of sensors and a two-node MCU. The fire fighting robot is wirelessly connected to the node MCU. If the node fires, it notifies the central node to the MCU. The Central Node MCU sends information to fire safety officers and launches a robot to carry out fire fighting operations and launches pumps to extinguish fires. MHZ-14 CO2 sensor values are analysed using graphs. If the CO2 value is in the lower limit on the graph, the robot will start pumping CO2. Otherwise, it uses a water pump.

Table:1 is listed the required of components, material and quantity for fabricating and assembling of GSM IoT Based Fire Fighting Robot.

Table 1. Components required

Components	Material	Quantity
L-Angle	Mild steel C 40	1
Ball Bearing	HSS	2
LDR Sensor Unit	Electronics	1
Gear wheel	Cast iron	2
D.C motor	Aluminium	1
Front Wheels	Nylon	2
Rear Wheels	Rubber	2
Microcontroller	Electronics	16F84 IC
Battery	Lead-acid	12V D.C - 1
Water D.C pump	Mild Steel	1

Journal of Physics: Conference Series **2040** (2021) 012045 doi:10.1088/1742-6596/2040/1/012045

Due to light intensity the LDR light has different resistance. It is much simpler for light intensity.

Resistance to LDR during dark burns up to Meg Om ranges. When sunlight falls on LDR it illuminated then the resistance of the LDR decreases abruptly (less than 10 kPa). White stone bridge is connected with LDR as shown in figure 1. The resistance is converted to different voltage and voltage is amplified and fed to comparator.

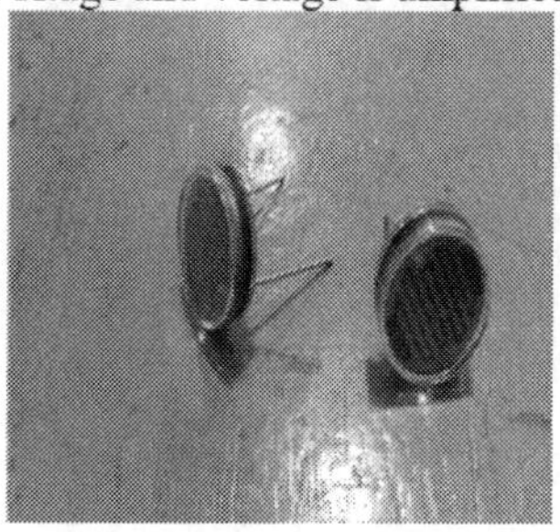

Figure 1 LDR Sensor Unit

Figure 2 L Angle

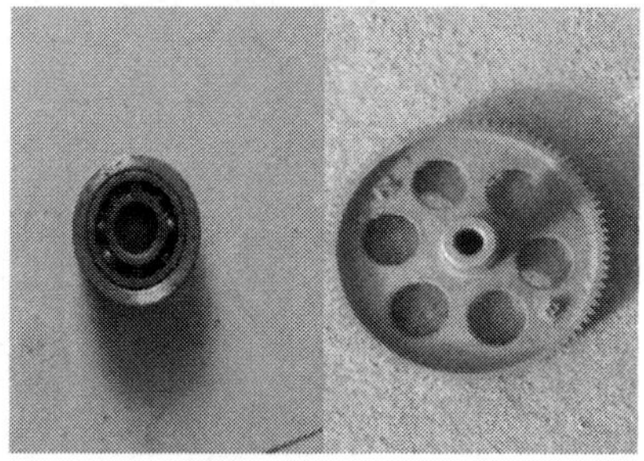

Figure 3 Ball Bearing & Gear Wheel

Angle steel is steel which is in 'L' shape and its angles are at 90 degree. It is one of the most common types of steel. The Shape of the 'L' angle can be equal to unequal at the bottom of the legs as shown in Figure 2. Ball bearings use to separate two different "races" or bearing rings to reduce surface contact and friction. Gears of different sizes change torque by their gear ratios, creating mechanical advantages and can be considered a simple machine. Rotational speed and two-mesh gears vary in proportion to their diameter as shown in figure 3.

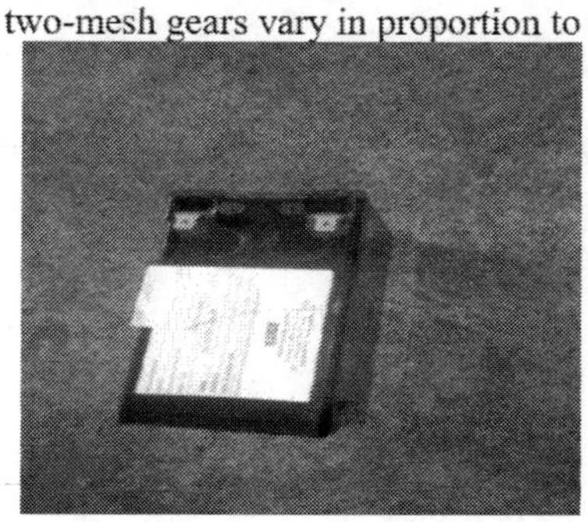

Figure 4 D.C. Battery

Figure 5 D.C. Motor

Figure 6 D.C. Pump

The batteries are the only storage device that is technically and economically available as shown in figure 4. Since both the photovoltaic system and the battery are high in cost. The DC machine can perform as generator or motor as shown in figure 5. While running the generator, it works mechanically and develops a voltage and it has ability to send electricity through voltage load resistance. Motor action develops a torque. Figure 6 is a 12 volt DC water pump will be used. Microcontroller d. C. Provides water supply so that during firing, D.C. Run the water pump directly.

The 89C52 microcontroller will be used as the control unit in our project. Since the sensor unit is giving the control signal to the microcontroller unit, the microcontroller d. C. Pump and rotating permanent magnet D.C. Give a signal to the motor. Block of memory is a Program memory which is continues program codes (using specialized erasable programmable read-only memory. / Programmable read-only memory programmers). It is not possible to read and write only under normal operating conditions.

Input storage known as read / write memory. So, that can perform, read and written programs. The Microcontroller (AT-89c52) have 128 pixel of inner input storage and 64K outward input storage. The Microcontroller has four 8-bit parallel ports (so 8 * 4 = 32 I / O lines are available). Each line contains the latch, the production driver and the Data buffer.

Port 0 (PO), Port 1 (P1), Port 2 (P2) and Port 3 (P3) were designated as four ports. External memory is used to access four ports, port 0 and port 2. Multifunctional process takes place in all port 3 pins. 8-binary digit is a port 3 pin of two-way with internal pull-ups.

Figure 7 Micro Controller

Figure 8 Flame Detector

2.2 Flame Detector
Flame Detector includes,

- Flame detector circuit
- Voltage divider
- Comparator
- The circuit is changing

2.2.1 Voltage divider
The potential or voltage divider provides a convenient way to obtain variable voltage from a constant voltage supply. In general, if the passive two-terminal electrical component of R1 and R2 are connected in sequence across a supply voltage V and the voltages developed in each are V1 and V2 respectively, I can say that the current is:

I x R1 (1)=V1
I x R2 (2)=V1
V1 + V2 = I (R1 + R2)=V ... (3)
Divided by (1) by (3) we get:
V1 / V = (I * R1) / (I * (R1 + R2))
Multiplying both sides by V:
V1 = (R1 * V) / (R1 + R2)
Similarly from (2) and (3) we get:
V2 = (R2 * V) / (R1 + R2)

2.2.2 Comparator
The comparator is a device used when different signal reaches a certain input range and this comparator output is used to drive the logic circuit. OPAMP is used to construct a comparison. This is a open-loop op-amp of output (Vsat (= Vcc) as (A) Ideal transfer properties. However, in commercial op-amp, have transition characteristics. Following types of comparators are available, there are

- Non-inverter comparator
- Comparator

3. Design and Fabrication

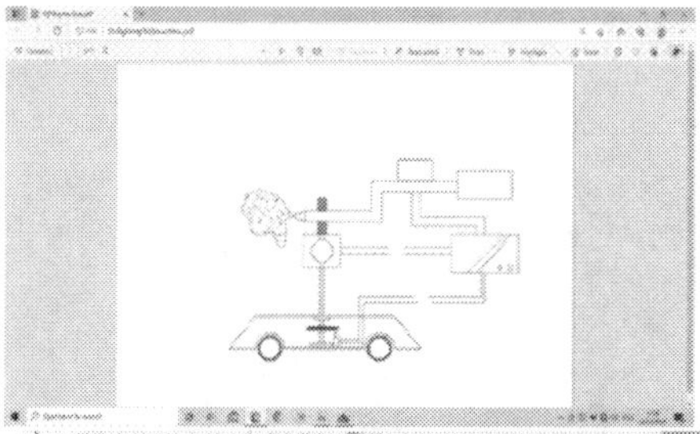

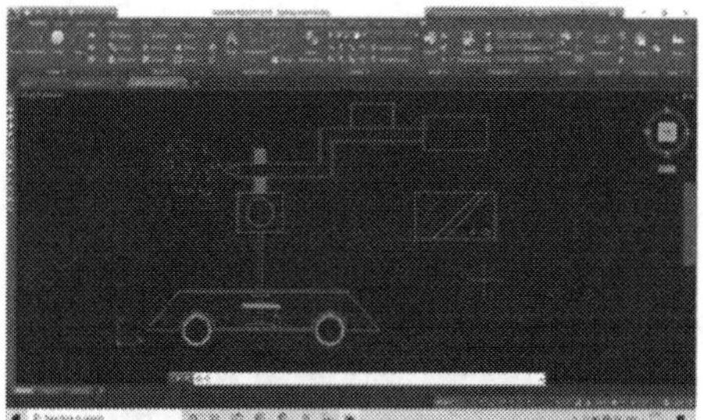

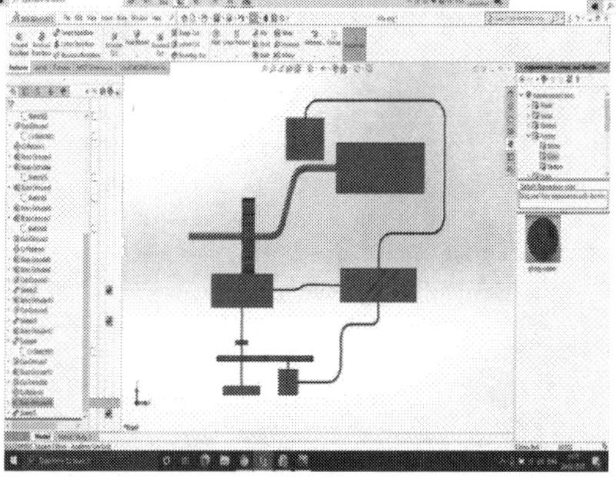

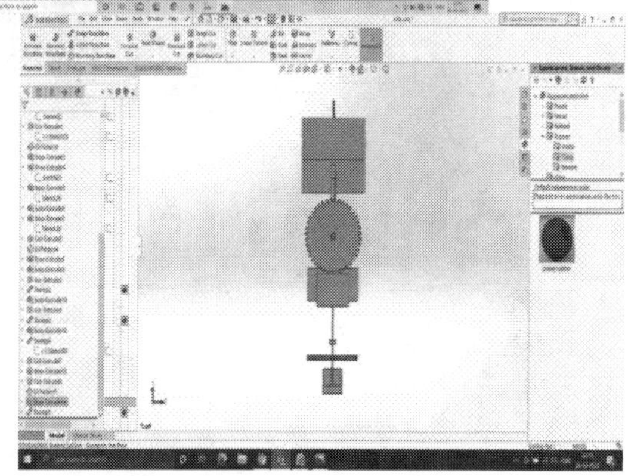

Figure: 9 3-D Front View of FFR **Figure: 10** 3-D Side View of FFR

The Fire Fighting intelligence can detect fires in an area of one foot square. The Fire detection limit in all directions is 360°. Nozzle adjustments i.e. for extinguish the fires water and any type of chemicals are used here. The project is based upon the AT89C52, it is composed and fast and responsive. No outward device is used for control this. This setup is very compact and easy to move.

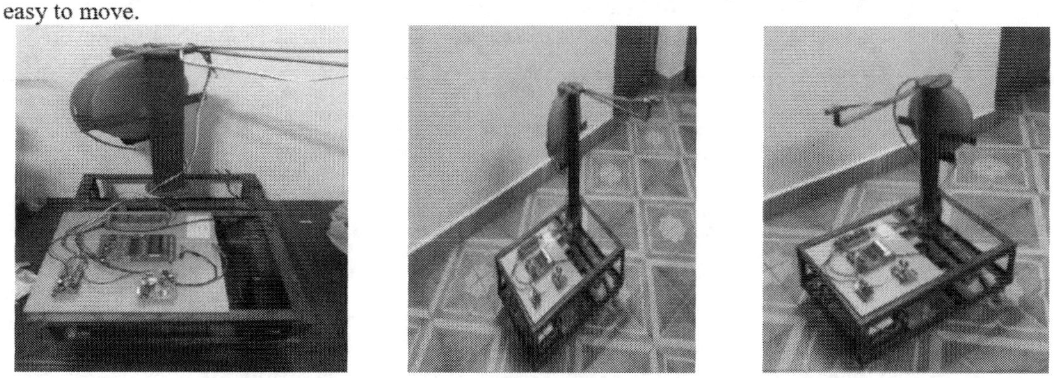

Figure: 11 Images of FFR Working Model with different Positions

IR sensor is used to detect flames. And it helps to find the direction of the fire area. Microcontroller receives the signal which is sent to the sensor it was basic information for the amplifier. Later having the signal of amplifier the microcontroller runs a program based on a signal, Such as permanent DC motor on stage to stage and D.C Pump Activation.
The rotary motion of the sensor and nozzle is fixed by a DC motor on rotating shaft of the fire fighting robot. The current is supplied by the electrochemical cells. D.C. motor control circuit receives power which is already stored in the electrochemical cells. DC motor operates

according to a program written on the microcontroller chip. (I.e.) Rotating motor ON and D.C water pump in off position.

4. Results and Discussion

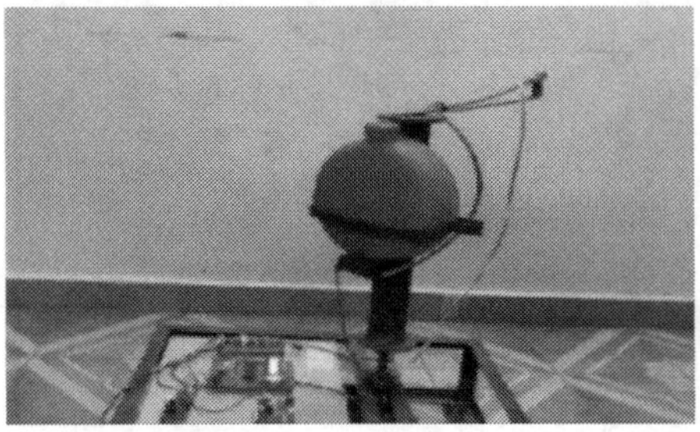

Here we have successfully developed a fire fighting robot. The robot detects temperature, smoke and flame in the area where the robot is located. The movement of this robotic vehicle is controlled by IoT according to the program. The robot assists in areas where natural disasters have occurred and fire explosions have occurred. If a fire is detected with the help of a sensor, the MCU operates the water pumping system through the relay circuit. A fire is detected with the help of sensors.

5. Conclusions

Therefore the firefighting robot can be used to avoid the human damage and helps to take precautionary actions to avoid the complication of fire. Fabrication and assemble of fire fighting robot using GSM and IOT based, which helps to extinguish the fire and simultaneously alerts the fire station before it gets complicated in the absence of the people.

References

[1] Jeelani, S., et al., Robotics and medicine: A scientific rainbow in hospital, *Journal of Pharmacy & Bioallied Sciences,* 2015. **Vol. 7,** pp. S381-S383.

[2] Aliff, M., S. Dohta, and T. Akagi, Simple Trajectory Control Method of Robot Arm Using Flexible Pneumatic Cylinders, *Journal of Robotics and Mechatronics,* 2015. **Vol 27,** pp. 698-705.

[3] Aliff M, D.S., and Akagi T, Control and analysis of simple-structured robot arm using flexible pneumatic cylinders. *International Journal of Advanced and Applied Sciences,* 2017. **Vol. 4**, pp. 151-157.

[4] Aliff, M., S. Dohta, and T. Akagi, Control and analysis of robot arm using flexible pneumatic cylinder. *Mechanical Engineering Journal*, 2014. **Vol. 1**, pp. DR0051-DR0051.

[5] M. Aliff, S. Dohta and T. Akagi, Trajectory controls and its analysis for robot arm using flexible pneumatic cylinders," *IEEE International Symposium on Robotics and Intelligent Sensors (IRIS*), 2015, pp. 48-54.

[6] M. Aliff, S. Dohta and T. Akagi, Trajectory control of robot arm using flexible pneumatic cylinders and embedded controller, *IEEE International Conference on Advanced Intelligent Mechatronics (AIM),* 2015, pp. 1120-1125.

[7] C. Xin, D. Qiao, S. Hongjie, L. Chunhe and Z. Haikuan, Design and Implementation of Debris Search and Rescue Robot System Based on Internet of Things, *International Conference on Smart Grid and Electrical Automation (ICSGEA)*, 2018, pp. 303-307.

[8] Yusof, M., and Dodd, T., Pangolin: A Variable Geometry Tracked Vehicle With Independent Track Control, Field Robotics, pp. 917-924.

[9] Day, C.-P., Robotics in Industry—Their Role in Intelligent Manufacturing. *Engineering,* 2018. **Vol. 4** pp. 440-445.

[10] J. Lee, G. Park, J. Shin and J. Woo, Industrial robot calibration method using denavit — Hatenberg parameters, *17th International Conference on Control, Automation and Systems (ICCAS),* 2017, pp. 1834-1837.
[11] Sani, N. S., Shamsuddin, I. I. S., Sahran, S., Rahman, A. H. A and Muzaffar, E. N, Redefining selection of features and classification algorithms for room occupancy detection, International Journal on Advanced Science, *Engineering and Information Technology*, 2018, **Vol. 8,** pp. 1486-1493.
[12] Holliday, J. D., Sani, N., and Willett, P., Calculation of substructural analysis weights using a genetic algorithm, *Journal of Chemical Information and Modelling*, 2015, **Vol. 55(2)**, pp. 214-221.
[13] Holliday, J. D., N. Sani, and P. Willett, Ligand-based virtual screening using a genetic algorithm with data fusion, Match: *Communications in Mathematical and in Computer Chemistry*, **Vol. 80**, pp. 623-638.
[14] SamsiahSani, N., Shlash, I., Hassan, M., Hadi, A., and Aliff, M, Enhancing malaysia rainfall prediction using classification techniques, *J. Appl. Environ. Biol. Sci*, 2017, ***Vol. 7***, pp. 20-29.
[15] Sani, N.S., Rahman, M.A., Bakar, A.A., Sahran, S. and Sarim, H.M, Machine learning approach for bottom 40 percent households (B40) poverty classification, International Journal on Advanced Science, Engineering and Information Technology, 2018, **Vol. 8**, pp.1698-1705.
[16] Kim, J.-H., S. Jo, and B.Y. Lattimer, Feature Selection for Intelligent Firefighting Robot Classification of Fire, Smoke, and Thermal Reflections Using Thermal Infrared Images. *Journal of Sensors*, 2016. pp. 13

Influence of Nozzle Hole Geometry of CI Engines Performance and Emission Parameters Fuelled by Diesel and Biodiesel-A Review

A.Imthiyas[1], M.Saravanan [2], Sanjaykumar B M [3], Sujith Raj P [3] and Udhaya Kumar M [3]
1,2 Assistant Professor, Department of Mechanical Engineering, Aarupadai Veedu Institute of Technology, Vinayaka Mission's Research Foundation, Deemed to be University, Tamilnadu, India.
3 UG Student, Department of Mechanical Engineering, Aarupadai Veedu Institute of Technology, Vinayaka Mission's Research Foundation, Deemed to be University, Tamilnadu, India.
E-Mail : imthiyas@avit.ac.in

Abstract In the quest to reduce hazardous pollutant emissions and environmental impact, various renewable alternative fuels derived from bio-based sources have entered the market. Biodiesel, a vegetable oil ester, has gained a lot of popularity and attention for diesel engine applications. The performance of diesel engine and emission characteristics are closely linked with fuel atomization and spray pattern of fuel injection nozzles. In modern engines various orifices with different designs are used. In a diesel engine, the shape of the diesel fuel injection nozzle and the fuel flow parameters in the nozzle have a considerable impact on the processes of fuel atomization, combustion, and exhaust emissions. The main focus of this study is to explore how number of nozzle hole and shape affects the performance of diesel engine, combustion, and emissions for both Biodiesel and diesel fuels.

Keywords: carbon monoxide, oxides of nitrogen, hydrocarbons.

I. Introduction

The demand for alternate fuels are increased day by day due to the increased number of vehicles .Recent days the ultimate objective of engine design is to reduce the hazardous gases such as carbon monoxide (CO), oxides of nitrogen (NOx), hydrocarbons (HC), and particulate matter (PM) emissions into the atmosphere. Given the challenges of properly controlling the combustion process, several actions aimed at mitigating its impacts, such as cleaning exhaust gases, have been attempted. The fuel injection system is the furthermost important part in the engine and it regulates the supply of fuel depends on the needs of the load and also maintain the injection pressure and fuel atomization, air-fuel mixing etc. Subsequently, it considerably disturbs engine performance and emissions. The mechanisms of the fuel injection system need exact plan measures, material determination, and manufacturing methods. Along with traditional pump-line nozzle arrangements, new ideas arose, such as distributing pumps, accumulation systems, common-rail systems, unit pump and unit injectors and so on. Attempts are made to reuse the waste cooking oil to use as a fuel in diesel engines rather than to dispose it. [14] According to the application's requirements, combustion may be tuned for the highest performance, emissions, and smooth operation. Exhaust emissions such as HC, CO, NOx, and ash are the maximum genuine worries about the diesel engines, because of non-uniform fuel appropriation, which makes the ignition blend be non-stoichiometric. Accordingly, the DI diesel engines ignition interaction is heterogeneous. It leads to an increase in air emissions. At the completion of the compression stroke the fuel is injected in to the cylinder in the form of liquid through the nozzle and blend with air present in the cylinder. The liquid jet exit the nozzle gets turbulent and expands out. The droplets are formed from fuel jet's outer skin. Once the in-cylinder fluids temperature reaches beyond the self-ignition temperature, the ignition started spontaneously and begin burning at a variety of areas when the appropriate conditions are present. Spray contact becomes a significant occurrence in high-speed Direct injection diesel engines as a result. Spray pattern has a significant impact on fuel jet distribution and evaporation. The spray patterns of fuel inside the cylinder are generated by the injection system which influence the performance and emission controlling processes. The main issue is the small size of the nozzle channel flow regions, which are smaller than one millimetre in diameter and can withstand the pressures of up to 200 bars. Increasing the nozzle holes inturn increases the atomization during combustion. [16,17] This makes quantities and other interpretations of the flow in the nozzle really challenging. The estimations are generally bound to consistent state estimations of nozzle stream coefficients or observations and estimations of stream in increased direct models .Due to the introduction of bio diesel and design changes of diesel engines powers more than the conventional fossil fuels. [18,19]

2. Methodology

2.1. Transesterification Reaction

The process of trans esterification is replacing the ester of an organic group R″ with the organic group R′ of an alcohol in organic chemistry. An acid or basic catalyst is frequently used to catalyse these reactions. It is the most popular and effective method for reducing the viscosity of vegetable oils. A blend of unsaturated fats, alkyl ester, and glycerol is created when fatty oil consolidates with three atoms of liquor within the presence of a catalyst. In the presence of a catalyst, esterification is a method for completely removing glycerol and unsaturated lipids from a vegetable oil.

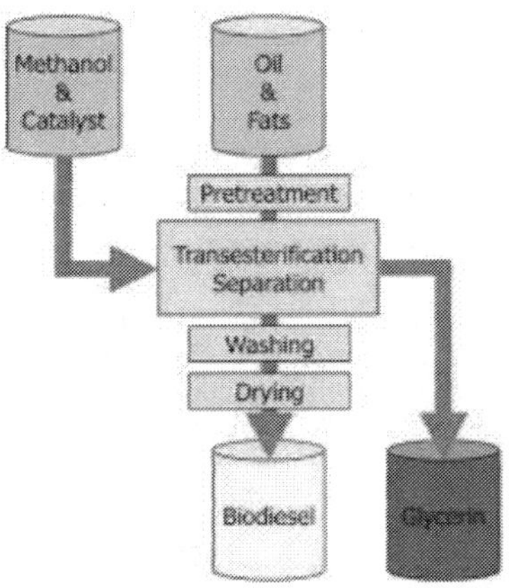

Figure 2.1 Transesterification process

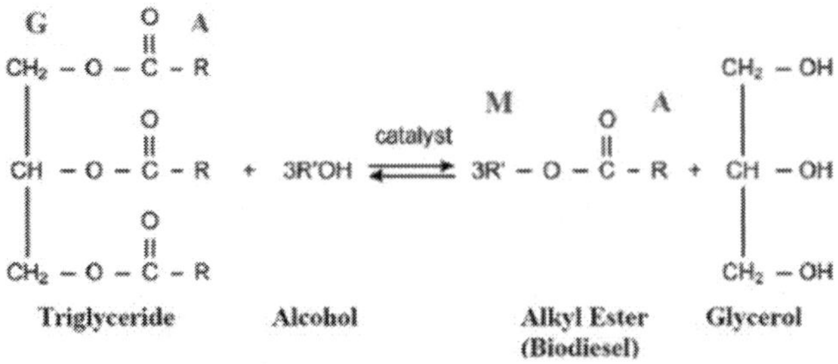

Figure 2.2 Transesterification Reaction

3. Results And Discussion

3.1. Performance Parameters

3.1.1. Brake Thermal Efficiency (BTE)

Sangeetha krishnamoorthi et al (2021) conducted experiments fuelled by preheated Spirulina methyl ester in VCR engine using 1-hole nozzle and 3-hole nozzle. When compared to Diesel and Preheated Spirulina, the BTE for unheated Spirulina was low. The main reason is that increased viscosity causes poor atomization and more fuel droplets owing to incorrect fuel and air mixing. Complete combustion happens in biodiesel due to the presence of oxygen, resulting in an increase in BTE value. The BTE of B20 Preheated Spirulina with a single or three-hole nozzle was found to be greater than unheated Spirulina and ordinary diesel. This might be because to the warmed Spirulina B20's improved viscosity, which allows for greater atomization and combustion. The BTE of unheated spirulina with a 3-hole nozzle is 26.3 percent, whereas plain diesel has a BTE of 28.73 percent and Spirulina B20 (preheated) at 60C has a BTE of 29.32 percent. At maximum load circumstances, with a 3-hole nozzle, the BTE of B20 Preheated Spirulina was found to be higher than diesel [1]

J. Deokar et al (2019) did experiments with 3, 4 and 5 hole nozzle. Because the B0 blend has a larger calorific value than the other blends evaluated, it has a enhanced brake thermal efficiency. Since the diameter of nozzle holes declines, the nozzle hole area also likewise reduces, the 5 hole nozzle has the highest braking thermal efficiency of all the nozzles examined. As a result, less fuel is used and the thermal efficiency is improved. Additionally, on the grounds that splash example of the 4-hole nozzle was uneven, the highest amount of fuel was encroached on the cylinder wall, bringing down BTE. [2]

C. Nandakumar et. al (2020) did experiments with different hole nozzles for a B50 blend. As number of hole increases in nozzle, the brake thermal efficiency also increases for B50 and its maximum for 5-hole nozzle. BTE may have increased as a result of finer fuel atomization, leading by improved air/fuel mixing and combustion. Also, because of the occurrence of oxygen in KME enhances combustion, the engine's BTE for B5O improves as the nozzle holes in number rises. In addition, it is predicted that the shortened combustion time would increase combustion efficiency; the B50 blend's reduced combustion time combined with an increase in the number of nozzle holes indicates an increase in performance and combustion efficiency [3].

Manjunath Channappagoudra et al (2018) did experiment in diesel engine using 5-hole nozzle that for Diesel and other DSOME mixes, there is a differential between BTE and braking power. Based on his research, because of its lesser calorific value, higher specific gravity, and larger viscosity, biodiesel has a lower BTE for all loads than petroleum diesel. Furthermore, increasing the biodiesel content in diesel reduces the BTE due to the lower calorific value. His research with several blends of biodiesel revealed that the DSOME-B20 blend outperforms the DSOME-B30 blend and DSOME-B100 blend. This is due to enhanced calorific value, reduced viscosity, and reduced density, which result in higher performance and a BTE that is extremely close to that of HS diesel. According to the data, diesel has the 31.32 percent BTE, after that DSOME-B10 with 30.42 percent BTE, which is slightly higher than DSOME-B20 with 29.93 percent BTE. However, even if it has more BTE, a 10 percent reduction in petroleum diesel is not a satisfactory result. When compared to the DSOME-B30 and DSOME-B100, the DSOME-B20 BTE is identical to the DSOME-B10. As a result, DSOME-B20 is determined to be the best mix among others [4].

Narayanan, S., et al. (2019) used a four, five, and six-hole nozzle to test various kapok methyl ester mixtures in a diesel engine. According to his research, As the IHN and mix ratio increase, the brake thermal efficiency improves. Higher brake thermal efficiency might be due to increased cylinder wall temperature as a function of load. Increase in IHN boosts brake thermal efficiency When the injection angle changes, the engine's thermal efficiency improves because more air and fuel are mixed together. BTE was found to be 15.2 percent, 17.6 percent, and 18.5 percent for B20, B30, and B40 correspondingly, at minimal load, while it was 20.3 percent, 23.4 percent, and 24.2 percent at maximum load (Jindal et.al, 2010; Syed alamet.al, 2016). Let's look at the reading for B30 with the smallest load using IHN. 15.2 percent, 17.2 percent, and 20.4 percent were recorded for three distinct IHNs[5].
Ranganatha Swamy L. et al. (2014) found that, the nozzle shape with 5 holes and an orifice diameter of 0.3 mm had the maximum braking thermal efficiency at fixed injection opening pressure (IOP). This might be because the atomization, spray properties, and air mixing were all enhanced, this results in better combustion. At 80 percent load, the BTE was found to be 28.5 percent, with the maximum value achieved with a 5HN and a 205 bar injection pressure. The BTE of 3HN and 4HN was 24.80 percent and 27.56 percent, respectively at 205 bar. The number of holes had no effect on ignition delay, but it did enhance the rate of fuel-air mixing, hence the 5HN was chosen as the superior. The holes in the 5HN were reduced in size from 0.3 to 0.2 mm, As a result, the combustion chamber's air and fuel mixture is enhanced and therefore increased BTE. When the thickness is smaller than 0.2 mm, the BTE is decreased because fuel droplets travel quicker than air, results poor in mixing and engine efficiency [6].

Khandal S. V. et al. (2015) found that, When the Hole count is raised from three to four for the same compression ratio and injection pressure, the BTE rises, but when the injection pressure is raised (240 bar), BTE decreases. Because Honge oil methyl ester (HOME) is nearly twice as viscous as diesel for the same injection pressure and compression ratio, the injected biodiesel droplets will be larger, resulting in a higher number of holes to enable appropriate mixing of the fuel injected (HOME) with the surrounding air. As a result, using a four-hole injector improves the process of fuel combustion. It also makes the charge more homogenous, which explains why the heat release rate (HRR) after premixed combustion is at an all-time high. The BTE increased as the number of holes was increased from three to four under similar injection time (IT) and compression ratio circumstances. This might be due to the fine biodiesel jets released by the 4hole injector, which provide ideal liquid fuel/air mixture, causing in better combustion. There is improved BTE as the number of holes was amplified from three to four for the similar injection time and injection pressure. As previously said, because HOME has a higher viscosity than diesel, a higher number of holes will allow the injected fuel to mix properly with the air, resulting in a more efficient fuel combustion process [7].

Thirunavukkarasu .R et al. (2018) found that, increasing the nozzle hole from single to 4 hole and 5 hole results in greater atomization of fuel , reduction in physical delay period of combustion which improves the BTE on a ceramic coated piston. The BTE of single and five hole nozzle gradually decreases by increasing the load but four hole nozzle BTE increases with increase of load. It's found that spray penetration , driblet size has greater influence on nozzle performance [11].The power output was increased by increasing the peak pressure of cylinder by employing doble and triple injection pulses. [13].

3.1.2. Brake Specific Fuel Consumption (BSFC)

Sangeetha krishnamoorthi et al.(2021) were conducted experiments by using Spirulina methyl ester using One hole & 3 hole nozzle for blend B20, as well as a 1HN and 3HN for B20 preheating up to 60o for blend B20. The value of BSFC for diesel is low when compared to unheated Spirulina among the several nozzles investigated due to the high calorific value of diesel. The BSFC of preheated spirulina with a 1HN was greater, but the BSFC of preheated spirulina with a 3HN was comparable to diesel. Preheated Spirulina at 60C with a 3HN yielded the lowest value. BSFC for Neat diesel was 0.29 kg/kw-hr with a 3HN, 0.34 kg/kw-hr for B20 Spirulina Preheated at 60C with a 1HN, and 0.30 kg/kw-hr for B20 Spirulina Preheated at 60C with a 3HN. At maximum load levels, the BSFC showed that the Preheated Spirulina outperformed the unheated biodiesel. The Brake Specific Energy Consumption (BSEC) is a measure that is used to compare the calorific content of different fuels. The BSEC denotes the effectiveness with which a certain fuel may be used to generate electricity. In addition, the BSEC values rise when the Blend value rises and the calorific value falls [1].

A. J. Deokar et al (2019) did experiments with 3, 4 and 5 hole nozzle. Because of the increased calorific content of the B0 blend, it had the lowest brake specific fuel consumption of all the blends evaluated. The area of the nozzle holes reduces as the diameter of the nozzle holes decreases, the 5HN has the lowest BSFC of the various nozzles examined. As a consequence, fuel consumption is reduced and thermal efficiency is improved. The BSFC increases for 4 hole nozzle because the spray pattern was uneven, allowing maximum amount of fuel that may be sprayed onto the cylinder wall, resulting in better BSFC. [2].

The engine's BSFC for B50 blend has been presented by C. Nandakumar et al (2020). For B50 with 5HN type, BSFC decreases as the number of nozzle holes increases, and eventually approaches diesel. The engine's BSFC drops for B50 with 3HN and 5HN from 250 g/kWh to 200 g/kWh respectively. The lower calorific value of KME required more fuel injection to reach the essential power yield, which cost the B50 with 3HN a higher BSFC. The BSFC, however, decreased as the count of nozzle holes increased because finely atomized fuel droplets favour evaporation of fuel and so expedite the combustion process. Previous study associating an increase in nozzle holes in diesel fuel to a drop in BSFC contradicts the results. [3].

As per Narayanan. S, et al (2019) findings the relationship between BSFC and primary factor load, as well as secondary variables for 1HN and blend ratio BSFC falls as 1HN rises, showing that the conversion of fuel's chemical energy into braking power is becoming more efficient. This might be attributable to greater air-fuel mixing in certain conditions [5].

3.2. Emission Parameters

3.2.1. CO Emissions

Sangeetha krishnamoorthi et al.(2021) investigated the relationship between carbon monoxide emissions and brake power for clean diesel with 1HN and 3HN, unheated Spirulina with 1HN and 3HN, and preheated B20 Spirulina with 1HN and 3HN. CO emissions were found to be reduced at the start load and even up to 75% load circumstances, but were shown to be higher at maximum load circumstances. When comparing unheated Spirulina to diesel and preheated Spirulina with 3HN, the CO emissions for unheated Spirulina were found to be higher. The reason for increased CO emissions at greater loads is that there is less oxygen supply when more fuel is injected. Co emission was 0.06 percent Vol for diesel with 3HN and 0.07 percent Vol for unheated Spirulina with 3HN. Even at maximum load, preheated B20 Spirulina with 3HN had a value of 0.04 percent. Vol, which is less than the value of diesel [1].

A. J. Deokar et al. (2019) discovered that the B0 blend emits the least carbon monoxide of all the blends examined due to its greater calorific value. Because the diameter of the nozzle holes reduces, the area of the nozzle holes likewise lowers, the 5-hole nozzle emits the least carbon monoxide of the various nozzles examined. As a result, fuel consumption is lower, thermal efficiency is improved, and carbon monoxide emissions are reduced. Carbon monoxide emissions rise with the 4-hole nozzle because the spray pattern was uneven, allowing the maximum quantity of fuel to impact on the cylinder wall, resulting in higher carbon monoxide emissions [2].

C. Nandakumar et al. (2020) discovered that increasing the number of nozzle holes, the fuel jet's finer dispersion lowers the fuel core inside the spray, allowing for lean mixing and burning of the well-mixed fuel/air mixture. As the nozzle holes in numbers increases for B50, the air-fuel equivalency ratio decreases, lowering CO emissions. Furthermore, the oxygen concentration in the fuel and greater highest heat output during combustion improve CO oxidation, lowering CO by 47.3 percent for B50 with 5-hole nozzle compared to 3HN. CO emissions for B50 with standard 3-hole nozzle geometry were determined to be higher than for diesel because of the dense mixture created by KME's greater viscosity and lower calorific content. CO emissions are reduced as the fuel mixing procedure improves, and the number of nozzle holes increases, bringing it closer to diesel [3].

Narayanan. S, et al (2019) found that the combustion quality in an internal combustion engine is measured by carbon monoxide emissions. CO is a result of incomplete combustion, hence the lower the CO emission, the better the combustion quality. Because the calorific value only lowers slightly with the fuel mix, the data does not suggest any major variance in CO emission with regard to blend ratio. The increase in the number of nozzles from four to six enables increased air and fuel miscibility, improving the odds of full combustion and thereby lowering CO emissions [5].

Ranganatha Swamy L. et al. (2014) discovered that emissions of CO followed the same pattern as HC emissions, with lesser CO emissions from 5HN than 3 and 4 HN. Incomplete combustion was the cause of the higher CO levels in the exhaust for 5 hole injectors with orifice sizes ranging from 0.3 to 0.2 mm. CO emissions were shown to be lower as injector hole sizes were reduced, owing to less wall impingement with diesel related to bigger hole sizes. Higher hole size injectors, on the other hand, result in fuel deposition on the combustion chamber walls. As a result, CO emissions from 0.3 mm hole injectors were discovered to be higher [6].

According to Khandal S. V. et al. (2015), For any given injection pressure, carbon monoxide decreases with compression ratio. CO, on the other hand, diminishes as the injection pressure is increased. At a higher compression ratio (17.5) and injection pressure of 230 bar, the quantity of CO was determined to be low. CO reduces as the compression ratio increases for a fixed value of injection time(IT), nevertheless as the IT grows, the quantity of CO rises. CO behaved similarly to HC in terms of process parameters (injection pressure and injection time). By increasing the number of holes in the injector from three to four, CO was decreased. The increased HRR accumulated for a four-hole injector may help to cut CO emissions. However, because higher non-homogeneity decreases combustion efficiency, a larger CO area was found close the piston wall with a 3HN [7].

3.2.2. HC Emissions

Sangeetha krishnamoorthi et al. (2021) investigated the HC emissions of plain diesel with 1HN and 3HN, unheated Spirulina with 1HN and 3HN, and preheated Spirulina B20 at 600 C with 1HN and 3HN. The value of HC emission for unheated Spirulina and diesel was found to be higher than for warmed Spirulina due to the large quantity of fuel stored in the combustion chamber because of the extended ignition delay. Diesel had an HC value of roughly 29ppm with a 3HN, whereas unheated Spirulina had an HC value of roughly 27ppm with a 3 HN. The HC emission amount of preheated spirulina B20 is 24ppm. This is due to the warmed Spirulina B20 blend's reduced igniting latency at extreme power output. Poor combustion is among the causes for the existence of unburned HC, and the HC value in biodiesel is low owing to full combustion due to the molecular oxygen present[1].

A. J. Deokar et al. (2019) discovered that of all the blends examined, the B0 blend emits the least amount of hydrocarbons due to its greater calorific value. Because the diameter of the nozzle holes reduces, the area of the nozzle holes also lowers, the 5 hole nozzle emits the least amount of hydrocarbons of all the nozzles examined. As a result, fuel consumption is lower, thermal efficiency is better, and hydrocarbon emissions are reduced. The hydrocarbon emission from the 4 hole nozzle rises because the spray pattern of the 4 hole nozzle is uneven, allowing the maximum quantity of fuel to impact on the cylinder wall, resulting in greater hydrocarbon emission [2].

For different nozzle hole geometries, C. Nandakumar et al. (2020) discovered HC emission of B50 mix. It indicates that when the engine load increases, the HC emission rises due to an increase in the equivalency ratio. The B50 mix emits more HC than a single diesel with a traditional 3HN. This is due to the B50 blend's increased viscosity and density when compared to diesel. These features have a negative impact on the fuel atomization process, resulting in bigger fuel droplets for B50 than pure diesel. The evaporation rate decreased as a result of this activity, the quantity of gasoline absorbed by the wall and via holes began to grow. The quantity of HC released is reduced as the nozzle geometry's hole count increased. The B50 emits HC emissions that are equivalent to diesel at lower loads when equipped with a 5-hole nozzle. The HC emission is greatly decreased with greater loads. At 80 percent and 100 percent load, the HC emissions of a B50 with 5HN are lowered by 11.66% and 8.86%, respectively, when compared to a diesel with a 3HN. As the count of nozzle holes rises, the atomization of the fuel stream gets thinner, resulting in a quicker vaporization rate and improved fuel-air mixing. Furthermore, the inclusion of extra oxygen molecules in the kapok biodiesel molecular structure reduces HC emission [3].

Narayanan. S, et al (2019) discovered that the unburnt hydrocarbons are commonly referred to as hydrocarbon emissions, meaning that they are a product created when fuel particles are not entirely burned. The graph shows a modest drop in unburned hydrocarbons as IHN increases, indicating that combustion improves. In relation to the biofuel blend, there is no substantial change in HC emissions[5].

Ranganatha Swamy L. et al. (2014) discovered that, the 5 hole nozzle shape resulted in a considerable reduction in HC emission due to improved combustion. A decreased ignition delay will result from improved atomization. The physical delay will be reduced as the spray quality improves. The 5 hole nozzle geometry was found to emit less HC than the 3 and 4 hole nozzle geometry. Because of enhanced atomization and correct combustion of diesel, there were less unburned hydrocarbons with the 5 hole nozzle [6]. Apart from injection strategy nozzle geometry like orifice diameter nozzle hole size, spray angle affects fuel
Distribution which have direct impact of HC emissions [20, 21].

Khandal S. V. et al (2015) found that, for any given injection pressure value, HC falls with compression ratio. For decreased HC, a grouping of 230 bar injection pressure and a elevated compression ratio (17.5) was shown to be effective. Increasing the number of holes in the injector from three to four resulted in lower HC when the injection pressure and compression ratio are the same. This might be attributable to a 4hole injector's improved mixing of air and fuel in the combustion chamber, resulting in full combustion. Furthermore, in these situations, increased BTE might be to blame for the tendency for a given injection time, HC reduces uniformly as the compression ratio increases, while HC tends to rise as the IT increases. For decreased HC emission, a lower IT with a greater compression ratio was critical. When compared to greater compression ratios, HC was more susceptible to lower compression ratios, especially with a 4-hole nozzle. Even Increasing the count of holes in the nozzle from three to four reduces the HC for the same injection pressure and injection timing [7].

3.2.3. NOx Emissions

Sangeetha krishnamoorthi et al.(2021) show how NOx emissions change with BP for diesel with1HN and 3HN, unheated Spirulina with with1HN and 3HN, and Preheated Spirulina B20 with 1HN and 3HN. The amount of NOx in unheated Spirulina is determined to be closer to that of diesel, and this is why because the insufficient vaporization in biodiesel, which resulted in lesser combustion, which resulted in lesser NOX emissions. The NOx amount for preheated B20 Spirulina with 1HN and 3HN are greater than the other two findings, owing to the premixed phase of combustion has the greatest temperature, resulting in higher NOx emissions. Diesel and unheated Spirulina had NOx levels of 1086 ppm and 1147 ppm, respectively. The Preheated SpirulinaB20 with1HN and 3HN was determined to be 1150ppm and 1228ppm at maximum load circumstances. [1].

A. J. Deokar and colleagues (2019) found among all blends, because the B0 blend has a larger calorific value than the other blends examined, it emits more nitrogen oxide. Because the width of the nozzle holes reduces as the area of the nozzle holes' decreases, the 5-hole nozzle emits the most nitrogen oxide of all the nozzles examined. This leads to increased fuel consumption and thermal efficiency, as well as increased nitrogen oxide emissions. Due to the pattern of the spray of the 4HN was uneven, the greatest quantity of fuel was impacted on the wall of the cylinder, resulting in lower NOx emissions [2]

C. Nandakumar et al. (2020) discovered smoke and NOx emissions for B50 with varied number of nozzle holes. Regardless of the presence of oxygen, NOx emissions for B50 were observed to be decreased than for diesel, owing to high viscosity, as a result, fuel atomization and combustion have been hampered, decreasing in-cylinder temperature and, as a result, NOx emissions. To offset the aforementioned problem, increase the number of nozzle holes, as finer atomization leads to more combustion, which boosts the in-cylinder temperature. As a consequence, NOx emissions increase steadily as increasing the number of holes in the nozzle and with a 5HN form, NOx emissions are determined to be equivalent to diesel. Furthermore, because the droplet size is projected to be lower, there will be insufficient momentum to drive the spray forward, resulting in a shorter fuel penetration length in order for it to evaporate and mix properly with the air. The heat loss would have decreased the in-cylinder temperature if the spray had penetrated farther to reach the wall. However, it is considered that such a phenomena did not occur, and that the homogenous mixing of air/fuel mixtures at the combustion chamber's limits increased the temperature within the cylinder, resulting in greater NOx emissions. However, when the number of holes in the nozzle rises, smoke production decreases because soot oxidation is favoured in the spray's flame regions. The fuel core will be short and lean as a result of the better air/fuel mixing, oxygen molecules will have an easier time identifying and oxidising the soot precursors formed during the premixed combustion phase, lowering smoke emissions. When compared to the typical 3-hole nozzle geometry, the smoke emission for B50 with 5HN geometry was reduced by 14.8% [3].

Narayanan. S, et al (2019) found at very high temperatures, the nitrogen molecule in the fuel reacts with oxygen to form nitrogen oxides (NOx). Fuel droplets impinging on the wall are thought to be a major source of NOx emissions. The graph below shows that increasing the nozzle hole number from 4 to 6 reduces NOx emissions. This means that as the nozzle hole gets bigger, the wall impingement gets smaller. Because kapok oil contains more oxygen than regular diesel fuel, increasing the fuel mixture improves combustion quality. The higher combustion quality raises the temperature within the cylinder, which raises NOx levels. Increased injector hole number resulted in a modest reduction in NOx, as seen by the statistics. NOx emissions appear to grow with increasing load in all configurations. NOx emissions grow as the proportion of biofuel increases across all fuel mixes[5].

According to Ranganatha Swamy L. et al. (2014), Because of the cycle's quicker combustion and greater temperatures, NOx emissions grow as the number of injector holes increases NOx levels vary in response to changes in adiabatic flame temperature. It's conceivable that increased NOx with larger holes is due to more heat created during premixed combustion and improved combustion inside the engine cylinder. These things change with the spray pattern of fossil fuels, implying that fuel composition has an impact on reaction zone stoichiometry and post-combustion mixing. With a 5 hole injector, reducing the injector orifice size cuts NOx emissions even further. Diesel had the greatest NOx emission of 963 ppm with a 0.3 mm opening, which might be related to the higher adiabatic flame temperature related with its combustion [6].

According to Khandal S. V. et al. (2015), NOx levels progressively rise as the compression ratio rises, regardless of injection pressure and NOx also rises with increasing injection pressure for a 3HN. The behaviour of NOx was shown to be non-linear, despite the fact that the results were the similar for 4HN. At delayed values of ITs, NOx grows linearly with the compression ratio for a 3HN (19-23 degree). However, at 25 degrees IT, NOx remains relatively constant, but for greater values of IT, NOx drops. For a specified value of injection time, NOx rises with an rise in the compression ratio, and NOx grows with more advanced injection time in the case of a 4-hole injector. The compression ratio had little effect on NOx. Advanced IT increases NOx for both 3 and 4 hole nozzles. Increased NOx was seen when the number of holes in the nozzle was increased from 3 to 4. The greater peak temperature in the combustion chamber may be the cause of increased NOx in the case of 4-hole injectors [7].

According to Shijun Dong et al. (2018),Nox emissions are considerably decreased by using reduced nozzle hole size on a dual fuel engine due to the lower combustion temperature . Soot emissions are increased slightly but maintained within the approved limits. [12]. Diesel engines fuelled with biodiesel emits NOx emissions in greater level than the fossil diesel fuels due to increased cylinder temperature [15].By using waste tire oil the HC and CO emissions are considerably decreased but NOx emission increased than conventional fuel . [23] From the literature it is evident that by adding butanol in the diesel fuel can reduce Nitrogen oxide emissions. [24,25]

4. Conclusion

We can conclude from this study that by decreasing the number of holes and size in the engine's nozzle, greater mixing of air and fuel in the combustion chamber is achieved, resulting in improved combustion and, as a result, higher BTE. Fuel particle sizes will grow as the hole's number in injector decreases, and the ignition delay period during combustion will lengthen. The BSFC rises as a result of this condition. Increased injector hole number, as a result the ignition delay time is reduced. As a consequence, the chances of homogenous mixing diminish while BSFC increases. CO emissions decreased as injector hole sizes were lowered, leading to less wall impingement with diesel as especially for larger hole sizes. Higher hole size injectors, on the other hand, result in fuel deposition on the combustion chamber walls. As a result, bigger hole injectors were found to produce greater CO emissions. If the number of nozzle holes was increased, which results in a drop in HC. This might be ascribed to a bigger hole injector allowing for better air fuel mixture to the combustion chamber, resulting in full combustion. Because of quicker combustion and greater temperatures achieved during the course of the cycle, NOx emissions increase as the number of injector holes rises. Greater NOx with bigger holes might be the result of enhanced combustion inside the combustion chamber and much more heat generated for the duration of premixed combustion.

Reference

[1] Sangeetha Krishnamoorthi , M. Prabhahar, V. Madhavan, T. Pragathish, F. Johnson, "Effect of variation in injector nozzle holes on the Performance of preheated Spirulina methyl ester used in VCR engine" Materials Today: Proceedings, 2020, 45, pp. 5840–5846

[2] A. J. Deokar1 and Dr. G. R. Selokar "effect of injection timing and nozzle hole geometry on compression ignition engine fuelled with thevetia peruviana biodiesel" International Journal od Advanced Research Int. J. Adv. Res. 7(4), 118-133, 2019

[3] C. Nandakumar, Vallinayagam Raman, C.G. Saravanan, M. Vikneswaran, S. Prasanna Raj Yadav and M.Thirunavukkarasu "Effect of nozzle hole geometry on the operation of kapok biodiesel in a diesel Engine" Fuel 276 (2020) 118114

[4] Manjunath Channappagoudra, K. Ramesh, and Manavendra G. "Effect of 5 hole nozzle and re-entrant toroidal piston bowl geometry on dairy scum biodiesel operated diesel engine" AIP Conference Proceedings 2039, 020003 (2018)

[5] Narayanan. S, Ramesh. K & Sakthivel. R (2019): Optimization of nozzle hole number for a diesel engine fueled with kapok methyl ester blend, Energy Sources, Part A: Recovery, Utilization, and Environmental Effects, DOI: 10.1080/15567036.2019.1675817

[6] Ranganatha Swamy L., T. K. Chandrashekar, N. R. Banapurmath, P. Nashipudi, "Effect of injection timing, combustion chamber shapes and nozzle geometry on the diesel engine performance", Universal Journal of Petroleum Sciences 2 (2014), 74-95

[7] Khandal S. V., Banapurmath N. R., Gaitonde V. N., Hosmath R. S, "Effect of number of injector nozzle holes on the performance, emission and combustion characteristics of honge oil biodiesel (HOME) operated DI compression ignition engine", Journal of Petroleum and Environmental Biotechnology 2015, 6:3.

[8] Praveen A. Harari, Akshatha D. S, Manavendra. G., "Effect of injection pressure and nozzle hole diameter on combustion parameters of CI engine fuelled with B20 neem blend", International Journal ofInnovative Research in Science, Engineering and Technology, Volume-4, Issue-5, Pages 2857-2870, 2015.

[9] Praveen A. Harari, Manavendra. G., "Experimental investigation on the combustion parameters of B20 neem blend in CI engine by varying injection timing and nozzle hole diameter", International Journal of Advance Research and Innovative Ideas in Education, Volume-1, Issue-4, Pages 503-510, 2015.

[10] Mansukh Pushparaj, Jadhav Vishal, Praveen A. Harari, Arun Pattanashetti, Shambhuraje Jagatap, Santosh Bhuimbar, Vijaykumar Meti, "Performance parameters of B20 neem blend in CI engine by varying injection timing and nozzle hole diameter", International Journal of Innovative Research in Science, Engineering and Technology, Volume-5, 434-1442, 2016.

[11] R.Thirunavukkarasu, M.Mahendran, R. Tamilselvan , Dr.S.Periyasamy "Investigation on Single, Four And Five Holes Fuel Injector Nozzle on Performance And Emission Characteristic of Diesel on A VCR Engine by Using Ceramic Coating Material on The Piston Crown"International Conference on Emerging Trends in Materials and Manufacturing Engineering (IMME17).

[12] Shijun Dong, Can Yang, Biao Ou, Hongguang Lu, Xiaobei Cheng "Experimental investigation on the effects of nozzle-hole number on combustion and emission characteristics of ethanol/diesel dual-fuel engine" https://doi.org/10.1016/j.fuel.2017.12.024

[13] Amin Maghbouli a, Wenming Yang , Hui An , Jing Li , Sina Shafee "Effects of injection strategies and fuel injector configuration on combustion and emission characteristics of a D.I. diesel engine fueled by bio-diesel "http://dx.doi.org/10.1016/j.renene.2014.11.092

[14] R. Dhanasekaran, S. Ganesan, B. Rajesh Kumar, S. Saravanan "Utilization of waste cooking oil in a light-duty DI diesel engine for cleaner emissions using bio-derived propanol , https://doi.org/10.1016/j.fuel.2018.08.093

[15] Shameer PM, Ramesh K. Experimental evaluation on performance, combustion behavior and influence of in-cylinder temperature on NOx emission in a D.I diesel engine using thermal imager for various alternate fuel blends. Energy 2017;118:1334–44. https://doi.org/10.1016/j.energy.2016.11.017.

[16] Cenk Sayin a, Metin Gumus a, Mustafa Canakci, Applied Thermal Engineering 61 (2013) 121e128.

[17] Federico Brusiania, Stefania Falfaria, Piero Pellonia, Energy Procedia 45 (2014) 749 – 758

[18] Subhash Lahane, K.A. Subramanian, Applied Thermal Engineering 64 (2014) 307e314.

[19] Zhixia He*, Wenjun Zhong, Qian Wang, Zhaochen Jiang, Zhuang Shao, International Journal of Thermal Sciences 70 (2013) 132e143.

[20]Montgomery D, Chan M, Chang C, Farrell P. et al., Effect of injector nozzle hole sizeand number on spray characteristics and the performance of a heavy duty D.I. diesel engine. SAE Technical Paper 962002; 1996, http://dx.doi.org/10.4271/962002.

[21] Kim Myung Yoon, Chang SL. Effect of a narrow fuel spray angle and a dual injection configuration on the improvement of exhaust emissions in a HCCI diesel engine. Fuel 2007;86(17–18):2871–80.

[22] Ağbulut Ü, Bakir H. The investigation on economic and ecological impacts of tendencyto electric vehicles instead of internal combustion engines. Duzce Univ J Sci Technol 2019;7(1):25–36.

[23] Mustafa Karagöz , Investigation of performance and emission characteristics of an CI engine fuelled with diesel – waste tire oil – butanol blends, https://doi.org/10.1016/j.fuel.2020.118872

[24] Yusri IM, Mamat R, Akasyah MK, Jamlos MF, Yusop AF. Evaluation of engine combustion and exhaust emissions characteristics using diesel/butanol blended fuel. Appl Therm Eng 2019;156:209–19.

[25] Rakopoulos DC, Rakopoulos CD, Hountalas DT, Kakaras EC, Giakoumis EG,Papagiannakis RG. Investigation of the performance and emissions of bus engine operating on butanol/diesel fuel blends. Fuel 2010;89(10):2781–90

Novel Enhanced Algorithm Based on DNA Cryptography to Secure Data Transfer over Network

[1] Samwal.Idris , [2] Anjali Mathur, [3] Md. Asdaque Hussain

[1]PG Student,[2] Associate Professor,[3]Associate Professor
Dept. of Computer Science and Engineering
KONERU LAKSHMAIAH EDUCATION FOUNDATION
Guntur,India
E.Mail id : 195034023@kluniversity.in

Abstract— As the crime rates are increasing, the process of ensuring security has become one of the important demands in wide networks. There is always a chance by third-party to threatened the data privacy and security, which led to continued development the information security. Cryptography defined as the process of encoding or encoding data to make it incomprehensible using mathematical operations. DNA cryptography is a new branch of DNA computation that represents a new era of technology in security for data transfer over the internet. In this paper, a Novel Algorithm Based on DNA Cryptography and Combination with One Time Pad and Caesar cipher has been introduced. The output of the algorithm is the tested using National Institute of Standards and Technology (NIST) statistical test suite much time to check the strength of an algorithm with the encrypted text. The algorithm has shown promising results.

Keywords— Cryptography, DNA Cryptography, DNA Computation, One Time Pad.

1. Introduction

As the crime rates are increasing, ensuring the security of data on a wide network getting crucial[1].There is always a chance by third-party to threatened the data privacy and security, which led to continued development in the information security [2]. Cryptography defined as the process of encoding or encoding data to make it incomprehensible using mathematical operations. The reverse process is called cryptanalysis. From a competitive standpoint, both of them are racing to get the best and latest technologies [3].

The researchers in the field of information security use many techniques and methods to improve security and provide many security services, which include confidentiality, integrity, and availability (CAI) triad in addition to the digital signature and non-repudiation to transfer data securely[4] [3][5].

Encryption began with classical Cryptography, when simple manipulations were performed on the letters, and ended with the entry of complex mathematical equations in what is known as modern Cryptography, and with the development oftechnology and the emergence of many techniques used in encryption such as Elliptical Curve Cryptography and Quantum Cryptography, these techniques makeCryptography more complex[5]. DNA cryptography is a new branch of DNA computation that represents a new era of technology in security for data transfer over the internet[6].

2. DNA

DNA is existing in every living cells. Deoxyribonculeotides make up DNA ,which is along polymer of deoxyriboncleotides . the number of nucleotides (or a couple of nucleotides , referred to as base ser) contained in a sugar-phosphate backbone and bases that protrude inside . The polarity of both nucleotide chains is antiparallel . There four types of nitrogen-containing regions called bases:

- Adenine (A)
- Cytosine (C)
- Guanine (G)
- Thymine (T)

These nitrogenous bases are combined to other by hydrogen bond where adenine is always combined with thymine and cytosine combined with guanine, also called a complementary rule.These repeated nucleotides are present in some specific region, known as telomeres, near the chromosome's end they keep chromosomal ends from being destroyed or merging with other chromosomes. One gram of DNA can store 10^8 terabytes of data[7].

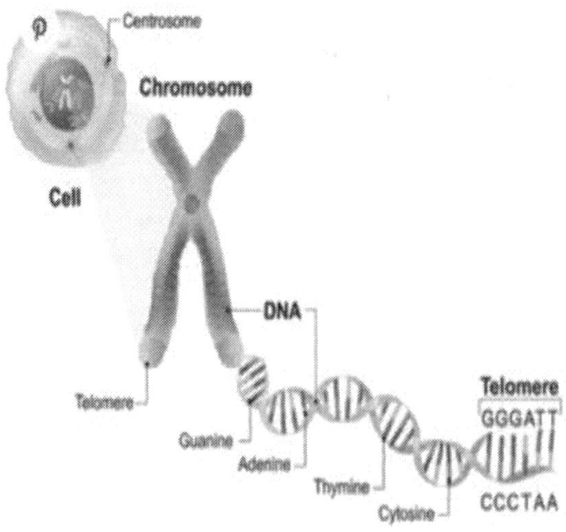

Figure 1

Telomere (Source: Google images)

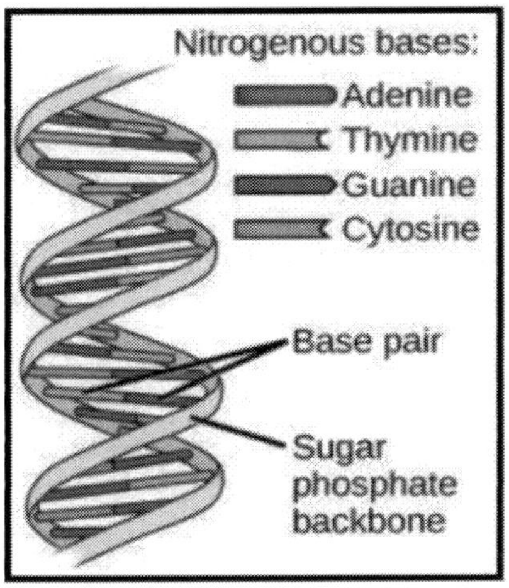

Figure 2 DNA Structure (source: Google images)

3. DNA COMPUTATION

Now a days, biological computations are playing important role in computer science and engineering field. DNA computing is a subfield of biological computation. With DNA computing the processes of DNA molecules for biological computing, can be processed. It is an area of common interest for researchers in the fields of computers and biology. DNA computing has the capabilities of parallelism and high density storage facility, that are extremely useful approaches for many computational problems in computer science and engineering. DNA also been explored as an excellent material and a fundamental building block for building large-scale nanostructures, constructing individual Nanomechanical devices, and performing computations[8].In DNA there are two parameters to consider. Another accuracy and time required for the separation of molecules. In the real system the separation will not be complete; there will be negative and false positive errors. It is also possible that in real-time applications the accuracy and precision of the partition may depend on the tube and the imaging coding [12].

4. DNA CRYPTOGRAPHY

DNA cryptography and information science were discovered after research of DNA computing by Adleman. According to him it is a process of securely hiding data in DNA sequences. It was explained by him,the use of DNA as an information carrierand the use of modern biotechnology as a measure to convert plaintext to cipher text[9].

5. PROPOSED METHODOLOGY

In this paper, we offer a novel algorithm based on DNA cryptography and combination with One Time Pad and Caesar cipher.

In general, cipher text foran OTP equation works like:

$C1 = (M1 \oplus K1)$ and $C2 = (M2 \oplus K2)$

if $K1 = K2$ then $C1 \oplus C2 = M1 \oplus M2$ where:

C1: ciphertext-1.

C2: ciphertext-2.

M1: Message1 plaintext.

M2:Message2 plaintext.

$\oplus$ indicate the XOR operation.

We proposed the algorithm that is much more secure, as, we proposed that:
C1 = (M1 ⊕ K1) and
C2 = (M2 ⊕ K2)
if K1 = K2
then C1 ⊕ C2 does not equal to M1 ⊕ M2.
It has been assumed that the implementation of Caesar cipher after the transformation to DNA sequence will reduce the frequency of letters in the properties of natural language, because of that the proposed algorithm becomes more efficient and harder to cryptanalysis.
The algorithm implemented here gives multilevel security also:
Level 1: DNA sequencing method.
Level 2: Implementing Caesar cipher.
Finally security is enhanced by using of One Time Pad algorithm.

Therefore, the method provided is a combination of those methods. Below is a detailed description of the situation:
Step 1: Input plain Text P.
Step 2: Convert P to 8-bit Format.

Step 3: Convert the resultsof Step 2 into the DNA Sequence (telomere)
Step 4: Apply Caesar cipher on the Telomere with Key (K), as fellow :

C = (input + key) Mod 4 eq. -1

(Here 4 indicate that the DNA Code is in 4 bases.)
Step 5: Apply One Time Pad with previously shared telomere as an OTP key.

Step 6: Convert the result in Step 5 to Bit Sequence and transfer it over to the network.

Let us assume that the k=3 and apply it to out given example HI: The DNA Sequence is "C A G A C A G C " and k1 = 3 which means each A in DNA will become T and C will be A in a same way, G will be replaced as C and finally, T will be replaced as G. so the output will be : ATCTATCA . And as a result, the original text is changed, so it reduced the reliance on the random key of the OTP cipher.

Table 1 :BIT ENCODING INTO DNA NUCLEOTIDES

Bits	Nucleotide
0000	A
0001	C
0010	G
0011	T

Table-2 Sample Key values:

ORIGINAL	ACGT			ACGT		
Key-0	Key -1	Key-2	Key-3	KEY-4	Key-5	Key-6
A--A	A--C	A--G	A--T	A-G	A—C	A—T
G--G	C--G	C--T	C--A	G--C	G—T	G—A
C--C	G--T	G--A	G--C	C--T	T—G	C—G
T--T	T--A	T--C	T--G	T--A	C—A	T--C

After applying the Caesar cipher next step is to apply the OTP cipher with Key k2 which is a DNA sequence and it has been shared between the two parties of the transmission process. As there is no XOR operation on DNA, instead of that, we use the below code sample to match the XOR operation result of binary to do this operation,assuming the k2 as a key and input is ATGCAGCA, the XOR operation for our example HI will be as follows:-

Table 3 XOR operation

KEY2	A	T	C	T	A	T	C	A
XoR	⊕	⊕	⊕	⊕	⊕	⊕	⊕	⊕
INPUT	A	T	G	C	A	G	C	A
RESULT	A	A	T	G	A	C	A	A

And by using table 1 we can convert the result into a bit and then to a text format as the final output of our algorithm
That is the cipher text. In this way we get the chipper text from the bits Sequence as the final output of encryption.

6. NIST STATISTICAL TEST

To check the randomness of any analysis, one of the most popular tool is NIST Statistical Test Suite (NIST STS). For our proposed algorithm, we use the NIST Statistical Test Suite to test the strength of the algorithm. The NIST Statistical test suite is a 16-component mathematical package designed to test several randomly generated binary sequences generated by hardware or software-based cryptographic random or pseudorandom generator. These tests focus on the various types of disorder that may occur in succession[10]. The 16 tests are listed in the table below.[11]

Table 4 shows the Result of NIST statistical test.

Test Name	Result
Frequency Test	Pass
Frequency Test within a Block	Pass
Run Test	Pass
Longest Run of Ones in a Block	Pass
Binary Matrix Rank Test	Fail
Discrete Fourier Transform (Spectral) Test	Pass
Non-Overlapping Template Matching Test	Pass
Overlap ping TemplateMatching Test	Fail
Maurer's Universal Statistical test	Fail
Linear Complexity Test	Fail
Serial test	Pass
Approximate Entropy Test	Pass
Cumulative Sums (Forward) Test	Pass
Cumulative Sums (Reverse) Test	Pass
Random Excursions Test	Pass

7. RESULTS AND CONCLUSION

The proposed algorithm is for data encryption, where the plain-text is first converted into a set 8-bits sequence and after that convert to a sequence of DNA. the data is encrypted by Applying Caesar cipher on the DNA Sequence with Key K can be shown like, C= (input + Key) Mod 4 bases. Finally, we apply One Time Pad with a previously shared DNA Sequence as the OTP key on both sides. The output of the algorithm is then tested using the NIST statistical test suite, that shows much time to check the strength and efficiency of the algorithm with the encrypted text while the algorithm has shown high and promising results in Randomness Testing .

REFERENCES

[1] A. Nath, A. Dodia, "Symmetric Key Encryption Algorithm using DNA Sequence," Int. J. Adv. Res. Comput. Sci. Manag. Stud**., vol. 6, no. 4**, pp. 108–115, 2018.

[2] K. C. Jithin, S. Sankar, "Colour image encryption algorithm combining Arnold map, DNA sequence operation, and a Mandelbrot set," J. Inf. Secur. Appl**., vol. 50, no. 102428**, pp. 1–22, Feb. 2020.

[3] A. Hazra, C. Lenka, A. Jha, M. Younus, "A Novel. Two Layer Encryption Algorithm Using One-Time Pad and DNA Cryptography," in Lecture Notes in Networks and Systems, **vol. 103, 2020**, pp. 297–309.

[4] P. B. Narasingapuram , M. Ponnavaikko, "DNA Cryptography Based User Level Security for Cloud Computing and Applications," Int. J. Recent Technol. Eng**., vol. 8, no. 5**, pp. 3738–3745, 2020.

[5] A. Kaushik, V. Thada, "VG1 Cipher – A DNA Indexing Cipher," Int. J. Innov. Technol. Explor. Eng**., vol. 9, no. 3**, pp. 221–226, 2020.

[6] Mohit, Manisha, P. Bangar, "Survey on DNA based Cryptography," Int. J. Res. Appl. Sci. Eng. Technol., **vol. 3, no. 3**, pp. 721–725, 2015.
[7] S. Kalyani, N. Gulati, "Pseudo DNA Cryptography Technique using OTP Key for Secure Data Transfer,"Int. J. Eng. Sci. Comput., **vol. 6, no. 5**, pp. 5657–5663, 2016.

[8] Rajni, "DNA Computing," Int. J. Eng. Comput. Sci**., vol. 6, no. 1**, pp. 19972–19976, 2017.

[9] O. Tornea, M. E. Borda, V. Pileczki,"Cryptographic Algorithm Based on Dna and RNA Properties," Int. J. Adv. Res. Comput. Eng. Technol**., vol. 7, no. 11**, pp. 237–241, 2018.
[10] J. K. M. S. U. Zaman, R. Ghosh,"A Review Study of NIST Statistical Test Suite: Development of an indigenous Computer Package,"
Aug. 2012.
[11] H. M. Mousa, "DNA-Genetic Encryption Technique," Int. J. Comput. Netw. Inf. Secur., **vol. 8, no. 7**, pp. 1–9, Jul. 2016.
[12] Byoung-Tak Zhang , Soo-Yong Shin, " Molecular Algorithms for Ecient and Reliable DNA Computing",Genetic Programming **vol .98**, pp. 735-74

A COMPARATIVE STUDY: PERFORMANCE AND OPTIMIZATION OF ARTIFICIAL NEURAL NETWORKS IN BLOOD CELL DATASETS

[1]Mr.K.Sivaraman, [2]Dr.V.Khanna
[1]Research Scholar, [2]Professor Department of Computer Science and Engineering
Department of Information Technology
1,2Bharath Institute of Higher Education and Research (BIHER)
Chennai-73, Tamil Nadu, India.
Mail id: sivaraman2006@gmail.com

Abstract: Artificial Neural Networks (ANN) generally referred to as Neural Networks are the statistical model, It is mainly focused on the biological neuron for the information or signal processing. A neural network is a dynamic system consisting of a community of interconnected neurons that offers a very exciting solution to complex problem solving and other applications that can play a crucial role in today's technology. In this paper we are using various process like pattern recognition, prediction and optimization of blood cells is usually undertaken to help clinicians determine a wide variety of haematics. Pathologies identify dengue, anemia, diabetes and other related diseases.

Keywords: Artificial Neural Network (ANN), Pattern Recognition, Prediction, Feedback Network

1. INTRODUCTION

Blood analysis has been an important field for quite a long time. With developments in the field of Computer science and informatics, it was thought that we can use this normal way of thinking blood cell process to develop some System of neural network. This which sound hyperbolic to suggest that the existing techniques of a whole research area are rapidly replaced by a new discovery, as if it were struck by a 'tsunami' science. But this devastating terminology is sufficient to explain Deep Learning's meteoric rise over the past several years-an increase marked by dramatic improvements over predominant approaches to the toughest problems.

Many jobs requiring intelligence or pattern recognition are exceedingly difficult to automate, but appear to be quite easy for people to do. Humans, for example, can distinguish between things and make sense of a large number of visual features in their environment with little effort. "Neural Nets," "Parallel Distributed Processing Networks," and "Connectionist Networks" are all terms that are used to describe computer systems that perform comparable tasks. A "Directed Graph" is made up of a set of "Nodes" (vertices) and a set of "Connections" (edges / links / arcs) linking node pairs, according to the theory. Each node performs a simple calculation.

2. Working of ANN:

The processing component of a neural network is the neuron. There are a few general abilities that this building block of human awareness possesses. A biological neuron simply gathers information from several sources, combines it in some fashion, applies a nonlinear operation to the result, and then outputs the final result. Multiple kinds of this type of neuron exist in humans, complicating man's attempts to electrically repatriate them. Natural neurons, on the other hand, have the same four basic components. Dendrites, soma, axon, and synapses are the biological names for these components. The different ways these neurons can be clustered together are discussed in the other sections that use neural networks. The human mind clusters information so that it can be understood in a complex, collaborative, and self-organizing fashion. In a three-dimensional cosmos, brain networks are built up from microscopic components. Currently, neural networks are nothing more than a collection of basic artificial neurons clustered together. This clustering is accomplished by forming layers that are then linked together. The connection between these layers to solve real-world problems is the second aspect of engineering networks.

A 'back-propagation cycle,' or often just a 'iteration,' is the name given to the complete process of input, processing, comparing output to correct response, and altering link strengths. The net will then be shown another image, and its response will be compared to the proper response, with the relation strengths adjusted as necessary. This approach can also handle hundreds or thousands of iterations.

3. Attributes of ANN:

Basically computers are good in calculations which primarily take the process of inputs and then give the result on the basis of the calculations performed on unique algorithms programmed in the software but ANN enhance their own The more rules, the better the choices they make. Basically the characteristics are those that should be represented in the following diagram.

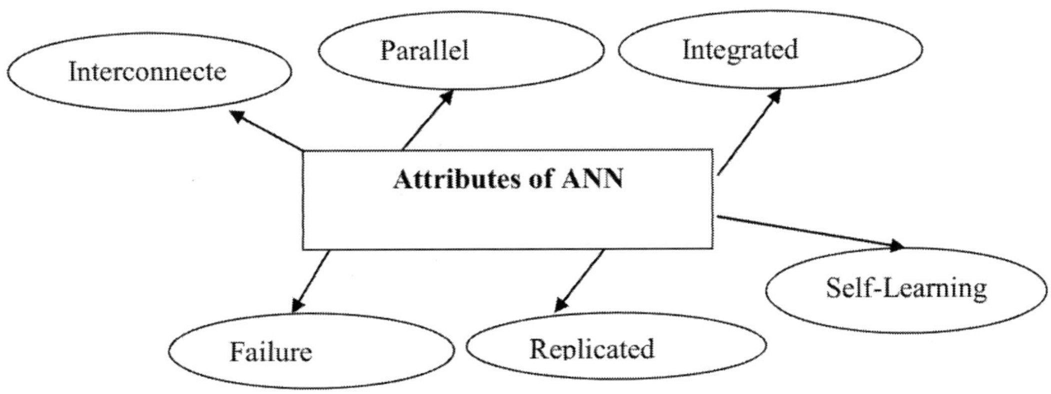

Fig.1 Attributes of ANN

4. Learning Process of ANN:

The Learning procedure of Artificial Neural Network Classified in to two pattern.

If the set of input units employs another pattern, the network learns to generate that pattern on the set of input units, which is known as associative mapping.

Detection of regularity in which units learn to respond to specific properties of the input patterns. While the network stores the relationships between patterns in associative mapping, each unit's response has a particular 'meaning' in regularity detection. This kind of learning mechanism is essential to the discovery of features and the representation of knowledge.

Each neural network possesses information found in the values of the weights of connections. Modification Information stored as a function of experience in the network implies a learning rule for changing the values of the experience Forces.

5. Applications of ANN

Medical: If the set of input units uses a different pattern, the network learns to generate that pattern on the input units, which is referred to as associative mapping.
Data Mining: Discovery from vast quantities of data, of relevant patterns (knowledge).
Expert Systems: A computer decisionmaking software that simulates a human expert's reasoning process.
Fuzzy Logic: Uncertain reasoning theory.
Artificial Life: Evolutionary Computation, Intelligence Swarm.
Artificial Immune System: A biological immune system based computer.

6. LITERATURE REVIEW

Artificial neural networks, which are non-linear mapping systems with a loosely defined structure based on biological nervous system notions, were investigated by O.A. Kurban. A typical real neuron has a branching dendritic tree that collects messages from many other neurons in a small area, as well as a cell body that integrates the impulses and generates a response. Manages metabolic functions); and along the branching axon, which disperses the reaction by interacting with many more neurons via dendritic tree interactions. Each neuron's response is a relatively simple non-linear function of its inputs, with the strength of its input connections having the biggest influence. Despite the relative simplicity of the individual components, structures contain a huge number of neurons. The performance of "neurons," or processing nodes, is entirely dependent on the information that is available locally at the node, whether it is internally stored or arrives via the weighted connection unit, which accepts input from many other nodes and transmits its output to other nodes. A single processing unit is unsuccessful when used alone; it produces a scalar output with a single number value, which is a simple nonlinear function of its inputs. The control of the system is sufficiently derived from a combination of numerous elements. A network can perform a number of functions by changing the topology of the connection and the values of the connecting weights. Complex functions can be built using links between units and weights. It has been proved that any function satisfying certain broad requirements may be estimated with arbitrary precision using a sufficiently large network with the suitable topology and property weights. The model is a highly simplified version of real nerve systems. The goal is to capture the important aspects of real-world information processing functions. Neurons are simple, highly interconnected processing units that perform two tasks: collecting inputs from other neurons or the external environment and producing an output from the aggregated inputs. It has been demonstrated that neural networks can accurately estimate most continuous functions by selecting the appropriate number of neuron units.

7. BIOLOGICAL INNOVATION

The human brain is made up of a network of neurons linked together via receptors and effectors. Dendrites are the receptors, and axons are the effectors. Dendrites receive signals from many other neurons in a small region; a cell body or soma integrates the inputs and generates a response signal; and a branching axon delivers the response by interacting with the dendrite trees of many other neurons. as illustrated in Figure

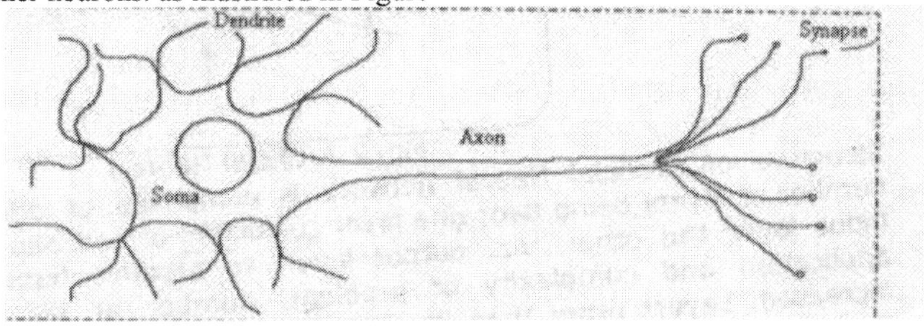

Fig 2: Biological Neurons

8. ANN METHODOLOGY

AANNs are massive parallel computational models that imitate the human brain's functioning. A weighted connection connects a large number of basic processors in an artificial neural network (ANN). On the other hand, processing nodes could be referred to as "neurons." The information accessible locally at the bulge, also internally stowed or received via biased nodes, is the only determinant of node performance. Each node receives data from a huge number of other nodes and relays it to the rest of the network. AANNs are massive parallel computational models that mimic the human brain's function. A vast number of basic processors are linked together via weighted connections in an artificial neural network (ANN). Processing nodes, on the other hand, could be referred to as "neurons." The only determinant of node performance is the information available locally on the bulge, moreover internally stowed or received via biased nodes. Separately node accepts data from a huge number of other nodes and relays it to the rest of the network. Ineffective on its own, a single processing component produces a scalar yield with a single numerical value that is a simple non-linear purpose of its ideas. A mistake occurs when the desired answer differs from the device's output. The machine receives this defect information and alters the system restrictions on a regular basis (the education rule). It goes through the process again and again pending the result is good. The result is clearly dependent on the data, as evidenced by this definition.

9. PROPOSED METHOD

The technique has been fine-tuned by the researchers utilising several well-known classifiers in machine learning. The most discriminative traits were effectively chosen in the proposed technique. It retrieves a high precision rate for disorder identification. WBCs were identified using a morphological technique based on blood pictures. Because it incorporates both wide and phantom features, the proposed technique attained a greater accuracy of more than 95%. To classify WBCs automatically, researchers created a deep learning algorithm-based technique. The Logistic Regression algorithm was chosen as a Deep Learning (DL) technique because it can handle a huge number of images.

10. RESULTS

The processing component of a neural network is the neuron. There are a few general abilities that this building block of human awareness possesses. A biological neuron simply gathers information from several sources, combines it in some fashion, applies a nonlinear operation to the result, and then outputs the final result. Humans have multiple forms of this type of neuron, confounding man's attempts to electrically reproduce brain processes. The four basic components of natural neurons, on the other hand, are the same. The biological names for these components are dendrites, soma, axon, and synapses. In the following parts that use neural networks, the various ways these neurons can be clustered together are explained. Information is clustered in the human mind in order to be interpreted in a complex, collaborative, and self-organizing manner. Brain networks are made up of microscopic components in a three-dimensional universe. Neural networks are currently nothing more than a collection of rudimentary artificial neurons arranged in a cluster. This clustering is achieved by generating layers that are then connected. The second part of engineering networks is the connectivity between these layers in order to solve real-world challenges.

11. CONCLUSION

Artificial neural networks are identified as an emerging technology in this research that can be utilised for a number of data analysis applications such as pattern identification and prediction. The network is able to identify the dataset using different experimental methods. The Neural network can capable of parallel processing .Our methods classify medical images based on attributes and we can achieve high performance and optimized result for identifying input.

REFERENCES

1. Laith Alzubaidi, Omran Al-Shamma et.al, Classification of Red Blood Cells in Sickle Cell Anemia Using Deep Convolutional Neural Network, Springer Nature Switzerland,2020.

2. Abdulkadir Şengür, Yaman Akbulut, White Blood Cell Classification Based on Shape and Deep Features,2020.

3. Roopa B. Hegde, Peripheral blood smear analysis using image processing approach for diagnostic purposes: A review, Nalecz Institute of Biocybernetics and Biomedical Engineering of the Polish Academy of Sciences. Published by Elsevier,2018.

4. Mazin Z. Othman, Thabit S. Mohammed, Neural Network Classification of White Blood Cell using Microscopic Images, International Journal of Advanced Computer Science and Applications, Vol. 8, No. 5, 2017.

5. Swaranjali Raskar, Gayatri Bhorkar, Neural Network Classification of Blood Cell Images, IJRECE ,Vol. 7 Issue 2 ,2019.

6. Su Yang, Bin Fang, Faster R-CNN Based Microscopic Cell Detection, International Conference on Security, Pattern Analysis, and Cybernetics (SPAC),2017.

7. K.Sona, C.Sriragavi, Detection Of White Blood Sample Cells Using CNN, International Research Journal of Engineering and Technology, Volume: 06 Issue: 03,2019.

8. Laith Alzubaidi, Mohammed A. Fadhel, Deep Learning Models for Classification of Red Blood Cells in Microscopy Images to Aid in Sickle Cell Anemia Diagnosis, Electronics 2020, 9, 427.

9. Abdullah Elen, M. Kamil Turan, Classifying White Blood Cells Using Machine Learning Algorithms, UMAGD, (2019) 11(1).

10. Nersu Kalyan, S S Mounika,, T Santhi Sri, Classification Of White Blood Cells Exploitation Convolutional Neural Networks, International Journal Of Scientific & Technology Research Volume 9, Issue 01, January 2020.

11. Thanh Tran Thi Phuong, Lam Binh Minh, Suk-Hwan Lee, Blood cell count using Deep Learning Semantic Segmentation,2019.

FABRICATION AND TRIBOLOGICAL STUDIES OF HYBRID COMPOSITE MATERIAL

K.Surendrababu[1] , Sathiyaraj S[2], M.Issac Premson[3], D.GuruMugilan[3], E.Sridhran[3]
Associate Professor[1], Assistant Professor[2] , UGScholar[3] ,Aarupadai Veedu Institute of Technology,Vinayaka Mission's Research Foundation, Deemed to be University,
Tamil Nadu, India,
Corresponding Author : surendrababu@avit.ac.in

Abstract. Many attempts have been undertaken to mechanically or chemically create various types of in organic filled polymers. These methods, however, necessitate the alteration of the filler material through intricate polymerization reactions, making them unsuitable for industrial use. Waste flyash recovery is challenging, and the utilisation rate is poor. This flyash is employed in novel composite materials as a filler or partial reinforcement. Flyash is employed as a filler material for polymer matrix composites in this study. Aramid and E-Glass fibres are utilised as reinforcement with epoxy resin to improve the bonding strength between the fibres and filler materials, resulting in a high load bearing capacity. The specimens are tested for mechanical properties such as tensile, flexural, and impact according to ASTM standards. The maximal impact strength was seen at a 5wt% flyash addition, however the tensile and flexural strength were steadily lowered with the addition of 5wt% flyash.

Keywords: flyash, polymer matrix, fibres, reinforcement, filler material, Aramid, E-Glass.

1. INTRODUCTION

For service longevity, Mechanical and tribological synergy is required for components that are in relative motion with their counterparts in any product.. Traditional materials frequently outperform modern materials in each of these area As a result, the part will fail prematurely. This has opened the door to new possibilities evolution of custom materials, i.e. composites, to fit specific applications. To synthesis and study the mechanical behaviour of Aramid/E-glass composite. This project comprises the creation of an a polymer composite reinforced with Flyash particles, as well as the study of its reaction to various mechanical behaviours. A fast and basic understanding of the following will aid in familiarising oneself with the present project and its goals.

Bheemappa Suresha et.al[1] Epoxy Hybrid Composite Fabrication and Tribological Behavior. Shubhajit Das[et.al [2] Fabrication and tribological analysis of SiC/B4C nanoparticle-reinforced AA6061 hybrid metal matrix composites. Jaswinder Singh[3] The features of Al/SiC/Gr hybrid composites that can be used in a range of tribological applications are discussed.. RManikandan[4] This study focuses on fabricating aluminium hybrid metal matrix composites with environmentally friendly agricultural waste, cow dung ash, and boron carbide using two-stage stir casting.D Lingaraju[5] Nanoparticle reinforced plastics are gaining popularity as a means of developing novel materials that maximise the efficient use of natural resources, particularly renewable resources.

2. MATERIALS AND METHODS

2.1. Materials Used

2.1.1. Thinner

Before and after combining Epoxyres with fillers, thinner is used to clean the beaker and other containers..Thinner is shown in figure 1.

FIGURE 1: Thinner

2.1.2. White Wax

The white wax aids in the removal of laminates from the moulding board without causing them to break. We can remove the manufactured laminate from the moulding board without any breakage or damage by applying white wax before removing it for one hour..White Wax is shown in figure 2 .

FIGURE 2 : White Wax

2.1.3. Moulding board

The base surface on which an o composite laminate is built is moulding board. Its dimensions are 400mm x 400mm.. In this moulding board the laminate is prepared with varying percentage nano by using handlay –up technique.

2.1.4. Mechanical stirrer

The mechanical stirrer is used for constant stirring of flyash with epoxyres in for more than 15minute .It makes uniformd is tributeion over the matrix.Mechnical stirrer is shown in figure 3.

FIGURE 3: Mechanical Stirrer

2.1.5. Epoxy Resin

Epoxy resin is matrix material and it is in the formo fliquid wax.Epoxy resin is shown in figure 4.

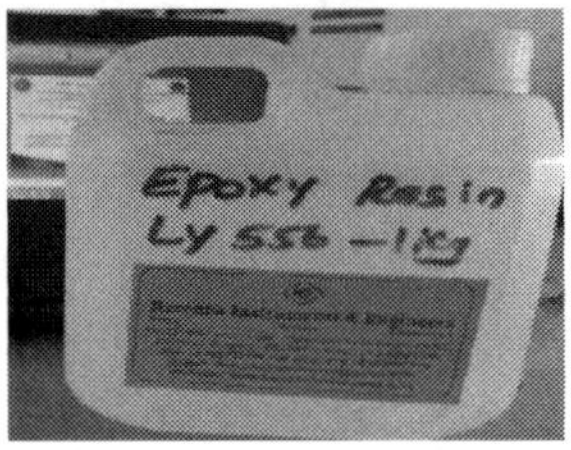

FIGURE 4: Epoxy Resin

2.2. Hardener

Hardener is the hardening and curing agent which is used to solidify the irepoxy resin material along with rein forcement. Because of the resin which is very less viscocity we are using hardener to solidify the resin. Here we are using the tri-ethylene tetra amineasa hardener fore poxy resin.Hardener is shown in figure 5.

FIGURE 5: Hardener

2.3. E-GLASSFIBER

Fibre glass is a type of fiber-reinforced plastic in which glass fibres are pushed into the material. Fiberglass is also known as glass reinforced plastic or glass fibre reinforced plastic for this reason..Fiberglas is a light, robust, and less brittle material. The best thing about fibreglass is that it can be moulded into a variety of intricate shapes. Fiberglass is frequently utilised in bathtubs, boats, aircraft, roofing, and other uses because of this..E Glassfiber is shown in figure 2.6.

FIGURE 6: E Glass Fiber

2.4. Aramid Fiber

Heat-resistant and strongsynthetic fibres are known as aramid fibres. They're employed in aerospace and military applications, as well as bicycle tyres, maritime cordage, and marine hull strengthening, and as an asbestos alternative..Aramid Fiber is shown in figure 7.

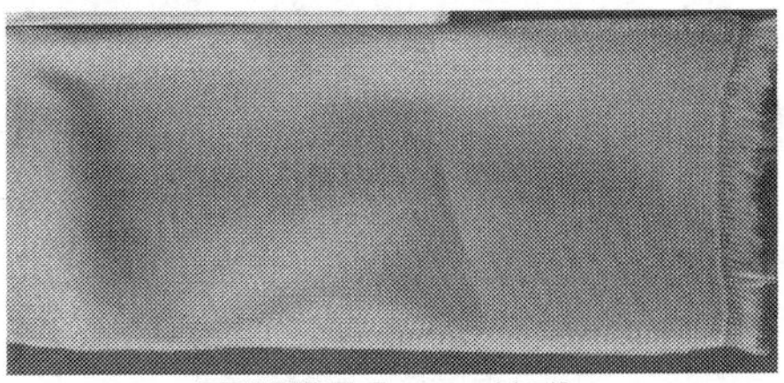

FIGURE 7: Aramid Fiber

2.5. Flyash

Flyash, also known as flue-ash, is one of the byproducts of combustion, and it contains the small particles that rise with the flue gases.. Bottom ash refers to ash that does not rise. Fly ash is the powder that is produced when coal is burned in a mechanical environment. All flyash in contains considerable quantity of silicondioxide ($SiO2$) (both amorphous and crystalline) and calcium oxide (CaO), both of which are endemic components in many coal-bearing rockstrata.Flyash is shown in figure 8

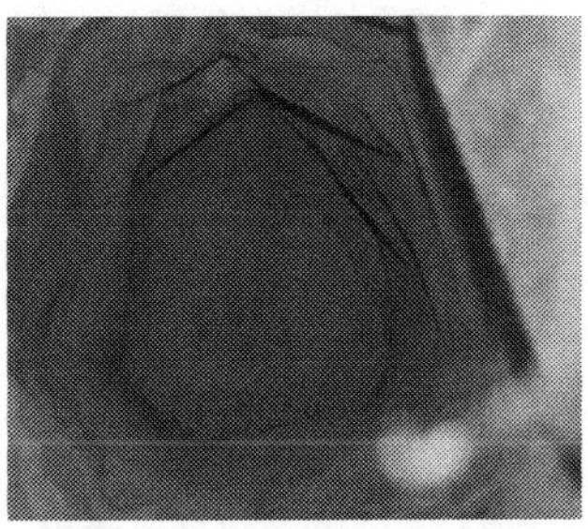

FIGURE 8 : Flyash

3. MANUFACTURING TECHNIQUE

hardener was the nano composite formation taking (1:1Ratio) the flyash dipresed through ashear mixing process. Then 10% hardener was added into the epoxy–fly ash mixture by weight and through mixed .Handlay-up is the most widel yused manufacturing technique .The square steel mold with the dimension of 400mm x 400mm was used to be prepared and the releasing agenta aqueous solution of polyvinyl lalcohol (PVA) is applied on the mouldevenly. The laminate are allowed to cure for 48hours and theres in was cool edtoroom temperature similarly the 0% laminate isprepared . Machine dasper dimensions of American Society for Testing and Material (ASTM) D3039, D790, and D256 were used to manufacture standard tensile, flextural, and impacts specimens from the curedlaminate of nano composites.

3.1. Handlay-Up Technique

Hand lay-up is the most basic way of composite processing. This approach also has a low infrastructure need. The processing procedures are straightforward in figure 9. To begin, a release gel is sprayed on the mould surface to prevent the polymer from clinging to it. To achieve a nice surface finish on the product, thin plastic sheets are used on the top and bottom of the mould plate. Woven mats or chopped strands can be used as reinforcement. Mats are cut to fit the mold's dimensions and laid on the mold's surface once the Perspex sheet has been applied. The rmosetting polymer in liquid form is then completely mixed with a prescribed hardener (curing agent) in a proper proportion and poured onto the surface of the matal for put in the mould. With the help of a brush, the polymer is spread evenly. The second layer of mat is then placed over the polymer surface, and a roller is pushed through the mat-polymer layer with little pressure to remove any trapped air and excess polymer. The method is continued for each polymer and mat layer until all of the layers have been restacked.

The 5w % of flyash is added to the resin while mixing at a rate of 1000 rpm, this was conducted for 20minutes until the

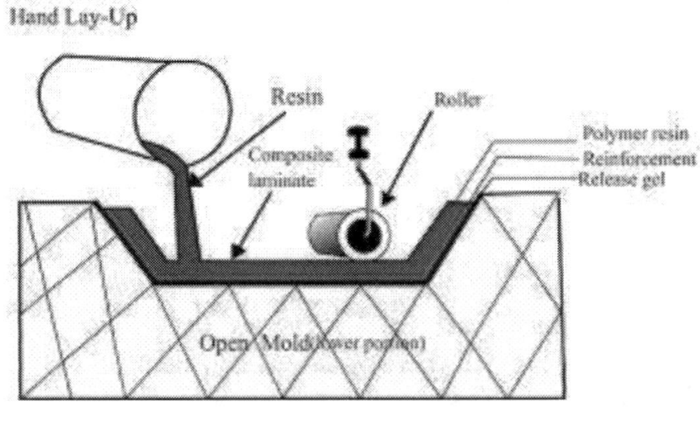

FIGURE 9: Handlay up Method

3.2. Experimental Procedure

STEP-1 Prepare the mould on the 400mm*400mm
STEP-2 Clean the mould using thinner
STEP-3 Apply the gel (PVA) on the mould using cotton.
STEP-4 In a digital weighing balance, weigh the Epoxy resin, hardener, syntactic foam, and fibre.
STEP-5 The preparation of a mixture of flyash and epoxyresin. A mechanical stirrer was used to mix the flyash particles with epoxy resin. The hardener was then added to the scattered suspension.
STEP-6 The maton is then placed in the mould and the resin is applied.
STEP-7 A roller is used to ensure that the resin is distributed evenly throughout the fibre..

STEP-8 After that, apply the glue and place the fibre on the foam.
STEP-9 Repeat the process and thesheet is allowed to dry for 48hrs
STEP-10 After drying the composite material is removed from the mould.
STEP-11 Machined the laminate togetthere quired size for testing.

FIGURE 10 : Aramid E Glass Composite

FIGURE 11: Aramid E Glass Composite 5% wt of Flyash.

4. RESULT AND DISCUSSIONS

4.1. Tensile strength of the polymer composite laminates

Tensile test result of the hybrid polymer nano composite laminates' were shown in table.1

TABLE. 1. Tensile test result

S.NO	Flyash content (wt%)	Maximum tensile load(KN)	Maximum tensile strength(N/mm^2)
1.	0%	18.09	289
2.	5%	5.47	97

4.2. Tensile test results

When comparing results of both sample there is a gradual decrease in tensile strength of the polymer lamina test while adding a 5wt% of flyash .The graphical representation of the results were shown in figure.12

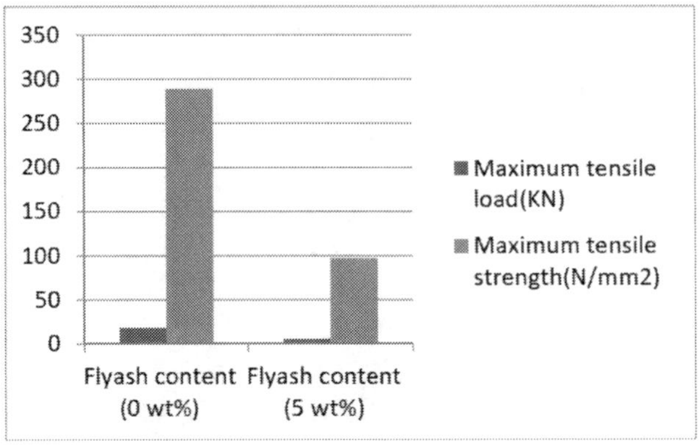

FIGURE 12: Tensile Strength

4.3. Flexural strength of the polymer composite laminates

Flexural test results of the hybrid polymer nano composite laminates 'were shown in table 2

TABLE. 2. Flexural Strenth

S.N O	FLY ASH content(wt %)	Maximum flexural strength(N/mm^2)
1.	0%	211
2.	5%	162

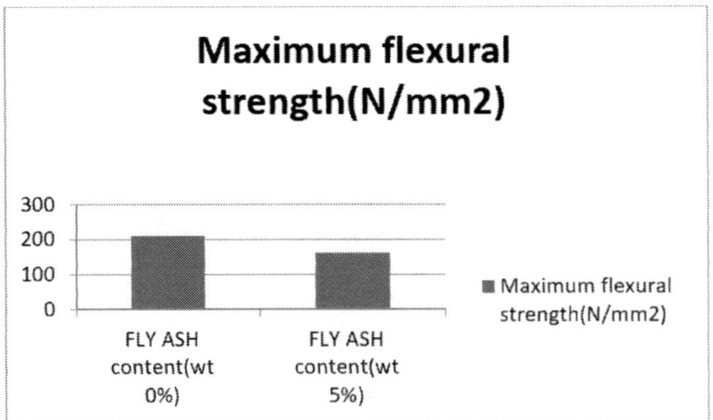

FIGURE 13: Flexural Strength

When comparing results of both sample there is a gradual decrease in flexural strength of the polymer laminates while adding a 5 wt% of fly ash. The graphical representations of the results were shown in fig 13 .

4.4. Impact strength of the polymer composite laminates

Impact test results of the hybrid polymer nano composite laminates'were shown in table.3

TABLE.3 Impacttestresult

S.NO	FLYASH content(wt%)	Test temperature(^{0}C)	Absorbedenergy -joules
1.	0%	24	06
2.	5%	24	14

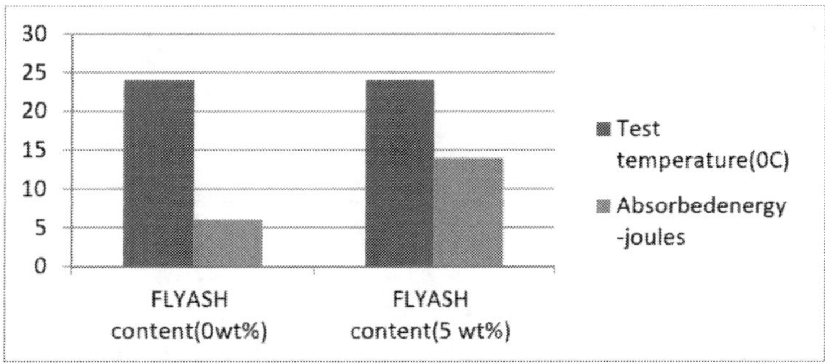

FIGURE 14: Impact Strength

When comparing results of both sample there is a gradual increase in impact strength of the polymer laminates while adding a 5 wt% of fly ash. The graphical representation of the results were shown in fig 14

5. CONCLUSION

As a comparative study, the current work aims to evaluate flyash as a filler material in polymer matrix composites and characterise the mechanical properties.Aramid/ E-Glass polymer composite with varying percentage of flyash have been successfully prepare dusing hand lay-up techniques. The created laminates were tested for mechanical qualities such as tensile, flexural, and impact.

The following findings are reported.The addition of flyash as fill ermaterial reduced the ten sile and flexural strength of the laminates. The strengths were decreased from 0w t% of flyash laminateto 5w t%of flyash laminate.But ,The Impact strength of the laminates was increased from 0w t% of flyash laminate to 5w t% of flyash laminate.

REFERENCES

[1] Bheemappa Suresha, and Rajashekaraiah Hemanth et.al, Fabrication and Tribological Behavior of Epoxy

[2] Hybrid Composites, Fabrication and Tribological Behavior of Epoxy Hybrid Composites, Published 2018 by Wiley-VCH

[3] Verlag GmbH & Co. KGaA. Shubhajit Das, Chandrasekaran M., Sutanu Samanta, Palanikumar Kayaroganam, Paulo Davim J., (2018) "Fabrication and tribological study of AA6061 hybrid metal matrix composites reinforced with SiC/B4C nanoparticles", Industrial Lubrication and Tribology, https://doi.org/10.1108/ILT-05-2018-0166.

[4] Jaswinder Singh et.al, Fabrication characteristics and tribological behavior of Al/SiC/Gr hybrid aluminum matrix composites: A review, Friction 4(3): 191–207 (2016) ISSN 2223-7690,DOI 10.1007/s40544-016-0116-8.

[5] RManikandan, T V Arjunan,et.al, Mechanical and tribological behaviours of aluminium hybrid composites reinforced by CDA-B4C, Mater. Res. Express 7 (2020) 016584 https://doi.org/10.1088/2053-1591/ab6b54.

[6] D Lingaraju, K Ramji, M Pramila Devi ,U Rajya Lakshmi,et.al, Mechanical and tribological studies of polymer hybrid nanocomposites with nano reinforcements, Bull. Mater. Sci., Vol. 34, No. 4, July 2011, pp. 705–712.

[7] Uzun, A., Asikuzun, E., Gokmen, U. and Cinici, H. (2017), Vickers Microhardness Studies on B4C Reinforced/Unreinforced Foamable Aluminium Composites, Transactions of Indian Institute of Metals, doi 10.1007/s12666-017-1163-1.

[8] Ravindran, P., Manisekar, K., Narayanasamy, R. and Narayanasamy, P. (2013), "Tribological behavior of powder metallurgy processed aluminium hybrid composites with the addition of graphite solid lubricant", Ceramic International, Vol. 39No. 2, pp. 1169-1182.

[9] Chen, H.S., Wang,W.X., Li, Y.L. and Zhang, P. (2015), "The design, microstructure and tensile properties of B4C particulate reinforced 6061Al neutron absorber composites", Journal ofAlloys and Compounds, Vol. 632, pp. 23-29.

[10] Dou, Y., Liu, Y., Liu, Y., Xiong, Z. and Xia, Q. (2014), "Friction and wear behaviors of B4C/6061 Al composite", Materials and Design, Vol. 60, pp. 669-677.

IMPROVING THE MECHANICAL PROPERTIES OF NATURAL FIBER REINFORCED SPONGE GOURD FIBER COMPOSITES

B Selva Babu[1], D Bubesh Kumar[2], K Arjun[3], A Arshad [3], Athul M Joy[3]

[1]Assistant Professor, [2]Associate Professor, [3]UG Scholars, Aarupadai Veedu Institute of Technology, Vinayaka Mission's Research Foundation, Deemed to be University, Tamil Nadu, India

Corresponding author: selvababu@avit.ac.in

Abstract. Natural fibers have emerged as a mainstay in the creation of Polymer Composites in recent years. Using a woven mat constructed from sponge fiber gourd (luffa), almond shell, and layered in an epoxy matrix, this work demonstrates the possibility of generating hybrid reinforced polymer composites using varied stacking sequences of composite layers. Natural fibers in an epoxy matrix were applied using hand-layup. The tests examined the possible influence of natural fiber reinforcement on the natural fiber hybrid composites.

Keywords; Luffa, Almond, Epoxy, Hand Lay-Up, Erosion

1. INTRODUCTION

When you combine two or more distinct materials together, you end up with a composite substance. In order for any composite material to be made, there are distinct components. Composites are made by combining two distinct materials: one is known as the matrix phase, and the other is a variety of various polymers. Composites are made by combining several natural or artificial materials to capitalise on their useful characteristics while taking steps to lessen their drawbacks. Reinforced plastic that incorporates glass fibers (strong yet brittle) is an example of an old composite (which is flexible). Composites are frequently used instead of metals of same weight and strength. The overwhelming majority of composites are created using a material known as a matrix that bonds fibers securely together[1]. As in an adhesive, matrices such as adhesives bond fibers together and cause them to be more resistant to outside influences. The reason for this is because matrices such as fibers, as well as other materials, enhance the stiffness and strength of the matrix, allowing it to bear fractures and cracks. common substance (though materials such as concrete may also be used)[2].

2. Constituent Materials

There are many types of composite materials. Common ones include fiber-reinforced composite materials, where the main components are fiber reinforcement, matrix, coupling agents, coating, and filler.

Fibers are the major load-bearing components, and the matrix surrounding them ensures they are properly orientated and placed. The shear stress applies the load on the fibers, which allows the load to transfer through the Matrix[3]. The Matrix protects the fiber from temperature and humidity-induced environmental issues. An excellent method for wetting fibers is to combine wetting agents with fiber coating additives. This will allow the fibers to be effectively wetted in the matrix, as well as promote a strong connection across the fiber-matrix interface[4]. Filling polymeric matrices with fillers is often done to decrease the cost while also improving dimensional stability.

2.1. Fibers

Bulk-form materials are softer and weaker than fibrous materials. Fibers contain most of the internal flaws, whereas bulk materials show the most outward mistakes. Fibers are tightly-aligned molecules, and their composition is made up of small crystals. Dense fibers are also helpful[5].

Fiber is the most important component of fiber reinforced composite materials. The majority of the composite reinforcement is taken up by the reinforcing fiber with the greatest volume %. Only the tensile load is absorbed by this reinforcing fiber[6]. With FRP composites' tensile, compressive, and flexural or shearing strength and stiffness enhanced by fiber reinforcement, FRP composites possess much greater tensile, compressive, and flexural or shearing strength and stiffness.

2.2. Lamina and Laminate

To make a laminate or substrate, several fibers are woven together to form a thin layer in a certain matrix. lamina fiber orientation: the location of the fibers inside the lamina may be continuous or discontinuous, and fibers can be orientated in one of many orientations, depending on the orientation of the lamina fibers[7]. On the major axis, unidirectional laminas feature fibers that run parallel to one another. The fibers have a comparable mechanical and physical performance in all directions, regardless of the direction they are located in the lamina[8].

Laminates are created by stacking several laminas on top of each other. Composite reinforced with fibers is most common, which has a specified degree of deformation and a specific load capacity. Orientation of the fibers in each lamina and the sequence in which many layers are layered determine the number of composite materials that may be created.

2.3. Natural Fibers Reinforced Composites

Research into natural fiber reinforced polymer composite materials has expanded dramatically in recent years owing to the increasing interest in such materials[9]. Recycling is important, as mentioned above, since it provides fiber and plant materials such as jute, cotton, flax, hemp, kenaf, sisal, ramie, pineapple, bamboo, banana, etc. as well as wood. Carbon, glass, and man-made fibers are low-density, renewable, abundant, and affordable alternatives. Compound that is synthesised using naturally occurring elements is far more ecologically aware than its artificially synthesised counterpart, which is used in many industrial sectors, including automotive, aerospace, construction, and building, as well as the military.

2.4. Application of Natural Fibers Reinforced Composites

When it comes the following applications, natural fiber composites might be a great material choice due to their inexpensive cost.

- Partition boars, panel sections for partition, flooring, wall framing, doors, windows, and roofs.
- They may also be utilised for various emergencies, such as hurricanes, floods, earthquakes, etc.
- Grain storage soils, bio-gas containers, post-boxes, and so on are examples of storage devices.
- Furniture: a table, chair, bath unit, and shower, for example.
- Pipes, electrical appliances, and so forth are known as electric devices.
- The more common applications: suitcases, helmets, lampshades, etc.
- Transportation: any type of boat, car, or train can be used for interior design.

2.5. Luffa Cylindrica as a Natural Fiber

A significant part of India's natural resources is derived from agriculture and forests.

FIGURE 1

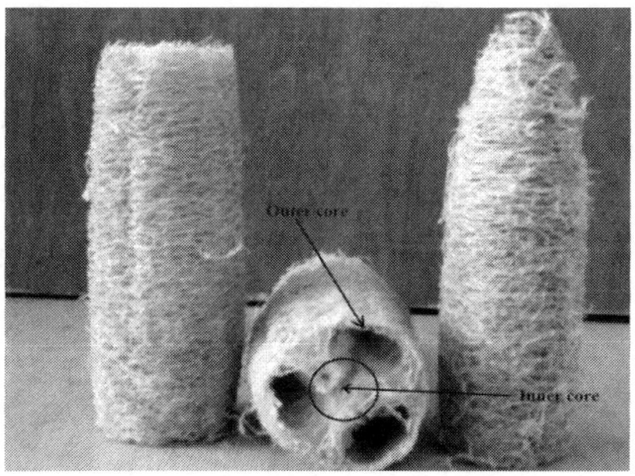

FIGURE 2

FIGURE 1 and 2 Luffa Cylindrica Plant with Fruit (1) the inner fiber core (2) and the outer core

One such natural resource whose potential for reinforcing polymer composites has not been studied too far is Luffa cylindrica, commonly known as Sponge-gourds. The cord-ligneous netting method uses the fibers to produce a mat-like structure that is free-form and multi-directional[10]. This vascular system is made up of fibers that are bonded together with plant-based natural resins. Of the 62% cellulose, 20% hemicelluloses, and 11.2% lignin, it comprises 62%. This fruit from the Sponge-gourd (L. cylindrical) plant, seen in Figure 1, has thick skin, while the sponge-gourd, which is part of the Cucurbitaceae family, has a multitude of fibers made up of a natural mat[11].

The major goal of this work is to make a PMC that is reinforced with luffa fiber and has a matrix material in the form of epoxy and which can be studied for tensile and flexural characteristics. For the production process, we have decided to use hand lay-up. After manufacturing the composites by different layers of fiber, the method of production included many fiber layer combinations, such as single, double, triple, etc.

Verma et al. assert that bagasse fiber is currently at the centre of a research-intensive era. Qualitative results and sources regarding current work on properties, processing, and application have been included in this assessment. The main aim of the research was to learn how to produce composite materials from bagasse fibers that use renewable resources. They found that using synthetic fibers combined with hybridization produced more suitably technical fabrics for interior automotive components. Bagasse, a low-cost, renewable source of sugars, has found many uses across numerous industries including packaging and the construction of electronic display stands. In order to get data for the characterisation of Luffa cylindrica, seven scientists collaborated and used Fourier transform infrared spectrophotometer, X-ray photoelectron spectroscopy (XPS), scanning electron microscopy SEM, and thermos gravimetric analysis. The hemicellulose, cellulose, and lignin content of Luffa cylindrica were examined as well. Polyester composite bearing an elongated cylinder which is reinforced with the seaweed Luffa cylindrica is subjected to water ageing in a tank with a sodium chloride concentration of 5 percent and water temperature of 500 degrees C for 170 hours. During the ageing process, a steady supply of steam is supplied by adding seawater from which sodium chloride has been added to the amount of 5 percent. And their values plummeted by 28%, 24%, 45% and 31% respectively. When water has aged, its characteristics change. Although tensile modulus and flexural modulus were not much impacted, it should be noted that According to the research conducted by Lassaad Ghali and his colleagues, flexural proprieties in luffa polyester composites are affected by fiber weight ratio, reinforcing structure, and chemical modification. The increase in the flexural strength and flexural modulus is accentuated by the action of acetylation and cyanethanolation. For flexural properties, composites with a greater fiber weight % perform better. When the fiber weight ratio rises, the amount of strain and strength it takes to absorb strain and strength increases They examined the different reinforcing structures, and then pondered about ways in which they could be put to use. According to Rokibi et al's study, alkalization has a positive effect on the flexural properties of Alfa fiber reinforced polyester matrix composites. The flexural properties, as determined by a measurement of the features of the materials, dictate how the alkaline treatment's peak results. Although Alfa 10% NaOH fiber processing was left untreated, the findings indicate that unprocessed Alfa 10% NaOH 24-hour fiber processing resulted in an increase in flexural strength and modulus of 23 MPa to 57 MPa, accompanied by a reduction in flexural strength and modulus of 1.16 to 3.04 GPa. However, notice that after being treated with a 5% NaOH solution for 48 hours, the flexural properties of composites decreased. Reduced lignin inside the cellulose fibrils keeps the cellulose together, causing this alteration. The lingo cellulosic fiber of Brazil, which was the subject of a research conducted by Satyanarayanan et. al, was investigated. As well as their current market circumstances, extraction methods, features, and applications, they spoke about the availability of cellulosic fibers from Brazil. In construction and automotive industries, non-renewable, costly petroleum-based synthetic fibers, synthetic fibers in composite materials, have proven themselves as viable rivals for cellulosic fiber.

3. MATERIALS AND METHODS

3.1. Raw materials

- Luffa fiber
- Almond Shell powder
- Epoxy resin
- Hardener

3.1.1. Luffa Fiber

One such natural resource that has the potential to serve as a fiber reinforcement in polymer composite is gourd sponge (known locally as Luffa cylindrica). As of yet, however, this resource's potential as a fiber reinforcement in polymer composite has not been explored. This kind of netting system utilises ligneous fibers, which are woven in a nested, multi-directional pattern. The resulting mat resembles grass. The vascular system is formed by long chains of plant-based polymers that are knitted together[12]. Luffaar hemicelluloses are mostly made up of cellulose and lignin. Hallow cellulose, a kind of hemicellulose, and hemicellulose, another form of hemicellulose, comprise about 82% of the total chemical components found in luffa. The chemical called lignin may also be found in luffa. Lignin, which is an energy storage system, also holds cellulose fibers together.

3.1.2. *Almond Shell Powder*

Natural goods made from agricultural waste. One of the agriculture wastages is non-used almonds shells, which are commonly referred to as almond shells. In this instance, we are focusing on these fibers and particles, and we are concentrating hemispheres to make a product.

3.1.3. *Epoxy Resin*

Aldrich's epoxy resin, Araldite LY 556, has these extraordinary properties.

3.2. Some Important Characteristics of Epoxy resin are

- The flexibility of the matrix results in a high-quality material.
- A good degree of handling ease.
- Tendency to shrink.
- Exceptional adhesion.
- resistive to flame
- Mechanical resistance is good.
- Toughness and overall mechanical characteristics
- high-mold-resin moulding (cold moulding rarely).
- Highest levels of smoke output.
- 120-175°C (250-350°F) for curing.

3.3. Properties of Epoxy Resin

Sl. No.	DESCRIPTION	SPECIFICATION
1	Company Code	LY556
2	Color	Transparent, Sticky liquid
3	Specific Gravity	1.14–1.19at25°C
4	Viscosity	9500-12500cgd@25°C
5	Density	1.15-1.2gm/cc
6	Compressive Strength	190N/mm

TABLE 1 Properties of Epoxy Resin

3.4. Hardener

In present work hardener (araldite) HY951 is used. This has a viscosity of 10-20 poise at 25°C.

3.5. Properties of Hardener

TABLE 2 Properties of Hardener

SLNO	DESCRIPTION	SPECIFICATION
1	Company code	HY951
2	Color	Transparent liquid
3	Specific gravity	0.95–1.05@25°C
4	Viscosity	2050cgs@25°C

4. EXPERIMENTAL WORK

4.1. Sample Preparation

The composite fiber is prepared by hand lay-up technique:

The practise of wet layups is one of the oldest and most frequently used methods for the manufacture of composite components. In the process of production, continuous fiber-reinforced composites are used at each stage. Woven or unidirectional composites are layered to produce a composite that may show features in many directions[13]. Each layer is oriented to maximise the features' potential. The laminated composite material may be made to maximise overall performance by layering multiple fibers in various orientations. By hand, we manually apply resins to woven, knitted, stitched, or bonded fabrics. When using a roller/roller combo with nip-type impregnators, we utilise rollers or rollers with nip-type impregnators to press resin into the textiles. Woven or laminated products may be left at room temperature to cure.

This research uses locally sourced luffa fiber and almond shell powder. It is cleaned and sized before it is cut to the correct size. Wooden moulds were used to create composite moulds that had dimensions of 200x200x5 millimetres. A specified volume % of the fibers was produced using a determined quantity of epoxy resin and hardener (ratio of 10:1 by weight). Figure 3.1 depicts the mould used to create the composite. Mould release may be used to make the removal of composite materials quick and easy. Simply apply a mould release sheet to the glass plate, followed by spraying the release within the mould. The liquid mixture was then thinly spread into the mould. The reaction began as the fiber lamina was placed into the liquid. This is true, in fact: during the fiber laminate process, resin was really added, and the operation was repeated to provide the proper thickness. The whole mix was poured into the mould once the process was complete. To prevent air bubbles from forming, conservatism was used. To pressurise the top of the mould, a solution of epoxy and hardeners was placed in the mould, and then the mould was allowed to cure at room temperature for 72 hours. To accommodate for this loss, much attention was taken throughout the manufacturing process of composite sheets. At the conclusion of the 72-hour period, the mould was removed from the samples.

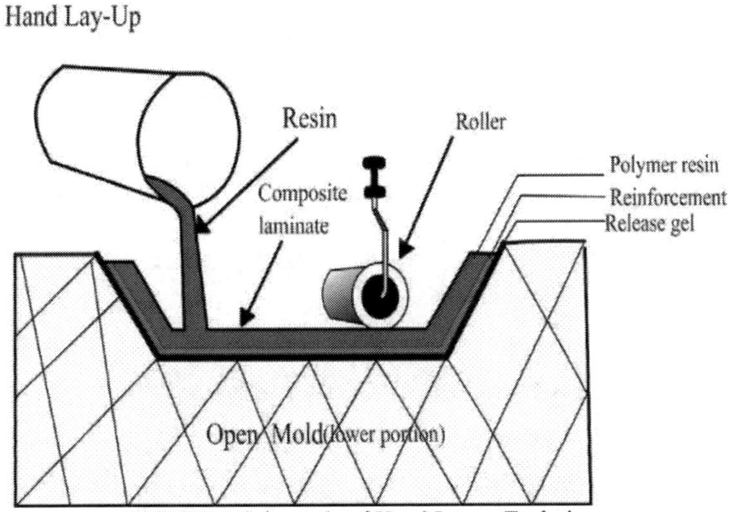

FIGURE 3 Schematic of Hand Layup Technique

4.2. Experimental Procedure

4.2.1. Measurement of Density

Using a pycnometer, the density of the sample is determined as follows:

ρ=(w1-w2/w3-w4) x Density of kerosene

In this example, the weights of the empty dry and clean pycnometer, the pycnometer with the sample, the pycnometer with the kerosene, and the pycnometer with the sample and kerosene are respectively W1, W2, W4, and W3.

The density of 0.540 g/cc was calculated using the formula.

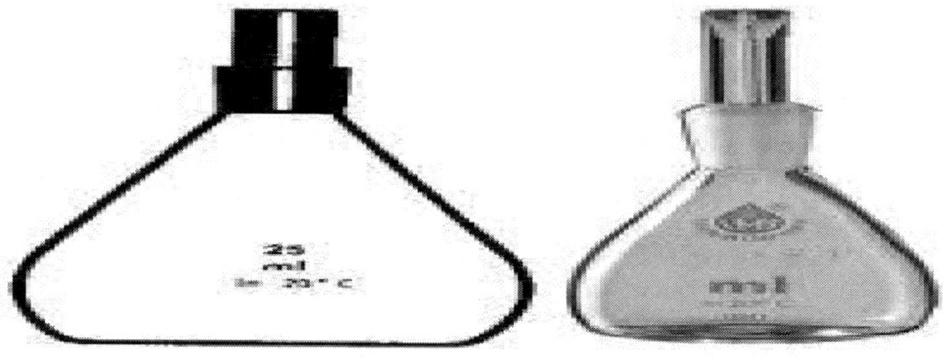

FIGURE 4 Pycnometer

Figure 4 shows the pycnometer used for density measurement of sample.

4.2.2. Cutting of Laminates into Samples of Desired Dimensions

For several tests, the Wire Hacksaw blade was utilised to cut each laminate into smaller pieces.

- Tensile test was performed on a dog bone-shaped sample (200mm x 25mm x 5mm).
- ASTM Standards specified that the sample be cut into a flat shape (150x15x5 mm) and FLEXURAL TEST applied.
- COMPRESSIVE TEST-A slab cut to the desired flat bar (150x25x5mm) was used
- IMPACT TEST-Sample was cut into flat bar V-not shape (125x15x5) mm.

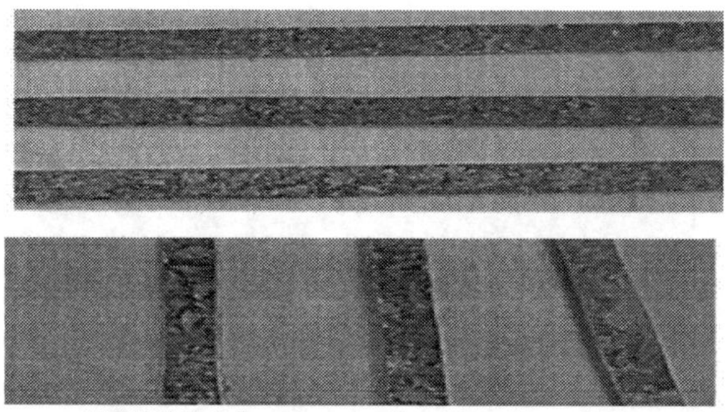

FIGURE 5 Specimen – For Tensile and Flexural Test

4.3. *Mechanical Testing of Sample*

4.3.1. Tensile Test

Flat specimens are typically tested using the tensile test. Dog-bone and the straight side type with tabs are the most often utilized specimen geometries. This case features the specimen in figure 6. The tensile tests were carried out using a computerized Universal Testing Machine INSTRONH10KS according to ASTMD 3039-76 standard. The specimen was 42mm long. For the experiments, the strain rate was kept constant at 2mm/min.

FIGURE 6 UTM Machine Sample Unloaded Condition for Tensile Testing

4.3.2. Flexural Test

To determine the flexural strength of the composites, a UTM machine was used and ASTMD 790-03 guidelines were adhered to. In figure 7, you can see the loading arrangement for the specimen and the image of the machine being utilized[14]. All the composite specimens had a form that was rectangular, with dimensions ranging from 100-125mm wide, 100-110mm long, and 4-8mm thick. In order to keep a crosshead speed of 0.5mm per minute, a 70mm span was utilised.

This equation established the flexural strength of composites.

FS=3FL/2bd2

When t is equal to the specimen thickness, l is the specimen gauge length, b is the specimen breadth, and τ is the flexural strength.

FIGURE 7 UTM Machine Sample Unloaded Condition for Flexural Testing

5. RESULTS AND DISCUSSION

5.1. Tensile Test

Tensile testing was conducted using an ASTMD 3039-76 UTM machine in compliance with the specification stated in ASTMD 3039-76. All the specimens were shaped like a dog bone, which was 140 mm long, 10 mm wide, and 5 mm deep. The findings are reported in the Table 3 for single, double, and triple layer samples.

TABLE 3 Tensile Test

Samp le	Volume fraction of fiber(%)	Tensile strength (MPa)
S_1	10	245
S_2	20	174
S_3	30	137

A clean epoxy sample is known as S1. A double layer luffa sample is called S2, and a triple layer luffa sample is known as S3.

The results of the tensile strength analysis revealed that S1 had the maximum tensile strength, which means it had 245MPa tensile strength.

Strength properties of composites are primarily affected by fiber content and fiber strength. Table 3 shows that single, double, and triple layer samples exhibit tensile strength variation, and Figure 8 shows the variance in tensile strength for single, double, and triple layer samples. The tensile strength of a composite increases with total tensile strength.

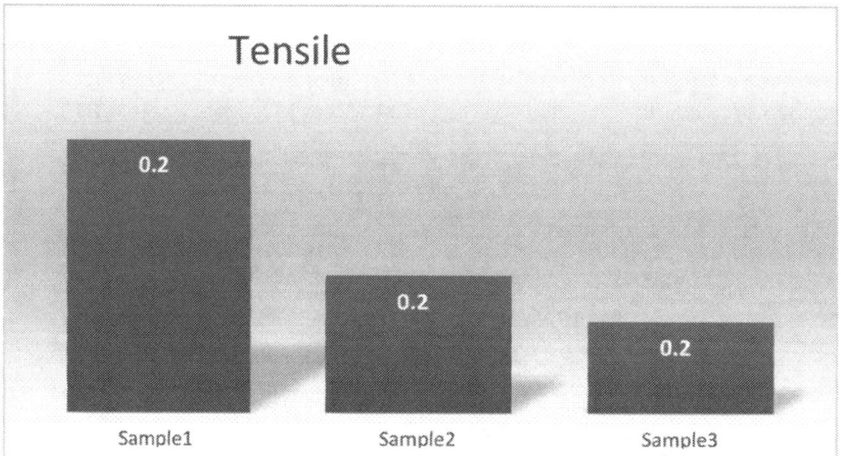

FIGURE 8 Variation of Tensile Strength with Different Layers of Samples

5.2. Compressive Test

The ASTMD 3039-76 standard-compliant compressive test was conducted using a UTM machine. The entire set consisted of canine-shaped samples of the aforementioned dimensions (150mm × 25mm x 5mm). The results are reported in the Table 4, which classifies single, double, and triple layer sample

TABLE 4 Compressive Test

Sample	Volume fraction of Fiber (%)	Compressive Strength (Mpa)
S1	10	1 7 3
S2	20	1 8 1
S3	30	1 9 3

For sample S1, the label says that it is a neat epoxy sample that contains 10% of the volume with fiber; however, for sample S2, the label says that it is a double-layer luffa and groundnut reinforced sample with 20% of the volume with fiber content; for S3, the label states that it is a triple-layer luffa and groundnut reinforced sample with 30% of the volume with fiber content. To solve this problem, it was discovered that S3 had the maximum compressive strength of all the samples, which was equal to 193 MPa.

Fiber content and fiber strength are commonly considered to be the key factors that determine the strength characteristics of the composite. Therefore, the compressive test results shown in Table 4 and in Figure 9 indicate a wide range of variance for the single, double, and triple layer samples. Increasing compressive strength is observed with single, double layer composite materials.

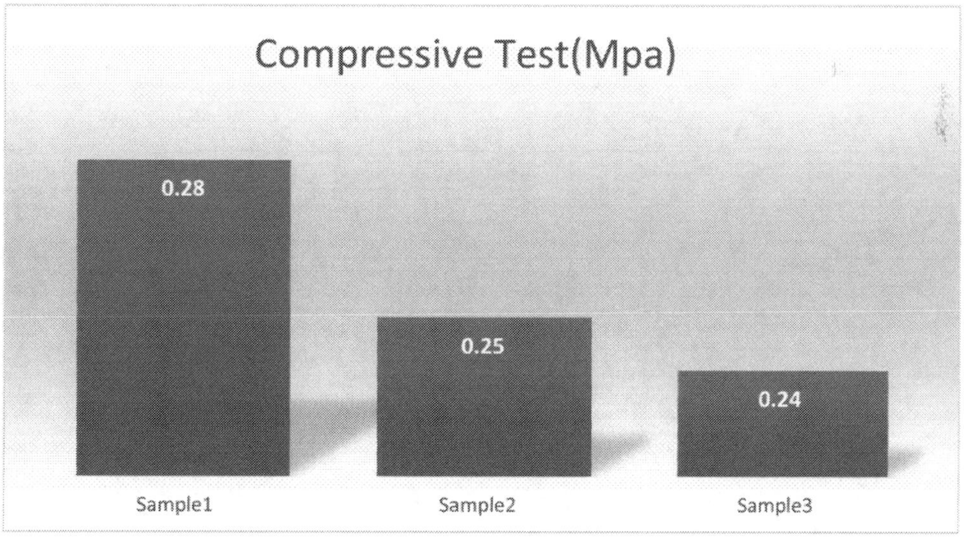

FIGURE 9 Variation of Compressive Strength with Different Layers of Samples

5.3. Flexural Test

In line with ASTM standard, a three-point bend test was conducted using a UTM machine to determine the flexural strength of the composites. All of the specimens (composites) were rectangular in shape and were 150x15x5 mm in length, width, and height. The Span length was 55 millimetres. All three samples of luffa fiber combinations were used in the experiment. Table 5 details the findings.

TABLE 5 Flexural Test

Sample	Volume fraction of fiber (%)	Flexural strength (MPa)
S_1	10	244
S_2	20	440
S_3	30	383

Where S1 denotes a clean epoxy sample with a total of 10% volume fraction of fiber, S2 represents a sample that consists of two layers of luffa, along with ground nuts that are used as reinforcements, and S3 represents a third layer of luffa, as well as ground nuts, that are used as reinforcements.
It was observed that S2 has the maximum flexural strength, that is, it has a flexural strength of 440 MPa.
Fiber content and fiber strength are commonly considered to be the key factors that determine the strength characteristics of the composite. Thus, as seen in Table 5 and Figure 10, the tensile strength variations within the single, double, and triple layer samples are represented. For single, double-layer composite composites, there is a slow but steady rise in flexural strength.

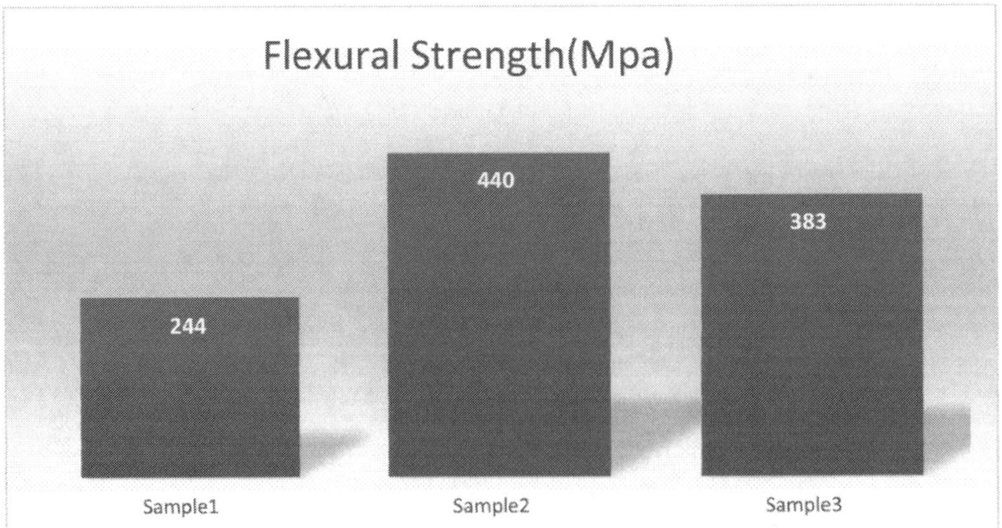

FIGURE 10 Variation of Flexural Strength with Different Layers of Samples

5.4. Impact Test
The Charpy impact test, known as the Charpy v-notch test, is a standardised high strain-rate test that evaluates the amount of energy absorbed by a material during fracture. When material is exposed to applied energy, it stores that energy. This energy, which is a measure

of material toughness, is a tool for studying temperature-dependent ductile-brittle transition. Industry use has expanded as a result, as it is straightforward to produce and run, which produces outcomes that are inexpensive and quick to get.

TABLE 6 Impact Test

Sample	Volume fraction of fiber(%)	Impact strength (Joules)
S_1	10	0.28
S_2	20	0.25
S_3	30	0.24

An epoxy sample, which has a volume fraction of fiber content of 10 percent, is known as S1; a Luffa and ground nut reinforced sample, which has a volume fraction of fiber content of 20 percent, is known as S2; and a third sample, which has a volume fraction of fiber content of 30 percent, is known as S3.
So, after determining the impact strength of all samples, it was determined that S1 has the maximum impact strength. This means it has a value of 0.28 joules. Table 6 and Figure 11 therefore demonstrate the tensile strength difference across single, double, and triple layer samples. The tensile strength of a single, double layer composite increases with time.

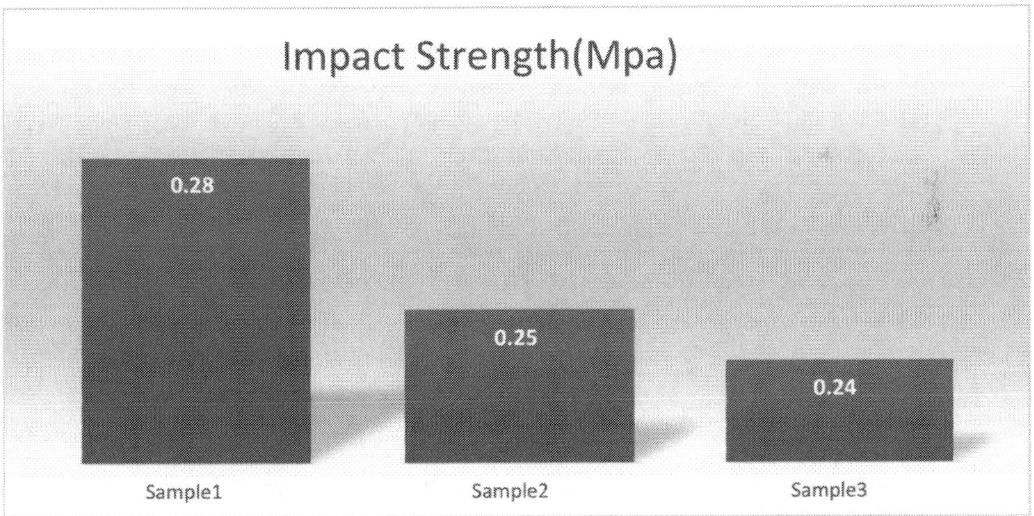

FIGURE 11 Variation of Impact Strength with Different Layers of Samples

6. CONCLUSION

The article presents a method of making an epoxy-bonded laminate composed of luffa fiber and almond shell reinforcement. Following are the general conclusions drawn from the composite lead's mechanical behaviour:

- To design and fabricate novel class of epoxy-based composites reinforced with luffa cylindrica and almond shell natural fibers, a series of successful fabrications have been done.

- From this investigation, it has been noted that the tensile strength is at its peak in layer S1 sample i.e., 245MPa, which is more than the final applied amount of plain epoxy. However, the tensile strength of triple-layer composite has dropped considerably (S3). Poor fiber-matrix adhesion may be behind this reduction in strength.
- Although flexural strength is found to be higher than plain epoxy for all layers of luffa and Almond shell reinforced composites, flexural strength is not seen in luffa or Almond shell reinforced composites. S2 is the maximum all layer (S2) sample which, for plain epoxy, is equal to 440MPa.
- For all layers of luffa and Almond shell reinforced composites, compressive strength was found to be higher than plain epoxy. For layer sample (S3), i.e., 193MPa is determined to be the maximum, more than plain epoxy.
- For all layers of the luffa and almond shell reinforced composites, impact strength was found to be greater than plain epoxy. It is shown to be the greatest among all S2 sample measurements, which is 0.28 Joules. In the building and construction business, composites might be used in such things as partition boards, panels for partition and fake ceiling, and flooring.

REFERENCES

[1] Verma,D., Gope, P.C., Maheshwari,M.K., Sharma,R.K., "Bagasse Fiber Composites-A Review", J. Mater. Environ.Sci.3(6), 2012, PP.1079-1092.

[2] Yoldas Seki, Kutlay Sever, Seckin Erden, Mehmet Sarikanat, Gokdeniz Neser, Cicek Ozes, "Characterization of Luffa cylindrica Fibers and the Effect of Water Aging on the Mechanical Properties of Its Composite with Polyester", 2011, DOI 10.1002/app.34744.

[3] Lassaad Ghali, Slah Msahli, Mondher Zidi, Faouzi Sakli, "Effects of Fiber Weight Ratio, Structure and Fiber Modification on to Flexural Properties of Luffa- Polyester Composites", Advances in Materials Physics and Chemistry,1, 2011, PP.78-85.

[4] Mansour Rokbia, Hocine Osmania, Abdellatiflmadc, Noureddine Benseddiqd, "Effect of Chemical treatment on Flexure Properties of Natural Fiber-reinforced Polyester Composite", ProcediaEngineering10, 2011, PP.2092–2097.

[5] K.G.Satyanarayana, J.L.Guimara, F.Wypych, "Studies on lingo cellulosic fibers of Brazil. Part I: Source, production, morphology, properties and applications, Composites", Part A38, 2007, PP.1694–1709.

[6] Andrzej, Bledzki AndrisChate., "Natural fiber-reinforced polyurethane micro foams", 61, 2009, PP. 2405–2411.

[7] Oboh I. O. Aluyor E. O., "Luffa cylindrica an emerging cash crop", African Journal of Agricultural Research Vol. 4(8), August 2009, PP.684-688.

[8] Ghali L, Msahli S, Zidi M, Sakli F.,2009, "Effect of pre-treatment of Luffa fibers on structural properties", Materials Letters 63, 2009, PP.61–63.

[9] Demir H., Atikler U., Balkose D., Tihminhoglu F. "The effect of fiber surface treatment on the tensile and water sorption properties of polypropylene-luffa fiber composites", Composites: Part A 37, 2006, PP.447–456.

[10] Valcineide O.A., Tanobe, ThaisH.D., Sydenstricker, MarildaMunaro, sandroc. Amico, "A comprehensive characterization of chemically treated Brazilian sponge-gourds (Luffa cylindrica)", Polymer Testing 24, 2005, PP.474–482.

[11] Herrera-Franco, P.J., Valadez-Gonza lez, A., 2004, "Mechanical properties of continuous natural fiber-reinforced polymer composites", Composites: Part A35, 2004, PP.339–345.

[12] Mukherjee P.S., & Satyanarayana K.G.,1986, "Structure and properties of some Vegetable fibers-II. Pineapple leaf fiber," J. Material Science21, pp. 51–56.

[13] Lassaad Ghali, Mourad Aloui, Mondher Zidi, Hachmi Bendaly and Faouzi Saki, "Effect of chemical modification of luffa cylindrica fibers on the mechanical and hydrothermal behaviours of polyester composites", bioresource.com.

[14] S. Suhas , M.J. Silvister Raju , D.S. Vijayan , "Natural fibre reinforcement experimental study in polymer composite", Materials Today: Proceedings, , https://doi.org/10.1016/j.matpr.2020.12.051

MACHINE LEARNING MODELS FOR PREDICTION OF CARDIOVASCULAR DISEASES

[1]K.Sivaraman, [2]V.Khanna

[1]Research Scholar, [2]Professor, 1Department of Computer Science and Engineering, 2Department of Information Technology, Bharath Institute of Higher Education and Research Chennai-73, Tamil Nadu, India.

Corresponding Author Mail id: sivaraman2006@gmail.com

Abstract: Support Vector Machines (SVM) [9], Ada Boost (AB) [10], and Gradient Boosting (GB). Maximum Relevance, Minimum Redundancy (mRMR), Relief, and Least Absolute Shrinkage and Selection Operator (LASSO) are examples of fast correlation-based filters. The authors tested all attributes as well as the selected attributes generated by the above feature selection methods on the Cleveland heart disease dataset (CHDD) and the Hungarian heart disease dataset (HHDD). For their suggested framework, the authors were able to attain the greatest feasible model accuracy of 92.09 percent (all features) and 94.41 percent (selected features). The early detection of cardiac disease was demonstrated by several additional researchers.

Keywords : Machine Learning World Health Organization, cardiovascular disease, Support Vector Machines

1. Introduction

The term "cardiovascular disease" refers to a group of heart and blood vessel illnesses (CVD). Every year, more people die from CVDs than from any other cause in the world. According to the World Health Organization (WHO), 17.9 million people die each year from cardiovascular disease (CVD). It is estimated that it accounts for 31% of all deaths globally. In a CVD fact sheet, the WHO reported on another statistical analysis. By 2030, about 23.6 million people will die from CVDs, primarily heart disease and stroke, which will be the leading cause of mortality [1]. The greatest and simplest method to lower your risk of heart disease is to adjust your lifestyle behaviors, such as eating healthy foods and engaging in regular physical activity. Several tests are used by doctors to diagnose heart disease and stress tests during exercise [2]. These tests, on the other hand, are prone to side effects, take a long time to complete, and are costly. As a result, the requirement for early diagnosis of cardiac disease is unavoidable in order to protect patients at high risk from dying prematurely and to aid physicians in their prognosis.

Machine Learning (ML) is a rapidly expanding interdisciplinary topic that has recently attracted the attention of academics. The number of machine learning applications in the health-care industry is continually growing. In recent years, researchers and academicians have published various approaches for detecting cardiac problems. [3] proposed an intelligent framework for early detection of heart diseases, in which authors combined feature selection methods with Decision Tress (DT) [4], Random Forest (RF) [5, Logistic Regression (LR) [6, Nave Bayes (NB) [7], Artificial Neural Network (ANN) [8, Support Vector Machines (SVM)

[9], Ada Boost (AB) [10], and Gradient Boosting (GB) [11]. Maximum Relevance, Minimum Redundancy (mRMR) [13], Relief [14], and Least Absolute Shrinkage And Selection Operator (LASSO) [15] are examples of fast correlation-based filters. The authors tested all attributes as well as the selected attributes generated by the above feature selection methods on the Cleveland heart disease dataset (CHDD) and the Hungarian heart disease dataset (HHDD). For their suggested framework, the authors were able to attain the greatest feasible model accuracy of 92.09 percent (all features) and 94.41 percent (selected features). The early detection of cardiac disease was demonstrated by several additional researchers. Table 1 summarizes and reviews these state-of-the-art frameworks.

The state-of-the-art frameworks for cardiac disease detection are listed in Table 1. Table 1 shows the following observations.

- Several important datasets were used in the research, including CHDD, HHDD, Single Proton Emission Computed Tomography (SPECTF), and StatLog. Researchers like CHDD because it has more instances than any other publicly available database.
- In the identification of heart disease, feature engineering (FE) techniques take a back seat to data and machine learning (ML) models.
- The taxonomy of feature engineering (FE) in the listed state-of-the-art frameworks includes dealing with missing values, pervasive domain knowledge, introducing / removing dummy features, dealing with categorical features, creation of interactive and new features from existing raw features, and removal of unused / unwanted features.
- Several researchers experimented with different combinations of FE approaches to increase the performance of ML models.
- Current state-of-the-art results show that combining (assembling) ML models with appropriate FE approaches can outperform independent ML models.
- Various performance measures are used to evaluate the performance of the machine learning models nand, as a result, the overall system.
- Accuracy, F1 score, recall, sensitivity, specificity, receiver operating characteristics curve (ROC) - area under the curve (AUC), true positive rate (TPR), and false positive rate (FPR) are the most commonly used metrics (FPR).
- In any of the aforementioned state-of-the-art frameworks, the Diagnostics Odds Ratio (DOR) and False Omission Rate (FOR) are not used to evaluate the performance of the models
- Despite the fact that various frameworks for early diagnosis of cardiac disease have been developed, improvements in the model's robustness in terms of performance are still needed.

Table 1: state of the art frameworks

Ref & Year	Dataset used for analysis	FE	ML Models used for comparative study	Performance metrics considered for analysis	Best Model (concluded)
[19] 2015	CHDD	Correlation, PCA	MLP, NB, J48, RF,SVM, AB, Boosted tree, and binary discriminant.	Classification accuracy, Specificity ,Recall Precision ,ROC, F-measure MAE,RMSE	Ensemble-NB, AdaBoost, boosted tree
[21] 2016	SPECT heart disease database	partitioning module, inner classifiers module and fuser module	AB, LR Boost, MLP, RF	sensitivity, specificity and classification accuracy	Hybrid ensemble model
[23] 2016	CHDD	CFS, PSO	MLP, MLG, FURI, DT (C4.5)	Accuracy	Hybrid method (CFS + PSO + k means Clustering + MLP)
[16] 2016	StatLog heart disease dataset	correlation	ANN SVM, kNN, CT, NB	Classification accuracy sensitivity, specificity precision, negative predictive value (NPV), false positive rate (FPR), rate of misclassification (RMC), F1 measure	LR
[20] 2017	Heart disease dataset, SPECTF dataset	BPSORS-AR	NB, SVM, ANN	Accuracy, Sensitivity and Specificity.	CFARS-AR
[19] 2018	CHDD	Relief, mRMR, LASSO	LR , kNN, ANN, SVM (RBF kernel), SVM (linear kernel), NB, DT, RF	Specificity, Sensitivity, MCC, AUC	LR SVM (RBF)
[22] 2018	CHDD	Relief, Correlation Filter , Info gain Consistency, Chi-squared, Gain ratio, PSO	RF, NB, MLP,SVM	Accuracy	PSO with SVM
[18] 2018	CHDD , HHDD, LBMC, and SUH.			Accuracy, misclassification error, sensitivity (or recall), precision, specificity, F-score, ROC curve, AUC, and the K-S measure.	adaptive Boosting algorithm
[24] 2019	Hospital in Beijing, China	XGBoost, one hot encoding	RF, GBDT and XGBoost	AUC INE	XGBSVM hybrid model (XGBoost and SVM)
[25] 2019	CHDD	L1 regularized SVM, L2 regularized SVM- linear and RBF	AB, RF,Extra tree	Accuracy, sensitivity, specificity, MCC, ROC charts, AUC	HGSA- stacks two support vector machine (SVM) models
[26] 2019	NSR-CAD, NSR-CHF and CAD-CHF	Nonlinear features and dimensional Reduction method as GDA.	GDA Method Having Kernel Function Like Gaussian, RBF and Polynomial + OS-ELM Having Sigmoid, Hardlim, RBF and Sine Activation	Mean ±SD, p Value	OS-ELM
[27] 2019	CHDD	DT entropy	NB, GLM, LR, DL, DT, RF, GBT and SVM	accuracy, classification error, precision, F-measure, sensitivity and specificity	Hybrid HRFLM RF and Linear Method (LM)
[28] 2019	CHDD, HHDD Switzerland SPECTF	MFSFSA, FFSA, RFSA and FSSA	A binary class RBF kernel-based SVM	accuracy, specificity and sensitivity	FSSA

[29] 2019	CHDD	random search algorithm (RSA)	Adaboost model, extra tree ensemble model, randomized tree model, SVM with linear kernel function and support vector machine with RBF kernel function	Accuracy, sensitivity, specificity and MCC	RSA based RF
[32] 2020	CHDD	Relief, Minimal redundancy maximal relevance, Least absolute shrinkage selection operator and Local learning	SVM, LR, ANN, kNN , NB, and DT	accuracy, specificity, sensitivity, MCC and processing time	feature selection algorithm - FCMIM-SVM
[33]] 2020	CHDD	Pearson correlation and chi-square features selection	RF,AB, XG Boost GB	Correctly Classified Instances, Incorrectly Classified Instances, Kappa statistic, Mean absolute error, Root mean squared error Relative absolute error Root relative squared error	XG Boost ensemble method
[30] 2020	CHDD	FAMD, PCA, MCA	LR, kNN, SVM, DT, RF	Accuracy Specificity, Sensitivity, Score, MCC, AUC	MIFH - Machine intelligence framework for heart disease diagnosis (FAMD +RF)
[31] 2021	Statlog and CHDD	DBSCAN SMOTE-ENN	NB, LR, MLP, SVM, DT and RF	Accuracy, precision recall/sensitivity/true positive rate (rec/sec/TPR), f-measure MCC, false positive rate (FPR), false negative rate (FNR), and true negative rate (TNR)	heart disease prediction model (HDPM) -XGBoost

BPSORS-AR-Binary particle swarm optimization and rough sets based attribute reduction, CFARS-AR- Chaos firefly algorithm and rough sets based attribute reduction, CFS-Correlation based Feature Selection, CT-Classification Tree, kNN - k-nearest neighbour, MLP – Multi Layer Perceptron, MLR- Multinomial logistic regression, FURIA - Fuzzy unordered rule induction algorithm, MCC- Matthews' Correlation Coefficient, AUC - Area Under the Curve, PSO – Particle Swarm Optimization, , mRMR- minimal redundancy-maximal-relevance, LASSO-Shrinkage and Selection Operator, FAMD-Factor analysis of mixed data, PCA-Principal components analysis, MCA-Multiple correspondence analysis , INE –Improved normalized entropy, GBDT - Gradient Boosting Decision Tree, HGSA-hybrid grid search algorithm, GLM- Generalized Linear Model, GBT- Gradient Boost Tree MFSFSA - Mean Fisher score-based feature selection algorithm, FFSA- Forward feature selection algorithm, RFSA- Reverse feature selection algorithm, FSSA-Feature subset selection algorithm, DBSCAN- Density-Based Spatial Clustering of Applications with Noise, SMOTE-ENN- Synthetic Minority Over-sampling Technique-Edited Nearest Neighbor,

In this work, we propose a new framework for early detection of heart disease from CHDD

2. Understanding Dataset

Throughout our research, we used the Cleveland database (information about the creators can be found in Appendix-1). It can be found in the UCI machine learning repository. Cleveland database is a multivariate dataset containing 303 instances and 76 attributes. All of the published [Refs] experiments, on the other hand, use 14 of the 76 attributes as the subset of the Cleveland databaseTable 2 contains descriptions of each attribute, the category of data contained in the attribute, the range of values, and the statistical summary. Table 2 shows that of the 13 CHDD attributes, A1, A4, A5, A8, and A10 are continuous data attributes, whereas the remainder are categorical data attributes (A2, A3, A6, A7, A9, A11, A12, and A13).

Table 2: The synopsis of CHDD

Attributes	Description	Attributes data category / Attribute's values and its interpretation	Mean ± Std.dev
sex (A2)	Gender	Categorical (Number of unique values =2) • 1 = male • 0 = female	
age (A1)	Age in years	Continuous	54.410596±9.040163
trestbps (A4)	Blood pressure level (mm/Hg) (at admission time in the hospital) at rest	Continuous	131.645695±17.612202
cp (A3)	Types of chest pain	Categorical (Number of unique values =4) • 1 = typical angina • 2 = atypical angina • 3 = non-anginal pain • 4 = asymptomatic	
fbs (A6)	On fasting, the level of blood sugar level (mg/dl)	Categorical (Number of unique values =2) Blood sugar on fasting > 120mg/dl • 1=true • 0=false	
chol (A5)	serum cholesterols (mg/dl)	Continuous	246.738411±51.856829
thalach (A8)	Exercise Max Heart Rate Achieved	Continuous	149.605960±22.912959
restecg (A7)	Resting Electrocardiographic	Categorical • 0 = normal • 1 = ST-T wave abnormal • 2 = left ventricular hypertrophy Estes)	
oldpeak (A10)	ST depression induced by Exercise relative to Rest	Continuous	1.035430±1.160723
exang (A9)	Exercise Induced Angina	Categorical (Number of unique values =2) • 1=yes • 0=no	
thal (A13)	Thalassemia	Categorical (Number of unique values =3) • 3= normal • 6= fixed defect • 7= reversable defect)	
ca (A12)	Major Vessels colored by Flouroscopy	Categorical (Number of unique values =4) • 0= fluoroscopy color 0 • 1= fluoroscopy color 1 • 2= fluoroscopy color 2 • 3= fluoroscopy color 3	
slope (A11)	Slope of Peak Exercise ST Segment	Categorical (Number of unique values =3) • 1=upsloping • 2=flat • 3=downsloping	
num (Outcome)	The diagnosis for heart disease	Categorical (Number of unique values =2) • 0 = absence of heart disease (class 0) • x > 0 = presence of heart disease (class 1)	

3. Problem Formulation

The attribute "num" (outcome) denotes whether the patient has cardiac disease or not. It's an integer attribute with values ranging from 0 (no heart illness (class 0) through 1, 2, 3, and 4. (presence of heart disease in different states). The presence of heart disease in various states is addressed in this study as a single group dubbed "presence of heart disease" (class 1). As a result, the problem of heart disease is classified as a binary classification problem. The dataset under examination is fully dependent on early identification of cardiac problems using machine learning techniques. The nature of the attributes in the dataset are not separable based on the classes, regardless of the datasets (used in the study of heart disease detection). Figure 1 depicts the population distribution of all CHDD traits in terms of classes. It is obvious from Figure 1 that all CHDD properties are not linearly separable. The difficulty in distinguishing between the presence of heart disease (class 0) and the absence of heart disease (class 1) is clearly demonstrated by the class wise data distribution of all CHDD features (figure 1). (class 0). Because of the nature of distribution data, detecting heart disease is a difficult task. We'll concentrate on supervised learning models as a data set with attributes and their corresponding classes in this research.

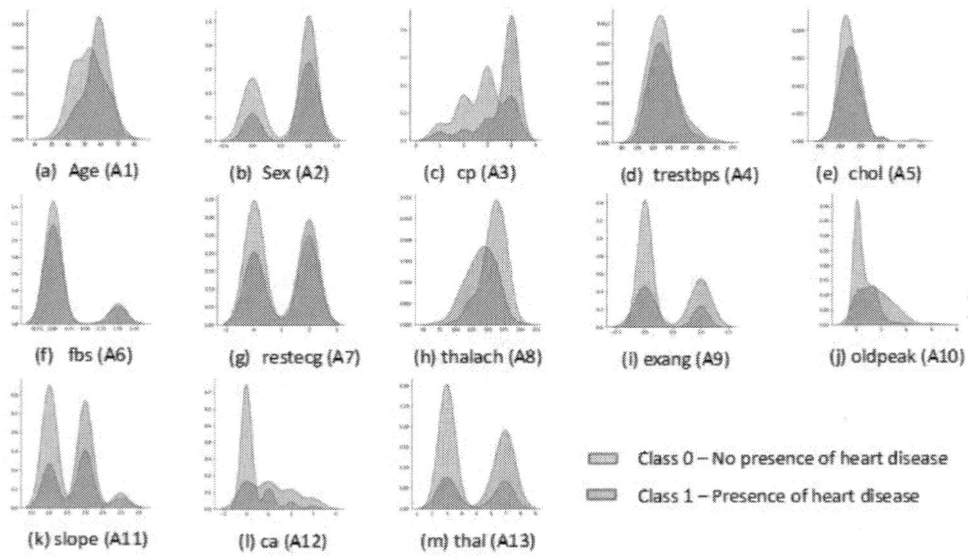

Figure1: Class wise Data distribution of all features of CHDD

The CHDD of observations D = {(X1, y1), (X2, y2)…, (Xn, yn)} where Xn is an attribute vector and yn is a class label (0 or 1), best subset (H) of D is arrived using different FE techniques and develop a binary classification model to learn from H to infer the value of yi given xi. FE includes dealing of outliers and missing values, data standardization, attributes selection and cross validation. LR, kNN, DT, NB, SVM, AB, XB and ensembling of different combination of the models are considered for performing evaluation of binary classification on the given dataset. In addition with sensitivity, specificity, AUC-ROC, F1 score and FOR, the independent prevalence metric DOR is also used to evaluate the performance of the models. The model providing more the number of best values of chosen metrics is considered as the best performing model.

4. Proposed Framework

The proposed framework is illustrated in Figure 2. Imputation of missing values and outlier removal processes are the integral part of the proposed framework. These two processes are inevitable steps to provide the quality data for the learning of ML model.

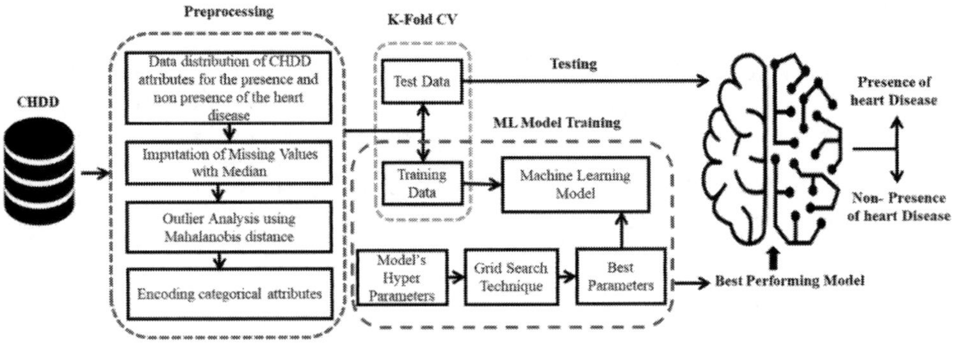

Figure 2: Proposed Framework

4.1. Dealing of missing values (IMV)

Missing values in a data set can be handled in one of two ways: they can be ignored or they can be accounted for. Dropping missing values results in a loss of information and precision for the data collected, as well as a reduction in the number of data available for training the model. Missing data is usually not fixed; rather, it is dealt with in order to deal with it. Impu does the missing data accounting. The task is to consider how to handle missing data using the imputation procedure that is provided for accounting missing data. When data is applied to the model, imputation by mean or median value progresses the model's performance metrics. The following is the formula for calculating the median for an odd number of observations of a feature in the data set (303 for CHDD).

$$\text{Median} = \frac{n+1^{th}}{2} \text{ term}$$

Where n is the number of observations While analyzing the unique values of all the features in the CHDD under consideration, it was discovered that missing values ('?') exist in ca (A12) and thal (A12) (A13). Four values with '?' in A12, the feature that describes the Major Vessels colored by Fluoroscopy, are found in the index [165, 191, 286, 301]. Only one data has a result of '1' among the four values (heart disease). A12 is the categorical value (0 represents fluoroscopy color 0, 1 represents fluoroscopy color 1, 2 represents fluoroscopy color 2, and 3 represents fluoroscopy color 3). Two values with '?' in A13, the feature that describes Thalassemia, are discovered in the index [86, 265]. Only one data has a result of '1' among the two values (heart disease). The category value for A13 is 3 (normal), 6 (fixed defect), and 7 (reversible defect). Because both are categorical data, the missing values are imputed with median values.

4.2. Dealing of outliers

Outliers are measurements that are markedly different from the rest of the data. These outliers are highly sensitive to data distribution and have a direct impact on the learning process of machine learning classifiers. Figure 1 depicts the positive or negative skewness of the characteristics A1, A4, A5, A8, and A10 (asymmetry). These are continuous data qualities, which is intriguing. As a result, the continuous data variables in this study are subjected to outlier analysis. A thorough understanding of data is critical in health care domain data analysis. Data are rarely treated as outliers in health care measurement in this context, despite the fact that they are statistically defined as such. Outlier data, on the other hand, has a direct impact on the performance of machine learning classifiers. Blood pressure levels are classified as normal (120/80 mm/Hg), high (120-129 /80 mm/Hg), Stage 1 (130-139 / 80-89), Stage 2 (140/90 mm/Hg), and hypertensive crisis (> 180/>120 mm/Hg) by the World Health Organization (WHO). Human blood pressure should be tested if it is greater than 180 mm Hg, according to the standards (hepertensive). The attribute A4 in particular has a value of

200 mm/Hg at the 125th incidence of CHDD, which is statistically regarded an anomaly. Imputing this with any of the statistical parameters of the appropriate property will be meaningless in this case. Also, retaining this value in the dataset will affect the performance of the ML classifier. To overcome the aforementioned scenario, the outlier rejection (dropping) process is preferred in the proposed framework rather imputation of outliers with any of the statistical parameters. The process of dealing outliers used in the proposed framework algorithmically framed using Mahalanobis Distance (MD) metric [34] as given below.

Algorithm: Dealing of Outliers

Input: The data [Ai] MxN , where M is the number of rows containing outlier data and N is the number of columns describing the multivariate features

Output: The data [Ai] PxN where P is the number of rows of outlier data and N is the number of columns describing the multivariate features

Load data X and compute its mean $\bar{A}_j = \frac{1}{M}\sum_{i=1}^{M} A_i$ = (μ1,μ2,μ3….μN) and j =1 to N (1)

Compute N x N covariance matrix, C NxN = $\sum_{i=1}^{M}(A_i - \bar{A})\,(A_i - \bar{A})^T$ (2)

Compute inverse of covariance matrix C-1 = $\frac{1}{|C|}\,Adj(C)$ (3)

Compute Mahalanobis Distance (MD) for the input data Ai using formula

$$D_i = \sqrt{(A_i - \bar{A})^{T} \cdot C^{-1}\,(A_i - \bar{A})} \quad (4)$$

Obtain threshold required for D_i for statistical outlier detection using chi-square percent point function (ppf) as Di has a chi-square distribution with degrees of freedom (p).

ppf (lower tail probability, p), where p is degrees of freedom (number of attributes considered for distance measurement)

[Ai] PxN is stored with data containing $D_i > threshold$

The MD is a multivariate distance metric that measures the standard deviation of a data instance from the data distribution's mean. The attributes considered here are A1, A4, A5, A8, and A10, and the actual correlation relationship between them is displayed in Figure 3. The covariance of A8 (maximum heart rate) is high since it has a strong association with other qualities. MD adjusts the variance of the attributes to 1 and calculates the distance after transforming the attributes into uncorrelated variables. By dividing the large covariance (C-1) as shown in equation 4, the distance is significantly minimized.

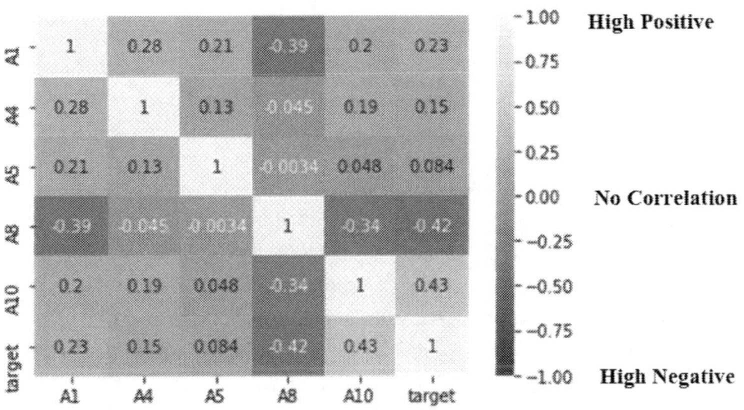

Figure.3 Correlation strength of the attributes with the target

The key to removing outliers is to fix the threshold. Because there are five qualities involved in the distance computation, the Chi-squared test statistics is used with p=5, where 5 is the degree of freedom. The number of occurrences to be deleted was also determined for different values of the probability (lower tail of the distribution). Table 3 shows the number of instances discarded in the dataset as a function of the ppf value (threshold).

Table 3: Lower Tail Probability and Pph Value

S.No	Lower Tail Probability	ppf value (threshold distance)	No. of instances dropped
1	0.99	15.086	6
2	0.97	12.375	13
3	0.95	11.07	18
4	0.93	10.191	21
5	0.91	09.521	26

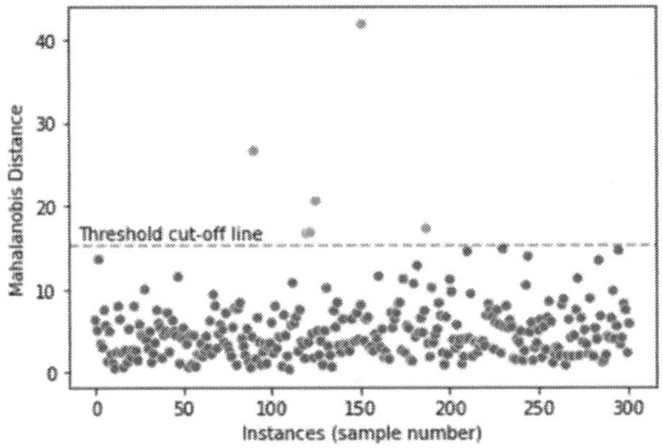

Figure 4.MD plot with threshold line (outliers for the experiment values presented in Table 3 (row~1) is highlighted in orange)

The threshold distance lowers as the lower tail probability drops (see Table 3). As a result, additional data instances are removed. Using 15.086 as a cut-off value for identifying outliers, it is discovered that 6 data instances are identified as outliers (see Figure 4) and those instances are only eliminated to avoid removing more data instances.

The existence of outliers is indicated by heavy tails in the data distribution (Figure 5(a)) in the continuous data attributes. The remaining CHDD attributes are categorical data attributes that do not have any positive or negative skewness. The length of the lower tail is shortened and the skewness of the attribute's distribution is adjusted to zero mean when outliers are removed (see Figure 5(b)). Because only 6 data instances are removed, other statistically defined outliers are still found. In this literature, statistically identified outliers are dropped in favor of imputation of outliers when considering health care data.

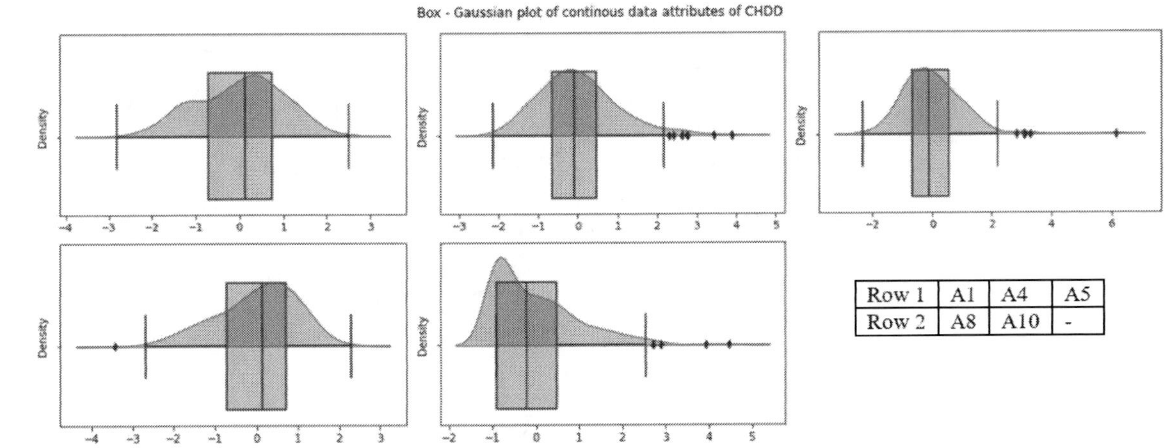

(a) Presence of outliers in the attributes

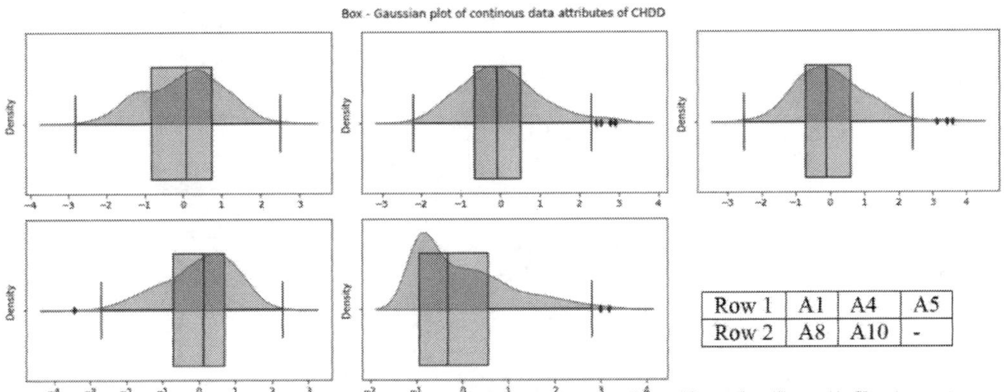

(b) After removal of outliers in the attributes

Figure 5: Data distribution of attributes with box plot (a) before outliers' removal (b) after outliers' removal

4.3. Cross Validation and Grid Search Technique

The technique of cross validation (CV) is used to generalize a machine learning model for an independent dataset. The key to the optimum performance of the ML model with fresh data is its generalisation. K-fold CV is a popular CV strategy used by machine learning practitioners to manage the bias-variance trade-off of their models. Figure 6 depicts the fold CV, which is utilized in this literature. The CHDD is split in to K (K=5) folds and K-1 (4 folds) used in the training phase. The optimum hyperparameters of the ML model are found using a grid search strategy. The same percentage of samples for each class (presence of heart disease / non-presence of heart disease) was maintained using a Stratified K-fold CV, as shown in i. The model's performance is calculated in each fold, and the average of all five folds is used to calculate the final model's performance.

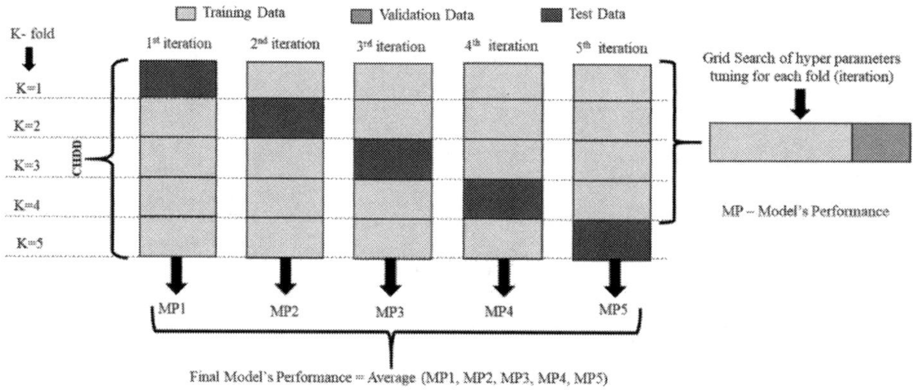

Figure 6: KCV and grid search of fine tuning of hyper parameters used for model's performance evaluation

The optimum hyper parameters of the ML model are found using a grid search technique. It considers all of the model's hyper parameter combinations in detail. Table 4 lists the hyperparameters that were considered in the extensive experiments for several ML models.

Table 4: ML Model's hyperparameters and its values considered in this study

ML Model	Hyperparameters [values] considered
LR	penalty=['l1', 'l2', 'elasticnet'], C (inverse regularization)=[0.2,0.3, 0.6, 0.8, 1] solver =['newton-cg', 'lbfgs', 'liblinear', 'sag', 'saga']
DT	criterion = ['gini','entropy'], splitter' = ['best'], min. sample leaf = [1,2,3,4,5] min. samples split = [0.1,0.2,0.3,0.4,0.5,0.6,0.7,0.8,0.9,1.0]
RF	criterion = ['gini','entropy'], number of estimators = [50,100,150,200,250]
NB	variance smoothing = [1e-01,1e-02,1e-03,1e-04,1e-05,1e-06,1e-07,1e-08, 1e-09,1e-10,1e-11,1e-12]
k-NN	leaf size = [5,10,15,20,25,30,35,40,45,50], distance = [1,2]. number of neighbors = [1,3,5,7,9,11,13,15,17,19,21,23,25,27,29,31, 33,35,37, 39, 41,43,45,47,49]
SVM	kernel= ['linear','rbf', 'sigmoid'], C (regularization) = [0.3, 0.6, 0.8, 1], gamma = ['scale', 'auto']
AB	learning rate = [0.1,0.5,1.0], number of estimators = [10,50,100,200], algorithm = ['SAMME','SAMME.R'],
XB	max. depth = [3, 4, 5], min. child weight = [1, 5, 10], gamma = [0.5, 1, 1.5, 2, 5] subsample= [0.5, 1.0] , colsample_bytree= [0.6, 0.8, 1.0],

5. Results and Discussions

5.1. Performance Evaluation – Environment and Metrics

The system utilized to develop the ML models and execute the lengthy model evaluation tests has an Intel (R) Core (TM) i7-4510U CPU @ 2.60 GHz, installed with RAM of 8.0 GB, and was running Windows 10 Operating System with Jupyter notebook and Python as the programming environment.

Experiments were carried out to assess the ML models. The ML models are assessed using a variety of measures, each of which provides a unique viewpoint on the study and evaluation of the models. Because the topic at hand is a binary classification problem with an unbalanced dataset, judging ML models just on accuracy (the proportion of accurate classifications) will lead to erroneous conclusions. The confusion matrix, which reports True Positive (TP), False Negative (FN), True Negative (TN), and False Positive (FP), is used to calculate the majority of the metrics (FP). Other measures like Accuracy (Acc), F1 score (F1), Precision (Pre), Specificity (Spe), Sensitivity (Sen), Diagnostics Odds Ratio (DOR), and False Omission Rate

(FOR) are calculated and reported using these four metrics. The parameters stated above are provided in their absolute values. In addition to these measures, the model's Area Under the Curve (AUC) with Receiver Operating Characteristics (ROC) is reported to evaluate the model's prediction ranking. These measures demonstrate the ML models' diagnostic capacity.

5.2. Performance Evaluation Results for ML model

In sections 4.1, 4.2, and 4.3, the outcomes of CHDD preprocessing are discussed. The results and robustness of the ML model are given in this section. The quantitative values of performance measures derived for several ML models are provided in Table 5, together with their standard deviations. The best performing classifier model is decided based on the performance metric in the last column of Table 5. The performance statistic with the highest value is highlighted (green colour). The two important metrics for the binary classification problem are accuracy and AUC, where accuracy is based on a threshold and AUC is a test of the quality of the internal value generated by the classifier. As CHDD is imbalance dataset, AUC is preferred over accuracy which helps in determine the robust classifier by providing insights over TPR and false FPR. Therefore, Table 6 is derived from Table 5, in a way to summarize AUC of the different models for different preprocessing. The best value of the performance metric is highlighted (green colour).

Table 5: Performance metrics obtained for ML models under different pre-processing

Preprocessing	Metric	LR	DT	RF	NB	kNN	SVM	AB	XB	Best Model
IMV (Exp #1)	**AUC**	**0.916±0.027**	0.826±0.060	0.894±0.034	0.914±0.031	0.743±0.050	0.912±0.026	0.907±0.029	0.901±0.025	**LR**
	Acc	0.844±0.052	0.748±0.077	0.824±0.032	**0.848±0.053**	0.679±0.057	0.834±0.048	0.821±0.030	0.827±0.049	**NB**
	Sen	0.805±0.074	0.690±0.094	0.783±0.072	**0.834±0.095**	0.532±0.086	0.791±0.082	0.769±0.080	0.776±0.076	**NB**
	Spe	**0.877±0.056**	0.797±0.102	0.859±0.033	0.859±0.032	0.804±0.055	0.871±0.037	0.865±0.032	0.871±0.038	**LR**
	Pre	**0.849±0.062**	0.752±0.110	0.826±0.031	0.833±0.040	0.698±0.084	0.839±0.043	0.831±0.027	0.836±0.045	**LR**
	FOR	0.157±0.050	0.248±0.063	0.174±0.045	**0.137±0.069**	0.330±0.047	0.167±0.058	0.181±0.050	0.178±0.053	**NB**
	DOR	45.346±28.704	28.983±44.22	26.704±12.738	**75.642±97.794**	5.684±2.764	43.480±43.983	25.720±11.696	34.607±27.254	**NB**
	F1	0.825±0.062	0.716±0.088	0.803±0.043	**0.832±0.065**	0.602±0.079	0.813±0.058	0.796±0.043	0.804±0.059	**NB**
IMV+OR (Exp #2)	**AUC**	**0.911±0.026**	0.822±0.041	0.896±0.034	**0.911±0.032**	0.731±0.058	0.909±0.029	0.871±0.093	0.903±0.023	**LR & NB**
	Acc	**0.848±0.039**	0.733±0.043	0.834±0.036	**0.848±0.054**	0.689±0.052	0.834±0.046	0.804±0.055	0.831±0.028	**LR & NB**
	Sen	0.797±0.077	0.605±0.076	0.783±0.062	**0.827±0.099**	0.529±0.068	0.775±0.078	0.731±0.102	0.797±0.081	**NB**
	Spe	**0.889±0.032**	0.840±0.091	0.876±0.029	0.864±0.024	0.821±0.045	0.883±0.036	0.864±0.033	0.858±0.026	**LR**
	Pre	**0.856±0.040**	0.774±0.102	0.839±0.035	0.832±0.040	0.709±0.078	0.845±0.051	0.815±0.049	0.823±0.020	**LR**
	FOR	0.155±0.046	0.279±0.038	0.168±0.041	**0.136±0.069**	0.321±0.040	0.171±0.050	0.200±0.066	0.159±0.046	**NB**
	DOR	40.838±20.049	15.166±15.126	33.289±23.001	**72.677±90.178**	6.106±2.860	37.184±24.846	23.639±13.597	27.597±9.582	**NB**
	F1	0.824±0.054	0.672±0.048	0.809±0.046	**0.828±0.069**	0.606±0.071	0.807±0.060	0.768±0.073	0.808±0.045	**NB**
IMV+OR+SS (Exp #3)	**AUC**	**0.912±0.028**	0.819±0.043	0.893±0.026	0.910±0.033	0.909±0.033	0.910±0.029	0.871±0.093	0.905±0.025	**LR**
	Acc	0.841±0.038	0.733±0.043	0.807±0.047	**0.848±0.054**	0.818±0.031	0.834±0.043	0.804±0.055	0.834±0.035	**NB**
	Sen	0.790±0.077	0.612±0.078	0.753±0.058	**0.827±0.099**	0.723±0.060	0.783±0.076	0.731±0.102	0.790±0.079	**NB**
	Spe	0.882±0.037	0.834±0.092	0.852±0.042	0.864±0.024	**0.895±0.013**	0.877±0.027	0.864±0.033	0.870±0.031	**kNN**
	Pre	**0.848±0.043**	0.769±0.103	0.808±0.052	0.832±0.040	0.849±0.029	0.838±0.042	0.815±0.049	0.835±0.031	**LR**
	FOR	0.161±0.046	0.276±0.040	0.193±0.043	**0.136±0.069**	0.202±0.032	0.167±0.048	0.200±0.066	0.162±0.051	**NB**
	DOR	37.306±19.559	15.074±15.164	23.494±16.984	**72.677±90.178**	24.588±8.453	34.176±22.828	23.639±13.597	35.175±26.481	**NB**
	F1	0.816±0.053	0.674±0.048	0.779±0.054	**0.828±0.069**	0.781±0.047	0.809±0.058	0.768±0.073	0.810±0.046	**NB**

The investigation of Table 5 and Table 6 provides the following evidences. NB outperforms other models on 5 metrics (Acc, Sen, FOR, DOR, F1) and LR outperforms other models on 3 metrics (AUC, Spe, Pre) for IMV preprocessing (Table 5, Exp #1). NB outperforms other models on 4 metrics (Sen, FOR, DOR, F1), LR outperforms other models on 2 metrics (Spe, Pre) and LR & NB provides 2 metrics (AUC, Acc) with equal values for IMV+OR (Table 5, Exp #2) . NB outperform other models on 5 metrics (Acc, Sen, FOR, DOR, F1), LR outperforms other models on 2 metrics (AUC, Pre) and kNN outperforms other models on 1 metric (Spe) for IMV+OR+SS preprocessing (Table 5, Exp #3). LR and NB are outperforming all other models in respect of most of the performance metrics reported for different preprocessing methods applied on the dataset. Exceptionally kNN is outperforming the other models in respect of Spe for the preprocessing of IMV+OR+SS. LR, DT, NB, SVM and AB demonstrate the best AUC values for IMV preprocessing (Table 6). RF provides better AUC value for IMV+OR preprocessing (Table 6). kNN and XB exhibits the best AUC values for IMV+OR+SS preprocessing (Table 6). From the above discussion, it is evident that NB is performing better than all other models on CHDD. CHDD has more categorical attributes (see Table 2) which are uncorrelated and mutually exclusive. This is a key factor that driven NB to produce better performance.

Table 6: Summary of tuned best hyperparameters for the best preprocessing with best AUC values (highlighted)

Preprocessing / Model	IMV	IMV+OR	IMV+OR +SS	Best preprocessing	Best hyperparameters
LR	**0.916±0.027**	0.911±0.026	0.912±0.028	IMV	penalty=l2, C=1
					solver=newton-cg
DT	**0.826±0.060**	0.822±0.041	0.819±0.043	IMV	criterion= entropy
					min. samples leaf= 5
					min. samples split= 0.1
					splitter =best
RF	0.894±0.034	**0.896±0.034**	0.893±0.026	IMV+OR	criterion =entropy,
					number of estimators =150
NB	**0.914±0.031**	0.911±0.032	0.910±0.033	IMV	Variance smoothing = 1e-06
kNN	0.743±0.050	0.731±0.058	**0.909±0.033**	IMV+OR+SS	leaf size = 10, p =1
					No. of neighbors = 27
SVM	0.743±0.050	0.909±0.029	0.910±0.029	IMV	kernel=linear
					C = 0.6
					gamma = scale
AB	**0.907±0.029**	0.871±0.093	0.871±0.093	IMV	learning rate =0.1
					no. of estimators =200
					algorithm =SAMME
XB	0.901±0.025	0.903±0.023	**0.905±0.025**	IMV+OR+SS	max. depth = 4
					min. child weight = 5
					gamma = 1.5, subsample= 1.0
					colsample_bytree= 0.6

6. Conclusion

The tuned best hyperparameters (5-fold CV and grid search) are reported for the best possible AUC of all the models in Table 6. In OR process, data instances identified as outlier (statistically) are dropped. This majorly downgrade the performance of most of the models.

Subsequently, the addition of standardization of continuous attributes of CHDD is also not guaranteed the improvement of the models performance. It is observed from Table 4 and 5 that DT demonstrates underperformance when compared with the other models. Encoding categorical value (one-hot) is the major cause of DT's underperformance as the encoding categorical attributes introducing sparsity in the CHDD which is not desirable for DT. Since DT is considered as the base learner for the boosting classifier (AB), the performance of AB is also not appreciable.

References

[1] World Health Organization (2017), Cardio Vascular Diseases fact sheet, Available Online: https://www.who.int/cardiovascular_diseases/about_cvd/en/

[2] Canters for disease control and prevention, Know the fact about Hear Disease, Available Online: https://www.cdc.gov/heartdisease/docs/consumered_heartdisease.pdf

[3] Muhammad, Y., Tahir, M., Hayat, M. et al. Early and accurate detection and diagnosis of heart disease using intelligent computational model. Sci Rep 10, 19747 (2020). https://doi.org/10.1038/s41598-020-76635-9

[4] I. Jenhani, N. B. Amor, and Z. Elouedi, ``Decision trees as possibilistic classifiers," Int. J. Approx. Reasoning, vol. 48, no. 3, pp. 784-807, Aug. 2008. https://doi.org/10.1016/j.ijar.2007.12.002

[5] L. Breiman, ``Random forests," Mach. Learn., vol. 45, no. 1, pp. 5-32, Oct. 2001. https://doi.org/10.1023/A:1010933404324

[6] Irfana P. Bhatti, Heman D. Lohano , Zafar A. Pirzado and Imran A. Jafri , 2006. A Logistic Regression Analysis of the Ischemic Heart Disease Risk. Journal of Applied Sciences, 6: 785-788. https://dx.doi.org/10.3923/jas.2006.785.788

[7] G. I. Webb, J. R. Boughton, and Z. Wang, ``Not so naive bayes: Aggregating one-dependence estimators," Mach. Learn., vol. 58, no. 1, pp. 5-24, Jan. 2005. https://doi.org/10.1007/s10994-005-4258-6

[8] A. Reinhardt and T. Hubbard, ``Using neural networks for prediction of the subcellular location of proteins," Nucleic Acids Res., vol. 26, no. 9, pp. 2230-2236, May 1998. https://doi.org/10.1093/nar/26.9.2230

[9]C. Cortes and V. Vapnik, ``Support-vector networks," Mach. Learn., vol. 20, pp. 237-297, Sep. 1995.

[10] B. Kégl, ``The return of AdaBoost.MH: Multi-class Hamming trees," 2013, Available Online: http://arxiv.org/abs/1312.6086

[11] Kartik Budholiya, Shailendra Kumar Shrivastava, Vivek Sharma, An optimized XGBoost based diagnostic system for effective prediction of heart disease, Journal of King Saud University - Computer and Information Sciences, 2020.
https://doi.org/10.1016/j.jksuci.2020.10.013

[12] L. Yu and H. Liu. Feature Selection for High-Dimensional Data: A Fast Correlation-Based Filter Solution. In Proceedings of The Twentieth International Conference on Machine Leaning (ICML-03), 856-863, Washington, D.C., August 21-24, (2003). https://www.aaai.org/Papers/ICML/2003/ICML03-111.pdf

[13] Hanchuan Peng, Fuhui Long, and Chris Ding, "Feature selection based on mutual information: criteria of max-dependency, max-relevance, and min-redundancy," IEEE Transactions on Pattern Analysis and Machine Intelligence, Vol. 27, No. 8, pp.1226-1238, (2005). https://doi.org/10.1109/TPAMI.2005.159

[14] KIRA, K. and RENDEL, L. The Feature Selection Problem: Traditional Methods and a new algorithm. Proc. Tenth National Conference on Artificial Intelligence, MIT Press, 129-134, (1992).
https://www.aaai.org/Papers/AAAI/1992/AAAI92-020.pdf

[15] Tibshirani, Robert, "Regression Shrinkage and Selection via the lasso". Journal of the Royal Statistical Society. Series B (methodological). Wiley. 58 (1): 267–88, (1996). http://www.jstor.org/stable/2346178

[16] Nguyen Cong Long, Phayung Meesad, Herwig Unger, A highly accurate firefly based algorithm for heart disease prediction, Expert Systems with Applications, Volume 42, Issue 21, 30 November 2015, Pages 8221-8231
https://doi.org/10.1016/j.eswa.2015.06.024

[17] Verma, L., Srivastava, S. & Negi, P.C, A Hybrid Data Mining Model to Predict Coronary Artery Disease Cases Using Non-Invasive Clinical Data, J Med Syst., 40, 178 2016. https://doi.org/10.1007/s10916-016-0536-z

[18] Kathleen H. Miao, Julia H. Miao , George J. Miao, Diagnosing Coronary Heart Disease Using Ensemble Machine Learning, International Journal of Advanced Computer Science and Applications, Vol. 7, No. 10, 2016, pp- 30- 39
https://thesai.org/Downloads/Volume7No10/Paper_4Diagnosing_Coronary_Heart_Disease_using_Ensemble_Machine_Learning.pdf

[19] Ashok Kumar Dwivedi, Performance evaluation of different machine learning techniques for prediction of heart disease, Neural Comput & Applic, Springer, 2016, https://doi.org/10.1007/s00521-016-2604-1

[20] Gupta N, Ahuja N, Malhotra S, Bala A, Kaur G. Intelligent heart disease prediction in cloud environment through
ensembling. Expert Systems. 2017;e12207. https://doi.org/10.1111/exsy.12207

[21] Haq, A., Li, J., Memon, M., Nazir, S., & Sun, R, A Hybrid Intelligent System Framework for the Prediction of Heart Disease Using Machine Learning Algorithms. Mob. Inf. Syst., 2018, 3860146:1-3860146:21. https://doi.org/10.1155/2018%2F3860146

[22] J. Vijayashreea, H. Parveen Sultana, A Machine Learning Framework for Feature Selection in Heart Disease Classification Using Improved Particle Swarm Optimization with Support Vector Machine Classifier Programming and Computer Software, 2018, Vol. 44, No. 6, pp. 388–397 https://link.springer.com/article/10.1134/S0361768818060129

[23] Elham Nikookar, Ebrahim Naderi, Hybrid Ensemble Framework for Heart Disease Detection and Prediction, International Journal of Advanced Computer Science and Applications, Vol. 9, No. 5, 2018, pp- 243- 248
https://thesai.org/Downloads/Volume9No5/Paper_33-Hybrid_Ensemble_Framework_for_Heart_Disease.pdf

[24] Wenbing Chang , Yinglai Liu, Xueyi Wu, Yiyong Xiao,Shenghan Zhou , Wen Cao, A New Hybrid XGBSVM Model: Application for Hypertensive Heart Disease, IEEE Access, Vol 7, 2019, pp- 175248- 175258.
https://doi.org/10.1109/ACCESS.2019.2957367

[25] Liaqat Ali , Awais Niamat , Javed Ali Khan, Noorbakhsh Amiri Golilarz, Xiong Xingzhong , Adeeb Noor, Redhwan Nour, Syed Ahmad Chan Bukhari, An Optimized Stacked Support Vector Machines Based Expert System for the Effective Prediction of Heart Failure, IEEE Access, Vol 7, 2019, pp- 54007 – 54014 https://doi.org/10.1109/ACCESS.2019.2909969

[26] Sulekha Saxena, Vijay Kumar Gupta, P. N. Hrisheekesha, Coronary Heart Disease Detection Using Nonlinear Features And Online Sequential Extreme Learning Machine, Biomedical Engineering: Applications, Basis and Communications, Vol. 31, No. 6 (2019) 1950046 (15 pages) https://doi.org/10.4015/S1016237219500467

[27] Senthilkumar Mohan, Chandrasegar Thirumalai, Gautam Srivastava, Effective Heart Disease Prediction Using
Hybrid Machine Learning Techniques,IEEE Access, Vol 7, 2019, pp- 81542- 81554
https://doi.org/10.1109/ACCESS.2019.2923707

[28] Syed Muhammad Saqlain, Muhammad Sher, Faiz Ali Shah, Imran Khan, Muhammad Usman Ashraf, Muhammad Awais,Anwar Ghani, Fisher score and Matthews correlation coefficient-based feature subset selection for heart disease diagnosis using support vector machines, Knowl Inf Syst (2019) 58:139–167

https://doi.org/10.1007/s10115-018-1185-y

[29] Ashir Javeed, Shijie Zhou, Liao Yongjian, Iqbal Qasim, Adeeb Noor, Redhwan Nour, An Intelligent Learning System Based on Random Search Algorithm and Optimized Random Forest Model for Improved Heart Disease Detection, IEEE Access, VOLUME 7, 2019, pp- 180235 – 180243, https://doi.org/10.1109/ACCESS.2019.2952107

[30] Ankur Gupta, Rahul Kumar ,Harkirat Singh Arora , Balasubramanian Raman, MIFH: A Machine Intelligence Framework for Heart Disease Diagnosis, IEEE Access, Vol 8, 2020 pp 14659- 14674 , https://doi.org/10.1109/ACCESS.2019.2962755

[31] Norma Latif Fitriyani, Muhammad Syafrudin,Ganjar Alfian , Jongtae Rhee, HDPM: An Effective Heart Disease Prediction Model for a Clinical Decision Support System, IEEE ACCESS, Vol 8, 2020 pp- 133034- 133050
https://doi.org/10.1109/ACCESS.2020.3010511

[32] Jian Ping Li , Amin Ul Haq , Salah Ud Din , Jalaluddin Khan, Asif Khan, Abdus Saboor, Heart Disease Identification Method Using Machine Learning Classification in E-Healthcare, IEEE ACCESS, Vol 8, 2020 pp- 107562-107582
https://doi.org/10.1109/ACCESS.2020.3001149

[33] Dhyan Chandra Yadav, Veer Bahadur Singh , Saurabh Pal, Veer Bahadur Singh , Analysis of Heart Disease Using Parallel
and Sequential Ensemble Methods With Feature Selection Techniques:Heart Disease Prediction,International Journal of Big Data and Analytics in Healthcare, Volume 6 • Issue 1 • January-June 2021. https://orcid.org/0000-0001-9545-7481.

[34] Mahalanobis, Prasanta Chandra , "On the generalised distance in statistics" (PDF). Proceedings of the National Institute of Sciences of India.,1936, 2 (1): 49–55

Performance Testing of VCR Engine using Plastic Oil Produced by Pyrolysis Method

Samuvel Michael .B[1], K.G.Muthurajan[2], K.Surendrababu [3].

1.Samuvel Michael.B.- Research scholar, Department of Mechanical Engineering, Aarupadai Veedu Institute of Technology, Vinayaka Mission Research Foundation, sammic69@gmail.com

2.Dr.K.G.Muthurajan,Professor, Department of Mechanical Engineering, Vinayaka Mission's Kirupananda Variyar Engineering College , kgmuthurajan@gmail.com.

3. B. K.Surendrababu- Research Scholar, Department of Mechanical Engineering, Aarupadai Veedu Institute of Technology, Vinayaka Mission Research Foundation, surendrababu@avit.ac.in.

ABSTRACT : Recycling is the retrieval, reclamation, reprocessing or refining process of waste to produce new products. Recycling has always been a top priority in waste management since it enables us not only to maintain the health of the environment, but also to reuse garbage profitably. Many techniques are utilised to transform plastic waste into articles that are not present in virgin plastic with distinctive features. Polymers undergo molecular and structural changes in this process leading to simpler basic materials with superior thermal properties than original plastics. While gasification and pyrolysis processes break plastics down, fluid and semi-liquid products are also generated.

Keywords: Waste plastic pyrolysis oil, diesel Compression ratio feature, Output

1. INTRODUCTION

Endothermally irreversible thermochemical disintegration The disintegration of complex, long-chain organic polymer molecules into units of simple short-chain composites occurs with or without pressure at temperatures over 300 C in the absence of air oxygen. A plastic waste pyrolyse involves the breakdown into solid, liquid and gas components of polymers. Char is primarily carbon in the solid phase, with minor quantities of other components depending on plastics. Oil-like material is employed in a liquid product that has features akin to diesel, similar to those originally petroleum-based polymers. Petroleum products like plastics are produced from a liquid material similar to oil. The gaseous phase results from Syngas (synthetic gas) and has a high calorific value. While pyrolyses need a number of phases Temperature Residence time

- Presence of catalyst
- Catalyst loading
- Catalyst contact mode
- Reactor Type

2. DISCRIBTION

2.1PYROLYSIS TECHNOLOGY

Pyro: heat, lysis: pafis disintegration. Pyrolysis is a chemical process that divides big molecules into smaller ones. The simplest example is the boiling of complex food molecules into smaller and easily consumable ones.

Plastic waste and pneumatic is long-chain molecules or hydrocarbons of polymer. Pyrolysis technology is the industrial process for the breakdown of big plastic/tire molecules into smaller oil, gas and carbon black molecules. In the absence of oxygen, plastic or tyre waste is pyrolysed at 350 to 550 degrees C and the reaction time is about 15 to 90 minutes.

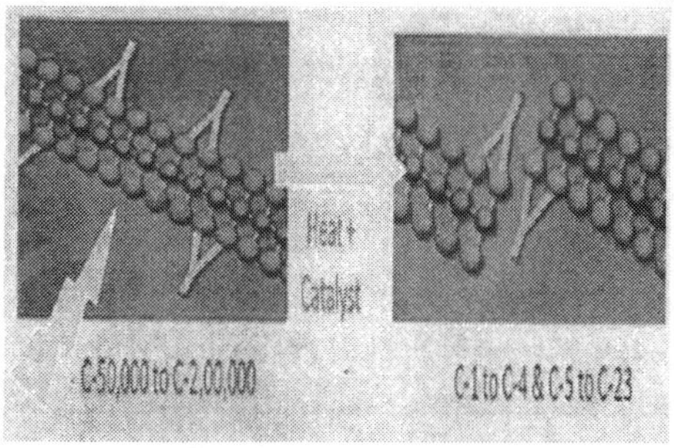

2.1 pyrolysis process

2.2 PYROLYSIS OIL

Pyrolysis oil is a sorney called biological crude oil or biological oil. It is a synthetic fuel under study as a replacement for petroleum. Biomass is extracted from dry biomass in a reactor with a temperature of about 500 degrees Celsius, with subsequent cooling, using liquid technique of destructive distillation. Pyrolysis oil is a kind of tar, which usually includes oxygen that is too high to be a hydrocarbon. This differs significantly from comparable petroleum products

2.3 CHARACTERIST PYROLYSIS

The pyrolysis-produced oil is acidic with a PH of 1.5-3.8. Acidity may be reduced by adding easily accessible basic components. There was little study on the stability of bio-oil acidity which has been changed with base components, whereas the precise composition of bio-oil relies on the bio mass source and circumstances. A typical falls of water is 20-28%, suspended particles and pyrolytic lignine 22-36yo, hydroxylacetaldehyde 8-120, levoglucosane 3-8o/o, acetyl acid 4-8%, acetol 3-6o/o, glycol 1-2 0/o, formic acid 3-6. During pyrolysis the water molecules are broken up and kept separately in other compounds with the pyrolysis liquid in the complex. The difference is important, since "water" does not separate as normal fossil fuels in pyrolysis oils

Batch Mode, Continuous Flow Mode, Fixed Bed, Fluidized Bed Reactor, and more variations are available.

Product yield at 430 degrees Celsius from actual waste plastics (from a German source) in a batch reactor was measured.

Product(s)	Thermal Pyrolysis Cat.	Pyrolysis (zeolite)
Oil	±78.7	±76.7 %
Gas	±7.6 %	±9.4
(char & CaO)	±17 %	±8.8 %

The inclusion of a catalyst increased the production of both gas and oil products while decreasing the yield of char product (char product).

2.4 PYROLYSIS BY CATALYTIC PYROLYSIS

Because of the ease with which heterogeneous catalysts may be recovered, they have become increasingly popular.

Lewis acids such as AlC13, FeC13, and other similar compounds are employed as homogeneous catalysts.

Solid acids, zeolites, silica alumina, and FCC catalyse reactions in heterogeneous catalysts.

Because of the acidity of their active sites and the crystalline microporous structure of their materials, they are particularly well suited for achieving high conversion at low temperature (temperature).

The acidic characteristics of these solids, particularly the strength and number of acidic sites, are responsible for the differences in their catalytic activity.

Properties of structures such as particle size, pore size distribution, and specific area have an important influence in the performance of these materials and composites.

3. EXPERIMENT SETTINGS FOR THE TEST:

The engine is directly connected to an eddy current dynamometer with a capacity of 3.5 kW (1500 rpm) and a direct drive. The engine and the air-cooled eddy current dynamometer are connected together by means of a tyre coupling system. The output shaft of the engine is linked to the dynamometer by way of a torque transducer, which measures the torque produced by the engine. A torque transducer generates an electrical signal that is proportional to the amount of torque applied to it. It is possible to transform a force into an electrical signal using a load cell, which is an electronic device (also known as a transducer). The load placed on the engine may be changed by manipulating the potentiometer on the control panel or by using the computer to control the load.

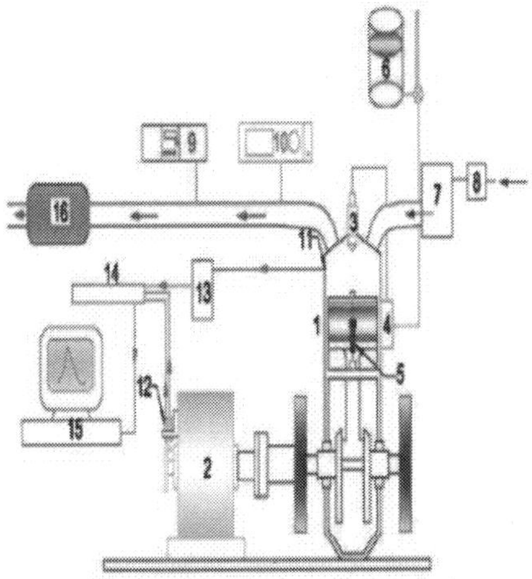

Fig.3.1 Experimental setup of VCR Engine

Fig.3.2 Photographic view of VCR diesel engine

4. EXPERIMENTAL PROCEDURE

Plastic oil poured into a tank and monitored the engine installations like coolant water, lubrication oil, fuel flow pipes, air lock and tested before engine experiments were started. And the test engine has been started and may be operated till steady conditions have been reached. The load of the engine was progressively raised to the maximum weight needed. At the same time, the dynamometer, all analyzers and measuring meters were activated, and manufacturers' instruction manuals carried out the proper preparation and measuring settings according to recommended methods. The testing were begun after the test engine was steady and the measurements were arranged and settings completed. The experiment technique is a continuous engine status test. Load settings were five phases with loads 0%, 25%, 50% and 75% correspondingly. Or each load-point, measured air intake, fuel consumption, air intake temperature, air intake temperature, gas exhaust temperature, engine cooling temperature, injection fuel speed, combustion pressure, crank angle, emission of hydrocarbon (HC), CO emissions, nitrogen oxides (NOX) emissions and sm.

5. RESULT AND DISCUSSION

Plastic oil tested in VCR engine and 5 gas analyzer has got readings. Based on that emission & performance reading I plat the graphs

5.1CARBON MONOXIDE VS LOAD

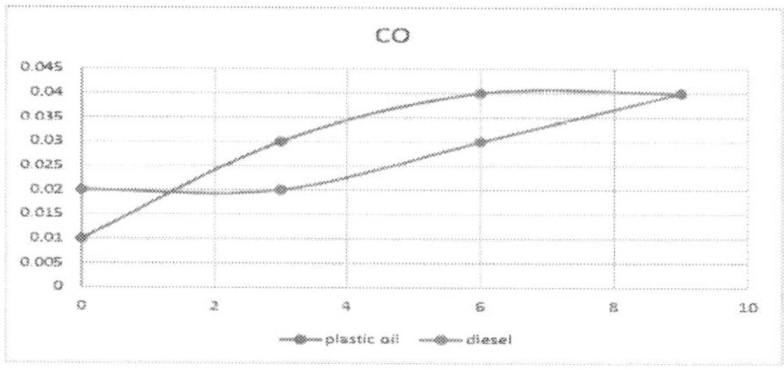

Fig.5.1 Variation of load Vs. CO

In this graph CO value shows the pyrolysis plastic oil compare with diesel with different load testing . The plastic oil value was higher then the diesel value in middle load but higher load plastic oil value is same to compare the diesel value
HYDROCARBON VS LOAD

5. 2. HYDRO CARBONS VS LOAD

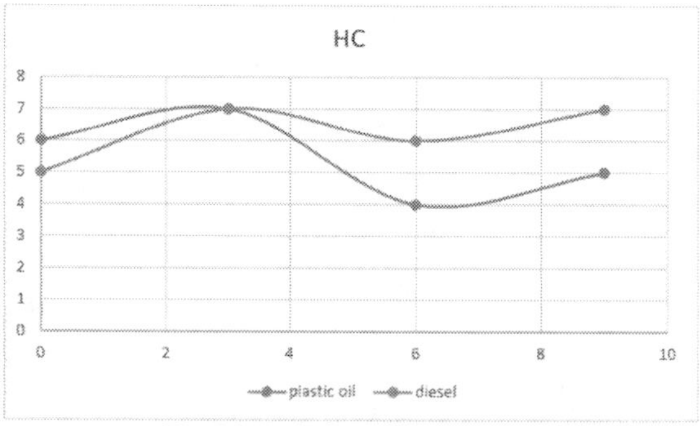

Fig.5.2 Variation of load Vs HC

In this graph HC value shows the pyrolysis plastic oil compare with diesel with different load testing . The plastic oil value washigher then the diesel value in initial load but higher load plastic oil value has lower to compare the diesel valuehigher engine load. From the results it is seen that the unburnt hydrocarbons is more at full load conditions, due to improper atomization of fuel.

5.3. CARBON DIOXIDE VS LOAD

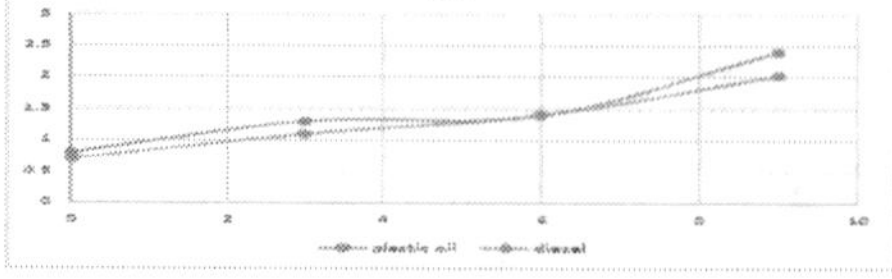

Fig.5.3 Variation of load Vs CO_2

In this graph CO value shows the pyrolysis plastic oil compare with diesel with different load testing . The plastic oil value was all most same like the diesel value

5. 4. NITRIC OXIDE VS LOAD

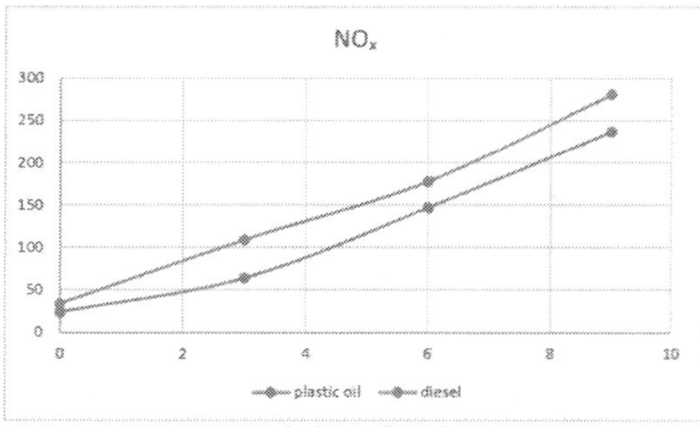

Fig.5.4 Variation of load Vs NOX

In this graph NO_x value shows the pyrolysis plastic oil compare with diesel with different load testing. The plastic oil value was shows lower than the diesel value

5.5 SPECIFIC FUEL CONSUMPTION VS LOAD

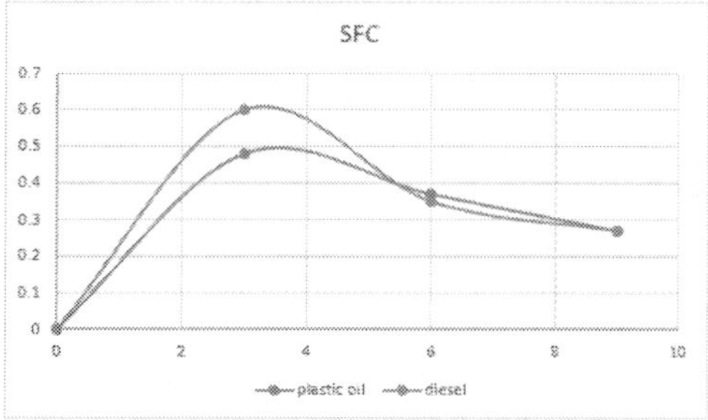

Fig.5.5 Variation of load Vs SFC

In this graph SFC value shows the pyrolysis plastic oil compare with diesel with different load testing. The plastic oil value was shows lower than the diesel value

5.6. BREAK THERMAL EFFICIENCY VS LOAD

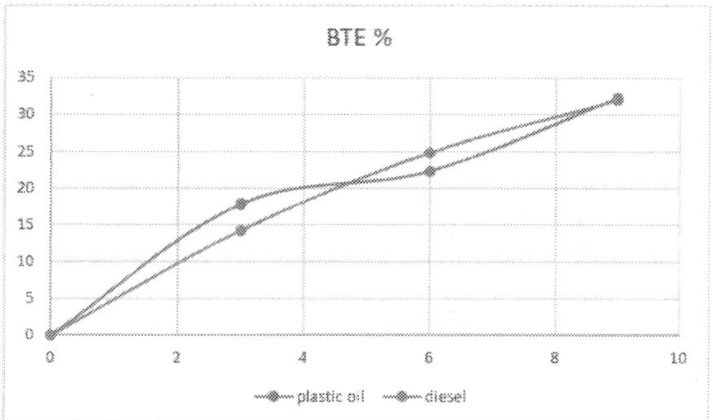

Fig.5.6 Variation of load Vs BTH

In this graph BTE vale shows the percentage of pyrolysis plastic oil compare with diesel with different load testing . The plastic oil value was all most same like the diesel value

6. CONCLUSION

Engine worked well without any significant changes to WPO. Thermal brakes The efficiency of the mixtures is reduced compared to diesel. The use of the mixture compared to diesel reduced CO emissions. The use of waste plastic oil as diesel has increased the emissions of NOx. In comparison to conventional diesel, the performance, emission and combustion characteristics of the multi-fuel variable compression engine fuelled by waste plastic oil and diesel mixes have been examined. The experimental findings indicate that the BTE, SFC of a variable compression motor, is based on the mixing waste, load and compression ratios of plastic oil. From this study the following findings are drawn: The thermal brake efficiency of the B30 mix is somewhat better than the regular diesel at a higher compression ratio. The particular fuel consumption of the B30 mixture is lower than that of all other combinations and diesel. This may be attributed to improved combustion and an increase in the mixture's energy content. For B30 and diesel the highest power generated at the compression ratio 18 is 2,07 kW and 2,12 kW. At greater compression rates, hydrocarbon emissions from different mixtures are increased. The increased compression ratio increases the HC emissions for the B30 mixture. The emission of nitrogen oxides (NOx) from the B30 waste plastic oil mix is greater than diesel. The CO emissions from the B30 combination are closer to the normal diesel and at the compression ratio 18 are much higher. NOx emissions have increased somewhat, but they are still similar to those of regular diesel fuel and are likewise acceptable. The experimental results also show that lower and medium percentages of waste petrol may be replaced by incomplete combustion of diesel fuel.

REFERENCE

1. K. Naima, A. Liazid, Waste oils as alternative fuel for diesel engine: A review, Journal of Petroleum Technology and Alternative Fuels, 4(3), 2013, pp. 30-43.
2. Kaustav Mukherjee, C.Thamotharan, Performance and Emission Test of Several Blends of Waste Plastic Oil with Diesel and Ethanol on Four Stroke Twin Cylinder Diesel Engine, IOSR Journal of Mechanical and Civil Engineering,11(2),2014,pp. 47-51.
3. Bridjesh P, PrabhuKishore, N, Mallikarjuna, M.V, and Alekhya, N., Performance
4. Analysis of Variable Compression Ratio Diesel Engine using
5. Calophylluminophyllum Biodiesel, Indian journal of science and technology, 9
6. (35),2016,pp.
7. Senthilkumar, Tamilkolundu, ChandrasekarMurugesan, The Evaluation of blend of Waste Plastic Oil-Diesel fuel for use as alternate fuel for transportation, 2nd International Conference on Chemical, Ecology and Environmental
8. Sciences,2012,pp. 66-70.
9. S. Madhu, A.V. Krishna Chaitanya and P. Bridjesh. Effects of Diethyl Ether on Performance and Emission Characteristics of a Diesel Engine Using Toroidal Profile Bowl Piston by Varying Injection Pressure. International Journal of Mechanical Engineering and Technology, 8(6), 2017, pp. 96–106.
10. Ravindra Randa and Aseem C. Tiwari. Experimental Investigation of Performance and Emission Parameters of CI Engine Using Karanja Oil Methyl Ester (KOME) and Biogas in Dual Fuel Mode. International Journal of Mechanical Engineering and Technology, 8(5), 2017, pp. 1010–1021.

Characteristics Study on Leachate Samples of Laloor, Thrissur

Ispara Xavier S[1*], Aravind Raj P S[2], Divahar R[2], Arun Mohan[3], Nithin Tr[3], Shamsad A C[3]
[1]Assistant Professor, [2]Associate Professor, [3]UG Student, Department of Civil Engineering, Aarupadai Veedu Institute of Technology, Vinayaka Mission's Research Foundation, Paiyanoor, Chennai
(*Corresponding Author, Email: isparaxavier.civil@avit.ac.in)

Abstract. Leachate generation is a biggest problem faced by municipal solid waste (MSW) landfills which causes significant hazard to surface water and groundwater. Leachate contains high concentration of pollution which can cause serious problems to the society. In this study, examining the characteristics of leachate generated from a landfill site situated in Laloor, In Thrissur, Standard methods were used to monitored, analyzed and examined the water and waste water. According to the standard methods for the Parameters were monitored included dissolved oxygen (DO), chemical oxygen demand (COD), biological oxygen demand (BOD), chlorides content, alkalinity, mpn, hardness and pH. After this study results are mapped by QGIS to confirm and show the impact of the leachate issue in laloor.

Key words: Leachate, solid waste, laloor, dump waste

1. Introduction

Leachate is the liquid that drains or 'leaches' from a disposal area. When water seeps through a litter, it results in the decomposition process of microorganisms. This activity releases decomposition by-products and quickly makes use of accessible oxygen, creating an oxygen deprived condition.

Its properties are based on factors like the decomposition degree, composition of the litter and its age, decomposition phase, the litter moisture content, water movement's rate, procedure of waste filling and temperature. Leachate can be categorized as a waste water type because of the inconsistency in its composition and this results not only in many parameters which are specific to the site like the area's geology, climate, geography, design of landfill but also from measures taken to control the same. Impacts on health and the environment of the generated wastewaters are also based on its composition. The natural diminishing levels of leachate pollutants in aquifers plays a key role in factors like understanding the dangers of leachates seeping into the groundwater, for the explanations of samples from observing wells and also for taking the necessary steps to mitigate the problem. The degradation's first step is characterized by the organic matter's degeneration. The anaerobic degeneration phase consists of two main processes of fermentation, the acid gene phase and methanogen phase which consists of biodegradable leachates and stabilized leachates respectively.

In this study, solid waste from different places in the Thrissur Corporation was collected and deposited in Laloor. This mixed waste consists of plastic, metals, glass, rubber, domestic hazardous waste, Food and vegetable waste, etc. During rainy season, the water along with waste get logged in the living area that creates a nuisance to the people. Due to these reasons mentioned above, it makes necessary for awareness to people about water quality. So, we found out the characteristic of waste water using GIS.

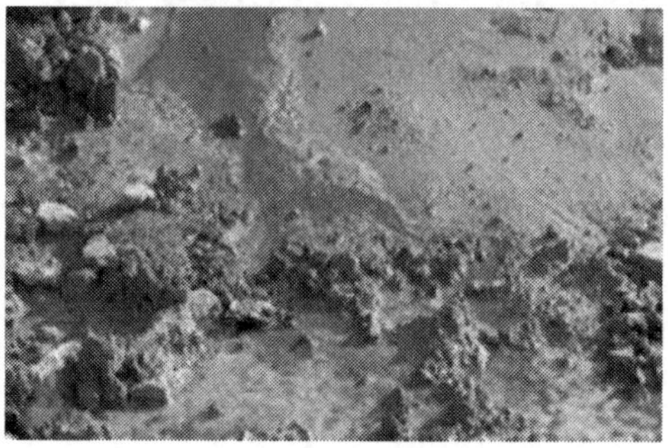

Figure 1. Leachate

1.1 QGIS

QGIS (formerly Quantum GIS) is a cross-platform, free and open-source desktop geographic information system (GIS) programme for viewing, manipulating, and analysing geographical data. QGIS interfaces with a variety of different open-source geographic information systems, including Post GIS, GRASS GIS, and Map Server. A GIS is any system that can store, modify, and display digital map data. A GIS system primarily makes use of two kinds of data. There are two types: vector and raster. Vector data is the most often used kind of map data, consisting of points, lines, and polygons that are used to represent physical features or places. The less often used format is raster data, which includes satellite images and aerial photography. Often, the aim of a map is to visualise a response to a query. Maps make it easier to detect patterns and clusters than spreadsheets or databases do.

2. Study Area

Laloor is a place in Thrissur city, Kerala. It is located in between the Ayyanthole, Olari, and Aranattukkara; it comes under the Thrissur Municipal Corporation limit. Thrissur is built around a hillock and situated in 10° 10'and 10° 46'North latitude and 75°55'and 77° 05'East longitude. The city is located on the National Highway 47 and is well connected to nearby cities and towns like Kochi (75 km), Kozhikode (144 km) and Palakkad. Thrissur Municipality was upgraded to a Corporation in October 2000 by joining the Panchayats of Ayyanthole, Ollukkara, Koorkkencherry, Ollur, and Vilvattom. In addition, parts of Nadathara were also merged with the Municipal Corporation and thereby increasing the total area of the Municipal Corporation from 12.65 sq. km to 101.42 sq. km. The population of the Thrissur Municipal Town steadily rose from 69,515 in 1951 to 317,526 in 2001.

2.1 Land Use Pattern:

The current land use pattern as shown in Table.1, indicates that 68.60 percent of the city land Is under residential use and includes houses in individual plots distributed all over the city. The Commercial area is less in comparison with residential area and consists of small establishments. Increased developments are seen in the MG road area. Industrial sector comprises mainly of service industries and small-scale industries; medium scale industries are concentrated in the city. Overall industries occupy 5.73 percent of the advanced area of the city. Of the total area available, 10.24 percent is occupied for public and semipublic use. 62 Percent of the total 113,967 of properties assessed in the city is residential. The dominance of land used for residential purposes is expected to continue almost at the same pattern in the Future also.

Figure 2. The burial ground and surrounding 1k m

Figure 3. collection of news about Laloor

The plot number 362 is an exact portion of Burial ground. At first, the land area was about 44 acres. After the public encroachment, the land reduced to about 24 acres. The tremendous increase in the population also increased the generation of waste enormously. A trenching ground and the treatment plant consist about 11 acres of land and the remaining land is scattered in small areas. The north of the burial ground is sasthri line. As well as south of the burial ground is Mathayipuramcolony. The west is assissibucker colony. As well as east is pulikal way.

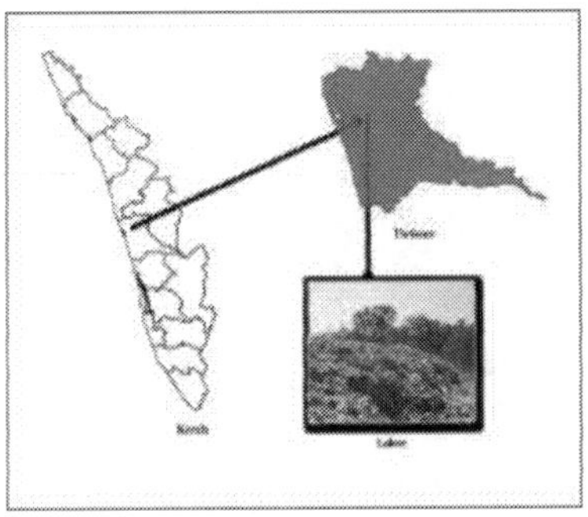

Figure 4. location of Laloor

Figure 5. Present situation in Laloor

After this strike the waste dumping was completely stopped in Laloor. The vehicle from outside not enter the dumping yard. The few people around the dumping yard deposited their waste in this dumping yard. Most of the waste removed by landfill method. However enough amount waste lying on ground. Now, the main problem is leachate.

3. Sampling and Materials

Water samples were collected from twelve well during the season of 2018 from 1 km surroundings on the landfill site. The containers were washed and collect samples in it. Physiochemical parameters such as pH were measured in site, while samples for either parameter analysis, they were transported into the laboratory. The sampling and analysis of various physico -chemical attributes were performed. The results obtained were then compared to the permissible limit of irrigation and drinking water quality standard (FAO, IS-10500, BIS). In this study, the water quality index was calculated by considering eight important parameters (pH, DO, BOD, TH, MPN, COD, alkalinity and Chloride).

Table 1. Standard of leachate

Physico-Chemical Parameters	Standard Value
Dissolved oxygen	5
pH	6.5-8.5
Total hardness	300
Ca	75
Mg	50
Cl	250
Biological oxygen demand	100
Chemical oxygen demand	10

In This study,15 samples were collected, then immediately transferred to the laboratory and stored in the cold Temperature (40c). Various physico-chemical parameters Analyzed in the contaminated water samples includes, pH, dissolved oxygen, chemical oxygen demand (COD), total alkalinity (TA), biological oxygen demand (BOD), total hardness (TH), chloride, collected regions are Assissibuckercolony, west of burial ground sasthrilanearea, north of burial ground,pulikkal way east of burial ground, mathayapuram colony south of burial ground

Table 2. Sample collection details

Sample 1	V.J. Chacko from sasthri line
Sample 2	John from sasthri line
Sample 3	Sheena from Assisi bucker
Sample 4	Ajitha from Assisi bucker
Sample 5	Vivitha from sasthri line
Sample 6	Omana from Pulikkal way
Sample 7	Thomas from Mathayapuram colony
Sample 8	Alwin from Mathayapuram colony
Sample 9	Lalitha from Assisi bucker
Sample 10	Aramukhan from Assisi bucker
Sample 11	Jancy from Pulikkal way
Sample 12	Leela from Assisi bucker
Sample 13	Mary from Mathayapuram colony
Sample 14	AlphonsafromMathayapuram colony
Sample 15	Elsa from Pulikkal way

4. Result and Discussion

Parameters were done in laboratory, including chemical oxygen demand (COD), dissolved oxygen (DO), biological oxygen demand (BOD), presence of chlorides, alkalinity, MPN, hardness and pH shown in Table 3

Table 3. Parameter's value obtained by laboratory tests

SAMPLES	PH	HARDNESS	COD	BOD	ALKALINITY	CHLORIDE	DISOLVED OXYGEN
S1	9	90	87	32	110	8.9	7.20
S2	6.7	161	79	70	214	5	9.8
S3	9.3	105	119	71	46	55	6.8
S4	6.8	123	161	81	52	50	8
S5	10	310	214	158	218	7	3
S6	10	314	160	57.5	26	35.9	12
S7	9	290	300	58	70	17	5.6
S8	9.1	274	302	42.5	66	36	6.3
S 9	9	362	70	43	58	42	4.2
S10	8	314	26	200	66	38	7.3

S11	10	318	400	180	30	21.9	8
S12	9.3	310	38	130	62	26	13
S13	6.3	320	350	50	74	30	6.5
S14	6.9	340	312	50	82	52	7
S15	8.3	362	110	30	34	32	8

PH value for drinking water ranges from 6.5 – 8.5. From graphical analysis, we found that sample 4, sample 2, sample 10, sample 14 and sample 15 are within this range. Other samples needed treatment methods for purification.

When the hardness is in the range of 0-75 then the degree of hardness is soft, if it is in the range of 75-120 then the degree of hardness is moderate, when it is in the range 120-300 then the degree is hard and when the hardness is greater than 300 the degree is very hard. From the graphical analysis, we found that sample 1, 3 are moderate, Sample 2, 4, 8 are hard and other samples are in the degree of very hard. Therefore, sample 1 and 3 used for drinking purpose after boiling because of its temporary hardness. Sample 2, 4, 8 are not useful for domestic purpose because it produces scale in water pipes.

According to WHO recommended Chemical Oxygen Demand is 10 ppm in drinking water. All samples are exceeding the limit.

Biological oxygen demand and chemical oxygen demand tests are helpful in finding the presence of biologically resistant of organic substances and also toxic level. The BOD level 1 – 2 ppm is said to be in a good range. A water having 3 – 5 ppm is considered moderately clean. Water to a BOD level of 6 – 9 ppm the water is considered somewhat contaminated. This is due to the presence of organic matter and bacterial decomposition of the waste. If the BOD level is 100 ppm or greater the water supply is considered very much contaminated with organic waste. Sample 5, 10 ,12, 11 are with a range of greater than 100 ppm, so these samples are very polluted with organic wastes.

According to Indian Standard Institution the recommended permissible value for alkalinity is 200 mg/l in drinking water. From the graphical analysis, we found that sample 2, 5 is not suitable for drinking purpose because its value 214 ,218 is greater than the limit.

The permissible limiting range for chlorides in drinking water is less than 250 mg/l. From graphical analysis, we found that all samples come within this range.

For drinking water, the allowed limit of dissolved oxygen is at least 5 mg/l. From graphical analysis, we found that sample 5 and 9 are not in the range

Membrane filters are porous, with pores of around 0.45m in diameter, which is large enough to hold microorganisms. A sample is initially fed over the membrane using a filter funnel and vacuum apparatus in this method. On the membrane surface, microorganisms present in the sample are maintained. The membrane is then moved to a specific plate that has a saturated pad with the appropriate media. During incubation, nutrients flow through the filter, facilitating organism development. Colonies of microorganisms develop on the membrane surface. After forming discrete colonies, they may be moved to a confirmation medium. The results are then calculated by using the conventional MPN standard methods. Within 24 hours the presence and absence of microorganism can be confirmed.

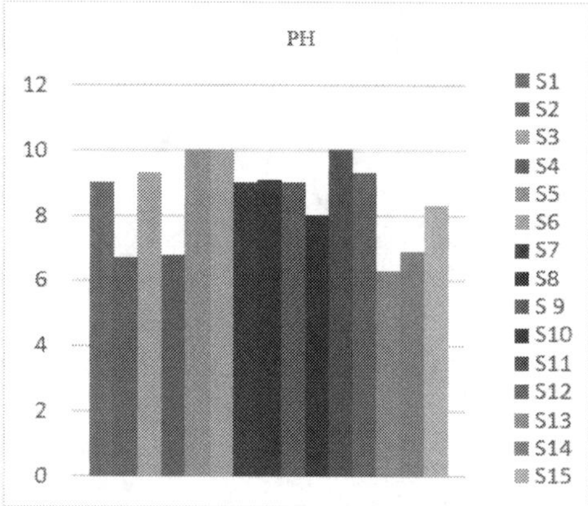

Figure 6. PH value of samples

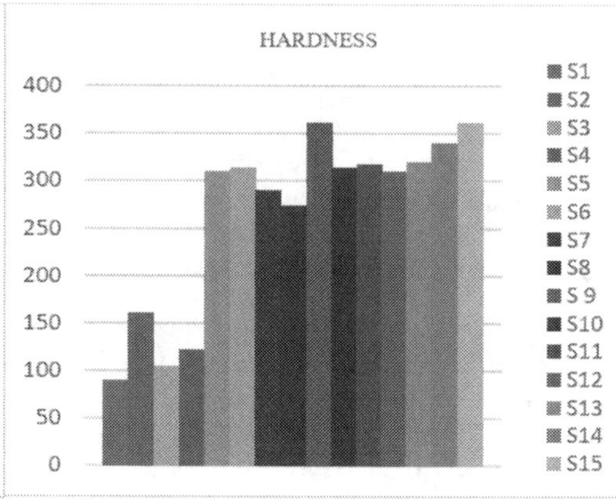

Figure 7. Hardness of samples

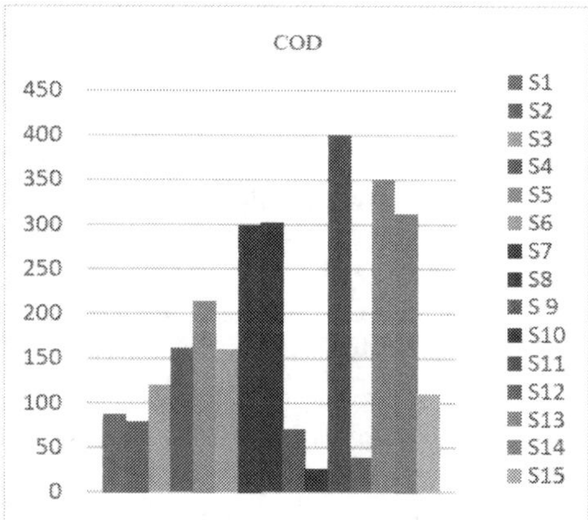

Figure 8. COD of samples

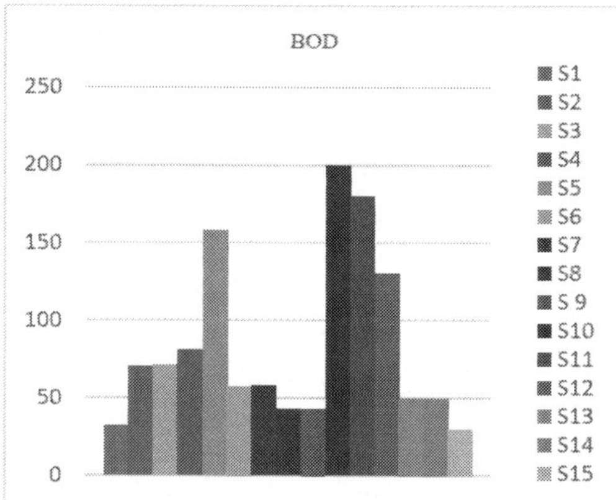

Figure 9. BOD of samples

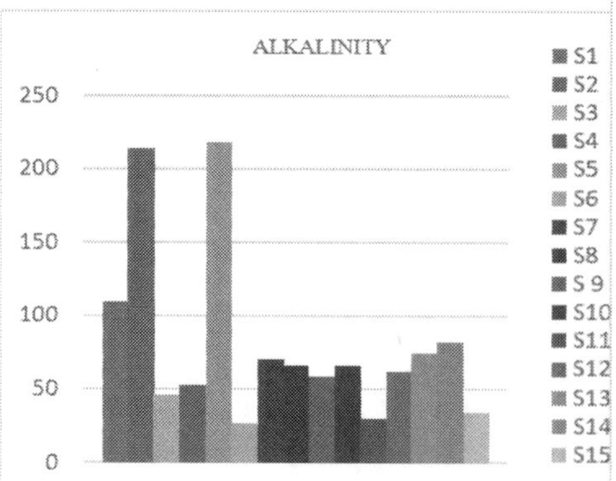

Figure 10. Alkalinity of samples

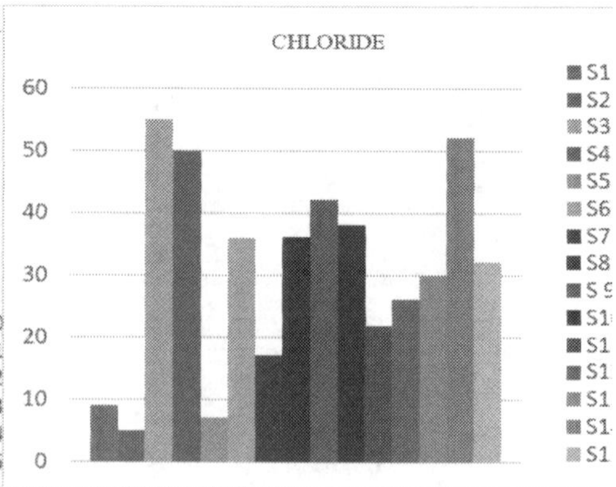

Figure 11. Chloride of samples

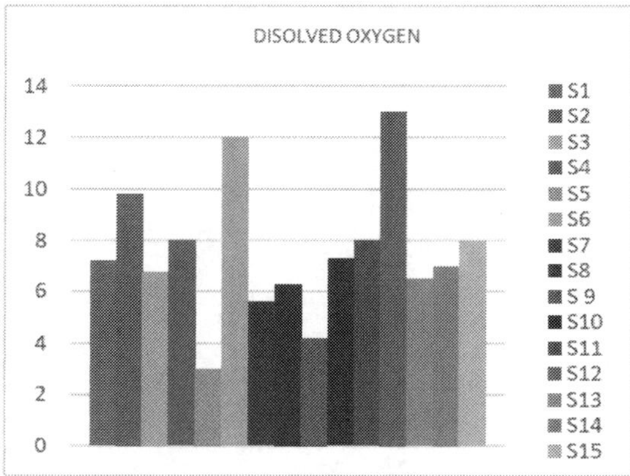

Figure 12. dissolved oxygen of samples

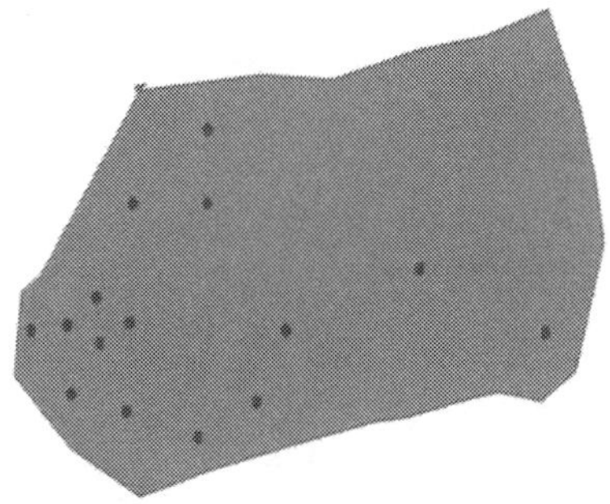

Figure 13. Geo referenced area

4.1 Geo-Referencing:

A geographic information system (GIS) is a framework for managing, gathering and data analysis. It analyzes spatial location and organizes layers of information into visualizations using maps and 3d scenes. Here the software used is Q-GIS 3.6 which performs advanced operations in Q-GIS. Firstly, the area including Azizibucker colony, Mathayipuram colony, Pulikkal way and Sasthri lane in Laloor derived from map and georeferenced then converted in to raster data as shown in figure.13

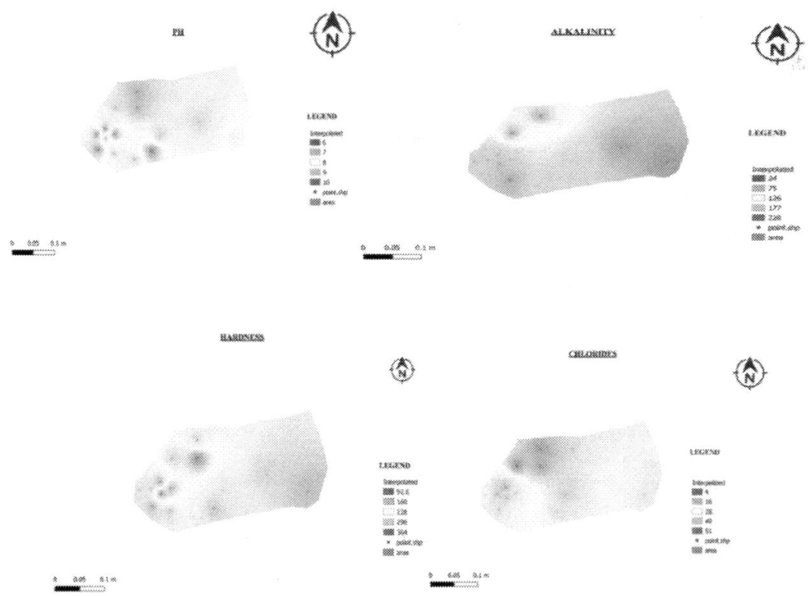

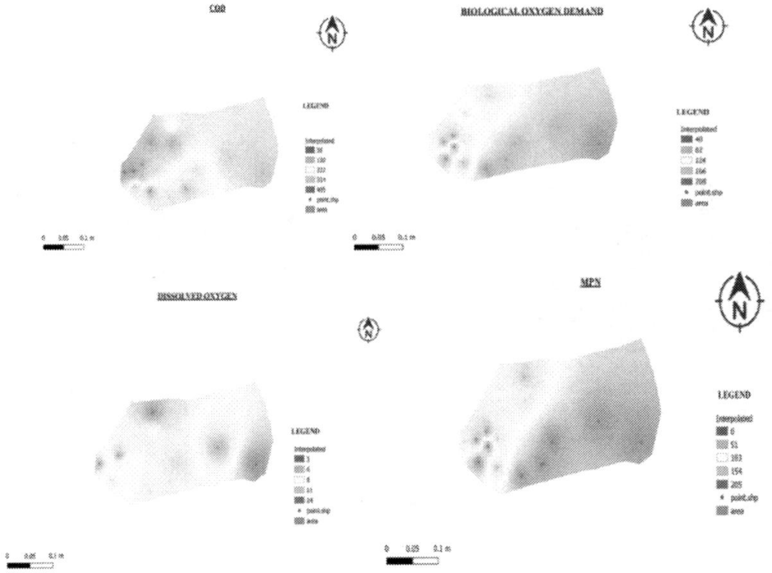

Figure 14. Graphical pictures of all the parameters

5. Conclusion

Leachate generation was the biggest problem in laloor, which is located in Kerala faced by municipal solid waste (MSW) landfills. Leachate contains a high contamination of pollutants which can cause serious problems to environment. This paper explains the impact of leachate peculation at the contaminated site at laloor. And on-groundwater quality was estimated from a contaminated site, water quality index mapped by using GIS. Parameters were checked by various tests which is confirms the impact of the leachate problem in the particular area. Membrane filters are recommended to analyze the waste water. As well as We can create the awareness program for the public about the Leachate problem, also can Educate the children concerning the waste management and the health hazards in schools. Thus, increasing the awareness of the matter in public.

REFERENCES

[1]. Aoyama, M. and Nagumo, T. 1996. Factors affecting microbial biomass and dehydrogenase activity in apple orchard soils with heavy metal accumulation. Soil Science and Plant Nutrition, 42: 821-831

[2]. Bandick, A.K. and Dick, R.P. 1999. Field management effects on soil enzyme activities. Soil Biol. Biochem., 31: 1471-1479.

[3]. Chandrakant Brookes, P.C.1991. Is the dihydrogen as assay invalid a same thod to estimate microbial activity in copper-contaminated soils? Soil Biology & Biochemistry, 23: 909-915.

[4]. Doelman, P. and Haanstra, L. 1979. Effect of lead on soil respiration and dihydrogen aseactivity. Soil Biol. Biochem., 11: 475-479.

[5]. Fuentes, B., Bolan, N., Naidu, R. and Mora, M.D.L.L. 2006. Phosphorus in organic waste-soil systems. J. Soil Sc. Plant Nutr., 6(2):64-83Glinski, J. and Stepniewski, W. 1985. Soil Aeration and its Role for Plants.

[6]. CRC Press, Boca Raton, Florida.Hayano, K., Takeuchi, M. and Ichishima, M. 1987. Characterization of a metalloproteinase component extracted from soil. Biol. Fert. Soils., 4: 179-183

[7]. Johansson, E., Krantz-Rülcker, C., Zhang, B.X. and Öberg, G. 2000.Chlorinationandbiodegradationoflignin. SoilBiologyandBiochemistry, 32: 1029-1032

[8]. Kiss, S., Dragan-Bilarda, M. and Radulescu, D.1978.Soilpolysaccharides: Activityandagriculturalimportance.Chapter4. In: Burns,R.G.(ed.)SoilEnzymes. AcademicPress,London

[9]. Ladd, J.N. andJackson, R.B.1982.Nitrogeninagriculturalsoils. Am. Scc.Agron., Wl.pp.173-228

[10]. Marzadori, C., Ciavatta, C., Montecchio, D. andGessa, C.1996.Effectsofleadpollutionondifferentsoilenzymeactivities. BiologyandFertilityofSoils,22:53-58.

[11]. Nannipieri, P., Ceccanti, B., Bianchi, D. andBonmati, M.1985.Fractionationofhydrolase-humuscomplexesbygelchromatography. Biol.Fert. Soils,1: 25-29

[12]. Pitchel, J.R. andHaynes, J.M.1990.Influenceofflyashonsoilmicrobialactivityandpopulations. J.Environ.Qual.,19:593-597

[13]. Raguotis, A.D.1967.Biological activity of sod-podzolic forest soils of the LithuanianSSR.Pochvovedenie.,6:51-56

[14]. Tyler, G.1974.Heavymetalpollutionandsoilenzymaticactivity.Pl. Soil,41:303-311.

[15]. Vekhrer, E.G.K.T. andShamshieva, 1968.ActivityofsomeenzymesinsoilsofcentralTeinShan. Pochvovedenie,3:94-100.

Design of Water Distribution System for Thirumitta code Grama Panchayat

Ispara Xavier S[1]*, Divahar R[2], Aravind Raj P S[2], Astoria Femy Roy[3], Athira P R[3], Maria Duna C B[3]
[1]Assistant professor, Department of Civil Engineering, Aarupadai Veedu Institute of Technology, Vinayaka Missions Research Foundation, Chennai, India
[2]Associate Professor, Department of Civil Engineering, Aarupadai Veedu Institute of Technology, Vinayaka Missions Research Foundation, Chennai, India
[3]Students, Department of Civil Engineering, Aarupadai Veedu Institute of Technology, Vinayaka Missions Research Foundation, Chennai, India
(*Corresponding Author, Email:isparaxavier.civil@avit.ac.in)

Abstract. The aim of this project is to supply the water to client with acceptable grade and amount. Jal Jeevan Mission (JJM) has been launched that aims at providing practical functional household tap connection (FHTC) to each village unit by 2025 in Republic of India. This project aims on designing a water distribution system for the Thirumittacode gramma panchayath situated in Pattambi Municipality, Palakkad district, Kerala. This project revolves around population prediction of the general practitioner by 2024 and 2054, water demand calculations, planning the water distribution system with associate degree acceptable water resource and value estimation. Use of loop4 software system adds on to the advancement of this project.

Keywords: Water distribution, Distribution system, deliver water, water source estimation, associate degree acceptable, designing water, revolves population, project.

1. Introduction

Water is the foremost necessary needs of life. With a Guaranteed handiness of moveable water is essential for human advancement. With the increasing population and increasing usage of water, there's a rise in carving for water in varied fields such as farming, industry, infrastructure development, agricultural activities etc. The aim of this project is to deliver water to client with acceptable standard, amount and enough force. A distribution system is one during which the water is transported from the centralized treatment plant or well provides to the service association or consumers´ faucets. These systems aim to preserve the standard and amount of water, also as maintain ample pressures within the distribution of water. Basically, these systems accommodate a network of taps, water carrying tanks, and different parts. Water distribution systems square measure aimed toward distributing water from reservoirs or aqueducts to the tip users. Jal Jeevan Mission (JJM) has been launched that aims at providing useful functional household tap connection (FHTC) to each rural family by 2025. The program aims on service Transportation at each family i.e., installation on well-organized basis in sufficient amount and of specified quality. These technology in designing and implementation of installation schemes, development of water sources, treatment and provide of water, authorization of Gram panchayat / area people, concentrate on service supply, partnership deals with different stakeholders, union with different programs, organized observation of program and to find service supply information mechanically to making certain standard services. this may facilitate in achieving the goal of Jal Jeevan Mission in its true spirit.

2. Water Distribution System

2.1 Methods Of Water Distribution System And Layout Of Distribution Network

There are some methods used for distribution of water that are Gravity system, pumping system, and combined gravity system. The operation of pumping stations where gravity pressure isn't enough, should be ensured. There are four principal ways to style a distribution system.

3. Results and Discussion

3.1 Population Forecasting of Thirumittacode GP

The most important element that calculate the amount of water to be supplied to a region is the population of that region. Population is forecasted based on design period. Various methods can be used for population forecast such as decadal growth rate method, Arithmetical Increase, Geometrical Increase, Incremental Increase etc. Here the method adopted is decadal growth rate method.
Total population = Urban population + Rural population
Decayed growth rate = 10%
Population in 2011: 31998

Population by 2021= 31998 * 1.1
= 35197.8 ≈ 35198

Population by 2024= 35198 + 35198 * (3/100)
= 36253.94 ≈ 36254

Population by 2034= (36254*1.1) 39879.4
= ≈ 39880

Population by 2044= (36254*1.1*1.1) 43867.34
= ≈ 43868

Population by 2054= (36254*1.1*1.1*1.1)
= 48254.074 ≈ 48255

3.1.1 Water Demand Calculation for Thirumittacode Grama Panchayath

Water demand is the total amount of water used by people or customers in an area. Hence it is a function of population. The domestic water demand is assumed as 130lpcd.

Estimated population by 2039 = 41693(15years from 2024)
Total domestic water demand = 41693*130 = 5420090 L/d= 5.42009 ML/d
Estimated population by 2054 = 48255 (30 years from 2024)
Total domestic water demand = 48255*130= 6273150L/d= 6.27315 ML/d

3.1.2 Water Resource Selection

BHARATAPUZHA River is selected as the water resource for the water distributing system in Thirumittacode grama panchayath. The supporting reasons for selecting Bharathapuzha River arethe river BHARATHAPUZHA is a perennial source of water and also the required head of the river is maintained in the catchment area with the help of regulator cum bridge at

Journal of Physics: Conference Series **2040** (2021) 012054 doi:10.1088/1742-6596/2040/1/012054

Vellyamkallu, Ottapalam Taluk, Palakkad and it has good water quality parameters.

3.2 Pump Design

This project proposes the design of centrifugal pump with the below computed horse power.

Horsepower or HP = ρ Q g H /η
P = density of fluid = 1000kg/(m³)
Q = flow of fluid = 8.62*10-3.m³/s
H = 73 m (total head)
η = efficiency of pump = 60% (assumed)
ρ Q g H = 1000*9.81*73*8.62*10-3.m³/s = 6.17 KN Pump Power = Water horse power/efficiency of pump water horse
power (WHP) =6.17*1000/746 = 8.27 hp
*1 hp=746 watts
Efficiency of pump assumed as 60%
Pump power = 10 hp

3.3 Location Data

Figure 1. satellite map of Thirumittacode

Sl.No.	Substance or Characteristics	Unit	Desirable limits	Permissible limits in the absence of Alternate source	Actual contents	
Physical and Chemical Analysis						
1	Turbidity	NTU	1	5	0.1	
2	pH		0.5-8.5	0.5-8.5	5.5	
3	Electrical Conductivity	mmhos/ cm			64	
4	Acidity	mg/l			50	
5	Alkalinity (Total)	mg/l	200	600	18	
6	Total Dissolved Solids (TOS)	mg/l	500	2000	35	
7	Total Hardness as (CaCo3)	mg/l	200	600	0.26	
8	Ca	mg/l	75	200	6.4	
9	mg	mg/l	30	100	0.97	
10	Cl	mg/l	250	1000	14	
11	Fluoride	mg/l	1	1.5	NIL	
12	Fe	mg/l	0.3	1	0.1	
13	Nitrate	Mg/l	45	45	25	
14	Sulphate	Mg/l	200	400	13	
15	Phosphate	Mg/l			NIL	
16	Ammonia	mg/l	0.5	0.5	NIL	
Bacteriological Analysis						
1	Number of Coliforms in 100ml				NIL	
2	E. coli				NIL	
Remarks: Low pH						

Table 1. water quality parameters of a water sample (from a well near Bharathapuzha River)

Journal of Physics: Conference Series **2040** (2021) 012054 doi:10.1088/1742-6596/2040/1/012054

3.5 Designed Intake Well

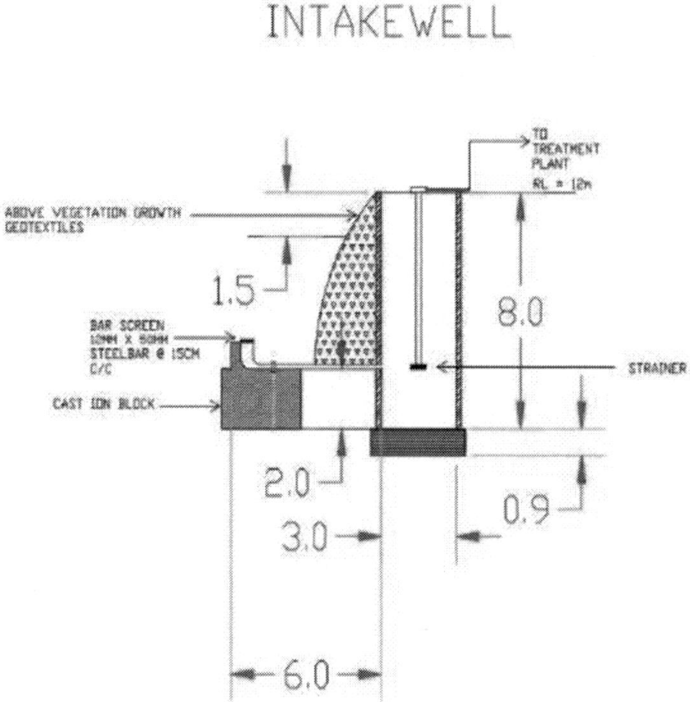

Figure 2. Designed intakes well in AutoCAD

4. Conclusion

By the end of this project, a water distribution system will be designed for the whole population of Thirumittacode Gram panchayat by 2024 and 30 years thereafter. The designed water distribution system will be including an intake well, storage tank, pumping system, water treatment plant and pipeline network. Then there won't be water scarcity in the respective Gram panchayat anymore. Ground water table regenerating methods can also add on to this project for more benefits. This project can also help as a reference to other water distribution projects as far as the data collection and designing strategies are concerned. This project can also be considered as a software driven project partially thus can be compared with other normal projects for further add on.

REFERENCE

[1] Bhavana Ajudiya, Shreyas Bhagde, Hitesh Hathaliya: "Designing of Rural Water Supply System

[2] ChavhanY, "Design of Distribution Network for Water Supply Scheme at Pindkepar Village by Branch Software", International Journal of Engineering Research and Applications, Vol.3, Page: 854-858, Issue: 5, 2013.

[3] Bao Y & Mays L W, "Model for water distribution system reliability".

[4] D.S. Vijayan, A. Mohan, J. Revathy, D. Parthiban , R. Varatharajan, "Evaluation of the impact of thermal performance on various building bricks and blocks: A review", Environmental Technology & Innovation 23 (2021) 101577, https://doi.org/10.1016/j.eti.2021.101577.

[5] Kay Chambers, john creasey and Leith forbes , " safe piped water: Managing microbial water quality in piped distribution systems", World health organization, UK

[6] Loan sarbu, "Design of optimal water distribution systems" INTERNATIONAL JOURNAL OF ENERGY, Issues 4, vol .3,2009.
[7] "Manual on water supply and Treatment" Third Edition, The GOI, Ministry of urban Development, New Delhi, March 1991,
[8] Development Forum, "water Decade Review" United Nations, Unnited Nations University, New York 1987.
[9] "Schedule of Rate" Technical cell- civil, Gujarat water supply and sewerage board (GWSSB) ,2008-2009
[10] Taylor,F.B" Functional design for effective operation of water supply systems in developing countries"J.AWWA, February ,1967.
[11]. Huang, d, vairavamoorthy, k, tsegaye, s,2010. Flexible Design of urban water distribution network System.
[12]. D. S. Vijayan, A. Mohan, J. J. Daniel, V. Gokulnath, B. Saravanan, and P. D. Kumar, "Experimental Investigation on the Ecofriendly External Wrapping of Glass Fiber Reinforced Polymer in Concrete Columns," vol. 2021, 2021.
[13]. Kirmeyer, G, W. Richards, and C. D. Smith, 1994. "An assessment of water distribution system.
[14]. Water supply engineering, Dr. B.C. Punmia
[15]. Quimpo, R.G., & Shamsi, U.M, RELIABILITY Based distribution system maintenance.
[16] Wagner,J.M., Shamir,U., &Marks water distribution reliability: simulation methods. Journal of water resources planning and management Division.
[17] D.S. Vijayan, D. Parthiban," Effect of Solid waste based stabilizing material for strengthening of Expansive soil- A review," Environmental Technology & Innovation, Volume 20, 2020, 101108, ISSN 2352-1864, https://doi.org/10.1016/j.eti.2020.101108
[18]. B.S.Jung, B.W.Karney "Fluid transient and pipeline optimization using GA and PSO: the diameter Connection (2004)
[19]. S. M. S. S, D. S. Vijayan, M. Anand, M. Ajona, and T. Jarin, " Biodegradation of P-nitro phenol using a novel bacterium Achromobacter denitrifacians isolated from industrial effluent water ," Water Sci. Technol., vol. 00, no. 0, pp. 1–12, 2021, doi: 10.2166/wst.2021.354.

Decolorization of Textile Wastewater with Activated Carbon made of Coconut Shell

Sangeetha S P[1], Susmita Bhowmick[2]*, Md Nawas Khan[2], Akash M[2]
[1]Vice Principal & Professor, Department of Civil Engineering, Aarupadai Veedu Institute of Technology,VMRF, Chennai
[2]UG Scholar, Department of Civil Engineering, Aarupadai Veedu Institute of Technology, VMRF, Chennai
(*Corresponding Author, Email: susmitags2130@gmail.com)

Abstract. This study focus on the removal of textile dyeing effluent by analyzing the physio-chemical parameters such as color, turbidity, pH, Chemical Oxygen Demand (COD). Activated carbon is used for removing color from the wastewater collected from silkworm textile industry, Bishnupur, West Bengal. We have used activated carbon made with coconut shell which is ecofriendly and low in cost. The results of the analysis were compared with the water quality standards of BIS (Bureau of Indian Standard) for further application of the treated water in agricultural field. Also, this paper included the case study on textile dyeing industry in Bankura district, West Bengal, India and its impacts of pollution load in the environment.
Keywords: Textile effluent, pH, turbidity, COD, color, activated carbon.

1. Introduction

Textile industry is one of the oldest and largest industries in India and it gives job to huge number of people in our country (Hussain et al., 2004). About 81 percent of all industries in India are situated in Tamil Nadu, Gujarat, Punjab, West Bengal, and Maharashtra (Husain et al., 2012), and this textile industry is diverse. According to the Ministry of Textiles, India now has 2,500 textile weaving facilities and 4,135 textile finishing industries (Elango et al., 2017), contributing 2.3 percent to the country's GDP, 13% to industrial output, and 12% to export profits in 2020. The textile sector is the world's third biggest user of water. In India, the usage of water for the textile industries increase with every year and was estimated to be about 63Billion Cubic Meter (Kabilashasundari et al., 2019). In India textile industry uses about 425000000 Gallons of water in daily basis and it takes about 500 gallons of water to produce enough fabric to cover one sofa (Rita Kant, 2012).Textile industry processes produce multi component wastewater and sludge in huge amount which can be difficult to treat (Paul &Dhas, 2008). The polluted water is affecting and damaging the environment vey severely. In textile waste water, the pollutants like color, chemical, pH, TDS, TSS etc. have destructed the environment and this physio-chemical textile industrial effluent was analyzed by different methods (Panwar et al., 2019).Activated charcoal is also preferred as an adsorbent for treating wastewater containing different classes of dyes for so long period (Himanshu Patel, 2018). It has the property to absorb the color. Peoples are trying to use low-cost materials to make and activated carbon and treat wastewater, such as agricultural by products and natural fibers, namely; waste coir pith (Namasivayam et al., 2001), corncob and barley husk (Robinson et al., 2002), sawdust (Khatmode et al., 2015), coconut shell, palmshell, pistachio shell, tropical wood and almond shell (Adib et al., 2018). In this project aggregates are used as filter medium and activated carbon is used as absorbent of color.

The wastewater was collected from a small textile industry in Bankura district in West

Bengal, were the silkworms are used for extracting the silk and used the various dyes to color the silk. Wastewater from dyeing units is rich in color, containing residues of reactive dyes and chemicals (Nasipude et al., 2015) and the water was get discharged in several water bodies like lakes and ponds without any treatment. Without proper treatment, disposal of wastewater can harm the aquatic species, ecosystem and environment as well (Adib et al., 2016). After the treatment, recycled water can be used for multi purposes like agriculture, car wash, gardening, etc. (Francics et al., 2015).

2. Materials Used

2.1 Coarse Aggregate

Coarse aggregate is often used as filtration medium to terminate the solids comes with wastewater and reduces the turbidity of water (Gnanaraj et al., 2019). In this project the two sizes of aggregate like, 6-8 mm and 12-16mm which is washed several times to remove all the dirt's from the surface of aggregates.

2.2 Fine Aggregate

Fine aggregates were works as physical strainer and biological renovator to help the pathogens to die. It also helps to decrease the COD of wastewater. The size of 0.5-1.0mm sand was sieved and washed 2 to 3 times to remove small particles and dried of (Gnanaraj et al., 2019).

2.3 Activated Carbon

Activated carbon has been used to clean water as a water filtration media. Activated carbon's primary function is to absorb colour and odour. Due to its great adsorption capability, granular activated carbon was extensively utilised to remove pollutants from water.

2.4 Physical phenomena

Possesses an exceptional ability to capture water-dissolved contaminates because of its rare quality. These contaminant-carbon surface interactions happen for Van der Waal forces.

2.5 Granular activated carbon

millimeter sized granular activated carbon can remove pollutants to concentrations below analytical detection limit and it requires about 1/4th of powdered activated carbon. But activated carbon never lasts long period, so periodically it's needed to be changed for proper removal.

3. Filter Setup

Table 1. Description of Material layers from bottom to top

Layer No.	Materials	Depth	Size
1st	Fine aggregate	150 mm	.5-1 mm
2nd	Coarse aggregate	100 mm	6-8 mm
3rd	Coarse aggregate	100 mm	12-16 mm
4th	Activated carbon	30 mm	Granular

Activated carbon is made of coconut shell which is heated and burnt up to 2/3rd part of volatiles out of shells, creating a carbonaceous mass full of pores. Then with help of pyrolysis process activated carbon was prepared. In 20 litre waste water bottle the filtration medium was set. The layers are shown below:

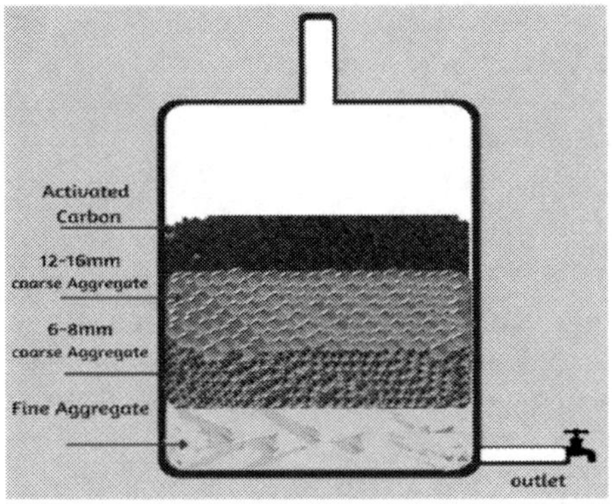

Figure 1. Filtration medium Setup

4. Tests & Results

Wastewater was collected first and treated, the standard values as per IS codes after treatment (Rice et al., 2017) are given in table 2.

Table 2: Reference values as per IS codes

No.	Parameter	Instrument	Methods
1.	pH	pH meter	IS 3025 Part 11-1983 (Reaffi:2012)
2.	COD	Standard Method	IS 3025 Part 58-2006 (Reaffi:2012)
3.	Turbidity	Turbidity meter	IS 3025 Part 10-1984(Reaffi:2012)
4.	Color	Spectro-photometer	IS 3025 Part 4-2006 (Reaffi:2002)

Table 3: Results before treatment

No.	Parameter	Result
1.	pH	9.1
2.	COD	1034 mg/l
3.	Turbidity	62.1 NTU
4.	Color	3685 Pt-Co

The wastewater now passes through the filter unit and collected in a glass jar for further treatment. Activated carbon is used as 2, 2.25, 2.5 g/l ratio for treating water.

Table 4: Post-treatment results

Journal of Physics: Conference Series **2040** (2021) 012055 doi:10.1088/1742-6596/2040/1/012055

No.	Parameter	Unit	Results		
			2(g/l)	2.25(g/l)	2.5(g/l)
1.	pH	-	8.7	8.8	8.8
2.	COD	mg/l	537	369	232
3.	Turbidity	NTU	70	21.8	9.5
4.	Color	Pt-Co	491	195	40

The results are presented in a chart.

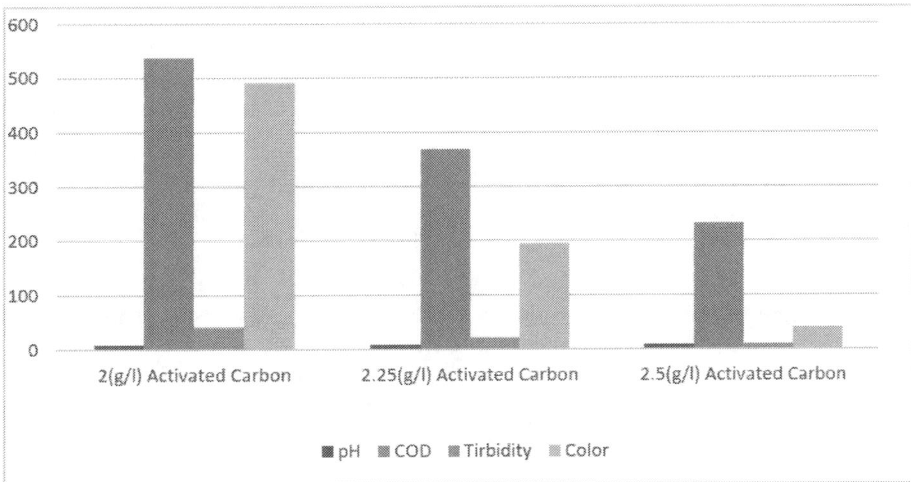

Figure 2. Comparison of results

Both before and after treatment parameters were compared with BIS (Bureau of Indian Standards).

No.	Parameter	Unit	Pre-Treatment Results	Post-Treatment Results			BIS Standards
				2(g/l)	2.25(g/l)	2.5(g/l)	
1.	pH	-	9.1	8.7	8.8	8.8	5.5-9
2.	COD	mg/l	1034	537	369	232	250
3.	Turbidity	NTU	62.1	41.6	21.8	9.5	10
4.	Color	Pt-Co	3685	491	195	40	25

Table 5: Comparison of results with BIS

5. Case Study and Discus

In Bishnupur, Dist. Bankura from where the water was collected has large number of small handloom industries where dyeing of silk materials happens. It generates huge amount of wastewater which is released in environment without treatment. So, the wastewater enters to the nearby lake and pollute those water bodies. Above method can be used for small scale industry very effectively to treat the wastewater and released in environment without any harm. With increasing the amount of activated carbon /lit more effective results could be achieved.

(a) (b) (c) (d)

Fig. 3: Waste water sample (a) before treatment, (b) treated with 2g/l activated carbon, (c) treated with 2.25g/l activated carbon, (d) treated with 2.5g/l activated carbon

The use of activated carbon as 2g, 2.25g, 2.5g per lit helps to remove the COD, Color, Turbidity, pH in acceptable range of BIS standards. 2g/l activated carbon is able to remove the COD and Color about 49.12% and 86.67%, 2.25g/l activated carbon is able to remove the COD and Color about 64.31% and 94%, 2.5g/l activated carbon is able to remove the COD and Color about 77.6% and 98%.

But the problem with activated carbon is that, it's never long lasts. So, the time frequency of changing the activated carbon could be done.

6. Conclusions

From the project we observed that waste water generated by textile industry has great impact in environment due to the presence of COD, color, turbidity and various high amount of chemicals. We can use activated carbon in 2.5g/l for the better results and it gives the results within the permissible limits of BIS standards. The discussed treatment can be helpful to decrease the waste water and manage the destruction of environment. Preparation of activated carbon with coconut shell will reduce the cost for treatment. Wastage of water needs to be prevented to overcome water scarcity. So, the treated water can be used for agricultural purposes, gardening and car washing etc.

REFERENCES

[1]. Abraham Francics A., Thenmozhi R., Sivakumar M., Sivakumar K., Sasikumar G.5, Thamaraimuthuayyanraj S. 6, "Waste Water Treatment Unit using Activated Charcoal", e-ISSN: 2395-0056, p-ISSN: 2395-0072, Volume: 05 Issue: 03 (2015)

[2]. Aijaz Panhwar, Khalida Faryal, Aftab Kandhro, Sofia Qaisar, "Assessment of Textile Industrial Effluent by Wastewater Quality Standards", International Journal of Scientific & Engineering Research,1659, ISSN 2229-5518, IJSER, Volume 10, Issue 6 (2019)

[3]. E.W. Rice, R.B. Baird, A.D. Eaton, editors, "Standard Methods for The Examination of
Water and Wastewater", 23rd Edition, ISBN: 9780875532875 (2017)

[4]. Gomathi Elango, Rathika Gand Santhini Elango, "Physio-Chemical Parameters of Textile
Dyeing Effluent and Its Impacts withCase study", International Journal of Research in Chemistry and Environment, ISSN 2248-9649 (2017).

[5]. Himanshu Patel, "Charcoal as an Adsorbent for Textile Wastewater Treatment", Separation Science and Technology, 53:17, 2797-2812, DOI:10.1080/01496395.2018.1473880 (2018)
[6]. Husain J. and Husain I., "Groundwater pollution by discharge of dyeing and printing Industrial wastewater in Bandi river", Rajasthan, India, Int J.Environment and Bioenergy, 2(2), 100-119 (2012)
[7]. J. Hussain, I. Hussain and M. Arif, "Characterization of Textile Wastewater", Jr. of Industrial Pollution Control 20 (1) pp. 137 -144© Enviromedia (2004).
[8]. Jaya Paul A/L Arumai Dhas, Removal of Cod and Colour Fromtextile Wastewater Using Limestone and Activated Carbon, University Sains Malaysia (2008)
[9]. Mohd Adib, Mohammad Razi1, Mimi Nur Attahirah Mohd Hishammudin1 and Rafidah Hamdan2, "Factor Affecting Textile Dye Removal using Adsorbent from Activated Carbon: A Review",MATEC Web of Conferences 103, 06015 (2017) DOI: 10.1051/matecconf/20171030, ISCEE (2016)
[10]. Mohd Adib, Mohammad Razi, Adel Al-Gheethi, Mohammed Al-Qaini & Anwar Yousef, "Efficiency of Activated Carbon from Palm Kernel Shell for Treatment of Greywater", Arab Journal of Basic and Applied Sciences, University of Bahrain, VOL. 25, NO. 3, 103–110https://doi.org/10.1080/25765299.2018.1514142 (2018)
[11]. Namasivayam C., Radhika R. and Suba S. "Uptake of dyes by a promising locally available agricultural solid waste: coir pith", Waste Manage., 21pp. 381–387 (2001)
[12]. Nitin P. Khatmode, Dr. Sunil B. Thakare, "Removal of pH, TDS, TSS & Color from Textile Effluent by using Sawdust as Adsorbent",International Journal of Sciences:Basic and Applied Research(IJSBAR)ISSN 2307-4531 (2015)
[13]. Rita Kant, "Textile dyeing industry an environmental hazard", Natural Science, 4, 22-26,DoI: 10.4236/ns.2012.41004. b (2012)
[14]. Robinson T., Chandran B. and Nigam P. "Removal of dyes from an artificial textile dye effluent by two agricultural waste residues, corncob and barley husk", Environmental Int.,28, pp. 29-33 (2002)
[15]. S. Christopher Gnanaraj, Ramesh Babu Chokkalingam, S. K. M. Pothinathan, R. Rekha, "Experimental Research on Treatment of Greywater using a Prototype", International Journal of Recent Technology and Engineering (IJRTE), ISSN: 2278-3075, Volume-9, Issue-2S2, December (2019)
[16]. S.Kabilashasundari, T.Nirmal Kumar, B.Naveen Karthik, J.Oliver Paul Nayagam, S.
Dhanasekar, "Decolorization and COD Removal from Textile Waste Water by Chemical, Adsorption and Biological Treatment Methods", International Journal of Engineering and Advanced Technology (IJEAT), ISSN: 2249 – 8958, Volume-8 Issue-4, April (2019).

PERFORMANCE OF PELLUCID CONCRETE BY USING OPTICAL FIBERS

P.Subathra[1*], S.P. Sangeetha[2]
[1]Research Scholar & Assistant Professor,2 Vice Principal(Academics) & Professor,Department of Civil Engineering, AarupadaiVeedu Institute of Technology,Vinayaka Mission's Research Foundation, India.
Corresponding Author Email : subathra@avit.ac.in

ABSTRACT: Energy efficient building and aesthetic view are the vital knot for infrastructure. Now engineers have developed smart translucent concrete(STC) using glass optical fibers (GOF).It can able to transmit light from one phase to the other phase of the wall. Smart translucent concrete is also known as pellucid concrete. Pellucid concrete reduces the consumption of electricity. Source of light can be taken from natural sunlight and artificial light source from LED having 20w. In this investigation ,incorporating optical fibers in various diameters are 1.5 mm and 2.5 mm. Concrete cube with provision of hole for providing glass optical fiber and arranged in equal spacing of about 15 mm and 25 mm. In this experiment ,analysis the performance of smart translucent concrete in the aspect of strength and luminance .The light intensity from the specimen can be measured with a help of photometer. It is an energy efficient building material innovated in recent years.

Keywords—Pellucid Concrete, smart translucent concrete(STC), Optical Fiber, Glass Optical Fiber (GOF). Strength, Photo Meter.

1. INTRODUCTION

Force is the premier valuable and supportive for man, when the utilization of control isn't so to speak used for the private explanation however for the age to use and solace in nature, in the two conditions of outside and inside trade also. By then the improvement of extension, moles, ports, roads and signals, by infers of fake force for use of trade and after that being developed and other mechanical assembly, it is also used in seepage for urban communities and towns. At the point when we are dealing with benevolent planning it has come in progress inside the field of advancement, hardly any people were having misconcept roughly charitable structure as a division of science which deals with Civilization. However, when the time passes the improvement of building structures, similar to underground roads, Bridge, raised Street, Point of interest structures, high rises and various other structure. At the point when we are thinking about monetary turn of events, in this time we have wide improvement that is the reason input, tall usage and tall pollution. The translucent cement was in any case made in 2001 by Hungarian plannerAron Lasancziat the particular College of Budapest. Translucent cement can be precast squares of assorted measure. In light Pellucid cement, which is usually known as direct concrete, Optical Fiber are projected into concrete

2. MATERIALS AND ITSPROPERTIES

2.1. Glass Optical Fiber

In this project we use glass optical fibers (GOF) of having diameter 1.5 and 2.5 mm,.The properties of glass optical fibers are given in table 1

Table 1. Properties of glass optical fiber

SL NO	PROPERTY	VALUES
1	Fiber diameter	1.5mm and 2.5mm
2	Transmissiondistance	400-700nm
3	Numerical aperture	0.25
4	Temperature in storage	70°C-400°C

2.2. CEMENT

Ordinary Portland cement grade 43 is used for casting, having specific Gravity 3.15 and fineness modulus value is lesser of 7% are given table 2

Table 2 : Physical Properties of Cement

SL.NO	PROPERTIES OF CEMENT	VALUES OBTAINED
1.	Fineness of cement as retained on 90 micron sieve	240
2.	Soundness	0.6
3.	Grade of cement	43
4.	Specific Gravity	3.15
5.	Consistency	34%
6.	Initial Setting time	30min
7.	Final Setting Time	560 mins

2.3. Fine Aggregates

River Sand is utilized as fine total in this think about which is coming beneath Zone-II ,Sand passing through IS 4.75mm Sifter is utilized for casting all the specimens are listed in table 3

Table 3: Physical Properties of Fine Aggregate

Sl.NO	Properties	Values obtained
1	Specific Gravity	2.6
2	Fineness Modulus	4.4
3	Water Absorption	5.6

2.4. COARSE AGGREGATES

Agreeing to IS guidelines, Coarse total passing through 20mm strainer is utilized for casting all specimens. sp. gravity ought to be 2.8 are provided in table 4

Table 4: Physical Properties of Coarse Aggregate

Sl.NO	Properties	Values obtained
1	Size of Aggregates	20mm
2	Fineness Modulus	6.3
3	Specific Gravity	2.8

2.5. Water

The water utilized for casting of tests ought to be free from natural substances and Conventional consumable water of ordinarily pH value between 6.5-8.5 is utilized for blending and curing the concrete specimen

3. TESTS AND RESULTS

3.1. Compressive strength :

Compressive strength test of the concrete sample was performed by compressive universal testing machine for 7 ,14 and 28 days curing time .The concrete samples were loaded gradually until failure occurred. During the test the load and displacement data of the concrete specimen was recorded using computer. The test result values of concrete sample and glass optical fiber samples are in the table 5 and table 6.From the results it is known that strength shows that 28 days and 2.5 mm dia ,25 mm spacing of glass optical fiber of curing has achieved its strength as per requirements compared to conventional concrete.

Table 5: Hardened Concrete cube Test

Conventional concrete Specimen Details	Average Compressive strength in N/mm^2		
	7 days	14 days	28 days
Specimen 1	16.8	19.5	24.3
Specimen 2	16.6	19.2	24
Specimen 3	16.5	19.4	24.2

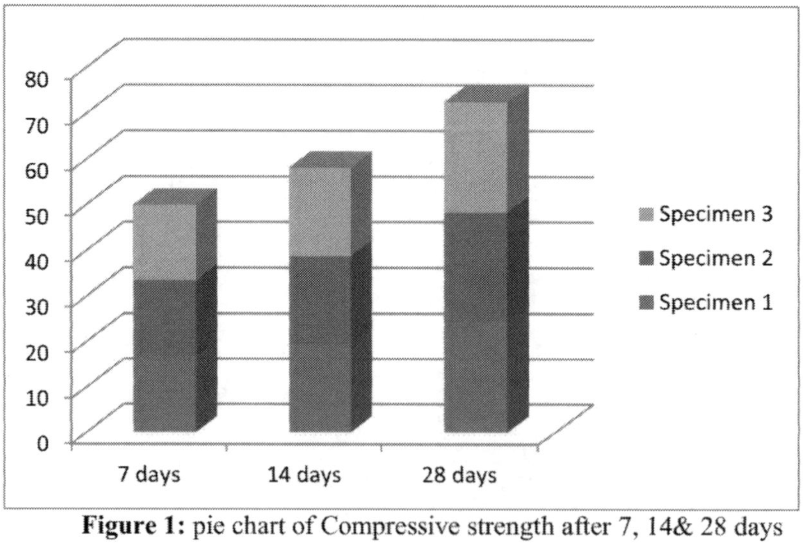

Figure 1: pie chart of Compressive strength after 7, 14& 28 days

Table 6: Compressive strength after 7, 14& 28 days

Glass optical fiber diameter (in mm)	Average Compressive strength (N/MM2)					
	7 Days		14 Days		28 Days	
Spacing	15 mm	25 mm	10 mm	15 mm	15 mm	25 mm
1.5	12.4	13.2	16.2	17.8	21.4	21.1
2.5	14.6	15.4	19.1	20.2	25.4	26.4

3.2. Average Compressive strength of GOF

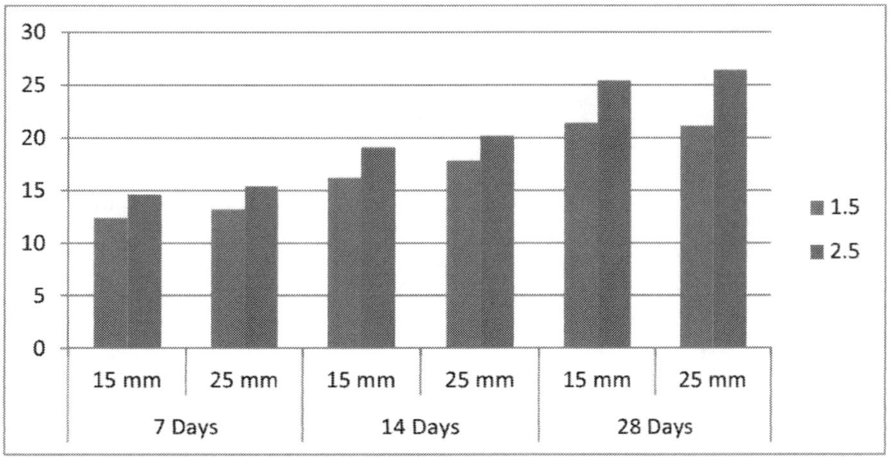

Figure 2 Average compressive strength of the concrete cube

3.3. ILLUMINANCE TEST :

It is a non-destructive test and concrete specimen from this test were reused for compression test. The luminance test was carried out by using photometer for 7 ,14 and 28 days of curing. The illuminance test was conducted after the use of glue gun for keeping glass optical fiber should be stable along with mould. The test setup of the illuminance test is given in fig 3 and illuminance result of glass optical fiber are in the table 7.The results shows that light from the artificial source gives maximum output in the 15mm spacing and 1.5 mm diameter of glass optical fiber.

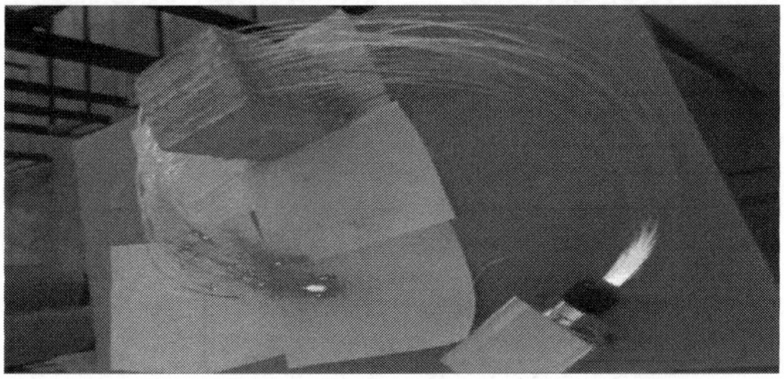

Fig 3 Test setup of illuminance test

Table 7: Light transmittance Results Of GOF

Glass optical fiber diameter (in mm)	Sunlight		Artificial light	
Spacing	15 mm (No of strands =64)	25 mm(No of strands =36)	15 mm (No of strands =64)	25 mm(No of strands =36)
1.5	2	2	700	650
2.5	2.5	2.5	1400	1300

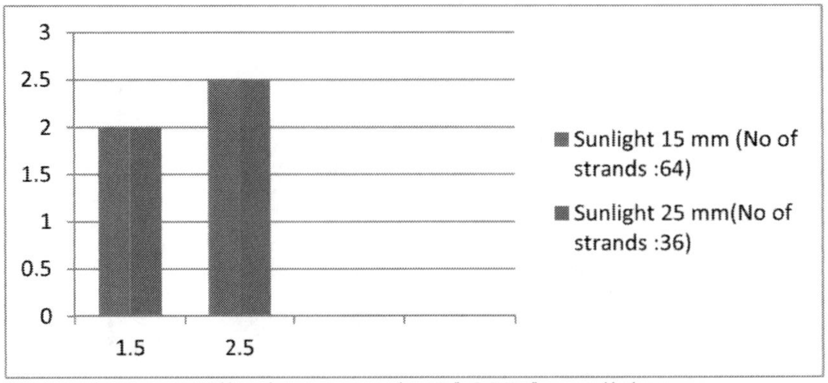

Fig 4 : Illuminance Results Of GOF for sunlight

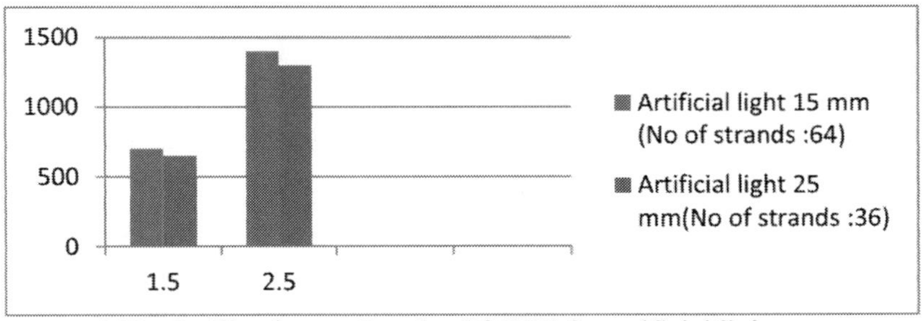

` **Fig 5 :** Illuminance Results Of GOF for artificial light

3.4. Analysis of Cost

The comparison cost of conventional concrete and glass optical fiber concrete of different diameter are in the table 8 and output of the glass optical fiber is shown in fig 4.When compared to conventional concrete samples ,glass optical fiber concrete specimen cost is high for 1.5 mm dia spacing of 15mm and 25 mm spacing .

Table 8: ForM25 GradeofConcrete (Normal&GOF) For1 Cube

	Cost			**Total cost**
	CC	**GOF**		
		15 mm(64 strands)	**25mm(36 strands)**	
CC	525	-	-	525
GOF- 1.5 mm dia optical fiber	525	1013	-	1538
GOF- 2.5 mm dia optical fiber	525	252		777
GOF- 1.5 mm dia optical fiber	525	-	1280	1805
GOF- 2.5 mm dia optical fiber	525	-	720	1245

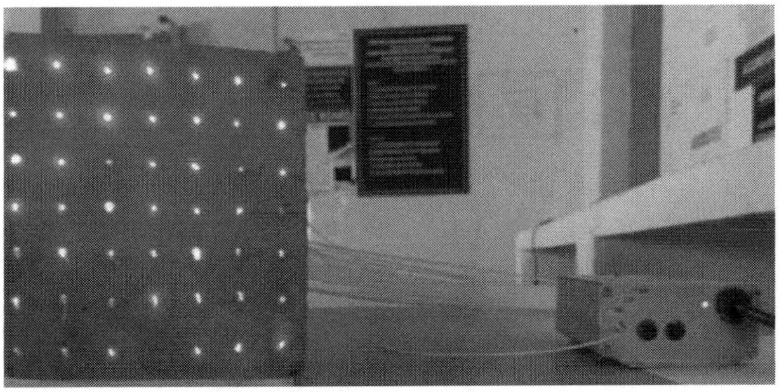

Fig 6 Illuminance test result

4. CONCLUSION

Transmission of light depends upon breadth of optical strands and dividing of strands. As we diminish the vertical and flat dispersing between filaments, it increments light. In GOF-64 were given 15 mm dispersing in both the vertical and level headings ,that appears increments light of light with regard to diminish in dividing. Last comes about appears that GOF 64 gives great light compared with GOF 36.The comes about appears that GOF-36 concrete 3d shape gives higher quality compared with GOF-64.By considering 15 mm and 25 mm dispersing ,25 mm dispersing of 2.5 mm breadth of glass optical fiber gives higher characteristics compressive quality in all the over diameters.

REFERENCES

[1] C. Amuthadevi, D. S. Vijayan, Varatharajan Ramachandran, "Development of air quality monitoring (AQM) models using different machine learning approaches", Journal of Ambient Intelligence and Humanized Computing, https://doi.org/10.1007/s12652-020-02724-2

[2] Arvindan Sivasuriyan et.al, "Development of Smart Sensing Technology Approaches in Structural Health Monitoring of Bridge Structures", Hindawi , Advances in Materials Science and Engineering Volume 2021, Article ID 2615029, 14 pages https://doi.org/10.1155/2021/2615029

[3] Arvindan Sivasuriyan, D.S. Vijayan, Wojciech Górski, Magdalena Daria Vaverková, Eugeniusz Koda, "Practical Implementation of Structural Health Monitoring in Multi-Story Buildings", Buildings 2021, 11, 263. https://doi.org/10.3390/buildings11060263

[4] D.S. Vijayan, A. Mohan, J. Revathy, D. Parthiban , R. Varatharajan, "Evaluation of the impact of thermal performance on various building bricks and blocks: A review", Environmental Technology & Innovation 23 (2021) 101577, https://doi.org/10.1016/j.eti.2021.101577.

[5] P. Subathra, et.al, Experimental investigation on concrete with marble dust and steel fiber, AIP Conference Proceedings 2271, 030016 (2020); https://doi.org/10.1063/5.0024779

[6] Kazuro Kageyama et.al, "Acoustic emission monitoring of a reinforced concrete structure by applying new fiber-optic sensors", June 2005, Smart Materials and Structures 14(3):52-59.

[7] P Jaishankar, C Karthikeyan, "Characteristics of Cement Concrete with Nano Alumina Particles", July 2017, IOP Conference Series Earth and Environmental Science 80(1):012005

[8] D. Zheren et al., "3D micro-concrete hybrid structures fabricated by femtosecond laser two-photon polymerization for biomedical and photonic applications," 2016 IEEE International Conference on Industrial Technology (ICIT), 2016, pp. 1108-1114, doi: 10.1109/ICIT.2016.7474909.

[9] B. Saravanan and D. S. Vijayan, Status review on the experimental investigation on replacement of red-mud in cementitious concrete, MaterialsToday:Proceedings, https://doi.org/10.1016/j.matpr.2020.05.500

[10] A. Leema Rose, D. Elango, D.S. Vijayan et al., Reliability analysis of corroded beams with FRP laminates, Materials Today: Proceedings, https://doi.org/10.1016/j.matpr.2020.09.410

AUTHOR INDEX